Northwest Community College
3 1512 00016 0439

ATMOSPHERIC CHEMISTRY

Atmospheric Chemistry

JULIAN HEICKLEN

Department of Chemistry
Center for Air Environment Studies
and
Ionosphere Research Laboratory
The Pennsylvania State University
University Park, Pennsylvania

The Casali Institute of Applied Chemistry
and
The Department of Environmental Health Sciences
The Hebrew University of Jerusalem
Jerusalem, Israel

ACADEMIC PRESS New York, San Francisco, and London 1976

A Subsidiary of Harcourt Brace Jovanovich, Publishers

ACADEMIC PRESS, INC.
111 Fifth Avenue, New York, New York 10003

United Kingdom Edition published by
ACADEMIC PRESS, INC. (LONDON) LTD.
24/28 Oval Road, London NW1

Library of Congress Cataloging in Publication Data

Heicklen, Julian
Atmospheric chemistry.

Includes bibliographies.
1. Atmospheric chemistry. I. Title.
QC879.6.H44 551.5'11 76-2944
ISBN 0-12-336740-9

PRINTED IN THE UNITED STATES OF AMERICA

To the future:
Judykins, Allie Pallie, and Debchick

And to those who made it possible:
my wife, Susan, and our parents

If I forget thee, O Jerusalem
Let my right hand forget its cunning
Let my tongue cleave to the roof of my mouth

CONTENTS

PREFACE xi
ACKNOWLEDGMENTS xiii

Chapter I. **Structure of the Atmosphere** 1

Physical Characteristics 1
Chemical Composition 6
Spectroscopy 30
Meteors 44
References 45

Chapter II. **Chemistry of the Upper Atmosphere** 51

Neutral Oxygen Atmosphere 58
Neutral Oxygen–Nitrogen Atmosphere 71
Oxygen–Nitrogen–Hydrogen Atmosphere 83
Carbon–Hydrogen–Oxygen Cycle 93
Overall Summary 97
Perturbations 112
References 120

Chapter III. **The Ionosphere** 122

E and F1 Positive-Ion Cycle 131
N and NO Chemistry 140
D Region 143
References 154

Chapter IV. **Atmospheric Pollutants** 156

CO and CO_2 157
Hydrocarbons 165
Oxides of Nitrogen 185
Oxidant 196
Halogenated Compounds 206
Sulfur Compounds 208
Particulate Matter 223
References 235

Chapter V. **Hydrocarbon Oxidation** 239

Slow Combustion 240
Rapid Combustion 262
References 271

Chapter VI. **Photochemical Smog** 274

Production 274
Hydrocarbon Reactivities 279
Chemistry of the Conversion of NO to NO_2 283
Initiation 286
Termination 290
Model Calculation 291
Oxidant Dosage 298
References 301

Chapter VII. **Reactions of O_3 and Singlet O_2** 304

NO_2 Photolysis 304
Rate Coefficients 310
Reactions of Ozone 312
Singlet O_2 325
References 327

Chapter VIII. **SO_2 Chemistry** 331

Photolysis of SO_2 331
Free Radical Reactions 341
Relative Importance of Homogeneous SO_2 Oxidation Reactions 347
Oxidation in H_2O Solution 348
References 351

Chapter IX. **Aerosol Chemistry** 354

Direct Introduction 356
Photochemical Smog 358
From SO_2 360
Fog Formation 364
References 365

Chapter X. **Control Methods** 367

Stationary Sources 367
Motor Vehicles 372
Chemical Control of the Atmosphere 388
Cost of Air Pollution 393
References 395

INDEX 397

PREFACE

The last comprehensive book on air pollution chemistry was the truly excellent monograph by Philip Leighton entitled "Photochemistry of Air Pollution," which appeared in 1961. Unfortunately it is now hopelessly out of date. Although a number of books have appeared recently on various aspects of upper atmospheric chemistry, they are not particularly comprehensive. In some cases they are restricted to the stratosphere or the ionosphere, or their emphasis is more on the meteorology or physics of the upper atmosphere. None of them relates tropospheric events with those in the upper atmosphere.

Over the past decade our knowledge of atmospheric chemistry has increased tremendously as a result of atmospheric observations and monitoring, laboratory determinations of reaction mechanisms and rates, and atmospheric modeling studies. The recently concluded Climatic Impact Assessment Program of the U.S. Department of Transportation has greatly increased our knowledge of the stratosphere. It seemed appropriate to attempt to correlate this material in order to obtain a comprehensive picture of the atmospheric chemical cycles. No attempt has been made to include details of engineering, meteorology, or health effects, though these have been discussed and included in an elementary manner when appropriate. The aim is to present the main features of the chemistry of the atmosphere in a manner that will prove useful to graduate students and to scientists in the field who wish to study atmospheric chemistry.

ACKNOWLEDGMENTS

This book was drafted and many of its threads were assembled while I was on sabbatical leave at the Casali Institute of Applied Chemistry at The Hebrew University of Jerusalem, 1973–1974. This material was used and developed as a course for the Department of Environmental Health Sciences at The Hebrew University. I sincerely thank Professors Gabriel Stein, Director of the Casali Institute, and Hillel Shuval, Head of the Department of Environmental Health Sciences, as well as members of the Physical Chemistry and Atmospheric Science Departments for making my stay possible and fruitful. In spite of the fact that Israel was engaged in a traumatic war during my visit and normal routines were disrupted, The Hebrew University and Israel continued to be an exciting and interesting place, both professionally and personally.

In developing this book, I drew heavily on the work of others. Extensive conversations with Professor Marcel Nicolet and Dr. Romualdas Simonaitis were especially helpful, as was correspondence with R. R. Baldwin, S. Braslavsky, R. Cadle, P. Crutzen, D. H. Ehhalt, G. Kockarts, F. S. Rowland, and C. F. H. Tipper. I particularly relied on the following publications:

Series of reports issued in 1969–1971 by what is now the Environmental Protection Agency—Rep. Nos. A.P. 49, 50, 52, 62–64, and 84.

G. Fiocco (1971). "Mesospheric Models and Related Experiments." Reidel, Dordrecht, The Netherlands.

D. Garvin and R. F. Hampson (1974). National Bureau of Standards Rep. NBSIR 74-430, "Chemical Kinetics Data Survey VII. Tables of Rate and Photochemical Data for Modelling of the Stratosphere (Revised)."

National Academy of Sciences (U.S.) Reports "Environmental Impact of Stratospheric Flight" (1975), "Report by the Committee on Motor Vehicle Emissions" (1973), and "Vapor Phase Organic Pollutants. Volatile Hydrocarbons and Oxidation Products" (1976).

F. S. Rowland and M. J. Molina (1975). *Rev. Geophys. Space Sci.* **13**, 1, "Chlorofluoromethanes in the Environment."

A. D. Danilov (1970). "Chemistry of the Ionosphere" (Engl. Transl.). Plenum, New York.

J. Garner (1972). Environmental Protection Agency Report, "Hydrocarbon Emission Effects Determination: A Problem Analysis."

D. Olsen and J. L. Haynes (1969). National Air Pollution Control Administration Publication No. APTD 69-43, "Preliminary Air Pollution Survey of Organic Carcinogens."

E. Robinson and R. L. Robbins. Stanford Research Institute Project PR-6755 Final Report, "Sources, Abundance, and Fate of Gaseous Atmospheric Pollutants" (1968) and Supplement (1969).

Advan. Chem. Ser. 76 (1968). "Oxidation of Organic Compounds,"

G. J. Minkoff and C. F. H. Tipper (1962). "Chemistry of Combustion Reactions." Butterworths, London.

M. Bufalini (1971). *Environ. Sci. Tech.* **5**, 685, "Oxidation of Sulfur Dioxide in Polluted Atmospheres: A Review."

American Chemical Society Report (1969). "Cleaning Our Environment. The Chemical Basis for Action."

J. W. Chamberlain (1961). "Physics of the Aurora and Airglow." Academic Press, New York.

Chapter I

I STRUCTURE OF THE ATMOSPHERE

The earth's atmosphere is necessary to support all of life. Its major constituents are nitrogen and oxygen, but many other species are also present, thus making it a complex and interesting chemical system. The other compounds present include all the known ones composed of H, N, and O (except for N_2H_4). In addition the natural atmosphere contains CO, CO_2, CH_4, and CH_2O. Various parts of the upper atmosphere contain ions and electronically excited species. When other species are introduced through meteor entry, volcanic eruption, urban air pollution, or aircraft exhaust, then atmospheric chemistry becomes so complex that all of its details have not been understood even in simplified schemes. In this book we shall attempt to present what is known about the composition and chemistry of the atmosphere.

PHYSICAL CHARACTERISTICS

The structure of the atmosphere has been considered in a number of ways, often by its physical properties.

Temperature

The most common division of the atmosphere has been by its temperature profile, as shown in Fig. I-1. At the surface of the earth, there is a large variation in temperature, but typically a daytime midlatitude temperature is about 290°K. As the altitude increases the temperature drops to a minimum value of about 210°K at about 15 km in the midlatitude regions. Specifically, at the equator the minimum temperature is 190°K at 17 km. At the poles the minimum temperature of 200°K is reached at about 8 km.

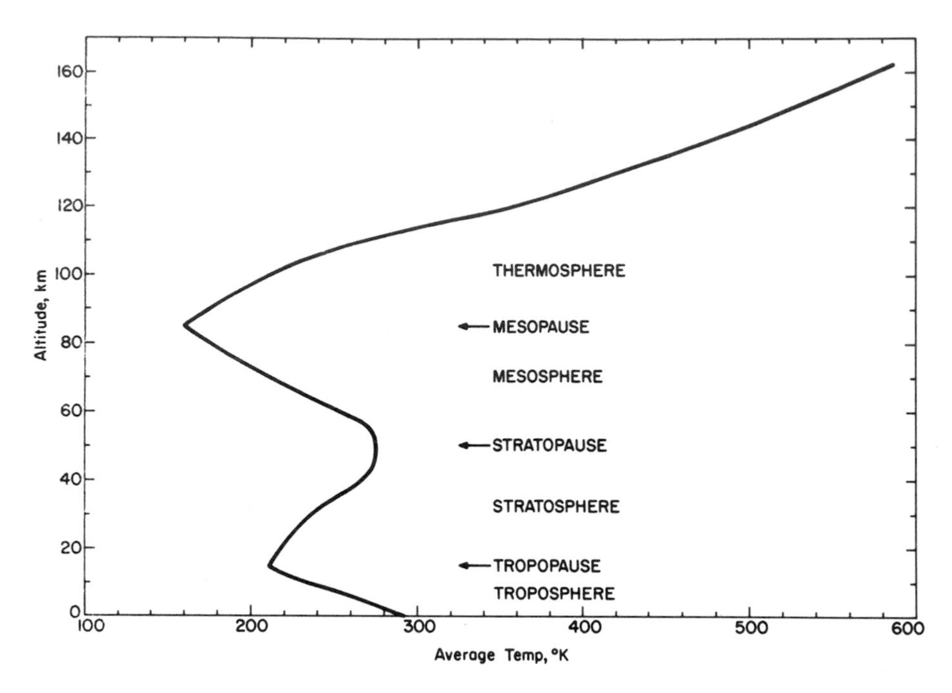

Fig. I-1. Plot of average temperature versus altitude.

This region of declining temperature with increasing altitude near the earth's surface is called the troposphere, and the region of the temperature minimum the tropopause.

Above the tropopause is the stratosphere, a region where the temperature rises with altitude until a maximum value of about 273°K is reached at 50 km at the equator and midlatitude regions. However near the poles there is a wide temperature variation. At Barrow, Alaska (71° N), the temperature maximum at 50 km varies from <240°K in the winter to >280°K in the summer (Theon and Smith, 1971). The altitude of maximum temperature is called the stratopause. Further increases in altitude are then accompanied by a falling temperature through the region known as the mesosphere until a minimum is reached again at 85 km (the mesopause). The minimum value at the mesopause is about 190–200°K at the equator, 170–210°K at midlatitudes, and 130–230°K at high latitudes (Theon and Smith, 1971). The higher end of the range is reached in summer, and the lower end in winter. Further increases in altitude lead to a rapid rise in temperature (5°/km at 150 km) through the region known as the thermosphere, which extends well beyond 150 km.

The reasons for the interesting temperature profile are related to the chemical composition of the atmosphere. At high altitudes, the sun's radiation in the vacuum ultraviolet region of the spectrum (<2000 Å) is being absorbed by atomic and molecular oxygen and nitrogen. This energy becomes heat and high temperatures result. As the altitude drops the vacuum ultraviolet radiation density drops because of this absorption and the temperature falls. However, at about 85 km and below another chemical species, ozone, becomes important. Its concentration increases as the altitude falls, reaching a maximum at about 25 km. It absorbs radiation between 2000 and 3000 Å, i.e., radiation that has not been removed at higher altitudes. Thus this absorption leads to heating, accounting for the mesosphere. At the stratopause (~50 km), enough of this radiation has been removed, so that the solar heating becomes less and less effective for further decreases in altitude, even though the ozone concentration is still rising. The increase in temperature below the tropopause to the earth's surface is due to absorption of the visible and near ultraviolet radiation (>3000 Å) at the earth's surface and reradiation.

Diffusion

Another way of structuring the earth's surface is through its diffusion characteristics. Since diffusion is a temperature- and pressure-dependent function it also changes with altitude. Table I-1 lists the average temperature, pressure, and density as a function of altitude. It can be seen that, although the temperature fluctuates up and down as the altitude is changed, the pressure and density fall continuously as the altitude increases.

Diffusion can arise from two sources. One of these is molecular diffusion, i.e., a molecule moves through the gas because of its individual velocity. This type of diffusion is readily computed from gas kinetic theory and varies inversely with the pressure. The computed value of the molecular diffusion coefficient for N_2 moving through N_2 is shown in Fig. I-2. This type of diffusion is unimportant at low altitudes, but increases in importance as the altitude is raised and the pressure drops. It becomes the dominant diffusion term above about 115 km.

At lower altitudes, the dominant diffusion term is that due to turbulent mixing (convection), referred to as eddy diffusion. The estimated eddy diffusion coefficients for vertical transport also are shown in Fig. I-2. There is considerable uncertainty in their values, so that a range of values is shown, as computed from different models (Hays and Olivero, 1970).

In the troposphere (<15 km) and mesosphere (50–85 km), the eddy diffusion coefficients are larger than in the stratosphere (15–50 km). This is because the temperature drops with rising altitude in the troposphere

TABLE I-1

Average Atmospheric Parameters at Various Altitudes

Altitude (km)	Temperature (°K)	Pressure (Torr)	Density (particles/cm³)
0	291	7.6×10^{2}	2.5×10^{19}
5	266	3.7×10^{2}	1.3×10^{19}
10	231	1.8×10^{2}	7.7×10^{18}
15	211	8.5×10	3.9×10^{18}
20	219	3.9×10	1.7×10^{18}
25	227	1.8×10	7.7×10^{17}
30	235	8.6	3.6×10^{17}
35	252	4.3	1.7×10^{17}
40	268	2.2	8.1×10^{16}
45	274	1.2	4.3×10^{16}
50	274	6.6×10^{-1}	2.3×10^{16}
55	273	3.6×10^{-1}	1.3×10^{16}
60	253	1.9×10^{-1}	7.2×10^{15}
65	232	9.4×10^{-2}	3.9×10^{15}
70	211	4.4×10^{-2}	2.0×10^{15}
75	194	1.9×10^{-2}	9.6×10^{14}
80	177	7.9×10^{-3}	4.2×10^{14}
85	160	2.9×10^{-3}	1.9×10^{14}
90	176	1.4×10^{-3}	7.6×10^{13}
95	193	6.4×10^{-4}	3.2×10^{13}
100	210	2.4×10^{-4}	1.1×10^{13}
105	235	—	—
110	265	5.8×10^{-5}	2.1×10^{12}
115	310	—	—
120	355	2.0×10^{-5}	5.4×10^{11}
125	390	—	—
130	420	8.5×10^{-6}	1.95×10^{11}
135	450	—	—
140	480	4.6×10^{-6}	9.3×10^{10}
145	503	—	—
150	530	2.7×10^{-6}	4.9×10^{10}
155	550	—	—
160	575	1.7×10^{-6}	2.9×10^{10}

and mesosphere; the warmer air at the bottom of these regions rises easily. The stratosphere is a region where the temperature profile rises with altitude, and the tendency for vertical mixing is very small. Thus the stratosphere is the most stable (to mixing) region of the atmosphere.

By the same argument, one might expect that the eddy diffusion coefficient would again be reduced in the thermosphere, where the temperature

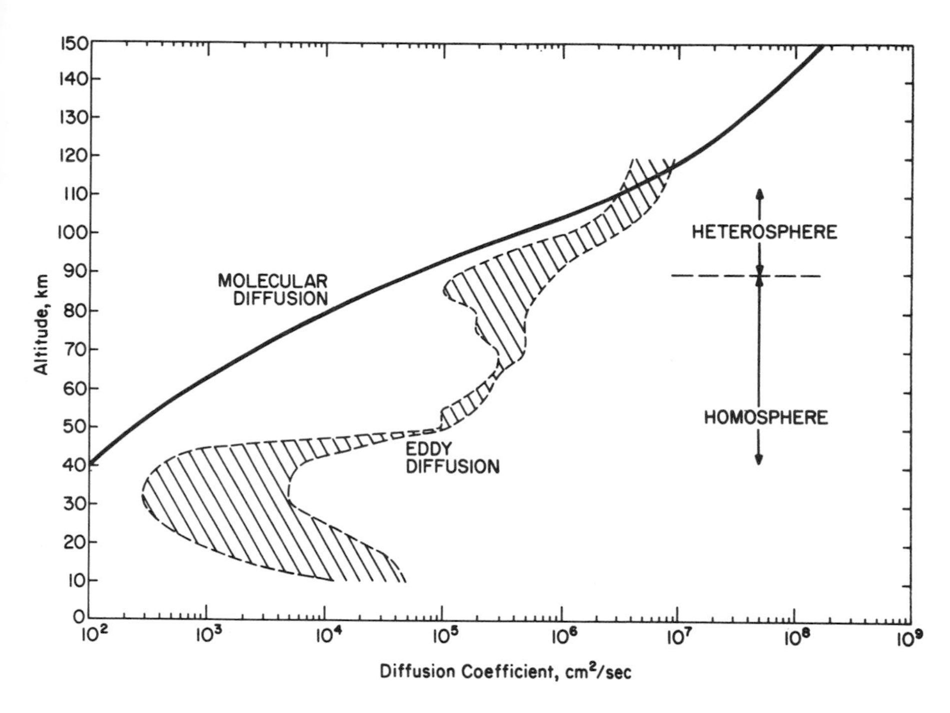

Fig. I-2. Semilog plots of diffusion coefficients versus altitude. Because of the uncertainty in the eddy diffusion coefficient, a range of values is shown. The molecular diffusion curve is for N_2 in N_2.

is again rising with altitude. However, in the region from 85 to 120 km, there are strong vertical shears in the horizontal wind induced by upward propagating gravity waves and atmospheric tides. These wind shears induce turbulence even in regions of increasing temperature so that vertical mixing by eddy diffusion continues to increase with altitude until about 120 km. Above this altitude, eddy diffusion becomes meaningless, because the molecular diffusion is more important, and turbulence is not a distinguishable phenomenon.

The region of the atmosphere below 90 km has a nearly constant mean molecular mass, and is therefore called the homosphere. The composition is essentially 78% N_2, 21% O_2, and 1% Ar. In this region molecular diffusion is negligible compared to eddy diffusion. Above 90 km, the O_2 and N_2 can be dissociated to a significant extent, and the mean mass of the species drops as the altitude increases. In this region, known as the heterosphere, molecular diffusion is significant and, above ~115 km, the dominant diffusion process.

Since eddy diffusion dominates below 115 km, there is thorough mixing of the atmosphere, and gravitational separation is unimportant. However, above 115 km molecular diffusion dominates and gravitational separation occurs. In particular, atomic H, the lightest neutral species, moves to the top of the atmosphere. The other species, all of similar molecular weight (14 for N to 32 for O_2) do not show nearly as much separation. In fact, the gravitational separation effect is not important below $\sim$160 km. We shall confine our discussion in this book to heights $\leq$160 km, since the chemistry is of interest only below this altitude. Therefore, gravitational separation can be ignored. Above 160 km, physical processes are more important than chemical processes in determining the structure of the atmosphere and they will not be considered here.

The subject of horizontal diffusion needs to be considered. Since the eruption of Krakatoa in 1883, it has been known that the mean speed of displacement around the earth is about 120 km/hr. In addition, winds bring matter northward and southward from the equator in the stratosphere, indicating that the meridional circulation of the stratosphere is different from that of the troposphere. Olivier (1942, 1948) has followed meteor trails to measure winds. He found that at 92, 61, and 36 km, the mean wind velocities were 48 m/sec by day, increasing to 56 m/sec at night. The effect of horizontal diffusion and winds is sufficiently small so that we will neglect it in our discussions.

Ionosphere

Another way of describing the atmosphere is through its ionization characteristics. Above about 60 km, ionization is important, and this region is called the ionosphere. It is subdivided into three regimes called the D, E, and F regions. The D region is from 60 to 90 km, and ionization here results primarily from the photoionization of NO. Between 90 and 120 km is the E region. Here the photoionization of O_2 is the most important ion-forming step. In the F region, above 120 km, O, O_2, and N_2 all photoionize, and there is a large amount of photoionization. A full discussion of the ionosphere is given in Chapter III.

CHEMICAL COMPOSITION

The principal species in the atmosphere are O_2 and N_2. They can absorb the sun's radiation and photodissociate. Thus atomic species can be produced that include nitrogen atoms and oxygen atoms in both their ground state, O(^{3}P), and first excited electronic state, O(^{1}D). In addition, O_3 is produced, and its photodissociation leads to the production of the first excited electronic state of molecular oxygen, $O_2(^1\Delta)$. In order to avoid confusion with this state, the ground electronic state of molecular oxygen is, when necessary, given its spectroscopic designation $O_2(^3\Sigma)$.

Several other species also exist in minor amounts in the atmosphere. These include the rare gases, and hydrogen and carbon compounds. The concentrations of a number of these species at the surface of the earth are given in Table I-2. The concentrations of N_2, O_2, CO_2, and the rare gases are constant throughout the homosphere ($\leq$85 km).

In addition to the species listed in Table I-2, there are also many free radicals and ions, as well as other minor constituents containing sulfur, halogens, and metals. Some of these species are present as aerosols. However, in this chapter we shall limit our discussion to neutral species containing only O, N, H, and C. Ionic species are discussed in Chapter III.

TABLE I-2

Relative Composition of Atmosphere at the Earth's Surface for Midlatitudes

Species	Concentration (ppm)	Reference
N_2	780,840	Nicolet (1960)
O_2	209,460	Nicolet (1960)
CO_2	325	Nicolet (1960)
He	5.24	Nicolet (1960)
Ne	18.18	Nicolet (1960)
Ar	9340	Nicolet (1960)
Kr	1.14	Nicolet (1960)
Xe	0.087	Nicolet (1960)
H_2O	Variable	Ehhalt and Heidt (1973a,b)
CH_4	1.4	Ehhalt and Heidt (1973a,b)
H_2	0.5	Ehhalt and Heidt (1973a)
NH_3	6×10^{-3} (estimate)	Robinson and Robbins (1968)
N_2O	0.25	Robinson and Robbins (1969)
CO	0.08	Goldman *et al.* (1973)
O_3	0.025	Hake *et al.* (1973)
NO	2×10^{-3}	Robinson and Robbins (1969)
HNO_3	$\leq$0.02	Ackerman (1975)
NO_2	4×10^{-3}	Robinson and Robbins (1969)
H_2O_2	?	—
CH_2O	?	—

Oxygen Species

In the thermosphere, O_2 dissociation is significant, and the O_2 mole fraction drops. A number of rocket measurements have been made. The values obtained are all about the same, to within a factor of 4, at any height. Mass spectrometric observations from 80 to 120 km made by Krankowsky *et al.* (1968) during daylight at 40–60° north latitude in the summer in the 1960s lie about a factor of 2 higher than daylight solar occultation measurements of Roble and Norton (1972) at 90–120 km made at the same latitudes in the winter of 1967. These measurements extended nicely to the nighttime data at 120–220 km of Hays and Roble (1973b) made by stellar occultation in August 1970 at 4–27° latitude and January 1970 at 48° latitude. There was also no dependence on longitude. However, Hays and Roble also made measurements in January, March, and September 1971 between −16 and 24° latitude and found these O_2 measurements to be about a factor of 2 lower at each altitude than their 1970 measurements. They associated this lowering to reduced solar intensity during the 1971 period. Since the O_2 concentration is almost independent of all parameters but altitude, for simplicity, we consider it stable and adopt the values of Hays and Roble (1973b) for August 1970 above 120 km and those midway between the data of Krankowsky *et al.* (1968) and Roble and Norton (1972) between 80 and 120 km. These values are listed in Table I-3 and are shown in Fig. I-3.

From the density and O_2 concentrations it is possible to compute the N_2 concentrations and, above 90 km, the $O(^3P)$ atom concentrations from the two relationships

$$[O] + [O_2] + [N_2] = \text{number density} \tag{1}$$

$$(\tfrac{1}{2}[O] + [O_2])/\,N_2] = 0.27 \tag{2}$$

These values have been computed and are also tabulated in Table I-3 and plotted in Fig. I-3.

The computed values for $O(^3P)$ can be compared with measured values from rocket flights. The data prior to 1973 were summarized by Offerman and Drescher (1973). The values were obtained between 94 and 102 km and were in the range 5.9×10^{11} atoms/cm^3, somewhat higher than our computed values, which reach a maximum of 4.2×10^{11} atoms/cm^3 at 97 km.

Offermann and Drescher made measurements themselves over the altitude range 90–160 km using two techniques. Below 120 km they measured the $O(^1S)$ emission at 5577 Å. Utilizing known rate constants for formation and removal of $O(^1S)$ they computed the $O(^3P)$ concentration, since it is produced by

$$3O(^3P) \rightarrow O_2 + O(^1S) \tag{3}$$

Above 120 km they measured $O(^3P)$ more directly by mass spectroscopy, since at these altitudes $[O_2] < [O]$ and the m/e 16 cracking peak of O_2 does not cause as great interference as at lower altitudes. They found two very interesting things:

1. The $O(^3P)$ profile showed a flat maximum during the summer.
2. The altitude of the peak concentration was lowest in the summer.

However these variations were not very great and average values of $[O(^3P)]$ were about

Altitude (km):	90	97	120	150
$[O(^3P)]$ (cm^{-3}):	3×10^{11}	6×10^{11}	1.0×10^{11}	2.5×10^{10}

Our computed concentrations are about 70% of their values.

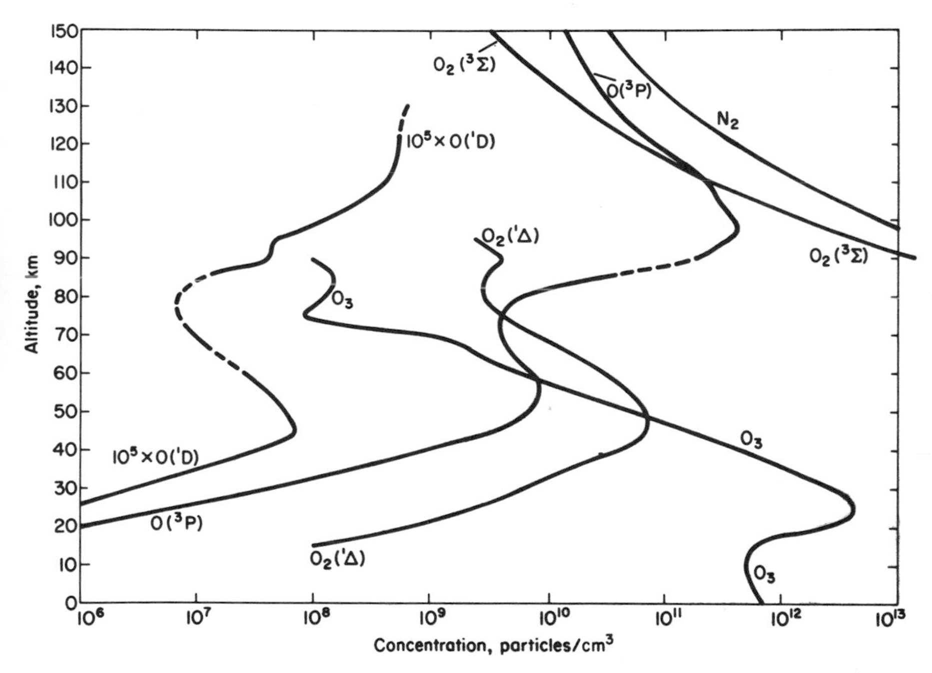

Fig. I-3. Semilog plots of typical concentrations of oxygen and nitrogen species versus altitude for an overhead sun at low latitudes. The $O_2(^1\Delta)$ concentration also includes any contribution from $O_2(^1\Sigma_g{}^+)$.

TABLE I-3

Typical Concentrations of Oxygen and Nitrogen Species at Various Altitudes for an Overhead Sun

Altitude (km)	$[O(^3P)]$ (particles/cm^3)	$[O(^1D)]$ (particles/cm^3)	$[O_2(^3\Sigma)]$ (particles/cm^3)	$[O_2(^1\Delta)]$[a] (particles/cm^3)	[N] (particles/cm^3)	$[N_2]$ (particles/cm^3)
0	—	2.2×10^{-2}	5.3×10^{18}	—	—	2.0×10^{19}
5	—	—	—	—	—	—
10	—	$<10^{-1}$	1.8×10^{18}	—	—	6.9×10^{18}
15	—	$>10^{-1}$	8.1×10^{17}	1×10^{8}	—	—
20	1.0×10^{6}	1	3.6×10^{17}	6×10^{8}	—	1.4×10^{18}
25	7.0×10^{6}	6	1.6×10^{17}	—	—	—
30	4.0×10^{7}	30	7.4×10^{16}	6×10^{9}	—	2.9×10^{17}
35	2.0×10^{8}	100	3.5×10^{16}	—	—	—
40	1.0×10^{9}	330	1.7×10^{16}	4×10^{10}	—	6.4×10^{16}
45	4.0×10^{9}	700	8.9×10^{15}	7×10^{10}	—	—
50	6.5×10^{9}	600	4.8×10^{15}	7×10^{10}	7.4×10^{4}	1.8×10^{16}
55	8.0×10^{9}	450	2.6×10^{15}	—	—	—
60	7.0×10^{9}	300	1.5×10^{15}	3×10^{10}	—	5.7×10^{15}
65	5.0×10^{9}	—	8.2×10^{14}	—	—	—
70	4.0×10^{9}	—	4.2×10^{14}	8×10^{9}	2.9×10^{5}	1.6×10^{15}
75	4.0×10^{9}	—	2.0×10^{14}	—	—	—
80	6.0×10^{9}	—	9.0×10^{13}	3×10^{9}	—	3.3×10^{14}
85	3.0×10^{10}	120	3.7×10^{13}	3×10^{9}	3.7×10^{6}	—
90	2.0×10^{11}	420	1.6×10^{13}	4×10^{9}	—	6.0×10^{13}
95	3.3×10^{11}	500	4.7×10^{12}	2.5×10^{9}	—	—
100	4.0×10^{11}	1200	1.9×10^{12}	—	—	8.6×10^{12}
105	2.9×10^{11}	2200	6.5×10^{11}	—	—	—
110	2.3×10^{11}	4100	3.0×10^{11}	—	7×10^{5}	1.57×10^{12}
115	1.5×10^{11}	5100	1.1×10^{11}	—	—	—

120	8.0×10^{10}	5100	6.0×10^{10}	—	2×10^{6}	3.7×10^{11}
125	4.8×10^{10}	—	3.3×10^{10}	—	—	—
130	3.5×10^{10}	—	2.0×10^{10}	—	3×10^{6}	1.4×10^{11}
135	2.7×10^{10}	—	1.1×10^{10}	—	—	—
140	2.1×10^{10}	—	7.1×10^{9}	—	5×10^{6}	6.5×10^{10}
145	—	—	—	—	—	—
150	1.3×10^{10}	—	2.6×10^{9}	—	6×10^{6}	3.3×10^{10}
155	—	—	—	—	—	—
160	8.0×10^{9}	—	1.2×10^{9}	—	—	2.0×10^{10}

[a] Includes $O_2(^1\Sigma)$, which may be significant at <40 km.

At the same time as Offermann and Drescher's experiments, an extensive study of measurements was made by Donahue *et al.* (1973) from 90 to 125 km. They found that the peak concentration occurred at 97 ± 2 km and that it varied somewhat, but was always $<5 \times 10^{11}$ atoms/cm^3, and that the concentration at 120 km was about 15% of that at 97 km. These values are somewhat lower than those of previous observations, but agree exactly with the values computed from Eqs. (1) and (2).

Donahue *et al.* (1973) also made observations similar to those of Offerman and Drescher (1973):

1. There are large latitudinal variations. The $O(^3P)$ concentration profile goes through a maximum in October in the Northern Hemisphere and in April in the Southern Hemisphere.
2. In addition to the seasonal variations with latitude, there is a very large semiannual variation (more than a factor of 2 at 100 km) in the global average abundance. Maxima occur in April and October, and minima in August and January.
3. The vertical distribution of $O(^3P)$ was independent of the time of day.

The implications of the above observations are striking, as pointed out by Donahue *et al.* (1973). Because of the need to balance production and loss of $O(^3P)$ on a global scale, it seems necessary to invoke a compensating variation in the global efficiency of vertical transport in the thermosphere, which means a variation in the average eddy diffusion coefficient. Also the large latitudinal variations in density require meridional flows of 10 to 50 m/sec at 100 km.

Actually there is a simpler explanation, at least in part, to the latitudinal variation. During the summer, more of the sun's energy is striking the summer hemisphere, more O_2 photodissociates, and $[O(^3P)]$ increases. As a consequence of lower O_2 density, the sun's radiation penetrates deeper and the $[O(^3P)]$ maximum value moves down in altitude.

The most recent measurement for $O(^3P)$ concentrations was made at night by Dickinson *et al.* (1974), using a rocket-borne oxygen atom resonance lamp and monitoring resonance fluorescence. The flight was made from South Uist, Scotland (57°20′ N, 7°20′ W), at 2237 local time on April 1, 1974. The results are shown in Fig. I-4. They agree well with the values of Offerman and Drescher (1973).

The O_3 profile has been measured in the upper atmosphere by a number of techniques. The one most common above 50 km is an occultation experiment in which an instrument carried aloft by a balloon or rocket measures the solar flux at various ultraviolet wavelengths absorbed by O_3.

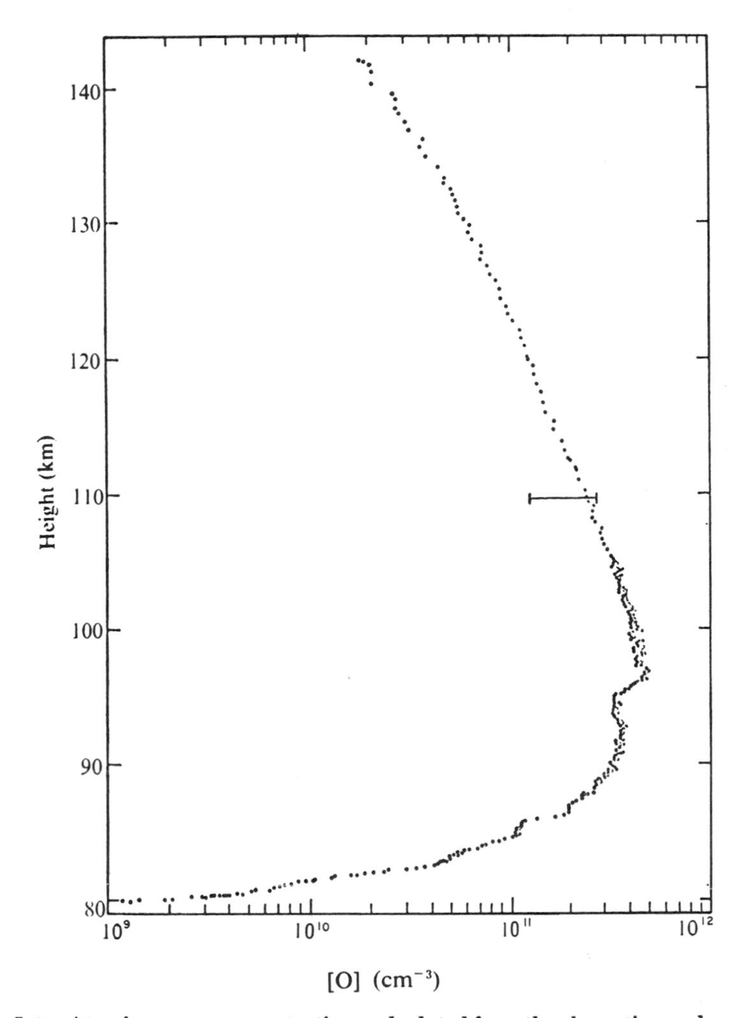

Fig. I-4. Atomic oxygen concentrations calculated from the absorption and resonance fluorescence results as a function of height (up-leg data only). The error bar indicates the maximum range of uncertainty in the absolute scale. From Dickinson *et al.* (1974) with permission of Macmillian Journals Limited.

As the instrument rises and gets closer to the spectral source, the intensity rises, being a function of the column height of O_3 between the source and detector. The method used most often below 50 km is a chemiluminescent method in which the ozone reacts with luminol to produce a chemiluminescence whose intensity is proportional to the O_3 concentration.

Below 50 km, the O_3 concentration profile does not vary during the day, and for <30° latitude there are not day-to-day variations. However, at larger latitudes, the O_3 concentration rises with latitude at any altitude and varies from day to day. In fact, there are seasonal variations in which [O_3] passes through a maximum in the spring and minimum in the autumn at each altitude. In a number of places in the world, the total atmospheric ozone has increased (up to 10% in some places) from 1960 to 1969 (Kulkarni, 1973). Detailed observations by Pittock (1974a,b) over Aspendale, Australia (38.0° S, 145.1° E) from 1965 to 1973 showed that the O_3 density decreased in the troposphere but increased in the upper stratosphere (>27 km). The total column density of O_3 decreased 1.1% over Aspendale during the same period.

It has been postulated that the seasonal variations at upper latitudes are due to a circulation transfer mechanism that moves O_3 from the low-latitude upper stratosphere to the high-latitude lower stratosphere during the winter. This "stored" O_3 is lost during the summer either by movement upward followed by photochemical destruction or movement downward followed by chemical reaction. This theory has not been substantiated as yet.

Mean O_3 concentrations at various latitudes for summer and winter for altitudes <50 km are tabulated in Table I-4, which is taken from Hake *et al.* (1973). The lower 34-km portions of the profiles are based on the systematic program of weekly ozonesonde ascents and are determined by the chemiluminescent method. The mean distribution above 34 km was determined from theoretical computations of the photochemical steady state and from measurements by rocket-borne spectrographs made over New Mexico in 1949 and Wollops Island in 1960. The ground level values of $\sim 7.0 \times 10^{11}$ molecules/cm^3 are somewhat higher than current best background values of 6.0×10^{11} molecules/cm^3 (see Chapter IV). Presumably the decrease with altitude shown in the first few kilometers is not real and [O_3] in this region should be constant or rise slightly.

The data from Table I-4 are also plotted in Fig. I-5 for the summer months. The latitudinal variation below 34 km is readily apparent. The maximum difference occurs at 16 km and is a factor of 7.1. As can be seen from the figure, the total O_3 at any altitude increases with latitude, becoming a maximum at 60°.

Above 50 km there are also a number of measurements. There are apparently no seasonal or latitudinal variations, although there may be diurnal variations. Daytime measurements were made by Johnson *et al.* (1952) and Weeks and Smith (1968). Their results are shown in Fig. I-6 and agree very well where they overlap. Both sets of data were from sun occultation experiments.

TABLE I-4[a]

Vertical Distribution of Ozone Density (molecules/cm³)

Altitude (km)	Tropical		Midlatitude		Subarctic	
	Summer	Winter	Summer	Winter	Summer	Winter
0	7.0×10^{11}	7.0×10^{11}	7.5×10^{11}	7.5×10^{11}	6.1×10^{11}	5.1×10^{11}
1	7.0×10^{11}	7.0×10^{11}	7.5×10^{11}	6.7×10^{11}	6.7×10^{11}	5.1×10^{11}
2	6.8×10^{11}	6.8×10^{11}	7.5×10^{11}	6.1×10^{11}	7.0×10^{11}	5.1×10^{11}
3	6.4×10^{11}	6.4×10^{11}	7.7×10^{11}	6.1×10^{11}	7.2×10^{11}	5.4×10^{11}
4	5.9×10^{11}	5.9×10^{11}	8.0×10^{11}	6.1×10^{11}	7.5×10^{11}	5.6×10^{11}
5	5.6×10^{11}	5.6×10^{11}	8.2×10^{11}	7.2×10^{11}	8.0×10^{11}	5.9×10^{11}
6	5.4×10^{11}	5.4×10^{11}	8.6×10^{11}	8.0×10^{11}	8.9×10^{11}	6.1×10^{11}
7	5.1×10^{11}	5.1×10^{11}	9.4×10^{11}	9.6×10^{11}	9.4×10^{11}	8.9×10^{11}
8	4.9×10^{11}	4.9×10^{11}	9.9×10^{11}	1.1×10^{12}	9.9×10^{11}	1.1×10^{12}
9	4.9×10^{11}	4.9×10^{11}	1.1×10^{12}	1.5×10^{12}	1.4×10^{12}	2.0×10^{12}
10	4.9×10^{11}	4.9×10^{11}	1.1×10^{12}	2.0×10^{12}	1.6×10^{12}	3.0×10^{12}
11	5.1×10^{11}	5.1×10^{11}	1.4×10^{12}	2.6×10^{12}	2.2×10^{12}	4.0×10^{12}
12	5.4×10^{11}	5.4×10^{11}	1.5×10^{12}	3.2×10^{12}	2.6×10^{12}	5.4×10^{12}
13	5.6×10^{11}	5.6×10^{11}	1.9×10^{12}	3.7×10^{12}	3.2×10^{12}	5.9×10^{12}
14	5.6×10^{11}	5.6×10^{11}	2.2×10^{12}	4.0×10^{12}	3.5×10^{12}	6.1×10^{12}
15	5.9×10^{11}	5.9×10^{11}	2.4×10^{12}	4.2×10^{12}	4.0×10^{12}	7.0×10^{12}
16	5.9×10^{11}	5.9×10^{11}	2.6×10^{12}	4.5×10^{12}	4.2×10^{12}	7.7×10^{12}
17	8.6×10^{11}	8.6×10^{11}	3.0×10^{12}	4.9×10^{12}	4.9×10^{12}	7.7×10^{12}
18	1.1×10^{12}	1.1×10^{12}	3.5×10^{12}	5.1×10^{12}	5.1×10^{12}	7.7×10^{12}
19	1.7×10^{12}	1.7×10^{12}	4.0×10^{12}	5.4×10^{12}	5.1×10^{12}	7.5×10^{12}
20	2.4×10^{12}	2.4×10^{12}	4.2×10^{12}	5.6×10^{12}	4.9×10^{12}	7.0×10^{12}
21	3.0×10^{12}	3.0×10^{12}	4.5×10^{12}	5.4×10^{12}	4.5×10^{12}	6.4×10^{12}
22	3.5×10^{12}	3.5×10^{12}	4.5×10^{12}	5.4×10^{12}	4.0×10^{12}	5.9×10^{12}
23	4.0×10^{12}	4.0×10^{12}	4.2×10^{12}	4.9×10^{12}	3.7×10^{12}	5.4×10^{12}
24	4.2×10^{12}	4.2×10^{12}	4.0×10^{12}	4.5×10^{12}	3.5×10^{12}	4.5×10^{12}
25	4.2×10^{12}	4.2×10^{12}	3.7×10^{12}	4.2×10^{12}	3.2×10^{12}	4.0×10^{12}
30	3.0×10^{12}	3.0×10^{12}	2.5×10^{12}	2.4×10^{12}	1.7×10^{12}	1.9×10^{12}
35	1.1×10^{12}	1.1×10^{12}	1.1×10^{12}	1.1×10^{12}	1.1×10^{12}	1.1×10^{12}
40	5.1×10^{11}	5.1×10^{11}	5.1×10^{11}	5.1×10^{11}	5.1×10^{11}	5.1×10^{11}
45	1.6×10^{11}	1.6×10^{11}	1.6×10^{11}	1.6×10^{11}	1.6×10^{11}	1.6×10^{11}
50	5.4×10^{10}	5.4×10^{10}	5.4×10^{10}	5.4×10^{10}	5.4×10^{10}	5.4×10^{10}

[a] From Hake *et al.* (1973) with permission. The data are taken directly from McClatchey *et al.* (1971).

Experiments were made at twilight and sunup on several days in 1967–1969 above White Sands and Churchill by Evans and Llewellyn (1970), who monitored $O_2(^1\Delta)$ emission to obtain the $O_2(^1\Delta)$ concentration profiles and then computed O_3 concentrations from the chemical reaction scheme

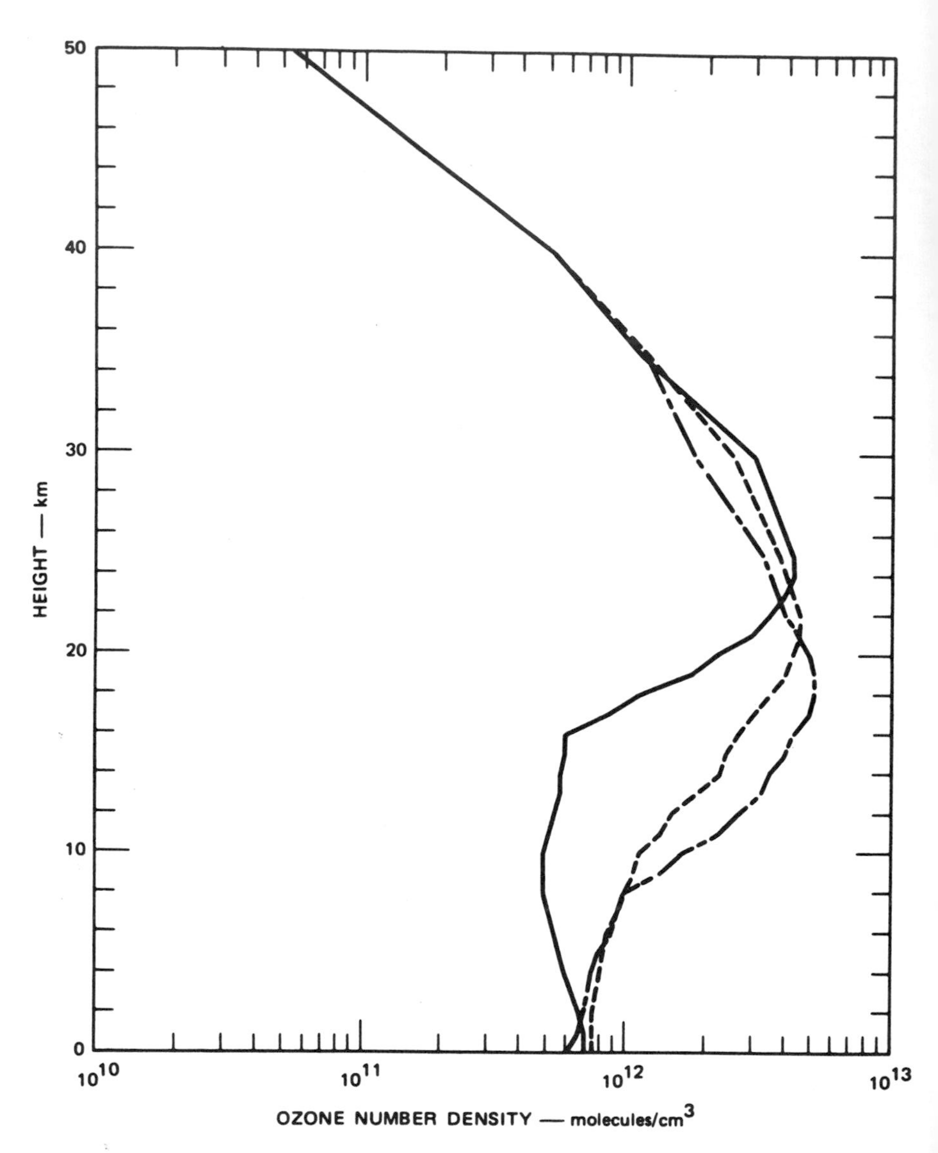

Fig. I-5. Verticle profiles of mean ozone density, 0–50 km. Data taken from McClatchey *et al.* (1971). ——, tropical; – –, midlatitude (summer); – - –, subarctic (summer). From Hake *et al.* (1973) with permission.

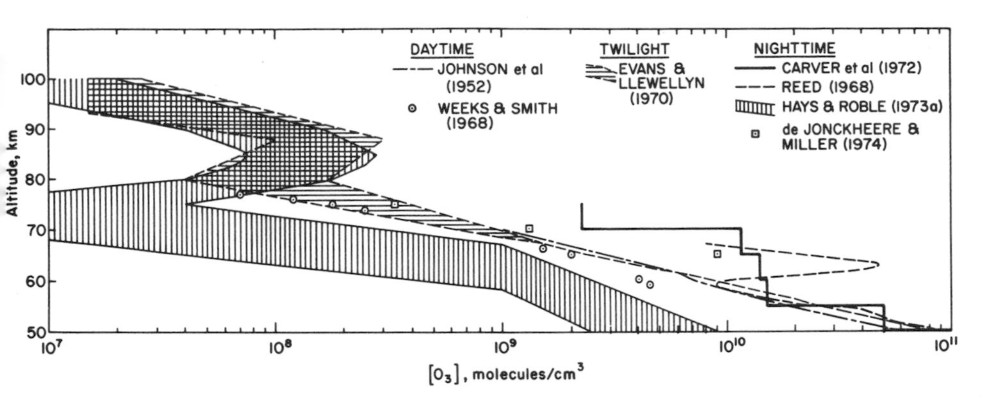

Fig. I-6. Vertical profiles of ozone density, 50–100 km.

(see Chapter II). Their data are also shown in Fig. I-6. They fit the daylight values exactly between 50 and 68 km. Above 68 km the data for different days spread and at higher altitudes encompass a factor of four spread. The lowest values fit those of Weeks and Smith at 74–77 km. However, there is a minimum at 80 and maximum at 88 km, before the downward trend with increasing altitude is again continued. This S-shaped curve was seen in all the flights.

Reliable nighttime observations were made in four studies. Carver *et al.* (1966) used moon occultation at 30°35′ south latitude on December 9, 1965, and above 50 km obtained the histogram shown in Fig. I-6. Reed (1968) used the upper atmosphere night glow as an occultation radiation source on a flight on May 27, 1960, from Wollops Island (37°51′ N). Her results are also plotted in Fig. I-6. Both sets of data lie above the daytime measurements between 60 and 75 km. The peak discrepancy is at 63 km, where Reed's data go through a maximum that is 12–16 times the daylight value. The Carver *et al.* data give $3\frac{1}{2}$–5 times the daylight value. De Jonckheere and Miller (1974) measured the attenuation of moonlight at 2570 Å on a rocket launched from Woomera on December 23, 1972. The observations were made at 65–75 km at 2217 hr local time at 30° south latitude and 136° east longitude. The results, shown in Fig. I-6, agree with those of Carver *et al.* (1972) at 65 and 70 km, but lie between those of Carver *et al.* and Hays and Roble (1973a) at 75 km.

More recently, a series of nighttime measurements at low latitudes were made from January 1970 to August 1971 by stellar occultation by Hays and Roble (1973a). They found that there was little or no variation of $[O_3]$ during the night, but they did obtain day-to-day variations by as much as a factor of four, which were unrelated to season or solar cycle change. Their data are shown in Fig. I-6 and are much lower than everyone else's.

Nevertheless, their results corroborate the observations of Evans and Llewellyn (1970) at high altitudes. In addition to the day-to-day variation they found a minimum of about 75 and a maximum at 85 km. The bulge moves upward to give a maximum at about 90 km at higher latitudes.

The difference between nighttime and daytime values is hard to understand for several reasons:

1. Some experiments give nighttime values higher than daytime values and some give nighttime values lower than daytime values.
2. The daylight and twilight values are the same, and the nighttime values do not vary during the night.
3. The Evans and Llewellyn data at 50 km (and lower) are too low to be compatible with >50 km data.

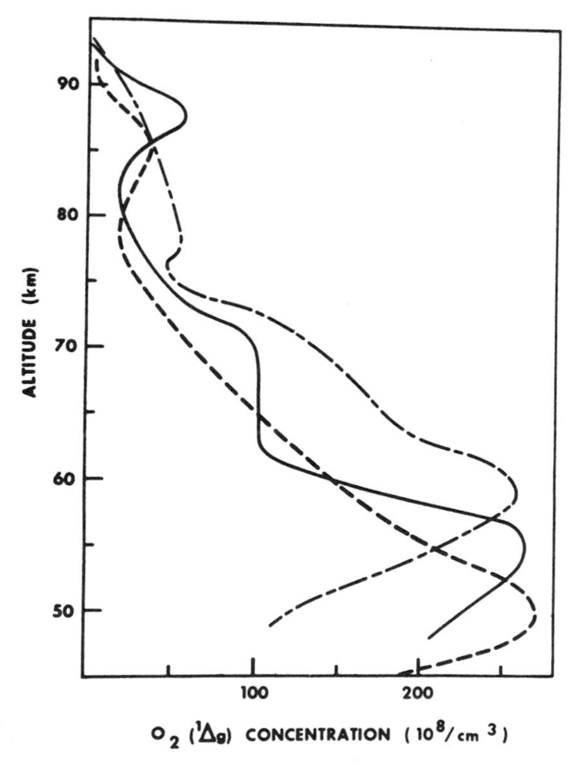

Fig. I-7. $O_2(^1\Delta_g)$ concentration profiles for large solar zenith angles χ. The July 1967 data were taken over Churchill, Manitoba, and the White Sands data were taken on October 11, 1966. ——, χ = 71.8°, P.M., July 11, 1967;- - -, χ = 80.8°, A.M., July 12, 1967; – –, χ = 75.5°, White Sands. From Evans and Llewellyn (1970) with permission of the Centre National de la Recherche Scientifique, Paris.

Thus we conclude that there really is no regular diurnal variation, although there are apparently day-to-day variations of as much as a factor of 4 above 70 km. There is a minimum, or at least a pause, at 75–80 and a maximum at 85–90 km. This maximum is at 85 km at low latitudes and moves upward at higher latitudes.

$O_2(^1\Delta_g)$ measurements have been made from its emission at 1.27 μm. The results to 1970 were reviewed by Evans and Llewellyn (1970) and are shown in Figs. I-7 and I-8. The observations have been made at large solar zenith angles (near sunrise or sunset) to minimize the interference of the sun's radiation. From Fig. I-7, for three sets of measurements made at solar zenith angles between 70 and 80°, the maximum values are all

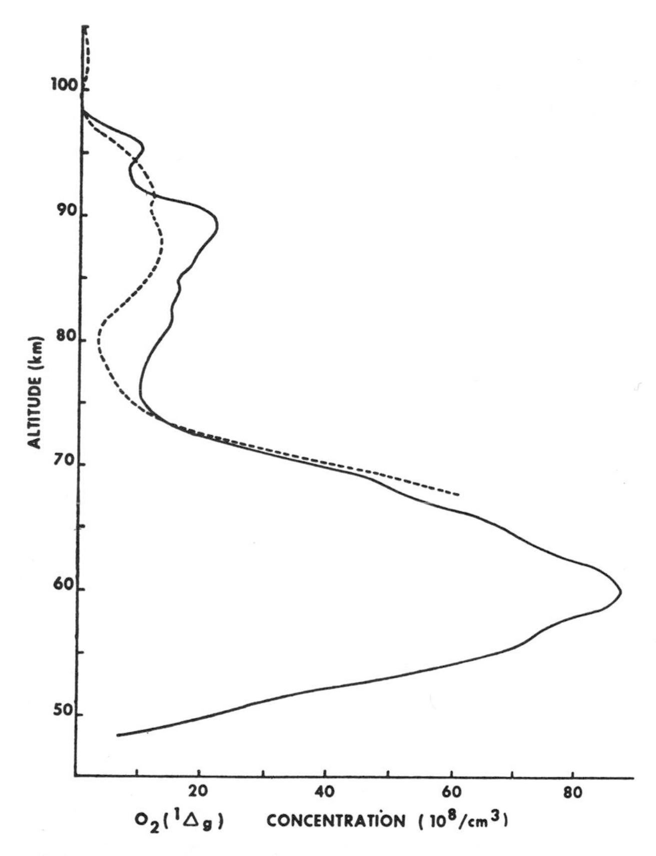

Fig. I-8. $O_2(^1\Delta_g)$ concentration profiles at sunrise and sunset at Churchill. χ = 90°; – – –, morning, March 29, 1969; ——, evening, March 28, 1969. From Evans and Llewellyn (1970) with permission of the Centre National de la Recherche Scientifique, Paris.

about $2.6 \times 10^{10}/cm^3$. However, the maximum in the profile occurs at higher altitudes at the higher latitudes. Also the peak is at a higher altitude for the Churchill measurements near sunrise than near sunset. In Fig. I-8, the observations are for a solar zenith angle $\chi = 90°$ in March at Churchill. The maximum in $[O_2(^1\Delta_g)]$ is about $8.5 \times 10^9/cm^3$, less than $\frac{1}{2}$ the maximum value in July for $\chi = 70–80°$. Thus we see that the $O_2(^1\Delta_g)$ concentrations and height profile distributions are functions of many variables: latitude, time of day, and season.

At night, the $O_2(^1\Delta_g)$ concentration drops to an immeasurably low value except at 70–110 km. Here Baker *et al.* (1974) found a peak value of 2×10^8 molecules/cm^3 at 90 km on two flights flown from Churchill. The results of a third flight depended on the azimuthal angle and showed a peak concentration of 4×10^8 molecules/cm^3. This high result was interpreted to reflect previous auroral activity.

The concentrations of all the oxygen species obtained from observations and chemical calculations (see Chapter II) are shown in Fig. I-3 for comparison purposes for an overhead sun at low latitudes. The solid lines represent observations and the dotted lines computations. Also for comparison the N_2 concentrations are shown.

At altitudes below 90 km, N_2 and $O_2(^3\Sigma)$ are by far the dominant species. However, above 90 km $O(^3P)$ becomes important, and its concentration exceeds that of molecular oxygen above 110 km. The $O(^1D)$ concentration profile more or less follows the same trend as that for $O(^3P)$, but $[O(^1D)]$ is very much smaller. Both $O(^3P)$ and $O(^1D)$ show maxima in the region of the stratopause. Also $O_2(^1\Delta)$ shows a maximum at the stratopause. Consequently, one might expect that a common process is responsible for $O(^3P)$, $O(^1D)$, and $O_2(^1\Delta)$ in the stratosphere and mesosphere, but that another process leading to $O(^3P)$ and $O(^1D)$, but not $O_2(^1\Delta)$, is more important in the thermosphere.

In contrast to the oxygen atom profiles, nitrogen atoms are not significant at all below 110 km. Even above 110 km their concentrations are at least three orders of magnitude less than those of $O(^3P)$. The other important homonuclear species in the atmosphere, O_3, shows a pronounced maximum at 25 km, an altitude much lower than that for the maxima for the other oxygen species.

Oxides of Nitrogen

In recent years a number of measurements have been made on NO, NO_2, N_2O, and HNO_3, principally in the stratosphere. For NO a number of measurements have been reported also for higher altitudes. Apparently the first reliable measurement for NO was made by Barth (1966), who measured the dayglow from the NO γ bands on four Aerobee rocket flights during 1963–1965 in different months at solar zenith angles of 9–64° and at 76–172 km. The NO γ-band intensity was converted to give $[NO] = 3.9 \times 10^7$ molecules/cm^3 at 85 km and a column density of 1.1×10^{14} molecules/cm^2 above 85 km. More recent corrections have raised these values by almost a factor of two.

Meira (1971) measured NO between 70 and 110 km for a solar zenith angle of 64° on January 31 and February 6, 1969, on rocket flights at 37.84° N, 75.48° W. Like Barth, he also measured the NO γ-band dayglow. His NO profile indicated a maximum of 1×10^8/cm^3 at 105 km and a minimum of about 1×10^7/cm^3 at 85 km.

Measurements by Tisone (1973) are more in line with those of Barth. His measurements over the 65–110 km region were taken at sunrise over Hawaii on May 26, 1971. He also looked at NO γ-band emission. The results of the three sets of experiments are shown in Fig. I-9. The three sets of data overlap, but the curve shape of the Meira data is considerably different from the other two.

More recently, Zalpuri and Samayajulu (1974) reported rocket measurements made on March 19, 1970, from a flight from Thumba, India (8.5° N, 76.9° E). The payload included propagation experiment, Lyman-α experiment, and a positive-ion mass spectrometer. From the measurements, derived values for NO concentrations at solar zenith angle of 53° were higher by a factor of 2 than those of Meira (1971) at 67–77 km, and lower by a factor of 2 at 84 km. For a solar zenith angle of 28°, the NO concentrations were lower by a factor of 2 to 4 from those of Meira at <75–84 km, but agreed with those of Meira at 85–90 km.

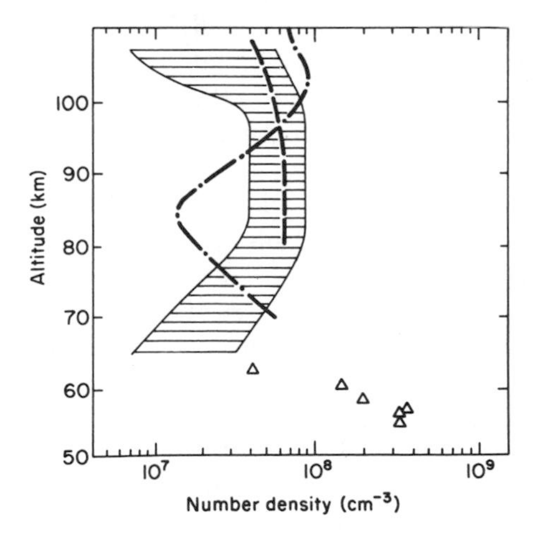

Fig. I-9. Experimental values of the number density of NO above 50 km. Data: – – –, Barth (1966); –•–, Meira (1971); hatched area, Tisone (1973); △, Pontano and Hale (1970). From Ackerman *et al.* (1973) with permission of Macmillan Journals, Ltd.

In the lower mesosphere, Pontano and Hale (1970) measured NO with a headlamp experiment in late 1968 just after sunset. In this experiment the lamp mounted on a rocket flown from the White Sands Missile range emits the Lyman-α line at 1816 Å to ionize NO, which is then measured as the total charge density. Their results, also given in Fig. I-9, show a rising NO profile as the altitude drops. They also report results at 42–55 km that give NO concentrations between 6×10^8 and 1.5×10^9 molecules/cm^3. In a later reevaluation of the data, Hale (1974) has lowered all of these values by a factor of 2.

No direct measurements have been made for NO in the thermosphere. However, Oliver (1974) deduced the NO concentrations above 130 km by observing the change in positive-ion concentrations during the 1970 eclipse of the sun. With this information and the rate coefficient for the charge exchange reaction he deduced the NO concentrations to be about 3×10^7 molecules/cm^3 at 130 km, falling to $<10^7$ molecules/cm^3 at 150 km.

The first measurement in the stratosphere was made by Ackerman *et al.* (1973, 1975). These measurements were taken in the afternoon of May 14, 1973, at 44° north latitude. Absorption spectra were taken from a balloon or Concorde aircraft between 16.5 and 37.5 km. The NO concentration varied with altitude between 10^8 and 10^9 molecules/cm^3, with a maximum occurring at about 25 km, i.e., at the same place as the O_3 maximum. Two other sets of experiments for stratospheric NO were performed at about the same time as those of Ackerman *et al.*

Toth *et al.* (1973) looked at infrared emission from NO at 11–26 km during a series of aircraft flights in March, 1973, in the early mornings over Albuquerque, New Mexico. They found the mean mole fraction of NO to be $(1.0 \pm 0.2) \times 10^{-9}$. On the other hand, Ridley *et al.* (1973, 1974) used the NO–O_3 chemiluminescent reaction to measure NO and found a constant mole fraction of 1.0×10^{-10} between 17.4 and 22.9 km for a balloon flight at 32°50′ N, 106° W, in midmorning March 16, 1973. This result was the same as that at 23.1 km on an earlier balloon flight on December 12, 1972.

The Ridley *et al.* data are exactly a factor of 10 lower than those of Toth *et al.* However, both indicate a drop-off in [NO] as altitude increases, in conflict with the data of Ackerman *et al.*, which show an increase in the same region. The data of Ridley *et al.* and Ackerman *et al.* agree at the tropopause, whereas those of Toth *et al.* and Ackerman *et al.* agree at 25 km.

Patel *et al.* (1974) used a spin–flip Raman laser to measure NO concentrations at 28 km by balloon flights on October 19, 1973, from Palestine, Texas (33° N). The concentration showed a marked dependence on the time of day, being $1.5 \times 10^8/\text{cm}^3$ or less at 7:00 A.M. local time and rising to $2 \times 10^9/\text{cm}^3$ at noon.

Savage *et al.* (1974) measured NO at midlatitudes by using a chemiluminescence technique. Their measurements extended from November 1973 to May 1974, and they found a seasonal variation of the NO concentration with a minimum in January and a maximum at the end of May. These minimum and maximum values are

Altitude (km):	18.3	19.8	21.3
$[NO]_{min}$ (molecules/cm^3):	3×10^8	2×10^8	2×10^8
$[NO]_{max}$ (molecules/cm^3):	1.4×10^9	1.2×10^9	1.2×10^9

Other measurements of NO in the stratosphere have been reported by Fontanella *et al.* (1974) at 43–51° north latitude for sunrise and sunset; by Lowenstein *et al.* (1974) at 38–49° north latitude for midday; and Murcray *et al.* (1968, 1973b, 1974). These results as well as those of the other stratospheric investigations are shown graphically in Fig. I-10.

It is clear that the NO profiles are not yet accurately known. There is no reason why they should depend on time of day, season, or latitude above the stratopause. In fact, different runs by the same experimenters gave the same results. However, in the stratosphere it is clear that there is a seasonal variation as well as a strong diurnal variation for the NO concentration. More work is needed to clarify the reasons for the seasonal variations in the stratosphere as well as to get more precise data in the mesosphere and thermosphere.

The diurnal variation in the stratosphere comes about because NO reacts rapidly with O_3 to form NO_2. During the day the NO_2 photodissociates so that a balance is maintained between NO and NO_2. However, at night the NO converts to NO_2. This conversion was shown most dramatically by Burkhardt *et al.* (1975), who found that between 20 and 28 km the NO concentration was at least an order of magnitude smaller in early morning and late afternoon than during midday, when the concentration was 2.0×10^9 molecules/cm^3 at 28 km. Realizing that NO and NO_2 interconvert, Ackerman *et al.* (1975) measured their sum and found it to correspond to a mole fraction of 1.3×10^{-9} at 20 km and 1.3×10^{-8} at 34 km, with a peak absolute concentration of $(4.2 \pm 1) \times 10^9$ molecules/cm^3 at 26 km.

Apparently, the first measurement of NO_2 in the upper atmosphere was made spectroscopically by Goldman *et al.* (1970). Since then measurements have been made by Ackerman and Muller (1973), Ackerman *et al.* (1973), Murcray *et al.* (1974), Brewer *et al.* (1973, 1974), Farmer *et al.* (1974),

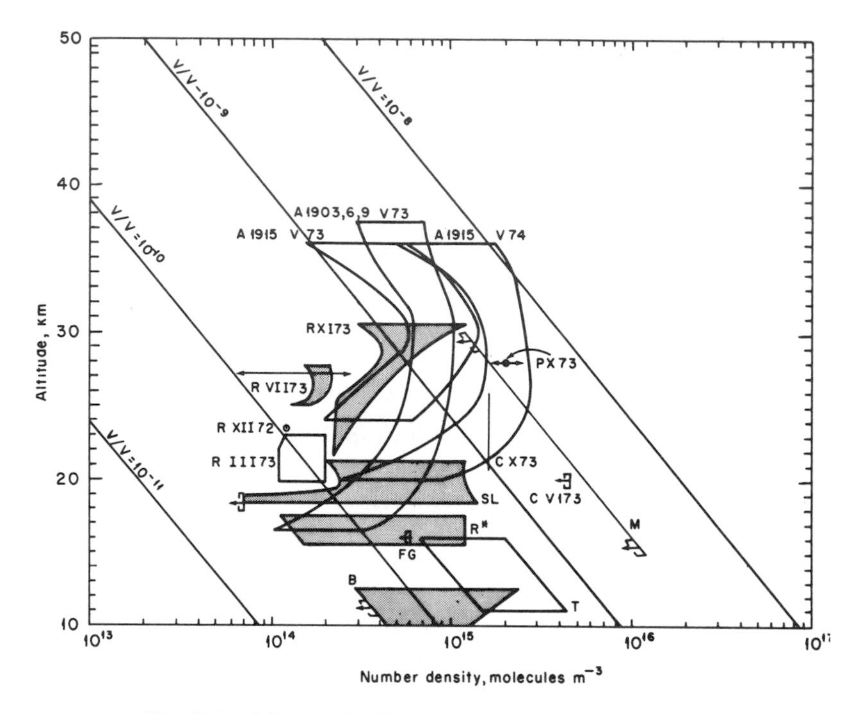

Fig. I-10 Measured NO concentrations in the stratosphere.

A 1903, 6,9 Ackerman *et al.* (1973, 1974), balloon 44° N, sunset V 73, solar absorption at 1903, 1906, and 1909 cm^{-1}, grille spectrometer, authors' uncertainty.

A 1915 Ackerman *et al.* (1975), balloon 44° N, sunsets V 73 and V 74, solar absorption at 1915 cm^{-1}, grille spectrometer, authors' uncertainties in measuring equivalent widths as amplified by inversion.

B Briehl *et al.* (1974), aircraft 25–37° N, midday I–II 74, chemiluminescence, range of observations, and limit of detectability.

C Chanin (1974), balloon 44° N, midday VI and X 73, solar absorption in NO γ-band.

FG Fontanella *et al.* (1974), aircraft 43–51° N, sunrise and sunset VI 73, solar absorption at 1891, 1909, and 1915 cm^{-1}, grille spectrometer, upper limit.

M Murcray *et al.* (1974), balloon 33° N, late afternoon I 72, solar absorption 1875–1915 cm^{-1}, grating spectrometer, upper limit derived from observed spectral features and possible interferences.

P Patel *et al.* (1974), balloon 33° N, midday X 73, spin–flip Raman laser absorption at 1888 cm^{-1}, acoustic detection, authors' uncertainty; temporal behavior from dawn to early afternoon also observed.

R Ridley *et al.* (1974), balloon 33° N, XII 72 and III, VI, IX 73, chemiluminescence. Ranges of data are shown for each date, and authors' uncertainty for a single measurement is shown as a double arrow.

R* Ridley *et al.* (1974), aircraft 9–34° N, afternoon IX 73, and 65–75° N, polar low sun XI 73, chemiluminescence. Range of data is shown.

SL Savage *et al.* (1974), aircraft 25–49° N, midday XI 73–V 74, chemiluminescence. Shown are envelope of data from 13 flights, one result below limit of detectability,

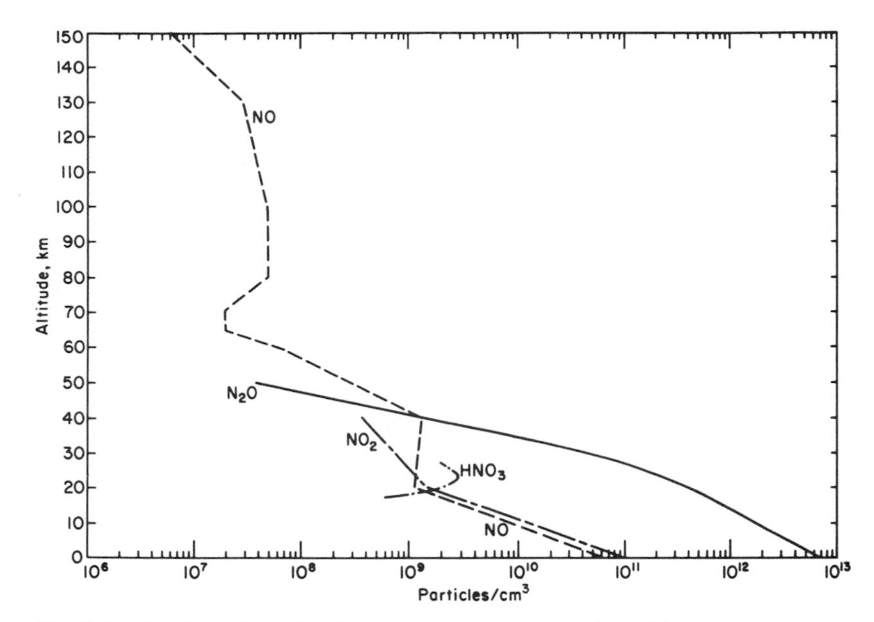

Fig. I-11. Semilog plots of estimated concentrations of the oxides of nitrogen versus altitude for an overhead sun.

and Fontanella *et al.* (1974). All the observations give about the same result, i.e., for midlatitudes the NO_2 concentration is between 10^9 and 10^{11} molecules/cm^3 between 10 and 30 km. The average results are shown in Fig. I-11.

A few measurements have been made for HNO_3 in the stratosphere. They are summarized in Fig. I-12. All the data are for midlatitudes, and they agree fairly well, although there is about a factor of 10 spread. The spread in the data of Murcray *et al.* represent data from four different flights in 1970. Flights in 1971–1972 from Fairbanks, Alaska, give a similar spread, but the average values are higher, almost by a factor of 2 at 23 km (Murcray *et al.*, 1974). Likewise, Lazrus and Gandrud (1974) found much lower concentrations near the equator than at large latitudes. Their equatorial results are shown in Fig. I-10 for altitudes between 15 and 28 km.

---

and double arrow for estimated uncertainty of a single measurement. Strong seasonal variation, with minimum in late winter.

T Toth *et al.* (1973), aircraft 43–51° N, sunrise and sunset X 73, solar absorption 1875–1915 cm^{-1}, interferometer, authors' uncertainty.

Courtesy of T. Hard, U.S. Department of Transportation Climatic Impact Assessment Program.

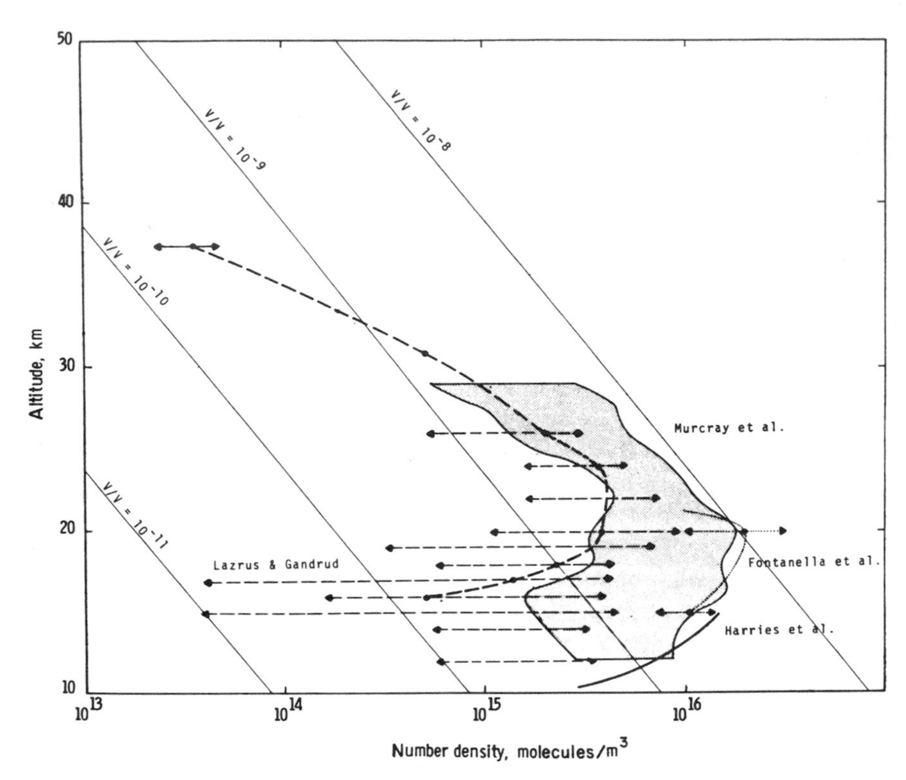

Fig. I-12. Measured nitric acid concentrations in the stratosphere. Data: . . . , Fontanella *et al.* (1974); —, Harries *et al.* (1974); - - -, Lazrus and Gandrud (1974); shaded area, Murcray *et al.* (1973a, 1974). Courtesy of T. Hard, U.S. Department of Transportation, Climatic Impact Assessment Program (1974).

The data also suggest higher concentrations of HNO_3 in winter and spring than in summer and fall for the midlatitude peak concentrations at 21 km.

A few observations on N_2O concentration have been made, all at midlatitudes (Ehhalt *et al.*, 1974; Farmer *et al.*, 1974; Harries *et al.*, 1974; Murcray *et al.*, 1973b; Schütz *et al.*, 1970). The results are all in good agreement and give an N_2O mole fraction of about (2–3) $\times 10^{-7}$ from 10 to 30 km, in good agreement with the concentration at the surface of the earth. However, at higher altitudes the mole fraction falls off to $\sim 3 \times 10^{-9}$ at 45 km and slightly lower at 50 km (Ehhalt *et al.*, 1974, 1975).

Hydrogen Compounds

Evans and Llewelleyn (1973) monitored the HO vibrational emission from $v = 9$ and $O_2(^1\Delta_g)$ emission at 1.27 μm to deduce H, O, and O_3

concentrations. They utilized the reactions

$$H + O_3 \rightarrow HO^{\dagger}\ (v \leq 9) + O_2 \tag{4}$$

and

$$O + O_2 + M \rightarrow O_3 + M \tag{5}$$

The vibrational energy in $HO^{\dagger}$ is removed mainly by radiation, so that its intensity is proportional to the rate of the first reaction above. The O_3 concentration was determined from the $O_2(^1\Delta)$ emission as discussed previously. Thus [H] and [O] could be computed from the known rate constants, respectively, of the above reactions. The results are shown in Fig. I-13. These results are specific for the particular experiment that occurred at twilight on September 12, 1969, above Fort Churchill. The values for [H] and [O] below 90 km are dependent on the solar intensity, and different results should be obtained under these conditions.

Anderson (1971) measured the 3064 Å HO radical electronic transition from resonantly scattered solar radiation on a rocket flight of April 22, 1972, over White Sands, New Mexico, at 1816 hr ($\chi = 86°13'$) between 45 and 70 km. He found $[HO] = 4.4 \times 10^6/cm^3$ between 45 and 50 km, rising to a maximum of $5.5 \times 10^6/cm^3$ at 60 km and then falling to $3.5 \times 10^6/cm^3$ at 70 km. The HO concentration should be dependent on latitude, season, and time of day, so that this reported result is specific for the particular flight.

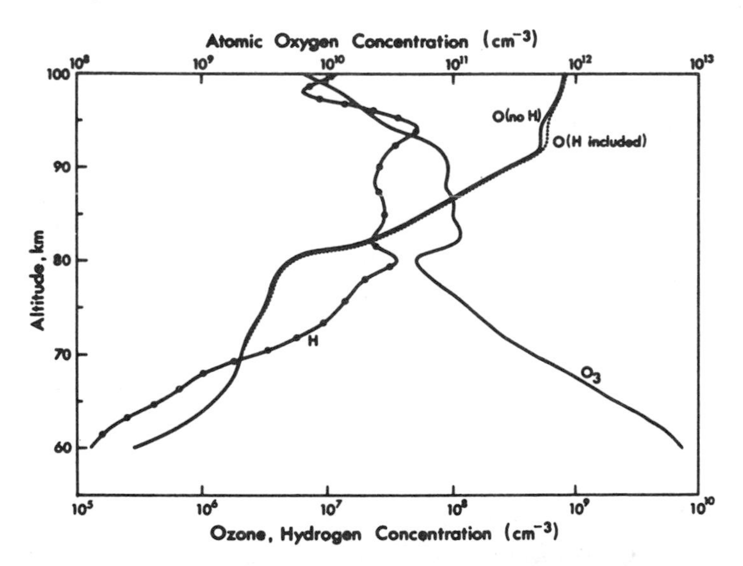

Fig. I-13. Concentrations of H, O, and O_3 in the mesosphere deduced from $O_2(^1\Delta)$ and $HO^{\dagger}$ emission at twilight on September 12, 1969. From Evans and Llewellyn (1973) with permission of the American Geophysical Union.

Ehhalt and co-workers (Ehhalt and Heidt, 1973a) have collected H_2 samples in balloon flights up to 35 km over Mississippi and Texas from 1967 to 1973. In a NASA flight over White Sands, New Mexico, on September 4, 1968, samples were collected between 44 and 62 km. The values are 0.5 ppm throughout the troposphere and about 0.4 ppm at the stratopause. Between these heights there is some variation among the five flights. However, all show a maximum value, the average maximum value being about 0.8 ppm at 27 km. A more recent report by the same group (Ehhalt *et al.*, 1974) shows a smaller maximum value at 27 km and gives a value of about 0.5 ppm for H_2 throughout the lower stratosphere. Ehhalt *et al.* (1975) used a cryogenic air sampler flown on an Aerobee rocket from White Sands Missile Range on May 23, 1973, and found $[H_2] = 0.47 \pm 0.02$ ppm at 40–50 km.

In the same NASA flight mentioned above, Scholz *et al.* (1970) found the H_2O concentrations to be 3–10 ppm at 42–62 km. In the low stratosphere, Mastenbrook (1974) found that for altitudes up to 30 km at midlatitudes (over Washington, D.C.), the ten-year (1964–1973) average mole fraction of H_2O vapor was $(2.6\text{–}2.7) \times 10^{-6}$. There is an annual cycle in the water vapor concentration ranging from 1 to 4 ppm. Furthermore, the average value increased during the first six years of measurements but then leveled off. Evans (1974) and Ackerman (1974) confirmed these findings. The former obtained a H_2O vapor concentration of 3.5 ppm from 10 to 30 km, but this increased to greater than 5 ppm at 45 km. The latter author obtained 3.4 ± 0.7 ppm from 20 to 37 km. A slightly higher value of 4 ppm was found by Hyson (1974) at 15–18 km in November 1972 at 23° and 34° south latitude. On the other hand Burkert *et al.* (1974) found a concentration of 1.5 ppm at 16 km, but 9.5 ppm at 30 km. The results of Murcray *et al.* (1974) also indicated H_2O vapor concentrations of 1.5 ppm or less between 15 and 20 km at midlatitudes, and these values rose to perhaps 6 ppm or more between 20 and 30 km. In the northern latitudes, above Alaska, the H_2O vapor concentration was <0.5 ppm from 17 to 29 km. On the flight of May 23, 1973, Ehhalt *et al.* (1975) found $[H_2O] = 4.0\ (+1.3, -0.9)$ ppm at 40–50 km.

We conclude that the average value for H_2O vapor concentrations at midlatitudes are 2 ± 1 ppm, 15–20 km; 3 ± 1 ppm, 20–30 km; 5 ± 1 ppm, 45 km.

Carbon Compounds

In the flights described above, Ehhalt and Heidt (1973a) also collected CH_4. They combined these data with the earlier data of Bainbridge and Heidt (1966) and Ehhalt *et al.* (1972) to obtain a uniform mole fraction of 1.41 ppm up to 12 km, followed by a linear drop to 0.25 ppm at 50 km. Ehhalt *et al.* (1975) confirmed this dropoff and obtained a CH_4 concentration of 0.37 ± 01 ppm between 40.8 and 50.6 km at 31° north latitude in May 1973. Ehhalt and Heidt (1973b) also made measurements from 0 to 8 km over Santa Barbara, California, and Scottsbluff, Nebraska. They found no systematic seasonal variation in the troposphere. Their one-year average gave 1.41 ppm CH_4.

Lowe and Cumming (1973) measured the infrared absorption of the ν_3 band of CH_4 on a balloon flight between 13 and 28 km on August 11, 1965. They found $[CH_4] = 0.9$ ppm at 13 km, and this value dropped to 0.7 ppm at 28 km. The former value is considerably below, but the latter value agrees very well, with the Heidt measurements.

More recent measurements by Ehhalt *et al.* (1974) as well as Farmer *et al.* (1974) at midlatitudes give about 1.0 ppm for CH_4 between 10 and 30 km. However, Ackerman and Muller (1973) get about 2.2 ppm at 16–25 km, which drops off to 1.0 ppm at 32 km. Burkert *et al.* (1974) report 1.4 ppm at 17 km and <0.1 ppm above 26 km.

In conclusion, the CH_4 concentration can be considered fairly constant at about 1.4 ppm up to 15 km, but then it falls off to <0.4 ppm at 50 km.

The CO distribution in the atmosphere is not well established. In 1961, Seeley and Houghton (1961) derived a constant mole fraction of 0.12 ppm throughout the troposphere by observing lines in the 4.7-μm band by an airborne infrared spectrometer. More recent measurements in the winter over West France (Sieler and Junge, 1969; Sieler and Warnick, 1972) give a constant value of 0.12–0.16 ppm in the troposphere and about 0.04 ppm just above the tropopause, which was at ~9 km. Goldman *et al.* (1973) flew several balloon flights and monitored the CO infrared absorption over Alamogordo in the January 1972 time period. They found $[CO] = 8 \times 10^{-2}$ ppm at 4 km, gradually decreasing to 4×10^{-2} ppm at 15 km. The most recent measurements of Ehhalt *et al.* (1974, 1975) and Farmer *et al.* (1974) show a variation between 0.5 and 1.1×10^{-7} ppm from 10 to 45 km. The highest values were obtained between 27 and 33 km.

The CO_2 mole fraction is constant at ~325 ppm throughout the atmosphere. The most recent measurement in the upper atmosphere at 40–50 km gave 316.2 ± 2.8 ppm (Ehhalt *et al.*, 1975). However, its concentration is rising each year (see Chapter IV).

Other Species

In addition to the species whose estimated concentrations have been measured, other minor neutral species are known to be present. These include N_2O_5, HNO_2, H_2O_2, NH_3, and CH_2O. Certain other free radical species such as HO_2, CH_3O_2, CH_3O, and HCO also must be present.

The concentration profiles of these species can depend on latitude and solar zenith angle. (The angle between the sun's rays and the perpendicular to the earth, i.e., an overhead sun has a solar zenith angle of 0°.) At night many of the species tend to disappear. These include all excited electronic states as well as the unstable intermediates easily removed in second-order processes (e.g., N, HO, HO_2). On the other hand, O_3, $O(^3P)$, and H tend to persist through the night at concentrations in great excess of their steady-state values.

The concentration profiles of the important positive ions and of the electrons are discussed in Chapter III.

SPECTROSCOPY

The energy source for all the chemistry in the atmosphere is the sun's radiation. The radiation intensity per 100 Å interval incident outside the earth's atmosphere (∞ km) and at the surface of the earth (0 km) is shown in Fig. I-14. In the visible region and above (>4000 Å) the sun's intensity is fairly constant. Below 4000 Å, the intensity falls with wavelength to a plateau between 2200 and 2500 Å. Further reductions in wavelength lead to a precipitous fall-off in intensity, with the exception of the spike at 1215.7 Å, which is due to the emission of the Lyman-α line by hydrogen atoms in the sun. However, at the earth's surface all the radiation below 2900 Å has been removed, as well as a significant fraction between 2900 and 3200 Å.

A more detailed estimate of the fall-off in the vacuum ultraviolet intensity is shown in Fig. I-15, which lists the solar flux below 2200 Å at 85, 50, and 30 km. The removal of the radiation below 3200 Å as the sun's rays penetrate the atmosphere is due to absorption by the species present in the atmosphere. A detailed discussion of the absorption processes and solar flux has been given by Ackerman (1971) and Kockarts (1971).

N_2 Absorption

N_2 does not absorb radiation above 1200 Å. It does remove radiation below this wavelength to produce N atoms or N_2^+, but exerts no effect on the longer wavelength radiation.

Nitrogen has been found to be transparent above 1200 Å, and even at wavelengths as low as 910 Å the absorption cross section is extremely low ($<4 \times 10^{-20}$ cm^2). Photoionization begins at 796 Å (14.46 eV) and rapidly

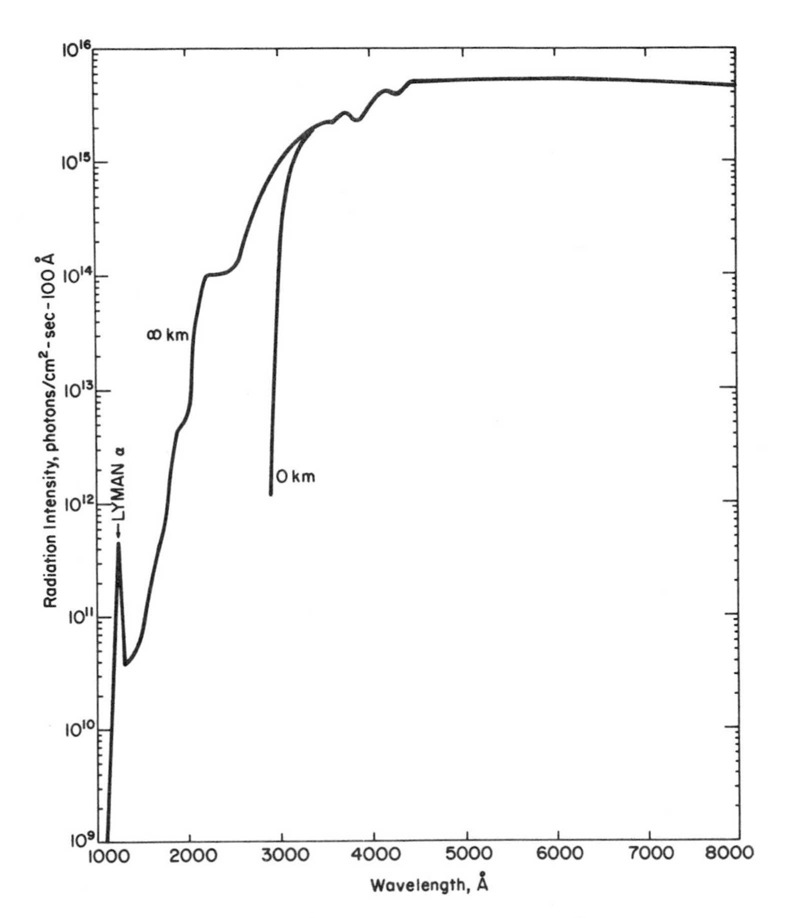

Fig. I-14. Semilog plot of the radiation intensity per 100 Å interval for an overhead sun versus the wavelength of radiation. Curves are shown for ground level and for a very large distance. The spike at 1216 Å is due to the hydrogen-atom emission line (Lyman-α).

reaches unit quantum efficiency. Therefore, N_2 photodissociation only occurs to any measurable extent at wavelengths between 910 and 796 Å. The actual amount of photodissociation due to N_2 absorption is negligible. The compounds responsible for the absorption of the radiation between 1000 and 3200 Å are principally O_2 and O_3.

O_2 Absorption

The absorption spectrum of O_2 from 1040 to 1800 Å is shown in Fig. I-16. There is structure below about 1400 Å, but at higher wavelengths to

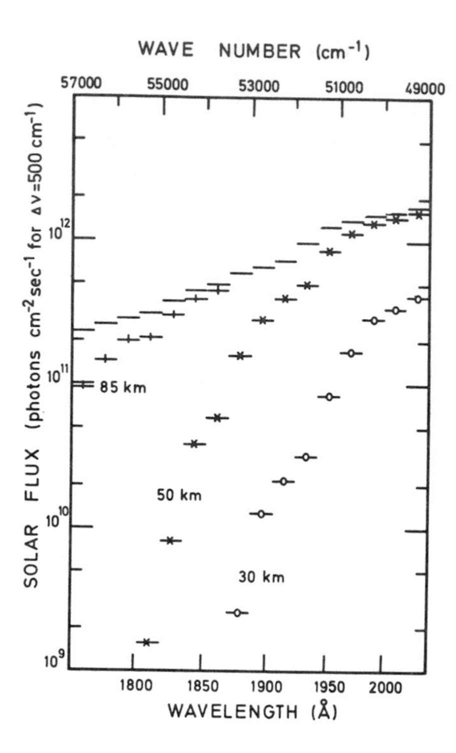

Fig. I-15. Semilog plot of the radiation intensity per 500 cm^{-1} interval for an overhead sun versus the wavelength of radiation. Curves are shown for 30, 50, and 85 km, and for a very large distance. From Kockarts (1971) with permission of Reidel Publ. Co.

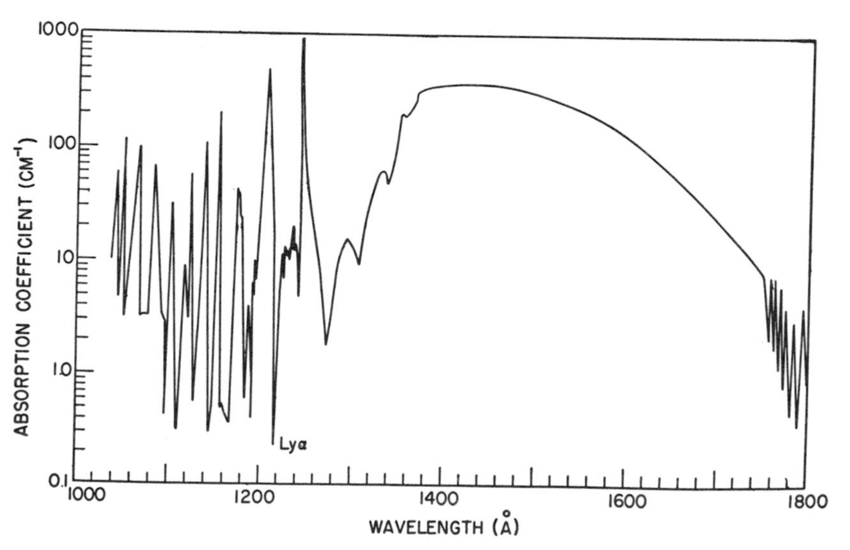

Fig. I-16. Room temperature absorption coefficients of 1 atm of O_2 at 0°C in the region 1000–1800 Å. From the data of Watanabe *et al.* (1953).

about 1740 Å, there is an absorption continuum, called the Schumann–Runge continuum. In this continuum, O_2 photodissociates to produce one ground state and one excited state oxygen atom:

$$O_2 + h\nu\ (<1750\ \text{Å}) \rightarrow O(^3P) + O(^1D) \tag{6a}$$

The energy necessary for this process corresponds to 1750 Å.

It should be noticed in Fig. I-16 that the O_2 absorption has a very pronounced minimum exactly corresponding to the Lyman-α emission at 1215.7 Å. This region is shown in much greater detail in Fig. I-17. This coincidence of the Lyman-α line with the O_2 minimum has great atmospheric significance, because it permits the Lyman-α radiation to penetrate well into the atmosphere (to about 65 km), whereas the other radiation below 1700 Å is effectively removed by O_2 absorption by 85 km.

Above 1800 Å there is a weak continuum superimposed on the Schumann–Runge absorption, which is the Herzberg continuum. It extends to >2400 Å and is shown in Fig. I-18. Absorption into the Herzberg continuum gives two ground state oxygen atoms, a process energetically possible for wavelengths below 2423 Å at 0°K:

$$O_2 + h\nu\ (<2423\ \text{Å}) \rightarrow 2O(^3P) \tag{6b}$$

Since this absorption is very much weaker than that of the Schumann–Runge continuum, the radiation penetrates very much deeper into the atmosphere, all the way to 20 km.

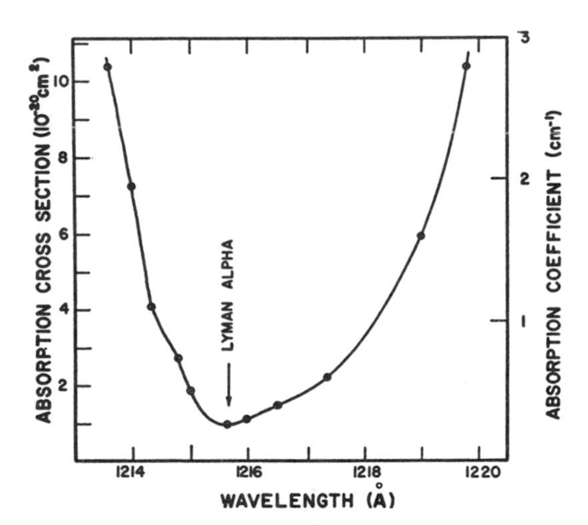

Fig. I-17. Absorption cross section of O_2 in the neighborhood of Lyman-α. From Watanabe (1958) with permission of Academic Press.

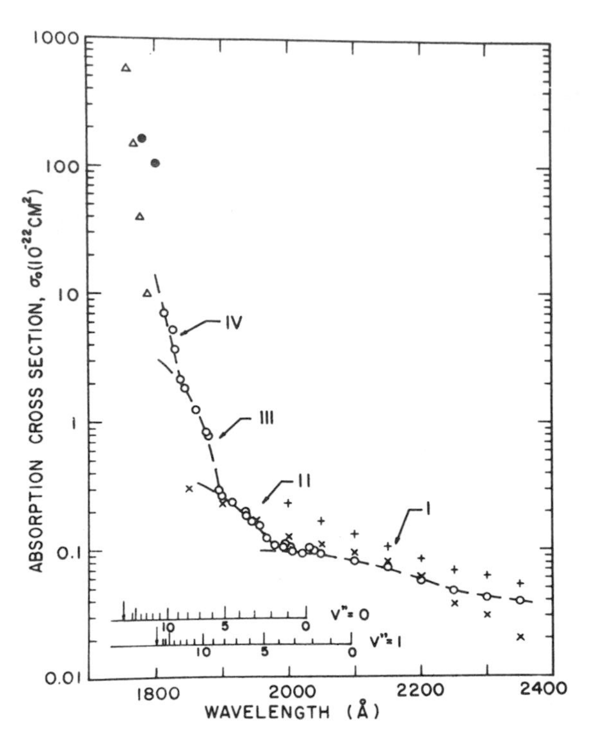

Fig. I-18. Absorption cross section of the O_2 continuum from 1800 to 2400 Å. Data: ●, Wilkinson and Milliken (1957); ×, Ditchburn and Young (1962); △, Hudson, Carter, and Stein [from Hudson and Carter (1968)]; +, Shardanand (1969); ○, Ogawa (1971). Positions of the calculated band origins of the Schumann-Runge absorption bands from $v'' = 0$ and 1 are also indicated by vertical lines. From Ogawa (1971), with permission of the American Institute of Physics.

O_3 Absorption

The radiation between 2200 and 3200 Å is absorbed by O_3 (which has its peak concentration at 25 km) as shown in Fig. I-19. The absorption is very strong and continuous below 3000 Å, a region is known as the Hartley band. Above 3000 Å, the absorption is much weaker and shows structure. This absorption is referred to as the Huggins bands.

The electronic energies of the various excited oxygen species are tabulated in Table I-5, and the wavelengths necessary energetically to produce these products from the photodissociation of O_3 are listed in Table I-6.

Below 3102 Å, there is sufficient energy to photodissociate O_3 to an electronically excited O_2 molecule and O atom at 0°K:

$$O_3 + h\nu\ (<3102\ \text{Å}) \rightarrow O_2(^1\Delta) + O(^1D) \tag{7a}$$

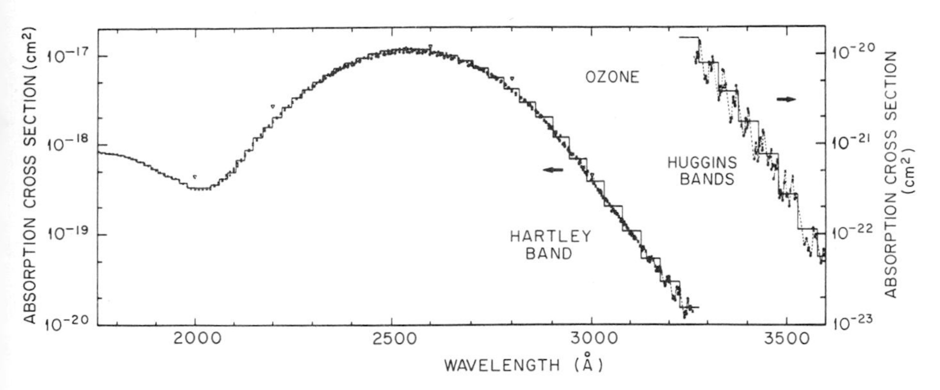

Fig. I-19. Absorption cross sections of O_3 versus wavelength. Data: •, Watanabe *et al.* (1953); ×, Inn and Tanaka (1953); ●, Vigroux (1953); ▼, Paetzold and Regener (1957); ▲, Hearn (1961); —, adopted average values. From Ackerman (1971) with permission of Reidel Publ. Co.

TABLE I-5

Electronic Energy of Excited States of Oxygen Species

Species	Electronic energy (kcal/mole)	Electronic energy (eV)	Wavelength (Å)
$O_2(^1\Delta)$	22.53	0.977	12,686.5
$O_2(^1\Sigma)$	37.51	1.626	7,621.4
$O(^1D)$	45.37	1.967	6,300
$O(^1S)$	96.62	4.189	2,958

TABLE I-6

Wavelength (Å) Needed to Produce the Various Products in O_3 Photodissociation at 0°K[a]

	$O(^3P)$	$O(^1D)$	$O(^1S)$
$O_2(^3\Sigma)$	11,788	4106	2365
$O_2(^1\Delta)$	6,110	3102	1993
$O_2(^1\Sigma)$	4,628	2668	1805

[a] Based on a dissociation energy of O_3 at 0°K of 24.25 kcal/mole and energy of $O(^3P)$ of 58.99 kcal/mole at 0°K (JANAF Tables).

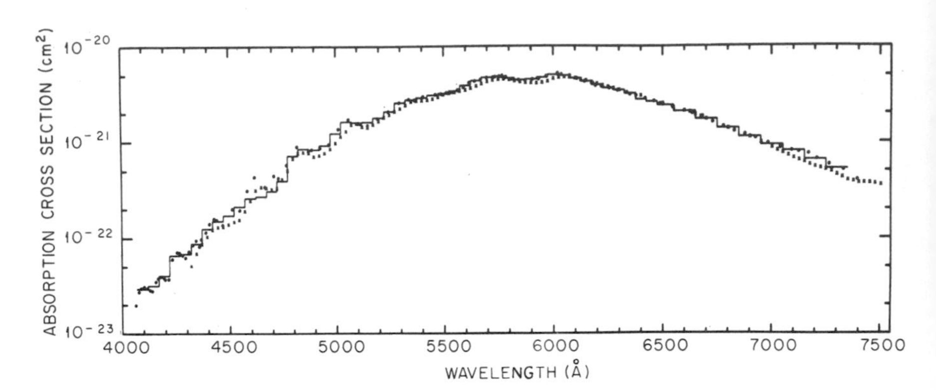

Fig. I-20. **Absorption cross section of ozone versus wavelength in the Chappuis bands. Data key as in Fig. I-19. From Ackerman (1971) with permission of Reidel Publ. Co.**

Reaction (7a) is known to occur with a quantum efficiency of 1 [i.e., every absorption results in reaction (7a) for wavelengths <3100 Å]. At higher wavelengths, the quantum efficiency for $O(^1D)$ drops, becoming zero at 3340 Å (Jones and Wayne, 1970; Castellano and Schumacher, 1972). At 3130 Å the $O(^1D)$ yield depends on the temperature, it being ~0.1 at 220°K and 0.28 at 293°K (Kuis *et al.*, 1974). This temperature dependence can be correlated with the rotational energy in the O_3 molecule.

Even though the quantum efficiency of $O(^1D)$ production drops to zero at 3340 Å, electronically excited O_2 is produced, and at 3340 Å its quantum efficiency of production is one or close to it. Reaction (7a) must be accompanied by the parallel process

$$O_3 + h\nu \rightarrow O(^3P) + O_2(^1\Delta \text{ or } ^1\Sigma) \tag{7b}$$

where either the first ($^1\Delta$) or second ($^1\Sigma$) electronically excited state of O_2 may be produced. At present, the fractions of $^1\Delta$ and $^1\Sigma$ are not known. There is no indication that the other energetically permissible paths

$$O_3 + h\nu \rightarrow O_2(^3\Sigma) + O(^1D) \tag{7c}$$

$$\rightarrow O_2(^3\Sigma) + O(^3P) \tag{7d}$$

occur to produce ground state O_2 for absorption into the Huggin's bands.

Ozone also has a very weak absorption in the visible region of the spectrum, known as the Chappuis band. This absorption, shown in Fig. I-20, always leads to photodissociation to ground-electronic state products

$$O_3 + h\nu\ (>4000\ \text{Å}) \rightarrow O_2(^3\Sigma) + O(^3P) \tag{7d}$$

Airglow

When radiation is absorbed and insufficient energy is available for photodissociation, then a bound excited electronic state can be produced. Such states can also be produced from the recombination of the photodissociation fragments. The energy level diagram showing these electronic levels for the three most important diatomic molecules in the atmosphere, N_2, O_2, and NO, are given in Figs. I-21, I-22, and I-23, respectively. Also shown are the energy levels of their ions. There are very many levels for each molecule, and a number of these play a role in the atmosphere because they can radiate light giving rise to the airglow.

The nighttime airglow, or nightglow, corresponds to the intensity of a candle at 100 m distance (Bates, 1960). The most prominent features come from the NO-O continuum afterglow and the banded structure from the Meinel emission from vibrationally excited HO radicals.

The NO-O afterglow, whose spectrum is shown in Fig. I-24, arises from the emission of electronically excited NO_2 (designated NO_2^*) formed in the combination of NO and $O(^3P)$:

$$NO + O(^3P) + M \rightarrow NO_2^* + M \tag{8}$$

$$NO_2^* \rightarrow NO_2 + h\nu \tag{9}$$

The light emission from reaction (9) appears greenish-yellow. Its spectrum starts at about 4000 Å, reaches a maximum at about 6200 Å, and continues well into the infrared. This emission originates in the region where the product of the NO and $O(^3P)$ concentrations is high. This is at 85–110 km at night, but at much lower altitudes (40–50 km) during the day.

The HO Meinel bands come from the emission of vibrationally excited HO with $v \leq 9$. The spectrum is a series of bands from 5500 to 20,000 Å. The vibrationally excited HO radicals are produced from the reaction of H atoms with O_3, and thus originate in the 56–100 km region during both the day and night. The reaction producing the emitting levels is

$$H + O_3 \rightarrow HO^\dagger\ (v \leq 9) + O_2 \tag{4}$$

where the dagger represents vibrational excitation.

Other features of the nightglow include emission from the bound electronically excited levels of O_2 and the so-called O(I) lines of electronically excited oxygen atoms. The characteristics of the various emissions are summarized in Table I-7. The excited O_2 molecules arise from the recombination of $O(^3P)$ atoms, and thus originate at 85–110 km, where the $O(^3P)$ concentration is a maximum. The reaction responsible is

$$2O(^3P) + M \rightarrow O_2^* + M \tag{10}$$

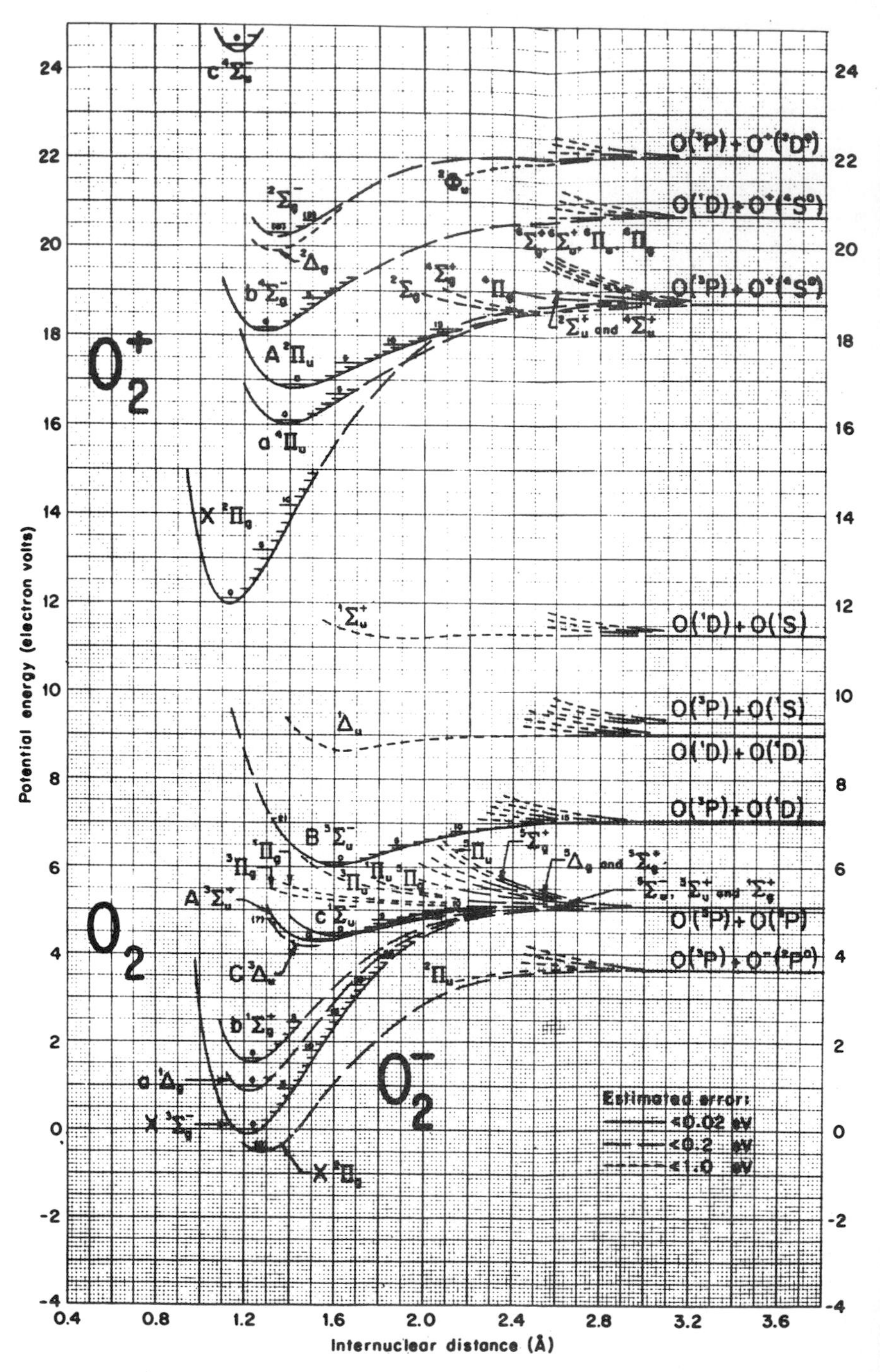

Fig. I-21. Potential-energy curves for O_2^-, O_2, and O_2^+. From Gilmore (1965) with permission of Pergamon Press.

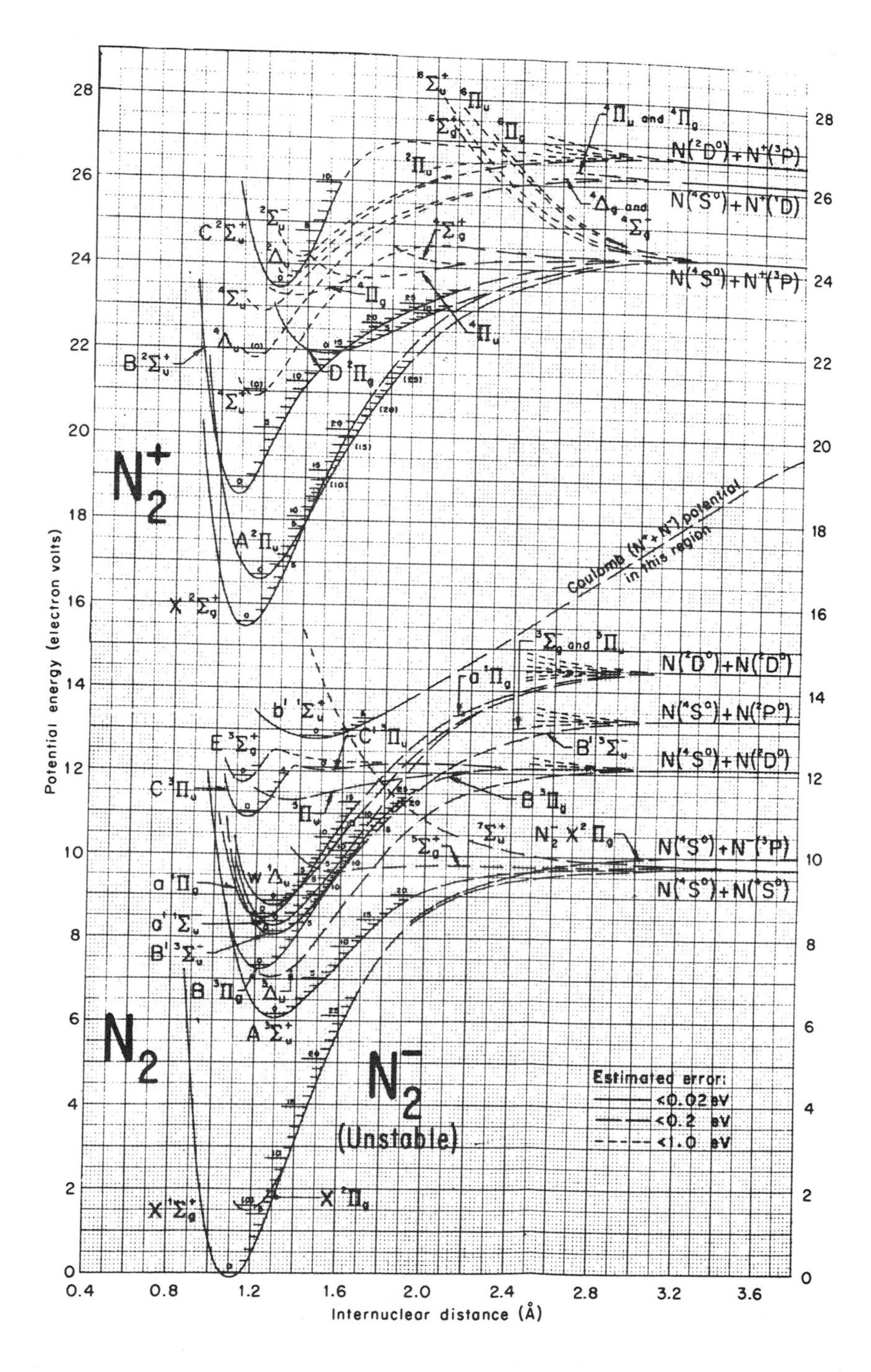

Fig. I-22. Potential-energy curves for N_2^- (unstable), N_2, and N_2^+. From Gilmore (1965) with permission of Pergamon Press.

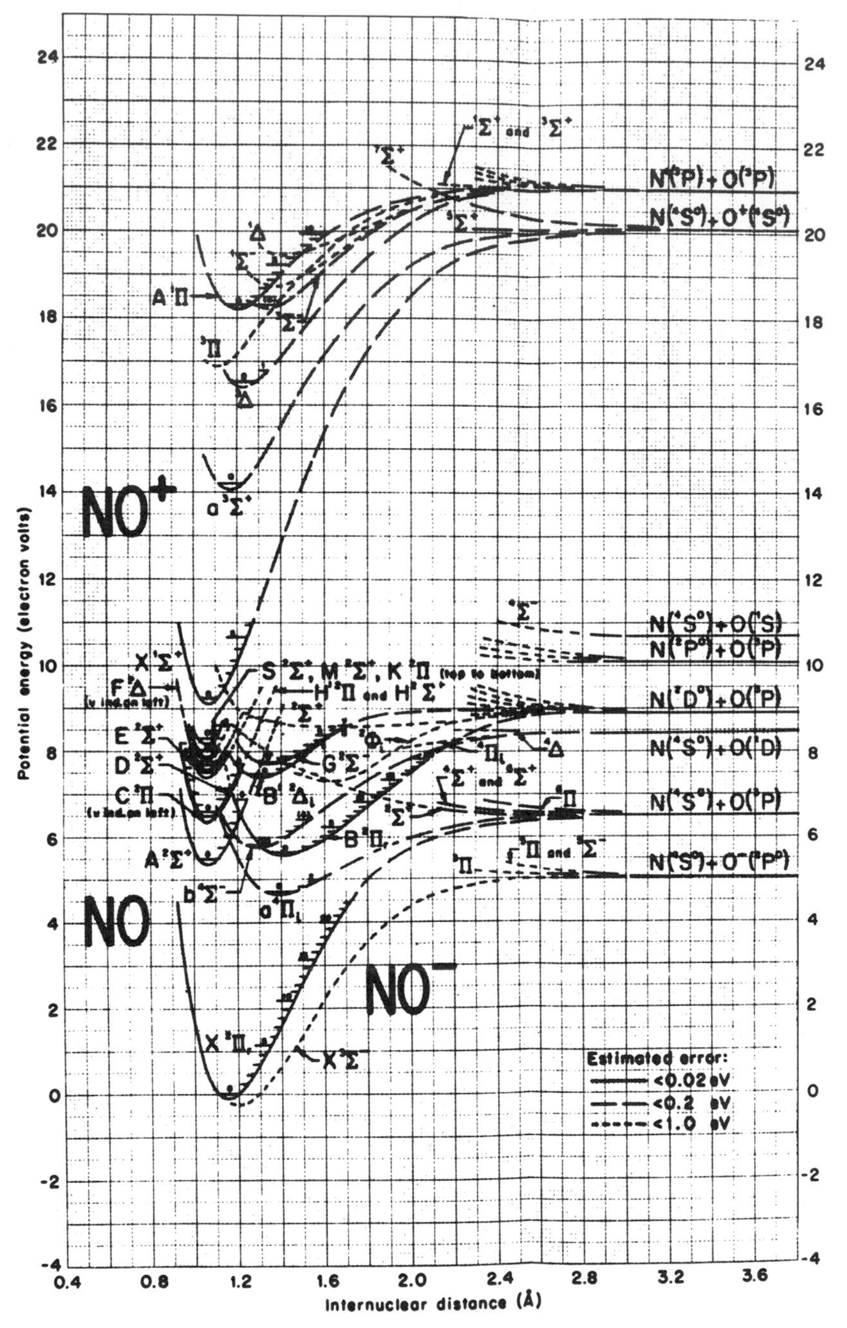

Fig. I-23. Potential-energy curves for NO^-, NO, and NO^+. From Gilmore (1965) with permission of Pergamon Press.

TABLE I-7

Atmospheric Emissions

Emission	Transition	Wavelength region (Å)	Altitude of peak intensity (km)	Reaction producing excited state	Comments
NO γ bands	$A\ ^2\Sigma^+ \rightarrow X\ ^2\Pi$	—	—	$N + O + M \rightarrow NO(A\ ^2\Sigma^+) + M$	Day
NO β bands	$B\ ^2\Pi \rightarrow X\ ^2\Pi$	—	—	$N + O + M \rightarrow NO(B^2\Pi) + M$	Day
O_2 Herzberg bands	$A\ ^3\Sigma_u^+ \rightarrow X\ ^3\Sigma_g^-$	3000–visible (3084–3832)	95 (85–110)	$2O(^3P) + M \rightarrow O_2(A^1\Sigma) + M$	Day, night
O_2 Kaplan–Meinel bands	$b\ ^1\Sigma_g^+ \rightarrow X\ ^3\Sigma_g^-$	8629, 8659	95 (85–110)	$O(^1D) + O_2(^3\Sigma) \rightarrow O(^3P) + O_2(^1\Sigma)$	Night
O_2 infrared bands	$O_2(^1\Delta) \rightarrow O_2(^3\Sigma)$	—	—	$O_3 + h\nu \rightarrow O_2(^1\Delta) + O(^1D)$	—
O(I) green	$O(^1S) \rightarrow O(^1D)$	5577	95 (85–110)	$3O(^3P) \rightarrow O_2(^3\Sigma) + O(^1S)$	Day, night, twilight
O(I) red	$^1D_2 \rightarrow {}^3P_{2,1}$	6300, 6364	>110	$O_2 + h\nu \rightarrow O(^1D) + O(^3P)$	
				$O_2^+(X^2\Pi_g) + e \rightarrow O(^3P) + O(^1D)$	Day, night, twilight
NO–O afterglow	$NO_2^* \rightarrow NO_2$	4000–14,000	85–110 (N)[a] 40–50 (D)[a]	$NO + O(^3P)(+M) \rightarrow NO_2^*(+M)$	Day, night
HO Meinel bands	From $v \leq 9$	5500–20,000	56–100	$H + O_3 \rightarrow HO\dagger + O_2(^3\Sigma)$	Day, night
Na D doublet	$^2P_{1,0} \rightarrow {}^2S_0$	5890, 5896	95 (85–110)	$Na(^2S) + h\nu \rightarrow Na(^2P)$	Night, twilight
N(I)	$^2D \rightarrow {}^4S$	5199	—	$N_2^+(X^2\Sigma_g^+) + e \rightarrow N(^4S) + N(^2D)$	Twilight
O_2	$^3\Delta_u \rightarrow a\ \Delta_g$	—		$2O(^3P) + M \rightarrow O_2(^3\Delta_u) + M$	Night
N_2^+ (first negative)	$B\ ^2\Sigma_u \rightarrow X\ ^2\Sigma_g^+$	—	>100	$N_2^+(X^2\Sigma_g^+) + h\nu \rightarrow N_2^+(B^2\Sigma_u^+)$	Twilight

[a] (N) = night, (D) = day.

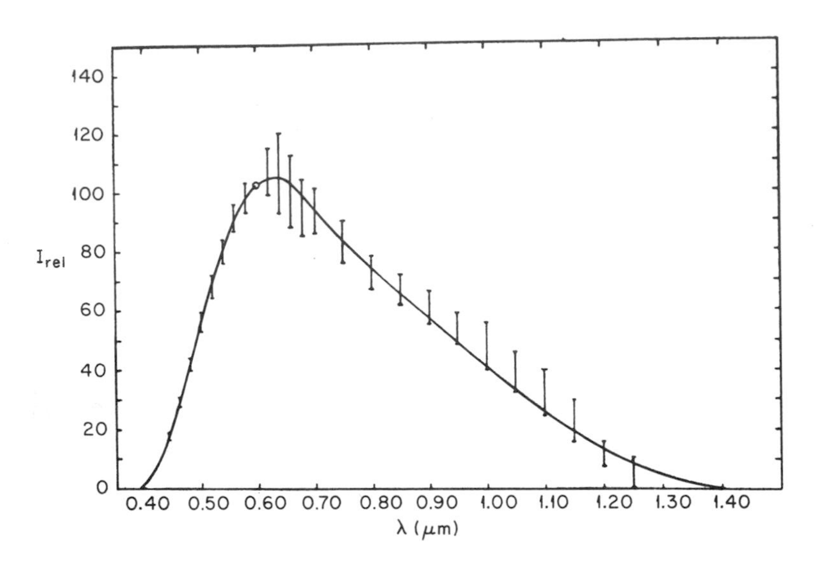

Fig. I-24. **Relative spectral distribution of the NO–O glow. From Fontijn *et al.* (1964) with permission of the American Institute of Physics.**

where O_2^* represents the first five electronically excited states of O_2 (See Fig. I-21). The combination reaction also produces $O(^1S)$ if the chaperone is $O(^3P)$ itself:

$$3O(^3P) \rightarrow O_2 + O(^1S) \tag{3}$$

Reaction (3) is the chief source of $O(^1S)$ production, and thus its emission at 5577 Å also originates at 90–110 km, with a peak emission at 97 ± 2 km. This green line comprises about 6 to 9% of the total *visual* brightness of the night sky (Chamberlain, 1961). However, the eye is particularly sensitive to green light, so that the actual photon intensity is much less.

The $O_2(^1\Delta)$ state arises primarily from the photodissociation of O_3. Thus one might expect its emission to originate at the height of the O_3 maximum, i.e., 25 km. However, $O_2(^1\Delta)$ is removed readily by reaction with O_3, so that its maximum is displaced upward to 50 km.

The $O(^1D)$ atom is produced by O_2 photodissociation for incident wavelengths <1750 Å, i.e., in the thermosphere, and by photodissociation of O_3 in the stratosphere. However, it is removed readily by collision with N_2 or O_2, so that its concentration is not large in the stratosphere. Its maximum concentration (and thus the location of the emission doublet at 6300 and 6364 Å) is >110 km during the day and >160 km at night. Because of the removal of $O(^1D)$ atoms by collision, their red lines slowly fade in brightness during the night and the maximum point of intensity moves upward. The removal of O (^{1}D) by O_2 leads to an additional source of O_2 ($^1\Sigma$):

$$O(^1D) + O_2 \rightarrow O(^3P) + O_2(^1\Sigma) \tag{11}$$

During the day this can lead to emission from O_2 ($^1\Sigma$) well into the mesosphere and even the stratosphere. The emission of O_2 ($^1\Sigma$) to O_2 ($^3\Sigma$) is known as the O_2 Atmospheric System. Meinel (1950) identified the 0–1 band whose origin is at 8645 Å with maxima in the P and R branches at about 8659 and 8629 Å, respectively. No emission has been observed from the 0–0 band, which is evidently reabsorbed in the lower atmosphere.

There is one other feature of the nightglow arising at 75–100 km with peak emission at 93 km. This is the emission from the Na D doublet at 5890 and 5896 Å. Presumably the electronically excited ^{2}P state of Na arises from electronic energy transfer from some other electronically excited species or from chemical reaction. The reaction proposed by Chapman,

$$NaO + O(^3P) \rightarrow Na(^2P) + O_2 \tag{12}$$

is not exothermic. Bates and Nicolet (1950) suggested

$$NaH + O \rightarrow Na(^2P) + OH \tag{13}$$

and Bates (1954) proposed

$$NaH + H \rightarrow Na(^2P) + H_2 \tag{14}$$

In addition to the above features, there is considerable infrared emission due to CO_2, O_3, H_2O, and at $\lambda > 15\ \mu m$, N_2O. For every strong emission there is a corresponding absorption minimum in the solar spectrum. The maximum intensity appears between 7 and 8 μm, but major bands appear at 4.3, 6.3, 9.6, and 15 μm.

All of the emissions of the nightglow are also present during the day (and at twilight where most observations have been made). However, a number of other emissions are also present during the day. These include the N_2^+ first negative emission ($B\ ^2\Sigma_u^+ \rightarrow X\ ^2\Sigma_g^+$), which originates at 90–125 km (closer to 125 km) from direct absorption by ground state N_2^+. The strongest lines are from the 0–0 band at 3914 Å and the 0–1 band at 4278 Å. There are several emissions resulting from the presence of N atoms (which are absent at night) during the day. These include doublet emission at 5199 Å (5198 and 5200) from the ^{2}D state of N atoms produced in ion annihilation,

$$N_2^+ + e^- \rightarrow N(^4S) + N(^2D) \tag{15}$$

$$NO^+ + e^- \rightarrow O(^3P) + N(^2D) \tag{16}$$

$$N_2^+ + O(^3P) \rightarrow NO^+ + N(^2D) \tag{17}$$

The reaction of N with O(^{3}P) gives rise to many electronic levels of NO, but the most important for emission are the A $^2\Sigma^+$ state, which gives rise to the γ-bands, and the B $^2\Pi$ state, which gives rise to the β-bands. In addition the 0–1 infrared Atmospheric (a $^1\Delta \rightarrow {}^3\Sigma$) O_2 band at 1.58 μm has been studied.

The Na D doublet at 5896 and 5890 Å is much stronger during the day than at night because of direct absorption of radiation as the primary source of Na(^{2}P). On some occasions, the Ca^{2+} resonance lines ($4s^2S_{1/2} - 4p^2P^0_{1/2,3/2}$) at 3968.5 and 3933.7 Å have been seen, as well as the analogous transition in Li^+, which gives a close doublet at 6708 Å (6707.89 and 6707.74 Å).

METEORS

Meteors that enter the atmosphere are small pieces of stone or iron, usually from comets' tails, which enter the atmosphere and evaporate at 80–120 km above the earth's surface. Presumably they are the source of the trace metals, e.g., Na, Ca, Li, found in the upper atmosphere. There are about 10^8 meteor impacts per day of meteors of sufficient size ($>10^{-3}$ gm) to be visible to the naked eye upon evaporation. The number of impacts increases as the size diminishes and is about 10^{15} per day for particles of 10^{-13} gm. For particles of mass $<10^{-10}$ gm, evaporation is not complete, and the residual unevaporated material falls to the earth as dust. Probably many millions of tons of dust are collected this way each year. From all meteors, 550 kg is believed to reach the earth each day (Greenhow and Lovell, 1960).

The mean free paths of the atmospheric atoms at 80–110 km are large compared to the size of the meteor. Consequently evaporation may result from the impact of only a few (or even one) gaseous molecule with the meteor. The modern theory of meteor evaporation has been worked out mainly by Whipple and his collaborators (Cook *et al.*, 1951; Thomas and Whipple, 1951; Thomas, 1952; Thomas and White, 1953; Smith, 1954; Cook, 1954; Whipple, 1955; Jacchia, 1955). The air molecules impacting on the meteor are trapped and the energy of impact heats the meteor until evaporation can occur. For very large meteors (>1 gm), evaporation is not complete and they fall to the earth, as do the micrometeorites. However, in the micrometeorite case the reason is that the large surface/volume ratio permits radiation of the energy before evaporation occurs.

During entry meteors can cause ionization. It is estimated that the electron production rate by meteors is $\sim 2.3 \times 10^{-4}$ electron/cm^3-sec (Kaiser, 1953, 1955). In the E region this is negligible during the day but may be significant at night, when solar radiation is negligible.

REFERENCES

Ackerman, M. (1971). *In* "Mesospheric Models and Related Experiments" (G. Fiocoo, ed.), p. 149. Reidel, Dordrecht, Holland, "Ultraviolet Solar Radiation Related to Mesospheric Processes."

Ackerman, M. (1974). *Planet. Space Sci.* **22,** 1265, "Stratospheric Water Vapor from High Resolution Infrared Spectra."

Ackerman, M. (1975). *Aeronom. Acta* No. 142, "NO, NO_2, and HNO_3 below 35 km in the Atmosphere."

Ackerman, M., and Muller, C. (1973). *Pure Appl. Geophy.* **106-108,** 1325, "Stratospheric Methane and Nitrogen Dioxide from Infrared Spectra."

Ackerman, M., Frimout, D., Muller, C., Nevejans, D., Fontanella, J. C., Girard, A., and Louisnard, N. (1973). *Nature (London)* **245,** 205, "Stratospheric Nitric Oxide from Infrared Spectra."

Ackerman, M., Frimout, D., Muller, C., Nevejans, D., Fontanella, J. C., Girard, A., Gramont, L., and Louisnard, N. (1974). *Can. J. Chem.* **52,** 1532, "Recent Stratospheric Spectra of NO and NO_2."

Ackerman, M., Fontanella, J. C., Frimout, D., Girard, A., Louisnard, N., and Muller, C. (1975). *Planet. Space Sci.* **23,** 651, "Simultaneous Measurements o NO and NO_2 in the Stratosphere."

Anderson, J. G. (1971). *J. Geophys. Res.* **76,** 7820, "Rocket Measurement of OH in the Mesosphere."

Bainbridge, A. E., and Heidt, L. E. (1966). *Tellus* **18,** 221, "Measurements of Methane in the Troposphere and Lower Stratosphere."

Baker, K. D., Bishop, R. H., and Megill, L. R. (1974). *J. Geophys. Res.* **79,** 243, "Rocket Measurements of $O_2(^1\Delta_g)$ Emissions in the Auroral Zone."

Barth, C. A. (1966). *Ann. Geophys.* **22,** 198, "Nitric Oxide in the Upper Atmosphere."

Bates, D. R. (1954). *In* "The Earth as Planet" (G. P. Kuiper, ed.), p. 576. Univ. of Chicago Press, Chicago, Illinois, "The Physics of the Upper Atmosphere."

Bates, D. R. (1960). *In* "Physics of the Upper Atmosphere" (J. A. Ratcliffe, ed.), Chapter 5. Academic Press, New York, "The Airglow."

Bates, D. R., and Nicolet, M. (1950). *J. Geophys. Res.* **55,** 235, "Theoretical Considerations Regarding the Altitude of the Layer Responsible for the Nocturnal Emission of the Sodium D Lines."

Brewer, A. W., McElroy, C. T., and Kerr, J. B. (1973), *Nature (London)* **246,** 129, "Nitrogen Dioxide Concentrations in the Atmosphere."

Brewer, A. W., McElroy, C. T., and Kerr, J. B. (1974). *Proc. Int. Conf. Structure, Composition, and General Circ. Upper and Lower Atmos. Possible Anthropogenic Pertubations, Melbourne, Australia,* p. 307, "Spectrophotometric Nitrogen Dioxide Measurements."

Briehl, D. C., Hilsenrath, E., Ridley, B. A., and Schiff, H. I. (1974). *Proc. Int. Conf. Envir. Impact Aerospace Ops. High Atmos., 2nd* 11, "In-Site Measurements of Nitric Oxide, Water Vapor, and Ozone from an Aircraft."

Burkert, P., Rabus, D., and Bolle, H. J. (1974). *Proc. Int. Conf. Structure Composition General Circ. Upper and Lower Atmos. Possible Anthropogenic Pertubations, Melbourne, Australia,* p. 267, "Stratospheric Water Vapor and Methane Profiles."

Burkhardt, E. G., Lambert, C. A., and Patel, C. K. N. (1975). *Science* **188,** 1111, "Stratospheric Nitric Oxide: Measurements during Daytime and Sunset."

Carver, J. H., Horton, B. H., and Burger, F. G. (1966). *J. Geophys. Res.* **71,** 4189, "Nocturnal Ozone Distribution in the Upper Atmosphere."

Castellano, E., and Schumacher, H. J. (1972). *Chem. Phys. Lett.* **13,** 625, "The Kinetics and Mechanism of the Photochemical Decomposition of Ozone with Light of 3340 Å Wavelength."

Chamberlain J. W. (1961). "Physics of the Aurora and Airglow," p. 350. Academic Press, New York.

Chanin, M. L. C. (1974). *Abstr. COMESA-COVOS Symp. Oxford,* "Observation of NO by Ultraviolet Absorption of Radiation."

Cook, A. F. (1954). *Astrophys. J.* **120,** 572, "The Physical Theory of Meteors. VI. The Light-Curve."

Cook, M. A., Eyring, H., and Thomas, R. N. (1951). *Astrophys. J.* **113,** 475, "The Physical Theory of Meteors. I. A Reaction-Rate Approach to the Rate of Mass Loss in Meteors."

Cumming, C., and Lowe, R. P. (1973). *J. Geophys. Res.* **78,** 5259, "Balloon-Borne Spectroscopic Measurement of Stratospheric Methane."

De Jonckheere, C. G., and Miller, D. E. (1974). *Planet. Space Sci.* **22,** 497, "A Measurement of the Ozone Concentration from 65 to 75 km at Night."

Dickinson, P. H. G., Bolden, R. C., and Young, R. A. (1974). *Nature* (*London*) **252,** 5481, "Measurement of Atomic Oxygen in the Lower Ionosphere Using a Rocket-Borne Resonance Lamp."

Ditchburn, R. W., and Young, P. A. (1962). *J. Atmos. Terrest. Phys.* **24,** 127, "The Absorption of Molecular Oxygen Between 1850 and 2500 Å."

Donahue, T. M., Guenther, B., and Thomas, R. J. (1973). *J. Geophys. Res.* **78,** 6662, "Distribution of Atomic Oxygen in the Upper Atmosphere Deduced from Ogo 6 Airglow Observations."

Ehhalt, D. H., and Heidt, L. E. (1973a). "Vertical Profiles of Molecular H_2 and CH_4 in the Stratosphere," presented at *AIAA/AMS Int. Conf. Environ. Impact Aerosp. Operations High Atmos., Denver, Colorado.*

Ehhalt, D. H., and Heidt, L. E. (1973b). *J. Geophys. Res.* **78,** 5265, "Vertical Profiles of CH_4 in the Troposphere and Stratosphere."

Ehhalt, D. H., Heidt, L. E., and Martell, E. A. (1972). *J. Geophys. Res.* **77,** 2193, "The Concentration of Atmospheric Methane between 44 and 62 Kilometers Altitude."

Ehhalt, D. H., Heidt, L. E., Lueb, R. H., and Roper, N. (1974). *Proc. Conf. C.I.A.P., 3rd,* p. 153 "Vertical Profiles of CH_4, H_2, CO, N_2O, and CO_2 in the Stratosphere."

Ehhalt, D. H., Heidt, L. E., Lueb, R. H., and Martell, E. A. (1975). *J. Atmos. Sci.* **32,** 163, "Concentrations of CH_4, CO, CO_2, H_2, H_2O and N_2O in the Upper Stratosphere."

Evans, W. F. J. (1974). *Proc. Int. Conf. Structure, Composition General Circ. Upper and Lower Atmos. Possible Anthropogenic Pertubations, Melbourne, Australia,* p. 249, "Rocket Measurements of Water Vapor in the Stratosphere."

Evans, W. F. J., and Llewellyn, E. J. (1970). *Ann. Geophys.* **26,** 167, "Molecular Oxygen Emissions in the Airglow."

Evans, W. F. J., and Llewellyn, E. J. (1973). *J. Geophys. Res.* **78,** 323, "Atomic Hydrogen Concentrations in the Mesosphere and the Hydroxyl Emissions."

Evans, W. F. J., Wood, H. C., and Llewellyn, E. J. (1970). *Planet. Space Sci.* **18,** 1065, "Ground-Based Photometric Observations of the 1.27 μ Band of O_2 in the Twilight Airglow."

Farmer, C. B., Raper, O. F., Toth, R. A., and Schindler, R. A. (1974). *Proc. Conf. C.I.A.P., 3rd,* p. 234, "Recent Results of Aircraft Infrared Observations of the Stratosphere."

Fontanella, J.-C., Girard, A., Giamont, L., and Louisnard, N. (1974). *Proc. Conf. C.I.A.P.*, 3*rd*, p. 217, "Vertical Distribution of NO, NO_2, and HNO_3 as Derived from Stratospheric Absorption Infrared Spectra."

Fontijn, A., Meyer, C. B., and Schiff, H. I. (1964). *J. Chem. Phys.* **40,** 64, "Absolute Quantum Yield Measurements of the NO–O Reaction and Its Use as a Standard for Chemiluminescent Reactions."

Gilmore, F. R. (1965). *J. Quant. Spectrosc. Radiat. Transfer* **5,** 369, "Potential Energy Curves for N_2, NO, O_2 and Corresponding Ions."

Goldman, A., Murcray, D. G., Murcray, F. H., Williams, W., and Bonomo, F. S. (1970). *Nature (London)* **225,** 443, "Identification of the ν_3 NO_2 Band in the Solar Spectrum Observed from a Balloon Borne Spectrometer."

Goldman, A., Murcray, D. G., Murcray, F. H., Williams, W. J., Brooks, J. N., and Bradford, C. M. (1973). *J. Geophys. Res.* **78,** 5273, "Vertical Distribution of CO in the Atmosphere."

Greenhow, J. S., and Lovell, A. C. B. (1960). *In* "Physics of the Upper Atmosphere" (J. A. Ratcliffe, ed.), p. 513. Academic Press, New York, "The Upper Atmosphere and Meteors."

Hake, R. D., Jr., Pierce, E. T., and Viezee, W. (1973). Stanford Res. Inst. Final Rep. 1724, "Stratospheric Electricity."

Hale, L. (1974). *Cospar Methods of Measurements and Results of Lower Ionosphere Structure,* Akademie Verlag, Berlin, p. 219 "Positive Ions in the Mesosphere."

Harries, J. E., Birch, J. R., Fleming, J. W., Stone, N. W. B., Moss, D. G., Swann, N. R. W., and Neill, G. F. (1974). *Proc. Conf. C.I.A.P., 3rd,* p. 197, "Studies of Stratospheric H_2O, O_3, N_2O, and NO_2 from Aircraft."

Hays, P. B., and Olivero, J. J. (1970). *Planet. Space Sci.* **18,** 1729, "Carbon Dioxide and Monoxide above the Troposphere."

Hays, P. B., and Roble, R. G. (1973a). *Planet. Space Sci.* **21,** 273, "Observation of Mesospheric Ozone at Low Latitudes."

Hays, P. B., and Roble, R. G. (1973b). *Planet. Space Sci.* **21,** 339, "Stellar Occultation Measurements of Molecular Oxygen in the Lower Thermosphere."

Hearn, A. G. (1961). *Proc. Phys. Soc.* **78,** 932, "The Absorption of Ozone in the Ultraviolet and Visible Regions of the Spectrum."

Hudson, R. D., and Carter, V. L. (1968). *J. Opt. Soc. Amer.* **58,** 1621, "Absorption of Oxygen at Elevated Temperatures (300–900 K) in the Schumann–Runge System."

Hyson, P. (1974). *Proc. Int. Conf. Structure, Composition General Circ. Upper and Lower Atmos. Possible Anthropogenic Perturbations, Melbourne, Australia,* p. 257, "Recent Measurements of Stratospheric Water Vapor over Australia."

Inn, E. C. Y., and Tanaka, Y. (1953). *J. Opt. Soc. Amer.* **43,** 870, "Absorption Coefficient of Ozone in the Ultraviolet and Visible Regions."

Jacchia, L. G. (1955). *Astrophys. J.* **121,** 521, "The Physical Theory of Meteors. VIII. Fragmentation as Cause of the Faint-Meteor Anomaly."

Johnson, F. S., Purcell, J. D., Tousey, R., and Watanabe, K. (1952). *J. Geophys. Res.* **57,** 157, "Direct Measurements of the Vertical Distribution of Atmospheric Ozone to 70 Kilometers Altitude."

Jones, I. T. N., and Wayne, R. P. (1970). *Proc. Roy. Soc.* **A319,** 273, "The Photolysis of Ozone by Ultraviolet Radiation. IV. Effect of Photolysis Wavelength on Primary Step."

Kaiser, T. R. (1953). *Phil. Mag. Suppl.* **2,** 495, "Radio Echo Studies of Meteor Ionization."

Kaiser, T. R. (1955). *J. Atmos. Terrest. Phys. Suppl.* **2,** 119, "The Incident Flux of Meteors and the Total Meteoric Ionization."

Kockarts, G. (1971). *In* "Mesospheric Models and Related Experiments" (G. Fiocco, ed.), p. 160. Reidel Publ., Dordrecht, Holland, "Penetration of Solar Radiation in the Schumann–Runge Bands of Molecular Oxygen."

Krankowsky, D., Kasprzak, W. T., and Nier, A. O. (1968). *J. Geophys. Res.* **73,** 7291, "Mass Spectrometric Studies of the Composition of the Lower Thermosphere during Summer 1967."

Kuis, S., Simonaitis, R., and Heicklen, J. (1974). *J. Geophys. Res.* **80,** 1328, "Temperature Dependence of the Photolysis of Ozone at 3130 Å."

Kulkarni, R. N. (1973). *Quart. J. Roy. Met. Soc.* **99,** 480, "Ozone Trend and Haze Scattering."

Lazrus, A. L., and Gandrud, B. W. (1974). *Proc. Conf. C.I.A.P., 3rd,* p. 161, "Progress Report on Distribution of Stratospheric Nitric Acid".

Loewenstein, M., Paddock, J. P., Poppoff, I. G., and Savage, H. F. (1974). *Proc. Conf. C.I.A.P., 3rd,* p. 213. "*In-Situ* NO and O_3 Measurements in the Lower Stratosphere from a U-2 Aircraft."

Mastenbrook, H. J. (1974). *Can. J. Chem.* **52,** 1527, "Water Vapor Measurements in the Lower Stratosphere."

McClatchey, R. A., Fenn, R. W., Selby, J. E. A., Volz, F. E., and Garing, J. S. (1971). Environmental Res. Papers No. 354, Air Force Cambridge Res. Lab., Bedford, Massachusetts, "Optical Properties of the Atmosphere" (Revised).

Meinel, A. B. (1950). *Astrophys. J.* **112,** 464, "O_2 Emission Bands in the Infrared Spectrum of the Night Sky."

Meira, L. G. Jr. (1971). *J. Geophys. Res.* **76,** 202, "Rocket Measurements of Upper Atmospheric Nitric Oxide and Their Consequences to the Lower Ionosphere."

Murcray, D. G., Kyle, T. G., Murcray, F. H., and Williams, W. J. (1968). *Nature (London)* **218,** 78, "Nitric Acid and Nitric Oxide in the Lower Stratosphere."

Murcray, D. G., Goldman, A., Csoeke-Poeckh, A., Murcray, F. H., Williams, W. J., and Stocker, R. N. (1973a). *J. Geophys. Res.* **78,** 7033, "Nitric Acid Distribution in the Stratosphere."

Murcray, D. G., Goldman, A., Murcray, F. H., Williams, W. J., Brooks, J. N., and Barber, D. B. (1973b). *Proc. Conf. C.I.A.P., 2nd,* DOT-TSC-OST-73-4, p. 86, "Vertical Distribution of Minor Atmospheric Constituents as Derived from Air-Borne Measurements of Atmospheric Emission and Absorption Infrared Spectra."

Murcray, D. G., Goldman, A., Williams, W. J., Murcray, F. H., Brooks, J. N., Stocker, R. N., and Swider, D. E. (1974). *Proc. Int. Conf. Structure, Composition and General Circ. Upper and Lower Atmos. Possible Anthropogenic Perturbations, Melbourne, Australia,* p. 292, "Stratospheric Mixing Ratio Profiles of Several Trace Gases as Determined from Balloon-Borne Infrared Spectrometers."

Nicolet, M. (1960). *In* "Physics of the Upper Atmosphere" (J. A. Ratcliffe, ed.), p. 17 Academic Press, New York, "The Properties and Constitution of the Upper Atmosphere."

Offermann, D., and Drescher, A. (1973). *J. Geophys. Res.* **78,** 6690, "Atomic Oxygen Densities in the Lower Thermosphere as Derived from *In Situ* 5577 Å Night Airglow and Mass Spectrometer Measurements."

Ogawa, M. (1971). *J. Chem. Phys.* **54,** 2550, "Absorption Cross Sections of O_2 and CO_2 Continua in the Schumann and Far-UV Regions."

Oliver, W. L. (1974). *J. Atmos. Terrest. Phys.* **36,** 801, "Determination of the Nitric Oxide Concentrations from Eclipse Variations of Ion Concentrations."

Olivier, C. P. (1942). *Proc. Amer. Phil. Soc.* **85,** 93, "Long Enduring Meteor Trains."
Olivier, C. P. (1948). *Proc. Amer. Phil. Soc.* **91,** 315, "Long Enduring Meteor Trains" (Second Paper).
Paetzold, H. K., and Regener, E. (1957). *In* "Handbook der Physik," Vol. 48. Springer Verlag, Berlin, "Ozon in der Erdatmosphäre."
Patel, C. K. N., Burkhardt, E. G., and Lambert, C. A. (1974). *Science* **184,** 1173, "Spectroscopic Measurements of Stratospheric Nitric Oxide and Water Vapor."
Pittock, A. B. (1974a). *Proc. Int. Conf. Structure, Composition and General Circ. Upper and Lower Atmos. and Possible Anthropogenic Perturbations, Melbourne, Australia,* p. 455, "Ozone Climatology, Trends and the Monitoring Problem."
Pittock, A. B. (1974b). *Nature (London)* **249,** 5458, "Trends in the Vertical Distribution of Ozone over Australia."
Pontano, B. A., and Hale, L. C. (1970). *Space Res.* **10,** 208, "Measurements of an Ionizable Constituent of the Low Ionosphere Using a Lyman-Alpha Source and Blunt Probe."
Reed, E. I. (1968). *J. Geophys. Res.* **73,** 2951, "A Night Measurement of Mesospheric Ozone by Observations of Ultraviolet Airglow."
Ridley, B. A., Schiff, H. I., Shaw, A. W., Bates, L., Howlett, C., LeVaux, H., Megill, L. R., and Ashenfelter, T. E. (1973). *Nature (London)* **245,** 310, "Measurements *in Situ* of Nitric Oxide in the Stratosphere between 17.4 and 22.9 km."
Ridley, B. A., Shiff, H. I., Shaw, A. W., Megill, L. R., Bates, L., Howlett, C., LeVaux, H., and Ashenfelter, T. E. (1974). *Planet. Space Sci.* **22,** 19, "Measurement of Nitric Oxide in the Stratosphere between 17.4 and 22.9 km."
Robinson, E., and Robbins, R. C. (1968). Stanford Res. Inst. Project PR-6755, Final Rep., "Sources, Abundance, and Fate of Gaseous Atmospheric Pollutants."
Robinson, E., and Robbins, R. C. (1969). Stanford Res. Inst. Project PR-6755, Supple. Rep., "Sources, Abundance, and Fate of Gaseous Atmospheric Pollutants Supplement."
Roble, R. G., and Norton, R. B. (1972). *J. Geophys. Res.* **77,** 3524, "Thermospheric Molecular Oxygen from Solar Extreme-Ultraviolet Occultation Measurements."
Savage, H. F., Loewenstein, M., and Whitten, R. C. (1974). Presented at AMS/AIAA Meeting, San Diego, "*In-Situ* Measurements of NO and O_3 in the Lower Stratosphere."
Scholz, T. G., Ehhalt, D. H., Heidt, L. E., and Martell, E. A. (1970). *J. Geophys. Res.* **75,** 3049, "Water Vapor, Molecular Hydrogen, Methane, and Tritium Concentrations near the Stratopause."
Schütz, K., Junge, C., Beck, R., and Albrecht, B. (1970). *J. Geophys. Res.* **75,** 2230, "Studies of Atmospheric N_2O."
Seeley, J. S., and Houghton, J. T. (1961). *Infrared Phys.* **1,** 116, "Spectroscopic Observations of the Vertical Distribution of Some Minor Constituents of the Atmosphere."
Seiler, W., and Junge, C. (1969). *Tellus* **21, 447,** "Decrease of Carbon Monoxide Mixing Ratio above the Polar Tropopause."
Seiler, W., and Warneck, P. (1972). *J. Geophys. Res.* **77,** 3204, "Decrease of the Carbon Monoxide Mixing Ratio at the Tropopause."
Shardanand. (1969). *Phys. Rev.* **186,** 5, "Absorption Cross Sections of O_2 and O_4 Between 2000 and 2800 Å."
Smith, H. J. (1954). *Astrophys. J.* **119,** 438, "The Physical Theory of Meteors. V. The Masses of Meteor-Flare Fragments."
Theon, J. S., and Smith, W. S. (1971). *In* "Mesospheric Models and Related Experi-

ments" (G. Fiocco, ed.), p. 131. Reidel Publ., Dordrecht, Holland, "The Meteorological Structure of the Mesosphere Including Seasonal and Latitudinal Variations."

Thomas, R. N. (1952). *Astrophys. J.* **116**, 203, "The Physical Theory of Meteors. III. Conditions at the Meteor Surface."

Thomas, R. N., and Whipple, F. L. (1951). *Astrophys. J.* **114**, 448, "The Physical Theory of Meteors. II. Astroballistic Heat Transfer."

Thomas, R. N., and White, W. C. (1953). *Astrophys. J.* **118**, 555, "The Physical Theory of Meteors. IV. Inquiry Into the Radiation Problem—A Laboratory Model."

Tisone, G. C. (1973). *J. Geophys. Res.* **78**, 746, "Measurements of NO Densities during Sunrise at Kauai."

Toth, B. A., Farmer, C. B., Schindler, R. A., Raper, O. F., and Schaper, P. W. (1973). *Nature (London) Phys. Sci.* **244**, 7, "Detection of Nitric Oxide in the Lower Atmosphere."

Vigroux, E. (1953). "Contribution à l'ètude expèrimentale de l'ozone." Masson, Paris.

Watanabe, K. (1958). *Advan. Geophys.* **5**, 153, "Ultraviolet Absorption Processes in the Upper Atmosphere."

Watanabe, K., Zelikoff, M., and Inn, E. C. Y. (1953). Air Force Cambridge Res. Center, Bedford, Massachusetts, Tech. Rep. No. 53-23, "Absorption Coefficients of Several Atmospheric Gases."

Weeks, L. G., and Smith, L. G. (1968). *Planet. Space Sci.* **16**, 1189, "A Rocket Measurement of Ozone Near Sunrise."

Whipple, F. L. (1955). *Astrophys. J.* **121**, 241, "The Physical Theory of Meteors. VII. On Meteor Luminosity and Ionization."

Wilkinson, P. G., and Mulliken, R. S. (1957). *Astrophys. J.* **125**, 594, "Dissociation Processes in Oxygen Above 1750 Å."

Zalpuri, K. S., and Somayajulu, Y. V. (1974). *J. Atmos. Terrest. Phys.* **36**, 1789, "Nitric Oxide Density Determination from Rocket Experiments at the Geomagnetic Equator."

Chapter II

CHEMISTRY OF THE UPPER ATMOSPHERE

The chemistry of the atmosphere can be attributed to the absorption of the sun's radiation giving rise to photodissociation. The reactive fragments produced further react, giving rise to the many different species observed. In this chapter we shall discuss the processes affecting many of the species in order to explain their altitude profiles. The processes important in influencing concentrations are the absorption of radiation, chemical reaction between species, and diffusion.

The effective rate coefficient for photon absorption, designated J, is the product of the photon flux (photons/cm^2-sec) and the absorption coefficient (cm^2/particle) and has the units of sec^{-1}, whereas the photoabsorption rate, designated I_a, is the product of J and the particle concentration and has the units of particles/cm^3-sec:

$$I_a\{X\} = J\{X\}[X] \tag{1}$$

Whereas $I_a\{X\}$ gives the photoabsorption rate of X, the photoreaction rate $I\{X\}$ also depends on the photoreaction efficiency $\phi\{X\}$:

$$I\{X\} = \phi\{X\}I_a\{X\} = \phi\{X\}J\{X\}[X] \tag{2}$$

The chemical reaction rate term (particles/cm^3-sec) is the product of the reaction rate coefficient and the concentrations of the species involved in the reaction. In many aeronomical discussions, these terms are designated a and n, respectively. However, we shall keep the usual chemists' notation of k for rate coefficients and brackets to denote concentrations. The symbol n will be reserved for the total concentration of all the species. Thus for the reaction

$$X + Y \rightarrow \text{products} \tag{3}$$

the reaction rate is

$$-d[X]/dt = -d[Y]/dt = k_{(3)}[X][Y] \tag{4}$$

For a given species X in a gas with n molecules/cm^3, the flux F_X with which X moves is

$$F_X = -D\left(\frac{d[X]}{dZ} - \frac{[X]}{n}\frac{dn}{dZ}\right) \tag{5}$$

and the change in concentration due to diffusion, $(\partial[X]/\partial t)_{diff}$ is

$$\left(\frac{\partial[X]}{\partial t}\right)_{diff} = \frac{d}{dZ}\left\{D\left(\frac{d[X]}{dZ} - \frac{[X]}{n}\frac{dn}{dZ}\right)\right\} \tag{6}$$

where Z represents the altitude and D is the diffusion coefficient. Equation (6) neglects mass separation effects due to gravity, i.e., diffusion due to differences in molecular weight, which are not important below 160 km.

Equation (6) is rather cumbersome to use because it requires knowing first and second height derivatives of the concentrations and first height derivatives of the diffusion coefficient. This detailed information is not available to us, so we shall assume that for a small enough region in space the concentration and diffusion coefficient dependences are exponential, viz.,

$$[X] = [X]_0 \exp\{-\xi_X Z\} \tag{7}$$

$$n = n_0 \exp\{-\xi_n Z\} \tag{8}$$

$$D = D_0 \exp\{-\xi_D Z\} \tag{9}$$

where the ξ parameters are the reciprocal scale heights. Then if the scale heights are sufficiently slowly varying, Eqs. (5) and (6) reduce to the simplified forms

$$F_X = D[X](\xi_X - \xi_n) \tag{10}$$

$$\left(\frac{\partial[X]}{\partial t}\right)_{diff} = D[X](\xi_X + \xi_D)(\xi_X - \xi_n) \tag{11}$$

In discussing the chemistry, there are very many species involved. To obtain a completely accurate picture of all that occurs, all possible photodissociations and chemical reactions must be included. This would be a formidable job, indeed, although it can be done with a high-speed computer. Our aim here will be more modest. We shall attempt to sort out only those terms of dominant importance for each species to understand the main features influencing the species concentration.

A separation of terms can be made by realizing that species with low concentrations will have little influence on species of large concentration (but the reverse is not true). Thus the ionic species, which are present at very low concentrations, have no influence on the chemistry of neutral species (with the exception of N and NO in some regions of space). A discussion of ion chemistry will be deferred until after a discussion of the neutral species. Further classifications can be made to simplify the discussion. Thus we shall discuss first the chemistry of the oxygen species, then introduce the oxides of nitrogen, then the hydrogen–oxygen species, and finally the carbon-containing compounds.

One other simplifying assumption will prove to be very useful. This is the steady-state assumption, which states that the overall rate of change of concentration of a species X is unimportant compared to the sum of the rates of the formation (or removal) processes for species X, i.e., $d[X]/dt \sim 0$, or that the total rate of formation is approximately equal to the total rate of removal. This approximation, of course, is not exact because the sun's position, and thus the photodissociation rates, is constantly changing. As a result, the species concentration change with the time of day. However, if the change is small enough, the approximation will be valid.

A knowledge of when the steady-state assumption should apply to a species X can be obtained from the lifetime of that species $\tau\{X\}$. The lifetime is computed as the ratio of the species concentration to the sum of its removal rates

$$\tau\{X\} = [X]/\sum_r R_r \tag{12}$$

where R_r is the rate of a reaction that consumes X. If $\tau\{X\}$ is small compared to the pertubation time of the system, the steady-state approximation is valid. The pertubation of the system is the change in the solar radiation intensity due to a change in solar zenith angle. The pertubation time is thus of the order of an hour or two at noon and somewhat less at twilight. Consequently, if $\tau\{X\}$ is less than these times, it is expected that the steady-state approximation should be useful. At night there is no pertubation term and the steady-state approximation will hold for even longer lifetimes $\sim 10^4$ sec. Regardless of the lifetime, the steady-state approximation holds if the concentration is passing through a maximum or minimum.

There is another useful application of the steady-state hypothesis that applies when $\tau\{X\}$ is very much greater than the pertubation time if the pertubation is cyclic. In the atmosphere the solar radiation intensity cycles on a 24-hr time scale (not exactly cyclic because due to change of seasons the peak intensity changes somewhat from day to day). Thus for $\tau\{X\} > 3$ days $(2.5 \times 10^5$ sec) (but less than 30 days so that seasonal variations do not become important) the daily average radiation intensity can be used. If we consider that the approximate daytime intensity variation is sinusoidal, then the average intensity for a 24-hr period at any location is the peak intensity $\mathcal{I}_{max}$ divided by 4:

$$\mathcal{I}_{av} = \mathcal{I}_{max}/4 \qquad \text{for daytime sinusoidal behavior} \tag{13}$$

where the factor $\frac{1}{4}$ represents the ratio of the earth's cross-sectional area to its total surface area. Thus if $\mathcal{I}_{av}$ is used, the steady-state approximation should be valid. It should also be realized that Eq. (13) gives the average intensity incident on the earth's surface for light not attenuated by the atmosphere, where $\mathcal{I}_{max}$ is the intensity for an overhead sun. The exact instantaneous intensity at any point on the earth's surface and for positive values of $\mathcal{I}$ is given by

$$\mathcal{I} = \mathcal{I}_{max}(\cos\delta \cos\phi \cos\omega + \sin\delta \sin\phi) \tag{14}$$

where δ is the sun's declination (0° at the spring and autumn equinoxes, +23° at the summer solstace, and −23° at the winter solstace), ϕ the latitude, and $\omega = 2\pi t/86{,}400$ if t is the time in seconds. $\mathcal{I}_{max}$ is the intensity for an overhead sun, which itself varies with the seasons:

$$\mathcal{I}_{max} = S/R^2 \tag{15}$$

where S is the mean solar constant (1.95 cal/cm^2-min) and R the distance from the earth to the sun in astronomical units $(93 \times 10^6$ mi = 1 astronomical unit). Since R varies by 3%, being a minimum in January and a maximum in July, $\mathcal{I}_{max}$ varies by 6% over the course of a year.

Equations (13) and (14) apply on the surface of the earth only for those wavelengths (>3400 Å) that are not significantly attenuated by the earth's atmosphere. For any region in space for any wavelength, $\mathcal{I}$ is given approximately by

$$\mathcal{I} = \mathcal{I}_\infty \exp\{-\zeta \sec\chi\} \tag{16}$$

where the subscript ∞ refers to the top of the atmosphere, χ is the solar zenith angle (0° for an overhead sun), and ζ is the optical depth defined by

$$\zeta \equiv \int_Z^\infty n\alpha \, dZ' \tag{17}$$

with n the number density, α the effective extinction cross section for the wavelength considered, and Z the altitude. Equation (16) is only approximate, but it gives good results for $\chi \leq 75°$.

Equation (16) gives $\mathscr{I}$ as a function of solar zenith angle. However the optical depth ζ is different for each incident wavelength. Thus Eq. (16) is not applicable as it stands. However at any altitude, we have found empirically that the photodissociation rate (which is what really interests us) as a function of χ can be given approximately by

$$I_z = I_z^\circ \exp\{-\zeta_z'(\sec \chi - 1)\} \tag{18}$$

where I_z is the photodissociation rate at altitude Z for any χ, and I_z° is the photodissociation rate for an overhead sun. This empirical formula is good to within a factor of 2 at any χ, and the averages found using this formula are very much better (probably $\pm 20\%$). Values of ζ_z' are listed in Table II-1.

Before examining the chemistry in detail, the importance of various types of processes can be deduced from the concentration profiles with altitude. These concentration profiles (see Chapter I) depend not only on the chemistry, but also on the fact that the atmosphere is being thinned as the altitude increases. To separate out this density decrease, we can examine the relative concentrations of the species. For a number of the

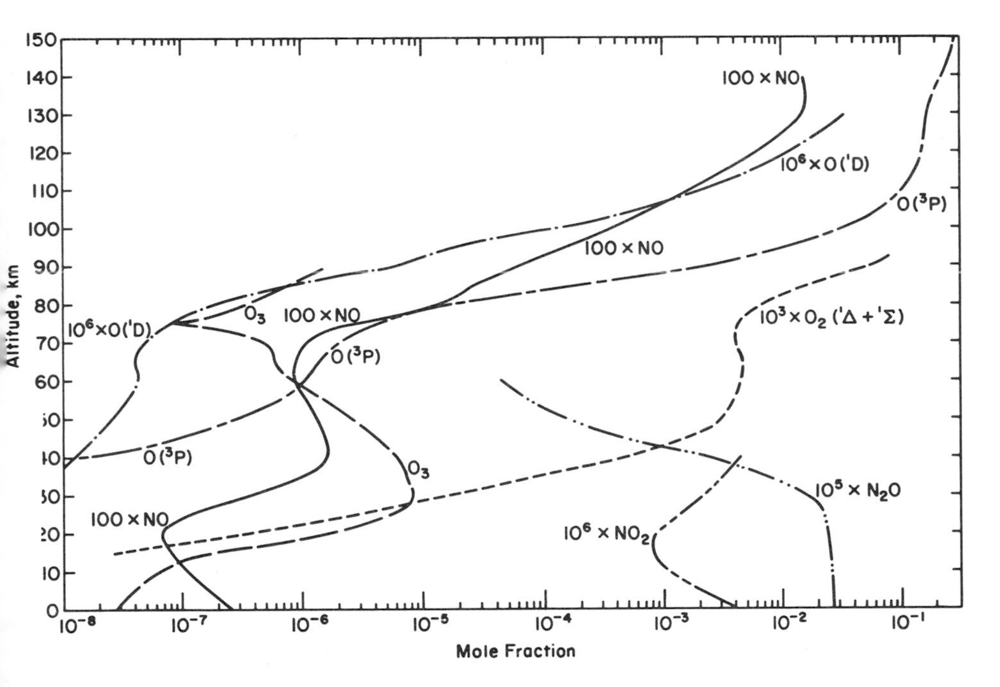

Fig. II-1. Semilog plot of typical relative concentrations of some atmospheric species versus altitude for an overhead sun based on absolute concentrations in Figs. I-3 and I-11.

TABLE II-1

Photodissociation Coefficient Parameters as a Function of Altitude[a, b]

Altitude (km)	$O_2 \rightarrow O(^1D) + O(^3P)$	$O_2 \rightarrow 2O(^3P)$	$O_3 \rightarrow O(^1D)$	$O_3 \rightarrow O_2(^1\Delta$ or $^1\Sigma)$	$N_2O \rightarrow N_2 + O(^1D)$	$NO \rightarrow N(^4S) + O(^3P)$
						J_z° (sec^{-1})
10	—	—	1.0×10^{-5}	1.0×10^{-4}	1.2×10^{-9}	—
20	—	1.7×10^{-12}	2.5×10^{-5}	1.4×10^{-4}	4.5×10^{-9}	—
30	—	7.0×10^{-11}	1.4×10^{-4}	3.2×10^{-4}	1.05×10^{-7}	3.4×10^{-8}
40	—	5.0×10^{-10}	1.25×10^{-3}	1.6×10^{-3}	4.6×10^{-7}	6.7×10^{-7}
50	—	1.2×10^{-9}	8.0×10^{-3}	8.0×10^{-3}	8.0×10^{-7}	2.7×10^{-6}
60	—	1.9×10^{-9}	1.0×10^{-2}	1.0×10^{-2}	—	5.4×10^{-6}
70	2.0×10^{-10}	3.0×10^{-9}	—	—	—	7.8×10^{-6}
80	2.0×10^{-9}	1.0×10^{-8}	—	—	—	9.3×10^{-6}
90	1.7×10^{-8}	4.0×10^{-8}	—	—	—	—
100	2.8×10^{-7}	9.0×10^{-8}	—	—	—	—
110	1.2×10^{-6}	1.2×10^{-7}	—	—	—	—
120	3.0×10^{-6}	1.2×10^{-7}	—	—	—	—
∞	3.7×10^{-6}	1.2×10^{-7}	1.0×10^{-2}	1.0×10^{-2}	8.0×10^{-7}	1.0×10^{-5}
						ζ_z'
10	—	—	5.0	0	1.39	—
20	—	5.3	2.30	0.392	1.93	—
30	—	1.93	1.26	0.385	1.33	2.94
40	—	0.478	0.488	0.415	0.441	1.23
50	—	0.185	0.261	0.261	0.165	0.561
60	—	0.128	0	0	—	0.264
70	9.29	0.231	—	—	—	0.111
80	7.50	0.231	—	—	—	0.045
90	5.14	0.187	—	—	—	—
100	0.392	0.198	—	—	—	—
110	0.392	0	—	—	—	—
120	0.308	0	—	—	—	—

[a] $J_z\{\chi\} = J_z^0 \exp\{-\zeta_z' (\sec \chi - 1)\}$.

[b] $J\{NO_2 \rightarrow NO + O(^3P)\} = 8.0 \times 10^{-3}$ sec^{-1} at ≤ 25 km, 1.0×10^{-2} at > 25 km. $J\{HONO \rightarrow HO + NO\} = 6.45 \times 10^{-4}$ sec^{-1} (Garvin and Hampson, 1974). $J\{O_3 \rightarrow O_2(^3\Sigma) + O(^3P)\} = 3.8 \times 10^{-4}$ sec^{-1}. $J\{NO_3\} = 1.0 \times 10^{-2}$ sec^{-1}.

more important species, relative concentration profiles are shown in Fig. II-1.

The relative concentrations fall into four main categories:

1. Relative species concentration rises with altitude. These include

$H_2O \rightarrow$ $HO + H$	$H_2O_2 \rightarrow$ $2HO$	$HNO_3 \rightarrow$ $HO + NO_2$	$CH_4 \rightarrow$ $CH_3 + H$	$CH_2O \rightarrow$ $CO + H_2$	$CH_2O \rightarrow$ $HCO + HO_2$
—	3.6×10^{-6}	7.2×10^{-7}	—	1.9×10^{-5}	7.8×10^{-5}
—	5.0×10^{-6}	1.1×10^{-6}	—	1.9×10^{-5}	7.8×10^{-5}
—	1.0×10^{-5}	2.1×10^{-6}	—	3×10^{-5}	$\sim 2 \times 10^{-4}$
—	4.0×10^{-5}	9.0×10^{-6}	—	$\sim 1 \times 10^{-4}$	$\sim 6 \times 10^{-4}$
—	9.3×10^{-5}	3.5×10^{-5}	—	1.4×10^{-4}	9.2×10^{-4}
1.5×10^{-8}	1.2×10^{-4}	—	—	1.4×10^{-4}	9.2×10^{-4}
4.4×10^{-7}	—	—	5.3×10^{-7}	—	—
3.1×10^{-6}	—	—	3.2×10^{-6}	—	—
5.6×10^{-6}	—	—	5.0×10^{-6}	—	—
6.5×10^{-6}	—	—	6.0×10^{-6}	—	—
—	—	—	—	—	—
—	—	—	—	—	—
6.5×10^{-6}	1.2×10^{-4}	3.5×10^{-5}	6.0×10^{-6}	1.4×10^{-4}	9.2×10^{-4}
—	0.478	0.630	—	—	—
—	0.408	0.488	—	—	—
—	0.408	0.347	—	—	—
—	0.347	0.381	—	—	—
—	0.519	—	—	—	—
0.770	0.036	—	—	—	—
1.07	—	—	2.78	—	—
0.462	—	—	0.478	—	—
0.126	—	—	0.0676	—	—
0.052	—	—	0	—	—
—	—	—	—	—	—
—	—	—	—	—	—

$O(^3P)$, $O_2(^1\Delta_g)$, and $O(^1D)$. This behavior results from the increasing importance of O_2 photodissociation with altitude. In the ionosphere [N] and [NO] also rise relatively with altitude, which is due respectively to the increasing importance of NO photodissociation and photoionization with altitude.

2. Relative species concentration has a maximum. O_3 and HNO_3 follow this behavior, which suggests a photoformation term with a maximum, with the species diffusing upward and downward from the maximum.

3. Relative species concentration has a minimum, such as for NO and NO_2. The indication is that important removal processes dominate the chemistry at the minimum. These species are produced both at the surface of the earth and in the upper atmosphere above the minimum, and they diffuse toward the minimum.

4. Relative species concentration is constant with altitude, at least for a considerable height, and then may drop with further increase in altitude. The species N_2O and CH_4 follow this behavior. They are formed at the earth's surface and diffuse upward until they reach a height at which removal processes become important.

NEUTRAL OXYGEN ATMOSPHERE

The chemistry of the atmosphere below ~160 km is caused principally by the absorption of the sun's radiation by O_2 and O_3. The radiation <1000 Å is not very intense even at the top of the atmosphere and is essentially completely removed by absorption by N_2 at altitudes >110 km.

As the sun's radiation penetrates the atmosphere, it photodissociates O_2, whose absorption spectrum was discussed in Chapter I. For radiation <1750 Å, there is sufficient energy to produce two oxygen atoms, the electronically excited $O(^1D)$ atom and the ground state $O(^3P)$ atom. This photodecomposition occurs with unit efficiency when absorption is into the Schumann–Runge continuum, which extends from 1740 to below 1400 Å:

$$O_2 + h\nu \text{ (Schumann–Runge continuum)} \rightarrow O(^1D) + O(^3P) \tag{19a}$$

At the top of the atmosphere for an overhead sun the intensity gives a photodissociation coefficient $J\{O_2\} = 3.7 \times 10^{-6}$ sec^{-1} for the Schumann–Runge continuum (Nicolet, 1971).

At wavelengths lower than for the Schumann–Runge continuum, a banded structure occurs in the O_2 absorption spectrum. Absorption here can lead to oxygen atom production and at $\lambda < 1027$ Å, photoionization of O_2. However, the sun's radiation is not very intense in this region (except for the Lyman-α line at 1215.7 Å) and no significant contribution is made to O_2 photodissociation.

Figure II-2 gives the photodissociation coefficients for O_2 for an overhead sun between 80 and 120 km. The Schumann–Runge continuum plays the most important role above 95 km, but this radiation is removed by ab-

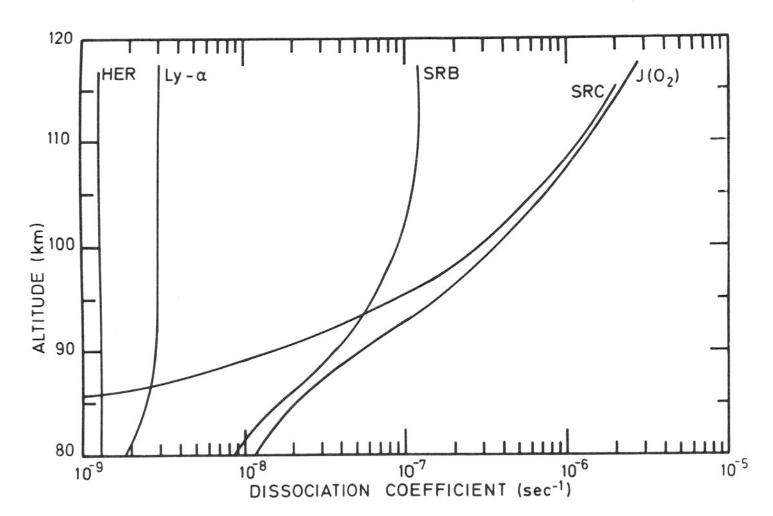

Fig. II-2. Photodissociation coefficients for O_2 in the region 80–120 km (lower thermosphere) for an overhead sun. The contribution to $J\{O_2\}$ of the Schumann–Runge continuum (SRC), the Schumann–Runge bands (SRB), the Lyman-α line (Ly-α), and the Herzberg continuum (HER) are shown. From Nicolet (1971) with permission of Reidel Publ. Co.

sorption, and its photodissociation coefficient becomes negligible below about 85 km.

For radiation >1750 Å, there is insufficient energy to photodissociate O_2 to produce electronically excited atoms. However, for radiation between 2400 and 1750 Å, two $O(^3P)$ atoms are produced. This can be done either by direct absorption into the Herzberg continuum

$$O_2 + h\nu \text{ (Herzberg continuum)} \rightarrow 2O(^3P) \tag{19b}$$

or by the Schumann–Runge band absorption into the $B\,^3\Sigma_u^-$ excited electronic state of O_2, which then predissociates:

$$O_2 + h\nu \text{ (Schumann–Runge bands)} \rightarrow O_2(B\,^3\Sigma_u^-) \rightarrow 2O(^3P) \tag{19c}$$

The photodissociation coefficients for these two absorptions also are given in Fig. II-2. The main contribution to the total photodissociation coefficient is from the Schumann–Runge bands between 90 and 60 km and from the Herzberg continuum below 60 km.

The contribution from the Lyman-α line is also shown in Fig. II-2. It gives a photodissociation coefficient of 3×10^{-9} sec^{-1} for an overhead sun at high altitudes. This value drops below 80 km, and Lyman-α is essentially removed at 65 km. It only makes a significant contribution to $J\{O_2\}$ in the region 70–85 km.

The total absorption coefficient for O_2, $J\{O_2\}$, continually falls with the altitude, until it is essentially zero at 20 km (Fig. II-3). However, the O_2 concentration is rising as the altitude diminishes, so that the rate of photodissociation $I\{O_2\}$ goes through a maximum at 42 km as shown in Fig. II-4.

Below 50 km, where the O_3 begins to be important, much of the radiation below 2100 Å has been removed. However, O_3 absorbs strongly in the region between 2100 and 3100 Å; this absorption is known as the Hartley band. In this region photodissociation proceeds with unit quantum efficiency to produce the electronically excited fragments $O(^1D)$ and $O_2(^1\Delta)$ (Lissi and Heicklen, 1972):

$$O_3 + h\nu \text{ (Hartley band)} \rightarrow O_2(^1\Delta) + O(^1D) \tag{20a}$$

This is the most important O_3 photodissociation process for O_3 above 40 km, but the radiation <3000 Å is rapidly removed, and $O(^1D)$ results mainly from absorption at 3000–3100 Å below the maximum in O_3 concentration at 25 km.

Above 3102 Å, there is insufficient electronic energy for process (20a), and if it occurs rotational–vibrational energy must be used. Thus at $\lambda > 3102$ Å, in the region of the Huggins bands, process (20a) decreases in importance and other photodissociations occur:

$$O_3 + h\nu \text{ (Huggins)} \rightarrow O_2(^3\Sigma) + O(^1D) \tag{20b}$$

$$\rightarrow O(^3P) + O_2(^1\Delta \text{ or } ^1\Sigma) \tag{20c}$$

In the region from 3100 to 3340 Å, it is known that either $O(^1D)$ or singlet O_2 is produced most of the time in the photodissociation act (Castellano and Schumacher, 1972; Jones and Wayne, 1969, 1970). It is not known whether the singlet O_2 is $^1\Delta$ or $^1\Sigma$. However, there are some stratospheric observations that suggest an excess of $O_2(^1\Sigma)$, which might result from process (20c). Kuis *et al.* (1975) have shown that all the $O(^1D)$ production can be explained as coming from process (20a), so that process (20b) is unimportant and can be neglected. At 3130 Å, they found the efficiency of process (20a) is temperature dependent, since rotational–vibrational energy is needed; its quantum efficiency is 0.29, 0.22, and 0.11, respectively, at 293, 258, and 221°K.

Finally O_3 can be photodissociated in the weak Chappuis absorption band, which extends from 4000 Å into the infrared region. Here the photochemical process always gives ground electronic state species:

$$O_3 + h\nu \text{ (Chappuis)} \rightarrow O_2(^3\Sigma) + O(^3P) \tag{20d}$$

The photodissociation coefficients at zero optical depth (>55 km) in the

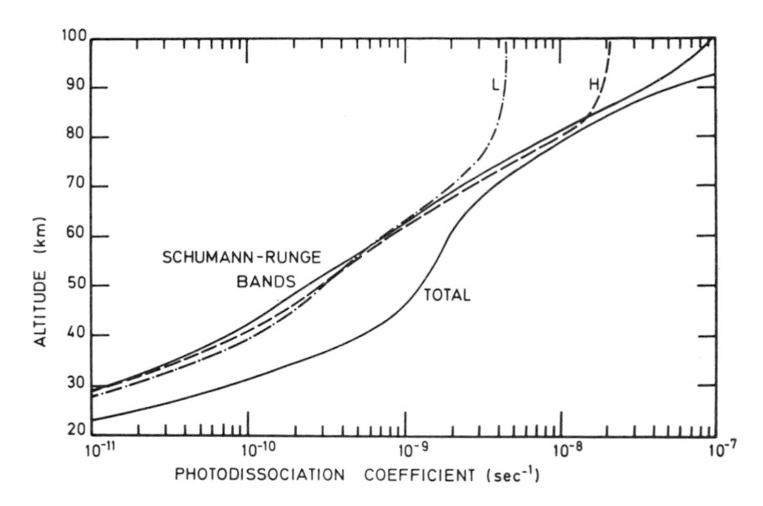

Fig. II-3. Photodissociation coefficients for O_2 versus altitude for an overhead sun. The plots give both the total coefficient and that due to absorption in the Schumann–Runge band for low (L) and high (H) values of the mean absorption coefficients. From Kockarts (1971) with permission of Reidel Publ. Co.

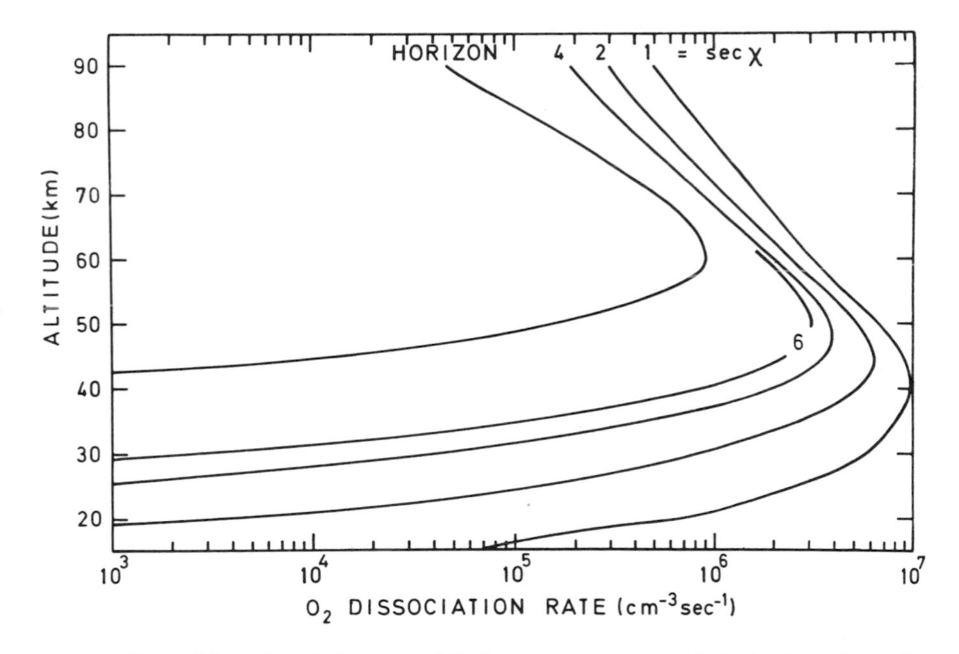

Fig. II-4. Photodissociation rate of O_2 in the mesosphere and stratosphere for various solar zenith angles. From Nicolet (1971) with permission of Reidel Publ. Co.

different regions for an overhead sun are

$$J\{\text{Chappuis}\} = 3 \times 10^{-4}\ \text{sec}^{-1}, \qquad J\{\text{Huggins}\} = 1 \times 10^{-4}\ \text{sec}^{-1}$$

$$J\{3000\text{–}3100\ \text{Å}\} = 2 \times 10^{-4}\ \text{sec}^{-1}, \qquad J\{\text{total}\} = 1 \times 10^{-2}\ \text{sec}^{-1}$$

Since the Chappuis bands are only very weakly absorbing (absorption cross section $= 5 \times 10^{-21}$ cm^2 at the absorption maximum near 6000 Å), $J\{\text{Chappuis}\}$ is essentially independent of altitude.

The photodissociation coefficients leading to the different photoproducts are shown in Fig. II-5. Above 40 km the Hartley photodissociation dominates and $O(^1D)$ and $O_2(^1\Delta)$ are produced nearly all the time. However, below the O_3 concentration maximum at 25 km, $O(^1D)$ is much less important than $O(^3P)$ production. Nevertheless, $O(^1D)$ production is still an important process [more important than $O(^3P)$ production from O_2 photodissociation]. The photodissociation rates for $O(^1D)$ and $O_2(^1\Delta_g)$ production from O_3 are shown for various solar zenith angles in Figs. II-6 and II-7, respectively. In Fig. II-7 it has been assumed that absorption into the Huggin's band gives exclusively $O_2(^1\Delta_g)$. However, some $O_2(^1\Sigma)$ may be produced, and thus the figure gives the photodissociation rate for total singlet O_2 production. The peak $O(^1D)$ and $O_2(^1\Delta_g)$ concentrations and the O_3 photodissociation rate appear at about 42 km for an overhead sun. For other solar zenith angles this peak moves upward, reaching close

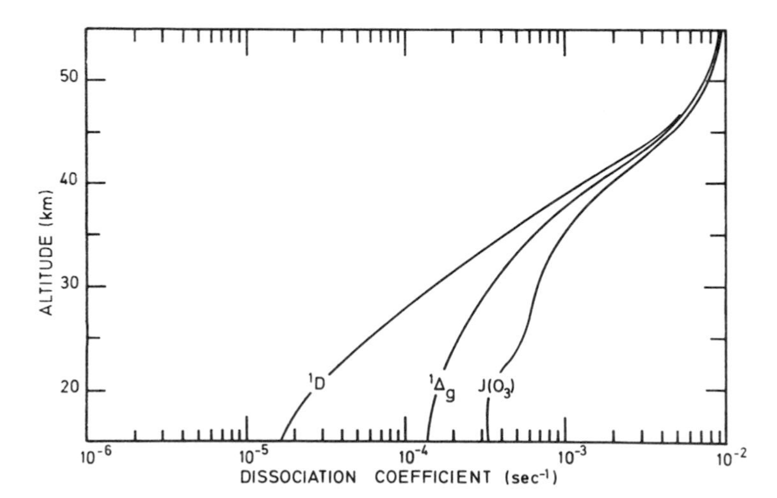

Fig. II-5. Photodissociation coefficient for O_3 for an overhead sun in the 15–55 km region. The curves give the total coefficient as well as the photodissociation coefficient for $O_2(^1\Delta_g)$ and $O(^1D)$ production. From Nicolet (1971) with permission of Reidel Publ. Co.

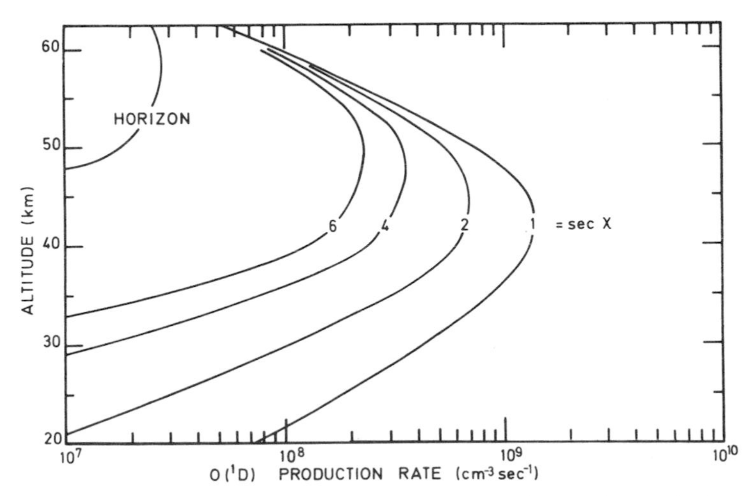

Fig. II-6. Production rate of O(^{1}D) from the photodissociation of O_3 for various solar zenith angles. From Nicolet (1971) with permission of Reidel Publ. Co.

to 60 km for sunrise and twilight. The large amount of energy from the O_3 absorption, and to a lesser extent that from the O_2 absorption, causes the temperature maximum at the stratopause around 50 km.

It is interesting to consider now the reasons for the temperature minima at the tropopause (8–17 km) and mesopause (~85 km). In the mesosphere as the altitude increases, the absorption of the sun's radiation falls and so does the temperature. However, above 85 km the O_2 photodissociation rate becomes nearly constant, so that the energy input is nearly constant. However, the gas density is still rapidly decreasing with increasing altitude, so that the energy must be distributed over fewer particles; the temperature rises throughout the thermosphere.

The temperature minimum at the tropopause can be attributed to absorption of the sun's radiation by the surface of the earth, which raises the surface temperature. The troposphere is then heated both by reradiation and upward convection of the warm surface air.

The photodissociation coefficients for an overhead sun are listed in Table II-1 as a function of altitude. In addition to photodissociation, chemical species are produced and removed by chemical reaction. Let us examine these species.

Odd Oxygen

First consider those species with an odd number of oxygen atoms, referred to as odd oxygen species. These are O_3, O(^{3}P), and O(^{1}D). The

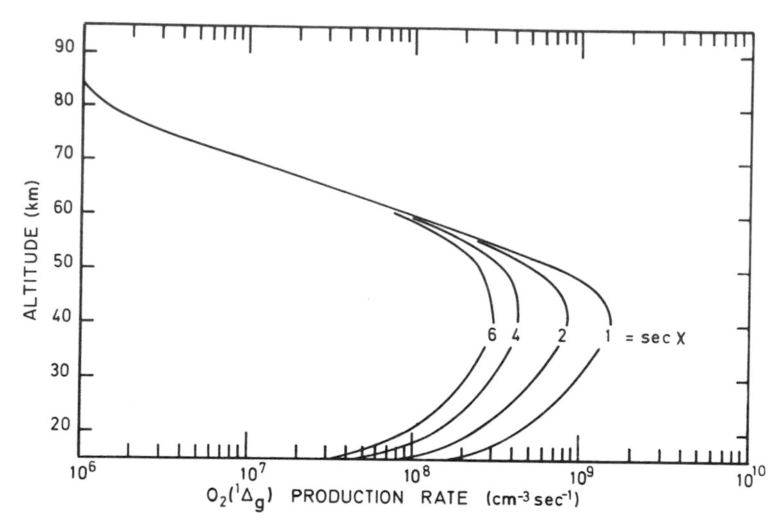

Fig. II-7. Production rate of $O_2(^1\Delta_g)$ from the photodissociation of O_3 at various solar zenith angles. Calculation assumes that absorption into the Huggin's band produces $O_2(^1\Delta_g)$ rather than $O_2(^1\Sigma_g{}^+)$. From Nicolet (1971) with permission of Reidel Publ. Co.

sum of the concentrations of these three species is called "odd oxygen." The reactions of importance for these species are

$$2O(^3P) + M \rightarrow O_2 + M \tag{21}$$

$$O(^3P) + O_2 + M \rightarrow O_3 + M \tag{22}$$

$$O(^3P) + O_3 \rightarrow 2O_2 \tag{23}$$

$$O(^1D) + M \rightarrow O(^3P) + M \tag{24}$$

The rate coefficients for these reactions are listed in Tables II-2 and II-3 Also for $O(^3P)$ in the thermosphere, diffusional transport is important, leading to removal of $O(^3P)$ in the lower thermosphere.

In principle, if the steady-state approximation is applicable, it can be applied to each individual species, and their concentrations can be calculated. In practice the steady-state approximation is valid during midday for each species, and the relative concentrations of the species can be computed in this way. Since the odd oxygen species interconvert into each other rapidly, but total odd oxygen is removed slowly, to obtain the absolute values of the species concentrations requires dealing with small differences in large numbers, and the steady-state approximation is not valid.

Reaction	Reaction number	k at 298°K[b] (cm^{3n}/sec)	A[b] (cm^{3n}/sec)	E (cal/mole)	Temp. range (°K)
$N + N + M \rightarrow N_2 + M$	—	—	8.3×10^{-34}	−1000	100–600
$N + O_2(^3\Sigma) \rightarrow NO + O(^3P)$	(53)	—	$1.1 \times 10^{-14}T$	6300	300–3000
$N + O(^3P) + M \rightarrow NO + M$	—	—	$1.8 \times 10^{-31}T^{-0.5}$	0	200–400
$N + O_3 \rightarrow NO + O_2$	(66)	5.7×10^{-13}	—	—	—
$2O(^3P) + M \rightarrow O_2 + M$	(21)	—	$3.80 \times 10^{-30}T^{-1}$ [c]	340	1000–8000
$O(^3P) + O_2(^3\Sigma) + M \rightarrow O_3 + M$	(22)	—	1.05×10^{-34}	−1020	200–346
$O(^3P) + O_3 \rightarrow 2O_2$	(23)	—	1.9×10^{-11}	4600	220–1000
$O(^3P) + CH_2O \rightarrow HO + HCO$	(132)	1.6×10^{-13}	3×10^{-11} [d]	2400[d]	300–1000
$O(^3P) + NO + M \rightarrow NO_2 + M$	(55)	—	4.2×10^{-33}	−1880	200–500
$O(^3P) + NO_2 \rightarrow NO + O_2$	(54)	9.1×10^{-12}	9.1×10^{-12}	0	230–350
$O(^3P) + CO + M \rightarrow CO_2 + M$[e]	—	—	5.5×10^{-33}	4100	298–472
$O(^3P) + H_2O_2 \rightarrow HO + HO_2$	(109)	—	2.75×10^{-12}	4250	283–373
$O_2(^1\Delta) + O_3 \rightarrow 2O_2(^3\Sigma) + O(^3P)$	(41)	—	6.0×10^{-11}	5700	296–360
$O_2(^1\Delta) + N_2 \rightarrow O_2(^3\Sigma) + N_2$	(39)	$<2 \times 10^{-20}$	—	—	—
$O_2(^1\Delta) + O_2 \rightarrow O_2(^3\Sigma) + O_2$	(40)	2.2×10^{-18}	$2.2 \times 10^{-18}(T/300)^{0.8}$	—	285–322
$N + NO \rightarrow N_2 + O$	(56)	2.7×10^{-11}	2.7×10^{-11}	0	300–5000
$NO_2 + O(^3P) + M \rightarrow NO_3 + M$	(60)	1.0×10^{-31}	—	—	—
$O_2(^1\Sigma) + N_2 \rightarrow O_2(^3\Sigma) + N_2$	(36)	2×10^{-15} [f]	—	—	—
$O_2(^1\Sigma) + O_2 \rightarrow 2O_2(^3\Sigma)$	(37)	1×10^{-16} [f]	—	—	—
$O_2(^1\Sigma) + O_3 \rightarrow 2O_2(^3\Sigma) + O(^3P)$	(38)	2.5×10^{-11} [g]	—	—	—
$2O_2(^1\Delta) \rightarrow O_2(^1\Sigma) + O_2(^3\Sigma)$	(42)	2×10^{-18} [h]	—	—	—
$O_2(^1\Sigma) \rightarrow O_2(^3\Sigma) + h\nu$	(34)	0.14[i]	—	—	—
$O_2(^1\Delta) \rightarrow O_2(^3\Sigma) + h\nu$	(35)	3.7×10^{-4} [i]	—	—	—

[a] From Garvin and Hampson (1974).

[b] $n = 1$ for second-order reactions; 2, for third-order reaction. $M = N_2$. [c] $M = O_2$. [d] Estimate.

[e] From Simonaitis and Heicklen (1972). This reaction third order above 15 km, but in the intermediate second–third order regime below 15 km. See Table II-9 for effective second-order rate coefficients.

[f] From Kearns (1971). [g] From Gilpin *et al.* (1971).

[h] From Arnold and Ogryzlo (1967) and Izod and Wayne (1968). [i] From Ogryzlo (1967).

TABLE II-3

Reactions of $O(^1D)$[a]

Reaction	Reaction number	k at 298°K (cm^3/sec)
$O(^1D) + N_2 \rightarrow O(^3P) + N_2$	—	5.4×10^{-11}
$O(^1D) + O_2 \rightarrow O(^3P) + O_2(^1\Sigma)$	(24′)	7.4×10^{-11}
$O(^1D) + O_3 \rightarrow O_2 + O_2^*$ [or $2O(^3P)$]	—	3.75×10^{-10}
$\rightarrow 2O_2$	—	1.25×10^{-10}
$O(^1D) + N_2O \rightarrow N_2 + O_2$	(45a)	1.1×10^{-10}
$\rightarrow 2NO$	(45b)	1.1×10^{-10}
$O(^1D) + CH_4 \rightarrow HO + CH_3$	(120a)	3.6×10^{-10}
$\rightarrow CH_2O + H_2$	(120b)	4.0×10^{-11}
$O(^1D) + H_2O \rightarrow 2HO$	(102)	3.5×10^{-10}
$O(^1D) + H_2 \rightarrow HO + H$	(94)	2.9×10^{-10}
$O(^1D) + H_2O_2 \rightarrow HO + HO_2$	(110)	$\sim 4 \times 10^{-10}$

[a] From Garvin and Hampson (1974).

To compute the absolute concentrations consider the rates of formation and removal of total odd oxygen. The only step producing odd oxygen is the photodissociation of O_2, reaction (19). Thus

$$\sum R_f\{\text{odd O}\} = 2I\{19\} = 2J_{(19)}[O_2] \tag{25}$$

The removal steps are reactions (21) and (23), and in the lower thermosphere, diffusion of $O(^3P)$:

$$\sum R_r\{\text{odd O}\} = 2k_{(21)}[O(^3P)]^2[M] + 2k_{(23)}[O(^3P)][O_3] + k_{\text{diff}}[O(^3P)] \tag{26}$$

where k_{diff} is the rate coefficient for diffusional loss. If the steady state were applicable (i.e., $\tau\{\text{odd O}\} < 5000$ sec), we could equate the noon values of the formation and removal rates, calculate $[O_3]$ from its steady-state equation

$$\sum R_f\{O_3\} = k_{(22)}[O(^3P)][O_2][M] = \sum R_r\{O_3\}$$
$$= I_{(20)} + k_{(23)}[O(^3P)][O_3] \tag{27}$$

and solve for $[O(^3P)]$ and $[O_3]$. The steady-state equation (27) is applicable during daylight, but the steady state on odd oxygen is not applicable. However, there is no change in odd oxygen over a 24-hr period. At all altitudes the concentrations of both $O(^3P)$ and O_3 are reasonably constant during daylight. Furthermore, at altitudes ≤ 60 km, $[O(^3P)]$ rapidly falls to a negligibly low value after sunset. Thus for these altitudes, $\sum R_r\{\text{odd O}\}$ is relatively constant during the day and can be equated to the daylight (12-hr) average value of $\sum R_f\{\text{odd O}\}$, since at night both $\sum R_f\{\text{odd O}\}$ and $\sum R_r\{\text{odd O}\} = 0$. Thus

$$2\bar{I}_{(19)} = 2k_{(21)}[O(^3P)][M] + 2k_{(23)}[O(^3P)][O_3] + k_{\text{diff}}[O(^3P)] \quad (28)$$

where $\bar{I}_{(19)}$ is the 12-hr daylight average value of $I_{(19)}$, the species concentrations are their average daylight values, and $[O_3]$ can be computed from Eq. (27). Furthermore, the 24-hr average lifetime for odd oxygen, $\tau\{\text{odd O}\}$, is

$$\tau\{\text{odd O}\} = ([O(^3P)] + [O_3] + [O(^1D)])/\bar{I}_{(19)} \quad (29)$$

Since $[O(^1D)]$ will always be negligibly small compared to the other species concentrations, it can be neglected in Eq. (29).

All that is necessary now is to determine $\bar{I}_{(19)}$. With the use of Eq. (18), $\bar{I}_{(19)}$ is computed. Its values are listed in Table II-4 (12-hr daylight averages of $I_{(19)}$ = 24 hr averages of $\sum R_f\{\text{odd O}\}$, since two odd oxygen species are produced with each photodissociation). Also listed are the computed daylight values for [odd O], $[O_3]$, $[O(^3P)]$, and $\tau\{\text{odd O}\}$.

The computed concentrations represent maximum values, since there are no other significant routes to odd oxygen production, but other minor species not yet considered can remove odd oxygen. Thus at 20 km the computed $[O_3] = 2.24 \times 10^{13}$ molecules/cm^3 is 3–10 times larger than the observed values; at 40 km, the computed value $[O_3] = 9.3 \times 10^{11}$ molecules/cm^3 is $\sim$80% larger than the observed value of 5.1×10^{11} molecules/cm^3. At 60 km, the value computed for $[O_3]$ is 1.75×10^{10} molecules/cm^3, and this also must represent a maximum daytime value. Observed daytime values have been reported for $[O_3]$ varying between 0.3×10^{10} and 1.5×10^{10} molecules/cm^3 (see Fig. I-6).

The lifetimes for odd oxygen are considerably longer than a day at 20 and 40 km, so that odd oxygen is essentially constant throughout the day. However, at 60 km the lifetime is about 9 hr; the nighttime values (all O_3) should be about a factor of 3 to 5 times lower than the daytime odd oxygen values. Thus at night $[O_3]$ should be lower than the daytime odd oxygen concentration.

If the same calculation is done at 80 km as was done at the lower altitudes then $[O(^3P)]$ becomes 4.6×10^{11} molecules/cm^3. However, at night the $O(^3P)$ does not vanish but falls exponentially with a lifetime of $\sim 1.4 \times 10^4$ sec. Also the nighttime average value for $[O_3]$ is about one-half that of the daytime value for $[O(^3P)]$. If the nighttime loss rate is also included

TABLE II-4
Calculated Daytime Concentrations for a Pure Oxygen Atmosphere for an Overhead Noonday Sun

Altitude (km)	Temp (°K)	ΣR_f {odd O}[a] (molecules/cm³-sec)	[odd O] (molecules/cm³)	τ {odd O}[a] (sec)	[O_3] (molecules/cm³)	τ {O_3} (sec)
20	219	1.93×10^5	2.24×10^{13}	2.3×10^8	2.24×10^{13}	1.9×10^3
40	268	5.79×10^6	9.34×10^{11}	3.2×10^5	9.32×10^{11}	5.16×10^2
60	253	2.41×10^6	7.95×10^{10}	3.3×10^4	1.75×10^{10}	3.28×10^1
80	177	7.50×10^5	$\sim 2 \times 10^{11}$	$\sim 3 \times 10^5$	$\sim 5 \times 10^8$	3.29×10^1
100	210	5.14×10^5	4.75×10^{11}	9.2×10^5	3.9×10^5	3.28×10^1

[a] 24-hr average value.

in the calculation, [O(^{3}P)] is reduced to 2.6×10^{11} cm³/sec during the day. If some diffusional loss is also considered ($k_{diff} \sim 1 \times 10^{-6}$ sec^{-1}) then [O(^{3}P)] is reduced to $\sim 2 \times 10^{11}$ molecules/cm³ and [O_3] $\sim 5 \times 10^8$ molecules/cm³. The typical daytime value listed for [O(^{3}P)] at 80 km in Table I-3 is 6.0×10^9 molecules/cm³, but the concentration gradient is very sharp at this altitude. Furthermore during summer (overhead noonday sun), the [O(^{3}P)] maximum shifts downward, so that the true value may be $\sim 6 \times 10^{10}$ molecules/cm³. Thus the calculated value is probably high by a factor of ~ 3 because of neglect of removal by the minor species. The lifetime for odd oxygen is $\sim 3 \times 10^5$ sec, so that total odd oxygen is relatively constant throughout the day, although O(^{3}P) gets converted to O_3 at night.

At 100 km, [O(^{3}P)] is constant through the whole 24-hr day, and removal of odd oxygen is primarily by reaction (21). Thus a 24-hr, rather than 12-hr, average must be used in the calculation. With diffusion neglected, [O(^{3}P)] is computed to be 1.7×10^{12} molecules/cm³. However, 100 km corresponds to near peak [O(^{3}P)] concentration and diffusional loss must be significant. The diffusional loss coefficient can be calculated from Eq. (11). The maximum value occurs when $\xi_{O(^3P)}$ lies midway between $-\xi_D$ and ξ_n. Since $\xi_D \sim -3 \times 10^{-7}$ cm^{-1} and $\xi_n \sim 17 \times 10^{-7}$ cm^{-1}, maximum diffusional loss will occur for $\xi_{O(^3P)} \sim 10 \times 10^{-7}$ cm^{-1}. With $D = 2 \times 10^6$ cm²/sec, $k_{diff} = 1 \times 10^{-6}$ sec^{-1}. Incorporation of diffusional loss into the calculation reduces [O(^{3}P)] to 4.75×10^{11} molecules/cm³, close to the observed value of 4.0×10^{11} molecules/cm³. The computed lifetime for odd oxygen is over 100 days, so that odd oxygen remains constant throughout the day.

$[O(^3P)]$ (molecules/cm^3)	$\tau\{O(^3P)\}$ (sec)	$[O(^1D)]$ (molecules/cm^3)	$\tau\{O(^1D)\}$ (sec)	$[O_2(^1\Sigma)]$ (molecules/cm^3)	$[O_2(^1\Delta)]$ (molecules/cm^3)
1.75×10^7	1.5×10^{-3}	5.48	9.8×10^{-9}	8.3×10^5	9.1×10^8
1.84×10^9	1.02	2.37×10^2	2.1×10^{-7}	4.1×10^6	3.1×10^{10}
6.2×10^{10}	1.15×10^2	4.18×10^2	2.4×10^{-6}	4.3×10^6	5.4×10^{10}
$\sim 2 \times 10^{11}$	1.37×10^4	2.11×10^2	4.1×10^{-5}	1.9×10^6	1.0×10^{10}
4.75×10^{11}	9.19×10^6	9.73×10^2	1.8×10^{-3}	8.8×10^5	1.0×10^7

The concentrations for $O(^1D)$ for an overhead sun can be computed from its steady-state rate law:

$$[O(^1D)] = (I_{(19a)} + I_{(20a)})/k_{(24)}[M] \tag{30}$$

These values are listed in Table II-4.

The lifetimes for each of the odd oxygen species for an overhead sun can be computed from their concentrations and removal rates:

$$\tau\{O(^1D)\} = [O(^1D)]/R_{(24)} \tag{31}$$

$$\tau\{O(^3P)\} = [O(^3P)]/(2R_{(21)} + R_{(22)} + R_{(23)}) \tag{32}$$

$$\tau\{O_3\} = [O_3]/(I_{(20)} + R_{(23)}) \tag{33}$$

These values are also listed in Table II-4. The lifetimes for O_3 and $O(^1D)$ are always <5000 sec, so that the steady-state approximation is valid. It should be noticed that at the lower altitudes the steady-state approximation is good for each of the odd oxygen species, but not their sum. This is because these species rapidly interconvert into each other via reactions (20), (22), and (24), but these reactions do not affect the overall odd oxygen balance.

Singlet O_2

There is only one known reaction that produces $O_2(^1\Delta)$ in the atmosphere, and that is reaction (20a):

$$O_3 + h\nu\ (<3102\ \text{Å}) \rightarrow O(^1D) + O_2(^1\Delta) \tag{20a}$$

At wavelengths longer than 3102 Å, singlet O_2 is produced, but it is not known whether it is in the $^1\Delta$ or $^1\Sigma$ state. Since there have been reports that there is apparently an excess of $O_2(^1\Sigma)$ from that expected from other sources in the stratosphere, we assume that $^1\Sigma$ is produced in the Huggin's band absorption,

$$O_3 + h\nu\ (3102\text{–}3400\ \text{Å}) \rightarrow O(^3P) + O_2(^1\Sigma) \tag{20c}$$

The recombination of $O(^3P)$ atoms should produce both $^1\Delta$ and $^1\Sigma$ O_2 some of the time. Unfortunately these fractions are not known, so we neglect this source. However, if the efficiency of singlet O_2 production is large, this could be an important source of singlet O_2 in the thermosphere, particularly for the $^1\Delta$ state. An important source of $O_2(^1\Sigma)$ is from O_2 quenching of $O(^1D)$, which produces $O_2(^1\Sigma)$ essentially every time $O(^1D)$ is quenched by O_2:

$$O(^1D) + O_2(^3\Sigma) \rightarrow O(^3P) + O_2(^1\Sigma) \tag{24'}$$

The removal terms for singlet O_2 are both by radiation and collisional quenching:

$$O_2(^1\Sigma) \rightarrow O_2(^3\Sigma) + h\nu \tag{34}$$

$$O_2(^1\Delta) \rightarrow O_2(^3\Sigma) + h\nu \tag{35}$$

$$O_2(^1\Sigma) + N_2 \rightarrow O_2(^3\Sigma) + N_2 \tag{36}$$

$$O_2(^1\Sigma) + O_2(^3\Sigma) \rightarrow 2O_2(^3\Sigma) \tag{37}$$

$$O_2(^1\Sigma) + O_3 \rightarrow 2O_2(^3\Sigma) + O(^3P) \tag{38}$$

$$O_2(^1\Delta) + N_2 \rightarrow O_2(^3\Sigma) + N_2 \tag{39}$$

$$O_2(^1\Delta) + O_2(^3\Sigma) \rightarrow 2O_2(^3\Sigma) \tag{40}$$

$$O_2(^1\Delta) + O_3 \rightarrow 2O_2(^3\Sigma) + O(^3P) \tag{41}$$

Quenching of singlet O_2 by $O(^3P)$ may also be important in the upper mesosphere or thermosphere. However, the rate coefficients for these reactions are unknown so that, for simplicity, quenching by $O(^3P)$ will be ignored.

Finally, the energy pooling reaction must be considered:

$$2O_2(^1\Delta) \rightarrow O_2(^1\Sigma) + O_2(^3\Sigma) \tag{42}$$

The rate coefficients for all the quenching reactions are listed in Table II-2; and that for reaction (24′) in Table II-3. It will turn out that under all conditions reactions (39) and (42) are negligible and can be ignored. Reactions (37) and (41) are not very important, but do make a small contribution to singlet O_2 removal.

The steady-state assumption is always valid for both singlet states of O_2 at noon, since their lifetimes cannot exceed their radiative lifetimes, which are 7 sec and 45 min, respectively, for $O_2(^1\Sigma)$ and $O_2(^1\Delta)$. The steady-state rate laws are

$$[O_2(^1\Sigma)] = \frac{J_{(20c)}[O_3] + k_{(24')}[O(^1D)][O_2]}{k_{(34)} + k_{(36)}[N_2] + k_{(37)}[O_2] + k_{(38)}[O_3]} \quad (43)$$

$$[O_2(^1\Delta)] = J_{(20a)}[O_3]/(k_{(35)} + k_{(40)}[O_2] + k_{(41)}[O_3]) \quad (44)$$

The computed steady-state concentrations for noon for a pure oxygen atmosphere are listed in Table II-4 for various altitudes. They are about 10^6 molecules/cm^3 for $O_2(^1\Sigma)$ and one to four orders of magnitude higher for $O_2(^1\Delta)$.

NEUTRAL OXYGEN-NITROGEN ATMOSPHERE

N_2 and N_2O

N_2O is the dominant oxide of nitrogen at the surface of the earth, its mole fraction being about 0.25 ppm. Its mole fraction is constant in the troposphere, but as can be seen from Fig. II-1, its relative importance drops quickly as the altitude increases in the stratosphere, until N_2O has become unimportant in the mesosphere.

In the atmosphere there are no chemical or photochemical sources of N_2O, only removal. Removal occurs by photodissociation, with a small contribution from the chemical term:

$$O(^1D) + N_2O \rightarrow N_2 + O_2 \quad (45a)$$

$$\rightarrow 2NO \quad (45b)$$

However, throughout the whole stratosphere the loss by reaction (45) is only about 4% of that estimated for photodissociation, with channels (45a) and (45b) being almost equally probable.

The photodissociation of N_2O produces N_2 and $O(^1D)$:

$$N_2O + h\nu \rightarrow N_2 + O(^1D) \quad (46)$$

The photodissociation coefficients for N_2O as a function of altitude and solar zenith angle are shown in Fig. II-8. N_2O photodissociation extends to wavelengths of 3150 Å, but most of the photodissociation in the upper stratosphere occurs at wavelengths <2500 Å.

Since N_2O is not produced in the atmosphere, but only destroyed, it must arise from the earth. In fact, N_2O is a waste product of life on earth and is brought into the atmosphere through upward diffusion. (In the troposphere the mole fraction is constant so that there is no net diffusional

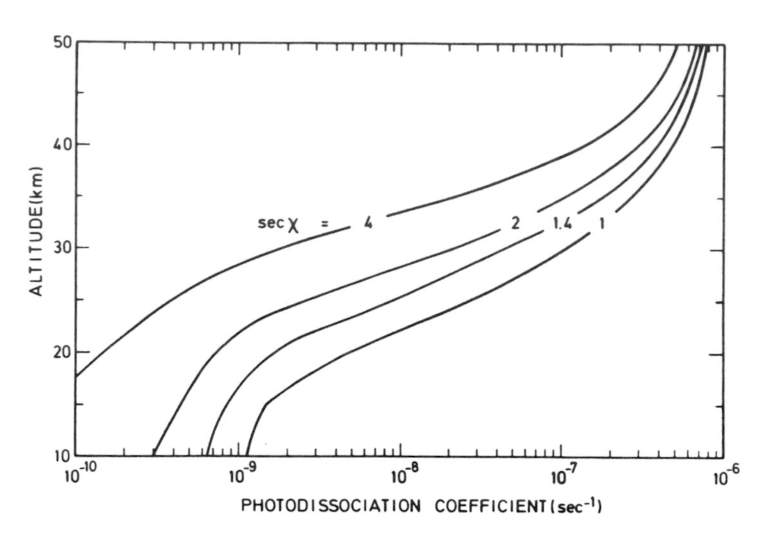

Fig. II-8. Photodissociation coefficient for N_2O in the 10–50 km region for various solar zenith angles. From Nicolet and Peetermans (1972) with permission of Centre National de la Recherche Scientifique.

gain or loss. The N_2O diffusion rate upward is constant, so that the inflow is offset by the outflow).

The 24-hr average removal rate coefficients for N_2O via reactions (45) and (46) must equal the production rate coefficient by diffusion, $k_{\text{diff}}\{N_2O\}$:

$$k_{\text{diff}}\{N_2O\} = k_{(45)}\overline{[O(^1D)]} + \bar{J}_{(46)} \tag{47}$$

The value for $[O(^1D)]$ at any time is given by Eq. (30), so that $\overline{[O(^1D)]}$ can be computed directly from $\bar{J}_{(19a)}$ and $\bar{J}_{(20a)}$. The values for $k_{(45)}\overline{[O(^1D)]}$ and $\bar{J}_{(46)}$ are given in Table II-5 as a function of altitude for a day with an overhead noonday sun.

Also $k_{\text{diff}}\{N_2O\}$ is related to the reciprocal scale height through the approximate relationship

$$k_{\text{diff}}\{N_2O\} = D(\xi_{N_2O} + \xi_D)(\xi_{N_2O} - \xi_n) \tag{48}$$

Thus if D, ξ_D, and ξ_n are known, ξ_{N_2O} can be computed and the N_2O concentration profile constructed. Unfortunately, values of D and ξ_D are not known accurately in the troposphere and stratosphere. However, the upper limit is $\sim 1 \times 10^4$ cm²/sec for D (see Fig. I-2). If values less than this are used, the computed N_2O concentration profile does not fit the observed values. Thus we take $D = 1 \times 10^4$ cm²/sec, the value adopted by most investigators (Grobecker *et al.*, 1974), and $\xi_D = 0$ for altitudes ≤ 40 km. At higher altitudes these parameters can be estimated from Fig. I-2 and

TABLE II-5

Calculated Concentrations of N_2O *as a Function of Altitude*

Altitude (km)	$k_{(45)}\overline{[O(^1D)]}$ [a] (sec^{-1})	$J_{(46)}$ [a] (sec^{-1})	D (cm^2/sec)	$10^7\xi_D$ (cm^{-1})	$10^7\xi_n$ (cm^{-1})	$10^7\xi_{N_2O}$ (cm^{-1})	$[N_2O]$ (ppm)
10	3.89×10^{-13}	3.09×10^{-10}	1×10^4	0	13.9	14.1	0.25
20	2.91×10^{-11}	1.04×10^{-9}	1×10^4	0	13.9	14.6	0.24
30	1.18×10^{-9}	2.74×10^{-8}	1×10^4	0	13.9	25.2	0.13
40	1.01×10^{-8}	1.60×10^{-7}	1×10^4	0	13.9	48.8	0.020
50	2.73×10^{-8}	3.26×10^{-7}	1×10^5	−10.4	13.9	31.0	1.7×10^{-3}
60	7.90×10^{-9}	4.0×10^{-7}	2×10^5	−6.9	13.9	25.1	4.3×10^{-4}
70	9.50×10^{-9}	4.0×10^{-7}	4×10^5	0	13.9	19.2	1.8×10^{-4}
80	9.25×10^{-9}	4.0×10^{-7}	4×10^5	0	17.0	21.7	1.1×10^{-4}

[a] 24-hr average for an overhead noonday sun.

they are listed in Table II-5 along with ξ_n. The computed values for ξ_{N_2O} are obtained from Eq. (48) and they are also listed in Table II-5. From these, the N_2O concentration profile curve is constructed and the estimated N_2O concentrations are tabulated in Table II-5.

The flux of N_2O from the surface of the earth can be related to $k_{diff}\{N_2O\}$ through

$$F_{N_2O} = \int_0^{\infty} k_{diff}\{N_2O\}[N_2O]\, dZ \tag{49}$$

With the values in Table II-5, the integration gives $F_{N_2O} = 2.66 \times 10^9$ molecules/cm²-sec. There is considerable uncertainty in this value since it is based on knowing the N_2O concentration profile. An alternate method is to combine Eqs. (10) and (11) to get

$$F_{N_2O} = \frac{k_{diff}\{N_2O\}[N_2O]}{\xi_{N_2O} + \xi_D} \tag{50}$$

At the surface of the earth, $k_{diff}\{N_2O\} = \bar{J}_{(46)}$, which is the same as at 10 km, i.e., 3.09×10^{-10} cm³/molecules-sec, $[N_2O] = 6.25 \times 10^{13}$ mole-

TABLE II-6[a]

Photodissociation Coefficients of NO *versus Height in the Mesosphere and Stratosphere*

Altitude (km)	Photodissociation coefficient (sec⁻¹)			
	δ(0–0) band		δ(1–0) band	
	sec χ = 1	sec χ = 2	sec χ = 1	sec χ = 2
85	6.80×10^{-6}	6.72×10^{-6}	2.84×10^{-6}	2.70×10^{-6}
80	6.64×10^{-6}	6.41×10^{-6}	2.61×10^{-6}	2.42×10^{-6}
75	6.33×10^{-6}	5.95×10^{-6}	2.36×10^{-6}	2.12×10^{-6}
70	5.80×10^{-6}	5.31×10^{-6}	1.94×10^{-6}	1.62×10^{-6}
65	5.22×10^{-6}	4.65×10^{-6}	1.54×10^{-6}	1.16×10^{-6}
60	4.44×10^{-6}	3.61×10^{-6}	9.08×10^{-7}	5.03×10^{-7}
55	3.54×10^{-6}	2.53×10^{-6}	4.65×10^{-7}	1.68×10^{-7}
50	2.53×10^{-6}	1.51×10^{-6}	1.75×10^{-7}	3.19×10^{-8}
45	1.52×10^{-6}	6.83×10^{-7}	3.61×10^{-8}	2.07×10^{-9}
40	6.64×10^{-7}	1.94×10^{-7}	2.44×10^{-9}	1.73×10^{-11}
35	1.70×10^{-7}	2.53×10^{-8}	1.80×10^{-11}	1.71×10^{-15}
30	3.36×10^{-8}	1.78×10^{-9}	2.97×10^{-14}	6.22×10^{-22}
25	6.92×10^{-13}	1.49×10^{-16}	2.93×10^{-22}	

[a] From Cieslik and Nicolet (1973).

cules/cm^3, and $\xi_{N_2O} = 14.0 \times 10^{-7}$ cm^{-1}. Assuming $\xi_D = 0$ gives $F_{N_2O} = 1.36 \times 10^9$ molecules/cm^2-sec. Thus we can say that $F_{N_2O} \simeq 2 \times 10^9$ molecules/cm^2-sec with an uncertainty of $\pm 33\%$. Of this N_2O introduced into the atmosphere, about 98% converts to N_2 and about 2% converts to NO.

NO and NO_2

The photochemical reactions of importance for NO and NO_2 are

$$NO + h\nu \rightarrow N + O(^3P) \tag{51}$$

$$NO_2 + h\nu \rightarrow NO + O(^3P) \tag{52}$$

The NO photodissociation is only important in the mesosphere and thermosphere; the photodissociation coefficients as a function of altitude have been given by Cieslik and Nicolet (1973). The absorptions of importance are into the ground and first vibrational levels of the C $^2\Pi$ state [the $\delta(0\text{–}0)$ and $\delta(1\text{–}0)$ bands, respectively] and are listed in Table II-6. The overall photodissociation coefficients are summarized in Table II-1. For NO_2, $J\{NO_2\} = 1.0 \times 10^{-2}$ sec^{-1} above 25 km and 8×10^{-3} sec^{-1} at 25 km and below.

The chemical reactions of interest are

$$N + O_2 \rightarrow NO + O(^3P) \tag{53}$$

$$NO_2 + O(^3P) \rightarrow NO + O_2 \tag{54}$$

$$NO + O(^3P) + M \rightarrow NO_2 + M \tag{55}$$

$$N + NO \rightarrow N_2 + O(^3P) \tag{56}$$

$$NO + O_3 \rightarrow NO_2 + O_2 \tag{57}$$

The rate coefficients for the reactions are listed in Tables II-2 and II-7. Under all conditions reaction (53) is negligible compared to reaction (56).

The diffusional term is only of importance in comparison to the photochemical and chemical rates in the mesosphere and thermosphere. Even here the term is only of importance for NO.

The steady-state expressions for N and NO_2 are

$$[N] = I_{(51)}/k_{(56)}[NO] \tag{58}$$

$$[NO_2] = \frac{(k_{(55)}[O(^3P)][M] + k_{(57)}[O_3])[NO]}{J_{(52)} + k_{(54)}[O(^3P)]} \tag{59}$$

During daylight, the steady-state expression for N is applicable since $(k_{(56)}[NO])^{-1} = \tau\{N\} < 2000$ sec at altitudes up to 150 km. Likewise for NO_2, $\tau\{NO_2\} \leq J_{(52)}^{-1} = 100$ sec.

TABLE II-7

Reactions of the Oxides of Nitrogen and Carbon[a]

Reaction	Reaction number	k at 298°K (cm^{3n}/sec)	A (cm^{3n}/sec)	E (cal/mole)	Temp. range (°K)
$2NO + O_2 \rightarrow 2NO_2 + O_2$	—	2.0×10^{-38}	3.3×10^{-39}	−1050	270–660
$NO + O_3 \rightarrow NO_2 + O_2$	(57)	1.7×10^{-14}	9×10^{-13}	2400	198–330
$NO_2 + O_3 \rightarrow NO_3 + O_2$	(61)	5×10^{-17}	1.1×10^{-13}	4900	220–340
$NO + NO_3 \rightarrow 2NO_2$	(62)	8.7×10^{-12}	—	—	—
$2NO_3 \rightarrow 2NO_2 + O_2$	—	—	5×10^{-12}	6000	293–309
$NO_2 + NO_3 + M \rightarrow N_2O_5 + M$	(63)	2.8×10^{-30}	—	—	—
$NO_2 + NO_3 \rightarrow N_2O_5$	(63)	3.8×10^{-12}	—	—	—
$N_2O_5 + M \rightarrow NO_2 + NO_3 + M$	(−63)	—	2.2×10^{-5}	19400	300–340
$N_2O_5 \rightarrow NO_2 + NO_3$	(−63)	—	5.7×10^{14}	21200	273–300
$CH_3 + O_2 + M \rightarrow CH_3O_2 + M$	(123)	2.6×10^{-31}	—	—	—
$CH_3 + O \rightarrow CH_2O + H$	(124)	$\sim 10^{-11}$ [b]	—	—	—
$CH_3 + O_3 \rightarrow CH_2O + O_2 + H$	(125)	$\sim 10^{-11}$ [b]	—	—	—
$CH_2O + O_3 \rightarrow CH_2O + O_2 + HO$	—	—	1.0×10^{-12} [b]	6000[b]	—

[a] From Garvin and Hampson (1974).
[b] Estimate.

The daylight ratio [NO_2]/[NO] can be obtained from Eq. (59) using the values for [$O(^3P)$] from Table I-3 and [O_3] from Table I-4 and Fig. I-6. These are listed in Table II-8. At 20 km there are about equal amounts of NO_2 and NO during the day, but the ratio drops rapidly as the altitude increases. At night when $J_{(52)}$ goes to zero and [$O(^3P)$] becomes markedly reduced, the NO tends to convert to NO_2; the NO concentration drops and the NO_2 concentration rises.

NO_3 and N_2O_5

Whenever NO_2 is present, it is possible to produce NO_3 via

$$O(^3P) + NO_2(+M) \rightarrow NO_3(+M) \tag{60}$$

$$NO_2 + O_3 \rightarrow NO_3 + O_2 \tag{61}$$

Chemical removal of NO_3 occurs via

$$NO + NO_3 \rightarrow 2NO_2 \tag{62}$$

$$NO_2 + NO_3(+M) \rightleftarrows N_2O_5(+M) \tag{63}$$

the reaction rate between two NO_3 molecules being too small to be of any significance. The rate coefficients for reactions (61) and (62) are listed in Table II-7, and that for reaction (60) in Table II-2. Reaction (63) and its reverse are partway between their second- and third-order regimes. Thus the effective second-order rate coefficients, corrected for temperature and pressure, are listed for the various altitudes of interest in Table II-9.

NO_3 can also be removed photochemically with a photodissociation coefficient $J\{NO_3\} = 1.0 \times 10^{-2}$ sec^{-1} at all altitudes. The products of

TABLE II-8

Relative Daylight Concentrations of the Oxides of Nitrogen

Altitude (km)	Temp. (°K)	[$O(^3P)$][a] (molecules/cm^3)	[O_3][b] (molecules/cm^3)	$\frac{[NO_2]}{[NO]}$	$\frac{[NO_3]}{[NO_2]}$
20	219	1.0×10^6	2.4×10^{12}	1.08	3.58×10^{-4}
40	268	1.0×10^9	5.1×10^{11}	0.265	1.38×10^{-3}
60	253	7.0×10^9	3.0×10^9	4.32×10^{-4}	5.06×10^{-4}
80	177	6.0×10^9	1.0×10^8	3.58×10^{-5}	2.52×10^{-5}
100	210	4.0×10^{11}	1.0×10^7	4.66×10^{-7}	4.40×10^{-5}

[a] From Table I-3.
[b] From Table I-4 and Fig. I-6.

TABLE II-9

Effective Second-Order Rate Coefficients for Reactions in the Intermediate Second- and Third-Order Regimes

			k, (cm^3/sec)				
Altitude (km)	Temp (°K)	Density (particles/cm^3)	$NO_2 + NO_3 \rightarrow N_2O_5$[a]	$HO + NO_2 \rightarrow HONO_2$[b]	$CO + O(^3P) \rightarrow CO_2$[c]	$N_2O_5 \rightarrow NO_2 + NO_3$[a]	$HO + NO \rightarrow HONO$[d]
0	291	2.5×10^{19}	—	7.0×10^{-12}	7.2×10^{-17}	—	—
5	266	1.3×10^{19}	—	—	2.4×10^{-17}	—	—
10	231	7.7×10^{18}	—	—	4.9×10^{-18}	—	—
15	211	3.9×10^{18}	1.9×10^{-12}	3.2×10^{-12}	1.2×10^{-18}	5.0×10^{-7}	2.4×10^{-12}
20	219	1.7×10^{18}	1.1×10^{-12}	2.5×10^{-12}	—	1.8×10^{-7}	1.5×10^{-12}
25	227	7.7×10^{17}	6.6×10^{-13}	1.6×10^{-12}	—	3.0×10^{-7}	8.9×10^{-13}
30	235	3.6×10^{17}	4.5×10^{-13}	9.6×10^{-13}	—	5.5×10^{-7}	4.8×10^{-13}
35	252	1.7×10^{17}	2.7×10^{-13}	5.5×10^{-13}	—	1.9×10^{-6}	2.6×10^{-13}
40	268	8.1×10^{16}	1.5×10^{-13}	2.7×10^{-13}	—	1.6×10^{-5}	1.1×10^{-13}
45	274	4.3×10^{16}	7.1×10^{-14}	1.4×10^{-13}	—	4.4×10^{-5}	5.9×10^{-14}

[a] From Johnston as reported in Garvin and Hampson (1974).
[b] From Tsang as reported in Garvin and Hampson (1974).
[c] From Simonaitis and Heicklen, (1972) assuming N_2 is one-third as efficient as N_2O.
[d] From Garvin and Hampson (1974).

this photodissociation are not known. They may be either $NO_2 + O$ or $NO + O_2$ or both. Fortunately the product distribution is of no consequence to the chemical balance in the atmosphere, so we need not be concerned with it.

The steady-state expression can be used to calculate $[NO_3]$ during the day since $\tau\{NO_3\} < J\{NO_3\}^{-1} = 100$ sec. Furthermore, photodissociation of NO_3 is the dominant removal process in the stratosphere and above. Thus an upper limit to the ratio $[NO_3]/[NO_2]$ can be computed from the steady-state expression

$$[NO_3]/[NO_2] < (k_{(60)}[O(^3P)][M] + k_{(61)}[O_3])/J\{NO_3\} \qquad (64)$$

These values are computed and listed in Table II-8. Under all daylight conditions $[NO_3]/[NO_2]$ is much less than one.

N_2O_5 is formed and removed only very slowly in chemical processes, some of which are not known. During the daylight, N_2O_5 gives NO_3, which immediately returns to the lower oxides of nitrogen. However, at night in the presence of NO_2, N_2O_5 can accumulate through reaction (63).

Because $[NO_3]/[NO_2]$ is small and because the chemistry of N_2O_5 is not completely understood, the species NO_3 and N_2O_5 will be ignored in our future discussion. Presumably both are unimportant. However, the concentrations of NO and NO_2 that we shall compute will really include NO_3 and N_2O_5 as well.

NO_x

Now that we have considered each of the oxides of nitrogen separately, it is useful to consider them as a group. For this purpose we will define a general oxide of nitrogen sum, NO_x, with a concentration computed as

$$[NO_x] \equiv [NO] + [NO_2] + [NO_3] + 2[N_2O_5] + [HNO_2] + [HNO_3] \qquad (65)$$

where the molecules HNO_2 and HNO_3 have included since they interconvert with the other oxides of nitrogen. Thus $[NO_x]$ is the total concentration of nitrogen tied up in the six molecules NO, NO_2, NO_3, N_2O_5, HNO_2, and HNO_3. Notice that N_2O has not been included.

In the troposphere NO_x is produced by the reaction of $O(^1D)$ with N_2O via reaction (45b). There are no other chemical or photochemical production or loss terms. The photodissociation of NO, reaction (51), is of no importance below 50 km. Even if it were important, it would not lead to NO_x loss, since the N atom produced would regenerate NO via

$$N + O_3 \rightarrow NO + O_2 \qquad (66)$$

Thus the NO_x production must be offset by diffusional loss, and the diffusional flux is downward below 50 km.

The NO_x concentration profiles can be obtained from the diffusion equations since the 24-hr average of $2R_{(45b)}$ must equal the diffusional loss at each altitude. Thus the flux can be computed from

$$F_{NO_x}\{Z\} = -\int_Z^{50\text{ km}} 2\bar{R}_{(45b)}\, dZ \qquad (67)$$

where $\bar{R}_{(45b)}$ is the 24-hr average rate of reaction (45b). The flux equation (5) can be rewritten

$$F_{NO_x}\{Z\} = \frac{-D_n\, d([NO_x]/n)}{dZ} \qquad (68)$$

Rearrangement and integration of Eq. (68) gives

$$\int_Z^{50\text{ km}} d([NO_x]/n) = -\int_Z^{50\text{ km}} (F_{NO_x}\{Z\}/D_n)\, dZ \qquad (69)$$

By numerical integration of Eq. (67), $F_{NO_x}\{Z\}$ is obtained and it is inserted into Eq. (69) for numerical integration of that equation. With the boundary condition that $[NO_x]/n = 3.0 \times 10^{-9}$ at the surface of the earth ($Z = 0$) (see Chapter IV), the concentration profiles for $[NO_x]$ can be obtained. The results are listed in Table II-10.

The relative concentration of NO_x increases with altitude to a maximum of 2.83×10^{-8} at 50 km, but the absolute value of $[NO_x]$ drops continually with increasing altitude. The relative maximum at 50 km corresponds to 6.5×10^8 molecules/cm^3, which agrees exactly with the observed value for daytime [NO] (Hale, 1974). (At 50 km during the day, all the NO_x is NO.)

The lifetimes for NO_x can be computed from the formula

$$\tau\{NO_x\} = [NO_x]/2\bar{R}_{(45b)} \qquad (70)$$

These are calculated and tabulated in Table II-10. The lifetimes are all much longer than a day, so that $[NO_x]$ is invariant to the diurnal cycle.

Above 50 km, reaction (45b) is unimportant, and there are no chemical or photochemical production terms for NO_x. However, NO_x is lost by photodissociation of NO, reaction (51), which removes 2NO molecules each time it occurs because the N atom produced reacts with another NO molecule. The NO loss must be offset by diffusional accumulation. Balancing the two process by realizing that during the day above 50 km, all

TABLE II-10

Concentration of NO_x *below 50* km *for a Day with an Overhead Noonday Sun*

Altitude (km)	$2\bar{R}_{(45b)}$ [a] (molecules/cm^3-sec)	$-10^{-6}F_{NO_x}\{z\}$ (molecules/cm^2-sec)	D (cm^2/sec)	n (molecules/cm^3)	$10^9[NO_x]/n$	$[NO_x]$ (molecules/cm^3)	$\tau\{NO_x\}$ (sec^{-1})
10	0.75	84.3	1×10^4	7.7×10^{18}	3.0	2.3×10^{10}	6.1×10^{10}
20	11.9	78.0	1×10^4	1.7×10^{18}	5.8	9.8×10^9	1.3×10^9
30	55.2	44.5	1×10^4	3.6×10^{17}	11.3	4.1×10^9	9.2×10^7
40	16.4	8.7	1×10^4	8.1×10^{16}	23.0	1.86×10^9	1.3×10^8
50	1.07	0.54	1×10^5	2.3×10^{16}	28.3	6.5×10^8	6.7×10^8
60	0.02	—	—	—	—	—	—

[a] 24-hr average value.

TABLE II-11

Calculated NO_x *Concentrations above* 50 km *for a Day with an Overhead Noonday Sun*

Altitude (km)	$2\bar{J}_{(51)}$ [a] (sec^{-1})	$10^{-5}D$ (cm^2/sec)	$10^7\xi_D$ (cm^{-1})	$10^7\xi_n$ (cm^{-1})	$10^7\xi_{NO_x}$ (cm^{-1})	$[NO_x]$ (molecules/cm^3)
50	1.78×10^{-6}	1.0	−10.4	13.9	54.4	7.2×10^8
60	4.12×10^{-6}	2.0	−6.9	13.9	55.9	5.3×10^6
70	6.64×10^{-6}	4.0	0	13.9	−34.4	4.0×10^6
80	8.42×10^{-6}	4.0	0	17.0	−38.2	1.8×10^8
90	9.3×10^{-6}	4.0	−6.9	17.0	−36.5	7.0×10^9

[a] 24-hr average value.

the NO_x is NO, and using Eq. (11) for diffusion gives

$$2\bar{J}_{(51)} = k_{diff}\{NO_x\} = D(\xi_{NO_x} + \xi_D)(\xi_{NO_x} - \xi_n) \tag{71}$$

where $\bar{J}_{(51)}$ is the 24-hr average NO photodissociation coefficient. The values for $\bar{J}_{(51)}$, D, ξ_D, and ξ_n can be obtained from previous considerations and are listed in Table II-11. Then ξ_{NO_x} can be computed, and from these the NO_x concentration profile deduced. The computed values for ξ_{NO_x} and $[NO_x]$ are also listed in Table II-11. In computing $[NO_x]$ it was assumed that its minimum value occurred at 65 km.

The values computed for $[NO_x]$ between 60 and 70 km lie lower than the daylight values for [NO] given in Fig. I-9. This is because our approximate diffusion Eq. (71) overemphasizes the minimum. At 80 km, the computed value is about a factor of 3 too high and at 90 km the computed value is about a factor of 100 too high. Clearly some other processes must be occurring at 80 km and above that produce NO_x, and reduce the need for diffusional input. These are the ionic processes, and they are deferred until Chapter III.

The lifetimes for NO_x are given by $(2\bar{J}_{(51)})^{-1}$ and these range from about 6 days at 50 km to $<1\frac{1}{2}$ days at 90 km. Thus in the upper mesosphere and lower thermosphere, $[NO_x]$ will not be quite constant during the daily cycle; it will tend to deplete during daylight and accumulate at night.

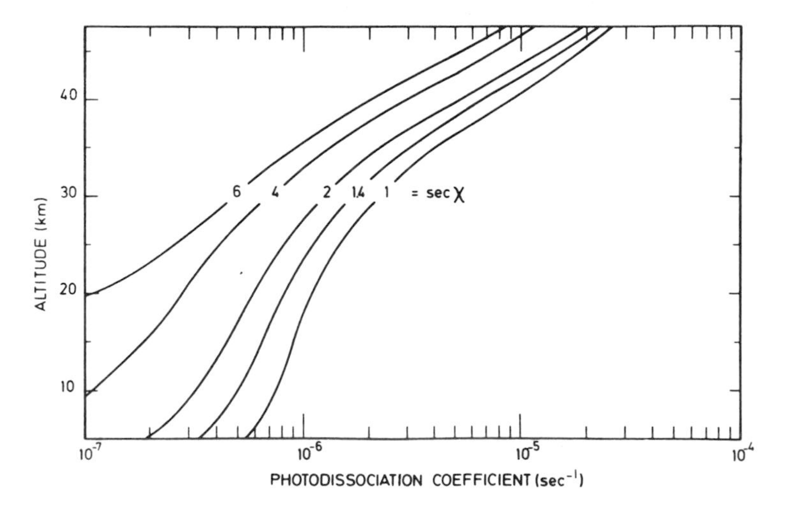

Fig. II-9. Photodissociation coefficient for HNO_3 in the 5–45 km region for various solar zenith angles. $J_\infty = 3.5 \times 10^{-5}$ sec^{-1}. From Brasseur and Nicolet (1973) with permission.

OXYGEN–NITROGEN–HYDROGEN ATMOSPHERE

The hydrogen-bearing species are of three types:

1. Those containing N but not C, called NH_x, which include NH_3, HNO_2, and HNO_3 as the important species, although NH_2 and HNO are present as transients. Possibly even other trace free radicals such as NH_2O_2, NH_2O, and NH are present. However, we shall limit our discussion to NH_3, HNO_2, and HNO_3.
2. The free radical species of hydrogen and oxygen: H, HO, and HO_2. These species contain one H atom each, and will be called HO_x.
3. The stable molecules of hydrogen and oxygen, which contain two hydrogen atoms each. These are H_2, H_2O, and H_2O_2.

NH_x

NH_3 arises from plant and animal wastes on the earth and diffuses upward, where it is removed in the troposphere and lower stratosphere. The main removal routes are probably combination with the acids HNO_3 and H_2SO_4, which are diffusing downward to produce their ammonium salts, which are solids and fall to the earth.

However, chemical reactions can also occur with NH_3. The principal routes for its destruction are photodissociation,

$$NH_3 + h\nu \rightarrow NH_2 + H \tag{72}$$

and attack by HO radical

$$NH_3 + HO \rightarrow NH_2 + H_2O \tag{73}$$

The Arrhenius parameters for $k_{(73)}$ are given in Table II-12. The NH_2 radical reacts rapidly with O_2 to produce H_2O, N_2, and N_2O (Jayanty *et al.*, 1976). Presumably the initial step is addition:

$$NH_2 + O_2 \rightarrow NH_2O_2 \tag{74}$$

In the lower stratosphere, reaction with O_3 might also occur. The details or rate coefficient of this attack are not known, but the products also include HNO_3. Possibly this process occurs through an NH_2O intermediate that is removed via attack by more O_3:

$$NH_2 + O_3 \rightarrow NH_2O + O_2 \tag{75}$$

$$NH_2O + O_3 \rightarrow NH_2O_2 + O_2 \tag{76}$$

$$NH_2O_2 + O_3 \rightarrow HO_2 + HNO_3 \tag{77}$$

TABLE II-12

Reactions of H, HO, *and* HO_2[a]

Reaction	Reaction number	k at 298°K, cm³/sec)	A (cm³/sec)	E (cal/mole)	Temp. range (°K)
$H + O_2 + M \rightarrow HO_2 + M$	(97)	—	2.1×10^{-32}	−580	203–404
$H + O_3 \rightarrow HO + O_2$	(98)	2.6×10^{-11}	—	—	—
$H + NO_2 \rightarrow HO + NO$	—	4.8×10^{-11}	5.8×10^{-10}	1480	298–633
$HO + O_3 \rightarrow HO_2 + O_2$	(105)	5.5×10^{-14}	1.6×10^{-12}	2000	220–450
$HO + CO \rightarrow H + CO_2$	(137)	1.4×10^{-13}	1.33×10^{-13}	0	200–400
$HO + CH_4 \rightarrow H_2O + CH_3$	(119)	—	2.95×10^{-12}	3540	240–370
$HO + NO_2 + M \rightarrow HONO_2 + M$	(85)	—	3.6×10^{-32}	−2200	273–400
$HO + NO + M \rightarrow HONO + M$	(78)	10^{-30}	2.2×10^{-32}	−2220	273–395
$HO + H_2 \rightarrow H_2O + H$	(95)	—	3.8×10^{-11}	5200	300–2000
$2HO \rightarrow H_2O + O(^3P)$	(104)	—	1.0×10^{-11}	1100	300–2000
$HO + O(^3P) \rightarrow H + O_2$	(96)	4.2×10^{-11}	4.2×10^{-11}	0	300–1000
$HO + HO_2 \rightarrow H_2O + O_2$	(103)	—	6×10^{-11} [b]	—	—
$HO_2 + O_3 \rightarrow HO + 2O_2$	(99)	1.5×10^{-15}	1×10^{-13}	2500	225–298
$HO_2 + NO \rightarrow HO + NO_2$	(100)	9.7×10^{-13} [c]	—	—	—
$HO_2 + NO_2 \rightarrow HONO + O_2$	(79)	1.8×10^{-13} [c]	—	—	—
$2HO_2 \rightarrow H_2O_2 + O_2$	(106)	3.3×10^{-12}	$\sim 3 \times 10^{-11}$	1000	300–1000
$HO + H_2O_2 \rightarrow H_2O + HO_2$	(111)	—	1.7×10^{-11}	1820	300–800
$HO + CH_2O \rightarrow H_2O + HCO$	(133)	1.4×10^{-11}	—	—	—
$HO + NH_3 \rightarrow H_2O + NH_2$	(73)	1.5×10^{-13}	3×10^{-11}	3200	—
$HO + HNO_3 \rightarrow H_2O + NO_3$	(86)	1.3×10^{-13}	6×10^{-13}	800	300–650
$HO_2 + O(^3P) \rightarrow HO + O_2$	(101)	5×10^{-11}	—	—	—
$HO_2 + H \rightarrow H_2 + O_2$	(107a)	—	4.2×10^{-11}	700	290–800
$HO + H \rightarrow H_2 + O(^3P)$	(114)	—	$1.4 \times 10^{-14}T$	7000	400–2000
$HO_2 + H \rightarrow 2HO$	(107b)	—	4.2×10^{-10}	1900	290–800
$HO + HNO_2 \rightarrow H_2O + NO_2$	(81)	6.8×10^{-12}	—	—	—

The photodissociation coefficient of NH_3 has not been reported in the literature. However, from the absorption spectrum of Watanabe (1954) and the radiation distributions in Fig. I-15, values for the photodissociation coefficient for an overhead sun can be estimated to be

$J\{NH_3\}$ (sec^{-1}):	3.3×10^{-5}	9.1×10^{-5}	11.8×10^{-5}	12.1×10^{-5}
Altitude (km):	30	50	85	∞

By extrapolation $J\{NH_3\} \sim 2 \times 10^{-5}$ sec^{-1} at 25 km.

Since the amount of NH_3 that rises to the stratosphere is small, we shall ignore it in our future discussion.

Let us now turn our attention to HNO_2. The chemistry of HNO_2 is still not completely elucidated, but certain general features can be summarized with some certainty. The main production terms for HNO_2 are surely

$$HO + NO + M \rightarrow HNO_2 + M \tag{78}$$

$$HO_2 + NO_2 \rightarrow HNO_2 + O_2 \tag{79}$$

The removal terms of importance are probably

$$HNO_2 + h\nu \rightarrow HO + NO \tag{80}$$

$$HO + HNO_2 \rightarrow H_2O + NO_2 \tag{81}$$

Other reactions that may play some role in HNO_2 removal are

$$O(^3P) + HNO_2 \rightarrow HO + NO_2 \tag{82}$$

$$H + HNO_2 \rightarrow H_2 + NO_2 \tag{83}$$

However in the stratosphere, where HNO_2 is important, reactions (82) and (83) can be ignored.

Since $J_{(80)}$ is 6.45×10^{-4} sec^{-1} (Garvin and Hampson, 1974) for an overhead sun, the steady state always applies to HONO during daylight. Thus

$$[\mathrm{HONO}] = \frac{k_{(78)}[\mathrm{HO}][\mathrm{NO}][\mathrm{M}] + k_{(79)}[\mathrm{HO_2}][\mathrm{NO_2}]}{J_{(80)} + k_{(81)}[\mathrm{HO}]} \tag{84}$$

Production of HNO_3 can occur by reaction of HO and NO_2 throughout the atmosphere:

$$HO + NO_2\ (+M) \rightarrow HNO_3\ (+M) \tag{85}$$

The only important chemical reaction for HNO_3 removal is

$$HO + HNO_3 \rightarrow H_2O + NO_3 \tag{86}$$

but also important is photodissociation,

$$HNO_3 + h\nu \rightarrow HO + NO_2 \tag{87}$$

The rate coefficients $k_{(85)}$ and $k_{(86)}$ are listed in Table II-12. Reaction (85) is a second-order reaction in its fall-off regime, so that equivalent second-order rate coefficients have been computed and are listed in Table II-9. The photodissociation coefficients for HNO_3 are shown in Fig. II-9 for the troposphere and stratosphere. $J\{HNO_3\}$ become constant at 3.5×10^{-5} sec^{-1} for altitudes >50 km.*

The removal rates for HNO_3 are slow so that the steady state does not apply, and diffusional loss terms must be included. The 24-hr average steady state gives

$$k_{(85)}\overline{[HO][NO_2]} = (k_{(86)}\overline{[HO]} + \bar{J}_{(87)} + k_{\text{diff}})[HNO_3] \tag{88}$$

Now HNO_3 is important only below 50 km, and for these altitudes HO disappears at night. Therefore $\overline{[HO]}$ is $\frac{1}{2}$ its daylight average value and $\overline{[NO_2]}$ is its daylight value. Furthermore, from the fact that the sum of the concentrations of the NO_x species equals $[NO_x]$, we obtain

$$\overline{[NO_2]} = ([NO_x] - [HNO_3])/(1 + [NO]/[NO_2]) \tag{89}$$

where the concentrations of HNO_2, NO_3, and N_2O_5 have been omitted because they are negligible. The ratio $[NO_2]/[NO]$ in Eq. (89) is the daylight average, values of which are listed in Table II-8.

Combination of Eqs. (88) and (89) gives

$$[HNO_3]/[NO_x] = \beta/(\tau\{HNO_3\}^{-1} + \beta) \tag{90}$$

where

$$\tau\{HNO_3\}^{-1} \equiv k_{(86)}\overline{[HO]} + \bar{J}_{(87)} + k_{\text{diff}} \tag{91}$$

$$\beta \equiv k_{(85)}\overline{[HO]}/(1 + [NO]/[NO_2]) \tag{92}$$

In order to compute $[HNO_3]/[NO_x]$ it is necessary to know $k_{(85)}$, $k_{(86)}$, $[NO_2]/[NO]$, $\overline{[HO]}$, $\bar{J}_{(87)}$, and k_{diff}. Values of $k_{(85)}$ and $k_{(86)}$ are listed in Table II-13, as are the daylight values of $[NO_2]/[NO]$ obtained from Table II-8. We anticipate later results and list values for $\overline{[HO]}$ that will be justified later. The values for $\bar{J}_{(87)}$ are obtained from the parameters in Table II-1. Values for k_{diff} are not known. However, in the stratosphere, they cannot exceed 2×10^{-7} sec^{-1}. This value is too small to play a role at 40 km and $[HNO_3]/[NO_x]$ is computed to be 0.087. At 20 km, if

* G. Kockarts has informed the author that $J\{HNO_3\}$ may be $\sim 10^{-4}$ sec^{-1} at ≥ 50 km.

TABLE II-13

Stratospheric Concentrations of HNO_3 *for a Day with an Overhead Noonday Sun*

Altitude (km)	$k_{(85)}$ (cm³/molecules-sec)	$\overline{[HO]}$[a] (molecules/cm³)	$[NO_2]/[NO]$[b]	$k_{(86)}$ (cm³/molecules-sec)	$\bar{J}_{(87)}$[a] (sec⁻¹)	$[HNO_3]/[NO_x]$	$\tau\{HNO_3\}$ (sec⁻¹)
20	2.5×10^{-12}	9×10^{5}	1.08	9.5×10^{-14}	3.7×10^{-7}	0.72	1.7×10^{6}
40	2.7×10^{-13}	7×10^{6}	0.265	1.34×10^{-13}	3.2×10^{-6}	0.087	2.4×10^{5}

[a] 24-hr average value.
[b] Daylight value.

$k_{\text{diff}} = 2 \times 10^{-7}$ sec^{-1}, $[HNO_3]/[NO_x] = 0.64$; but if $k_{\text{diff}} = 0$, $[HNO_3]/[NO_x] = 0.72$. Thus the exact value of k_{diff} is not very important. We expect diffusion to be negligible at 20 km and consider $[HNO_3]/[NO_x] = 0.72$ at 20 km.

The lifetimes for HNO_3 can be computed from Eq. (91) and they are listed in Table II-13. They are longer than a day, so that $[HNO_3]$ is essentially constant throughout the diurnal cycle.

HO_x

The three HO_x radicals H, HO, and HO_2 interchange among each other in the atmosphere, their ratios being determined by the rates of these interchange reactions. For H atoms the important production terms are

$$H_2O + h\nu \rightarrow HO + H \tag{93}$$

$$O(^1D) + H_2 \rightarrow HO + H \tag{94}$$

$$HO + H_2 \rightarrow H_2O + H \tag{95}$$

$$HO + O(^3P) \rightarrow H + O_2 \tag{96}$$

The important removal processes are

$$H + O_2 + M \rightarrow HO_2 + M \tag{97}$$

$$H + O_3 \rightarrow HO + O_2 \tag{98}$$

The rate coefficients for the chemical reactions are listed in Table II-12, whereas the photodissociation parameters for reaction (93) are listed in Table II-1. In the heterosphere, the removal terms are slow, and nearly all the available hydrogen is present as atomic hydrogen. However, in the homosphere reaction (97) is always sufficiently fast so that the steady-state approximation applies to H.

For HO, the important production terms are reactions (80), (87), (93), (94), (98), and

$$HO_2 + O_3 \rightarrow HO + 2O_2 \tag{99}$$

$$HO_2 + NO \rightarrow HO + NO_2 \tag{100}$$

$$HO_2 + O(^3P) \rightarrow HO + O_2 \tag{101}$$

$$O(^1D) + H_2O \rightarrow 2HO \tag{102}$$

The important removal terms are numerous. They include reactions (81), (86), (95), (96), and

$$HO + HO_2 \rightarrow H_2O + O_2 \tag{103}$$

$$2HO \rightarrow H_2O + O(^3P) \tag{104}$$

$$HO + O_3 \rightarrow HO_2 + O_2 \tag{105}$$

The rate coefficient for reaction (102) is listed in Table II-3, and those for the other reactions are listed in Table II-12. The daylight HO lifetime is always short and the steady-state approximation is always valid.

The reactions of importance forming HO_2 are (97) and (105), whereas those important in HO_2 removal are reactions (79), (99)–(101), (103), and

$$2HO_2 \rightarrow H_2O_2 + O_2 \tag{106}$$

$$HO_2 + H \rightarrow H_2 + O_2 \tag{107a}$$

$$\rightarrow 2HO \tag{107b}$$

The rate coefficients are in Table II-12. The daylight lifetimes for HO_2 removal are always sufficiently small for the steady-state approximation to be valid.

H_2, H_2O, H_2O_2

H_2, H_2O, and H_2O_2 are all produced and removed in the upper atmosphere. First let us examine the H_2O_2 cycle. H_2O_2 is produced by reaction (106) and removed by

$$H_2O_2 + h\nu \rightarrow 2HO \tag{108}$$

$$O(^3P) + H_2O_2 \rightarrow HO + HO_2 \tag{109}$$

$$O(^1D) + H_2O_2 \rightarrow HO + HO_2 \tag{110}$$

$$HO + H_2O_2 \rightarrow H_2O + HO_2 \tag{111}$$

The rate coefficients of interest are listed in Tables II-2, II-3, and II-12. The photodissociation coefficients for H_2O_2 are shown in Fig. II-10 for various solar zenith angles at altitudes below 60 km. The photodissociation parameters are summarized in Table II-1.

The steady state applies to the mesosphere. However, in the stratosphere, $\tau\{H_2O_2\}$ is large, but not large enough so that diffusion is important. The 24-hr average steady state gives

$$\overline{[H_2O_2]} = \frac{k_{(106)}\overline{[HO_2]}^2}{J_{(108)} + k_{(109)}\overline{[O(^3P)]} + k_{(110)}\overline{[O(^1D)]} + k_{(111)}\overline{[HO]}} \tag{112}$$

In the stratosphere all the terms on the right-hand side of Eq. (112) go to zero at night, so that only the daylight averages need to be considered. Also reactions (109) and (110) are negligible by comparison to reactions (108) and (111). Anticipating the average values of $\overline{[HO_2]}$ and $\overline{[HO]}$ permits us to calculate $[H_2O_2]$ from Eq. (112). The results are in Table II-14.

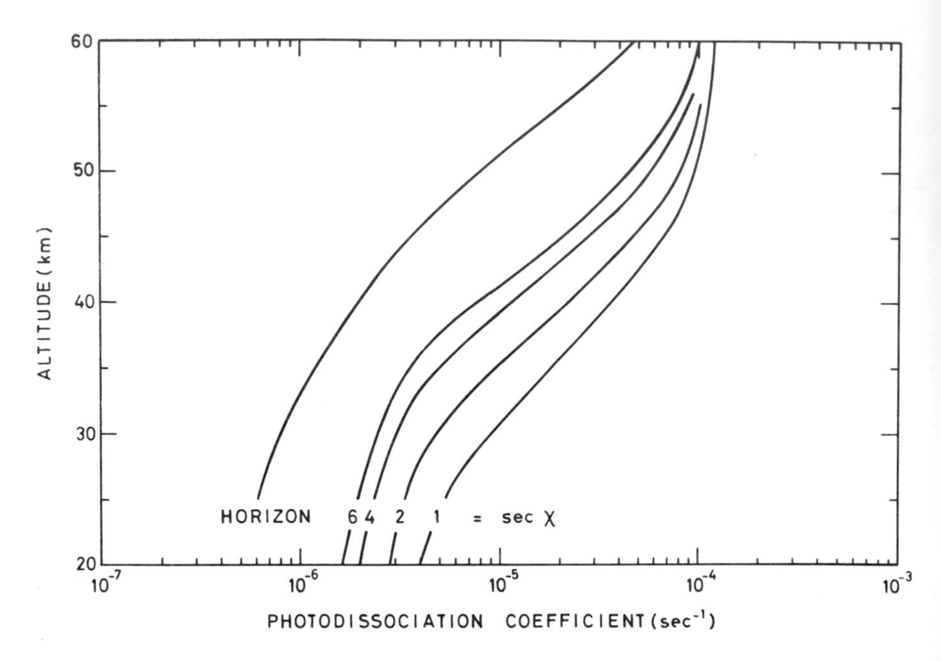

Fig. II-10. **Photodissociation coefficient for H_2O_2 in the 0–40 km region for various solar zenith angles. From Nicolet (1971) with permission of Reidel Publ. Co.**

The lifetimes for H_2O_2 have been computed from

$$\tau\{H_2O_2\} = (\bar{J}_{(108)} + k_{(111)}\overline{[HO]})^{-1} \tag{113}$$

They are also listed in Table II-14. At 20 km, $\tau\{H_2O_2\}$ is sufficiently long so that $[H_2O_2]$ should be constant throughout the day. However, at 40 km there will be some variation in $[H_2O_2]$.

The H_2 mole fraction is essentially constant in the troposphere at about 0.5 ppm (see Chapter I). In the mesosphere the important chemical production terms are reaction (107a) and

$$H + HO \rightarrow H_2 + O(^3P) \tag{114}$$

Chemical removal is mainly by reactions (94) and (95). All of these reactions are slow, and diffusion can also play an important role. It may lead to either production or removal of H_2, depending on the altitude.

The formation and removal processes for H_2 are slow all the way to the mesosphere. Furthermore they tend to offset each other. As a first approximation we can consider the mole fraction to be constant at 0.5 ppm up to 60 km.

TABLE II-14

H_2O_2 *Concentrations in the Stratosphere for a Day with an Overhead Noonday Sun*

Altitude (km)	Temp (°K)	$k_{(106)}$ (cm³/molecules-sec)	$\overline{[HO_2]}$[a] (molecules/cm³)	$\overline{J}_{108}$[a] (sec⁻¹)	$k_{(111)}$ (cm³/molecules-sec)	$\overline{[HO]}$[a] (molecules/cm³)	$\overline{[H_2O_2]}$ (molecules/cm³)	$\tau\{H_2O_2\}$[a] (sec)
20	219	3.0×10^{-12}	3.5×10^{7}	3.5×10^{-6}	2.7×10^{-13}	1.4×10^{6}	1.0×10^{9}	2.6×10^{5}
40	268	4.6×10^{-12}	1.7×10^{7}	2.8×10^{-5}	5.8×10^{-13}	1.4×10^{7}	3.6×10^{7}	2.8×10^{4}

[a] 12-hr daylight averages.

TABLE II-15

H_2O *Concentrations above 60* km *for a Day with an Overhead Noonday Sun*

Altitude (km)	Temp (°K)	$\overline{J}_{(93)}$[a] (sec⁻¹)	$k_{(102)}\overline{[O(^1D)]}$[a] (sec⁻¹)	D (cm²/sec)	$10^7\xi_D$ (cm⁻¹)	$10^7\xi_n$ (cm⁻¹)	$10^7\xi_{H_2O}$ (cm⁻¹)	$[H_2O]$ (ppm)
70	211	1.23×10^{-7}	1.52×10^{-8}	4×10^{5}	0	13.9	16.1	3.0
80	177	1.07×10^{-6}	1.47×10^{-8}	4×10^{5}	0	17.0	27.0	1.4
90	176	2.35×10^{-6}	4.30×10^{-8}	4×10^{5}	−6.9	17.0	36.9	0.33
100	210	2.92×10^{-6}	1.83×10^{-7}	2×10^{6}	−10.7	17.0	26.7	0.070

[a] 24-hr average values.

Above 60 km H_2 is produced readily by reaction (107a). Thus the relative concentration of H_2 rises to as high a value as possible near 80 km. Since we shall see that $[H_2O] = 1.4$ ppm and $[CH_4] = 0.015$ ppm at 80 km, and since $\sum H = 12$ ppm, $[H_2] \simeq 4.3$ ppm, where allowance has been made for some hydrogen in the minor species. Above 80 km, the relative importance of H_2 drops because removal by O (^1D) via reaction (94) becomes the dominant chemical process in regard to H_2. It is difficult to estimate the relative H_2 concentration at 100 km, but fortunately it is not of great importance. As an approximation we consider it to be $\simeq 1.0$ ppm.

The relative water concentration rises with altitude in the stratosphere, reaches a maximum in the mesosphere, and decreases with further increases in altitude. It is produced by reactions (81), (86), (95), (103), and (104) and is removed by reactions (93) and (102). The photodissociation coefficients for reaction (93) are shown in Fig. II-11 for altitudes of interest (60–100 km). The corresponding photodissociation parameters are listed in Table II-1. The lifetimes are large, diffusion is important, and the steady-state approximation does not apply.

The H_2O mole fraction increases with altitude in the stratosphere because of production via reaction (103). By 40 km, it has reached its maximum

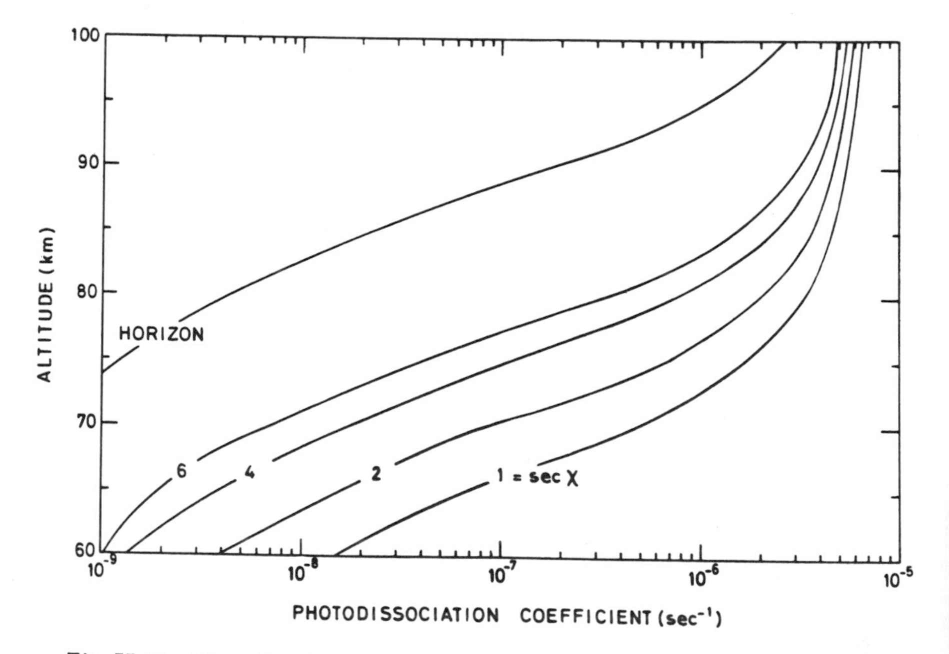

Fig. II-11. Photodissociation coefficients for H_2O vapor versus altitude for various solar zenith angles. From Nicolet (1970) with permission.

possible value of 4.78 ppm consistent with a hydrogen mass balance (since the sum of all the hydrogens =12 ppm) and the values for [H_2] and [CH_4]. Because [CH_4] drops with increasing altitude, [H_2O] reaches 4.9 ppm at 60 km.

Above 60 km, the production of H_2O by reaction (103) decreases rapidly in importance, and H_2O is removed by reactions (93) and (102). These processes must be offset by diffusional production, which must balance on a 24-hr cycle. Thus

$$k_{\mathrm{diff}}\{H_2O\} = \bar{J}_{(93)} + k_{(102)}\overline{[O(^1D)]} \tag{115}$$

The 24-hr average values of $J_{(93)}$ and $k_{(102)}[O(^1D)]$ are listed in Table II-15 for altitudes of 70 to 100 km. The production coefficient due to diffusion is related to the scale heights through

$$k_{\mathrm{diff}}\{H_2O\} = D(\xi_{H_2O} + \xi_D)(\xi_{H_2O} - \xi_n) \tag{116}$$

Equating Eqs. (115) and (116) allows computation of ξ_{H_2O}, and these values are listed in Table II-15. From them the H_2O concentration profile can be constructed, and the values obtained for [H_2O] are also listed in Table II-15.

The values for $[HO_x] \equiv [H] + [HO] + [HO_2]$ can be obtained by difference above 80 km since the sum of all the hydrogens must be 12 ppm. (There is no mass fractionation with altitude below 100 km.) For altitudes ≤ 80 km, [HO_x] varies during the daily cycle, and the concentrations of the individual species must be obtained from a detailed consideration of their chemical and photochemical production and loss terms, the diffusional terms being negligibly small by comparison.

CARBON–HYDROGEN–OXYGEN CYCLE

The carbon-containing compounds in the atmosphere include CH_4, CH_2O, CO, and CO_2. The free radicals of significance are CH_3O_2, CH_3O, and HCO. The CH_4 is oxidized to CH_2O, which in turn is oxidized to CO, which is converted to CO_2 in the troposphere and stratosphere. Above the stratospause, the CH_4 and CH_2O are fairly well exhausted, and an important process becomes CO_2 photodissociation to produce CO. This process becomes dominant in the thermosphere.

CH_4

CH_4 has been measured at mole fractions of 1.4×10^{-6} in the lower stratosphere, but this mole fraction drops at higher altitudes, being

TABLE II-16

Concentrations of CH_4 *for a Day with an Overhead Noonday Sun*

Altitude (km)	Temp (°K)	$\bar{J}_{(118)}$[a] (sec^{-1})	$k_{(120)}\overline{[O(^1D)]}$[a] (sec^{-1})	$k_{119}\overline{[HO]}$[a] (sec^{-1})
10	231	—	7.07×10^{-13}	—
20	219	—	5.29×10^{-11}	8.0×10^{-10}
30	235	—	2.15×10^{-9}	$\sim 6 \times 10^{-9}$
40	268	—	1.84×10^{-8}	2.93×10^{-8}
50	274	—	4.96×10^{-8}	$\sim 6 \times 10^{-9}$
60	253	—	1.44×10^{-8}	2.72×10^{-9}
70	211	1.08×10^{-7}	1.73×10^{-8}	$\sim 3 \times 10^{-9}$
80	177	1.09×10^{-6}	1.68×10^{-8}	4.29×10^{-9}
90	176	2.21×10^{-6}	—	—
100	210	3.0×10^{-6}	—	—

[a] 24-hr average values.

$<4 \times 10^{-7}$ at the stratopause. CH_4 is produced at the ground and diffuses upward. This is the only production process so that

$$\sum R_f\{CH_4\} = k_{diff}\{CH_4\}[CH_4] \tag{117}$$

The loss processes include both photodissociation and chemical attack:

$$CH_4 + h\nu \rightarrow CH_3 + H \tag{118}$$

$$HO + CH_4 \rightarrow H_2O + CH_3 \tag{119}$$

$$O(^1D) + CH_4 \rightarrow HO + CH_3 \tag{120a}$$

$$\rightarrow CH_2O + H_2 \tag{120b}$$

Thus the 24-hr average steady-state condition gives

$$k_{diff} = \bar{J}_{(118)} + k_{(119)}\overline{[HO]} + k_{(120)}\overline{[O(^1D)]} \tag{121}$$

The photodissociation coefficients for CH_4 are shown in Fig. II-12 for various altitudes and solar zenith angles, and the photodissociation parameters are summarized in Table II-1. The rate coefficient for reaction (119) is given in Table II-12, and those for reactions (120a) and (120b) in Table II-3. The appropriate 24-hr average rates have been computed and are listed in Table II-16. The diffusion coefficient is also given by

$$k_{diff}\{CH_4\} = D(\xi_{CH_4} + \xi_D)(\xi_{CH_4} - \xi_n) \tag{122}$$

The appropriate parameters are listed in Table II-16 from which ξ_{CH_4} can

k_{diff} (sec^{-1})	D (cm^2/sec)	$10^7\xi_D$ (cm^{-1})	$10^7\xi_n$ (cm^{-1})	$10^7\xi_{CH_4}$ cm^{-1})	[CH_4] (ppm)
—	1×10^4	0	13.9	—	1.4
—	1×10^4	0	13.9	—	1.4
8.15×10^{-9}	1×10^4	0	13.9	18.3	1.1
4.77×10^{-8}	1×10^4	0	13.9	29.9	0.37
5.56×10^{-8}	1×10^5	−10.4	13.9	19.8	0.093
1.71×10^{-8}	2×10^5	−6.9	13.9	15.0	0.044
1.28×10^{-7}	4×10^5	0	13.9	15.9	0.039
1.11×10^{-6}	4×10^5	0	17.0	27.2	0.015
2.21×10^{-6}	4×10^5	−6.9	17.0	36.0	0.0022
3.0×10^{-6}	2×10^6	−10.7	17.0	26.5	

be computed and the CH_4 concentration reconstructed. These values are also listed in Table II-16.

CH_2O

The cycle for CH_2O is fairly complex. It is produced primarily by the oxidation of CH_3 radicals, but reaction (120b) also plays a role. The

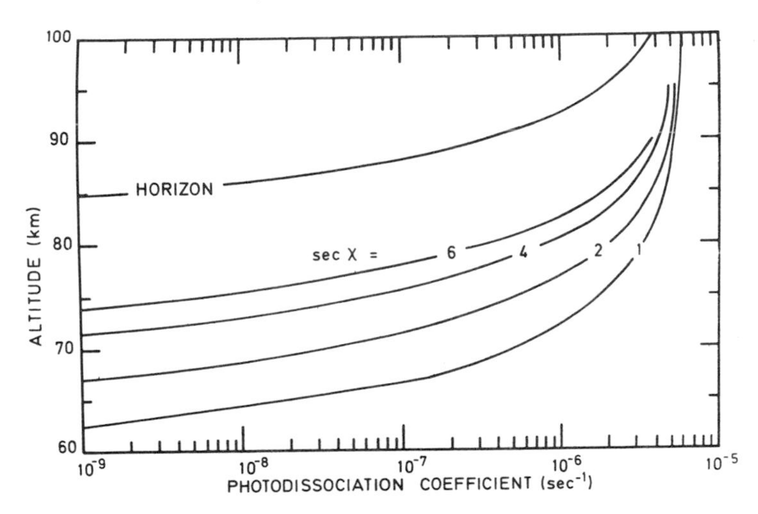

Fig. II-12. Photodissociation coefficient for CH_4 in the 60–100 km region for various solar zenith angles. From Nicolet (1971) with permission of Reidel Publ. Co.

sequence of events involved in CH_3 radical oxidation is

$$CH_3 + O_2 + M \rightarrow CH_3O_2 + M \tag{123}$$

$$CH_3 + O \rightarrow CH_2O + H \tag{124}$$

$$CH_3 + O_3 \rightarrow CH_2O + O_2 + H \tag{125}$$

for CH_3O_2,

$$CH_3O_2 + O(^3P) \rightarrow CH_3O + O_2 \tag{126}$$

$$CH_3O_2 + NO \rightarrow CH_3O + NO_2 \tag{127}$$

and finally for CH_3O,

$$CH_3O + O_2 \rightarrow CH_2O + HO_2 \tag{128}$$

$$CH_3O + HO \rightarrow H_2O + CH_2O \tag{129}$$

$$CH_3O + O \rightarrow CH_2O + HO \tag{130}$$

Many of the rate constants for the above reactions have not been measured. Fortunately, accurate rate coefficients are not usually necessary, since the radicals are in their steady states and their diffusion terms are not important. The total rate of CH_2O production $\sum R_f\{CH_2O\}$ is then equal to the rate of CH_4 loss.

The important removal terms for CH_2O are

$$CH_2O + h\nu \rightarrow H_2 + CO \tag{131a}$$

$$\rightarrow H + HCO \tag{131b}$$

$$CH_2O + O(^3P) \rightarrow HCO + HO \tag{132}$$

$$CH_2O + HO \rightarrow H_2O + HCO \tag{133}$$

The quantum efficiencies of the two photodissociation processes of CH_2O have been summarized by Calvert *et al.* (1972) for wavelengths between 2900 and 3600 Å. Based on these values, Nicolet and Peetermans (1972b) have estimated those at zero optical depth. With the new quantum efficiencies (Osif, 1975), the photodissociation coefficients for an overhead sun accepted here are listed in Table II-1. The rate coefficients for reactions (132) and (133) are tabulated in Tables II-2 and II-12, respectively. It can be seen that the lifetimes for an overhead sun are $<10^4$ sec; the steady state is nearly achieved:

$$[CH_2O] = \frac{(J_{(118)} + k_{(119)}[HO] + k_{(120)}[O(^1D)])[CH_4]}{J_{(131)} + k_{(132)}[O(^3P)] + k_{(133)}[HO]} \tag{134}$$

CO and CO_2

The CO is present at a constant mole fraction of about 10^{-7} throughout the homosphere. Therefore, production and removal processes offset each other. On the other hand CO_2, which has a mole fraction of 3.25×10^{-4} throughout the homosphere, falls in importance as it photodissociates in the thermosphere.

For CO the production processes include reaction (131a) and

$$CO_2 + h\nu \rightarrow CO + O(^3P \text{ or } ^1D) \tag{135}$$

$$HCO + O_2 \rightarrow HO_2 + CO \tag{136}$$

The rate coefficient of reaction (136) is not known, but it is fast. Thus HCO is always in a steady state and the total rate of CO production is independent of $k_{(136)}$, being

$$\sum R_f\{CO\} = I_{(131)}\{CH_2O\} + R_{(132)} + R_{(133)} + I_{(135)}\{CO_2\}$$

The only important chemical removal process is

$$HO + CO \rightarrow H + CO_2 \tag{137}$$

but diffusional loss also occurs in the thermosphere and upper mesosphere. The rate coefficient $k_{(137)}$ is listed in Table II-12.

OVERALL SUMMARY

Odd O

In order to complete our picture of the atmosphere we return to a consideration of odd oxygen, taking into account all the processes that can produce and remove it. For a pure oxygen atmosphere, we found the computed values of odd oxygen to be much higher than observed. Including all the reactions of consequence on the odd oxygen cycle and equating the 24-hr average rates of formation and removal gives

$$\begin{aligned} 2\bar{R}_{(19)} + \bar{R}_{(46)} + 2\bar{R}_{(51)} + \bar{R}_{(52)} + \bar{R}_{(104)} \\ = 2\bar{R}_{(21)} + 2\bar{R}_{(23)} + \bar{R}_{(54)} + \bar{R}_{(55)} + \bar{R}_{(57)} + \bar{R}_{(96)} + \bar{R}_{(98)} \\ + \bar{R}_{(99)} + \bar{R}_{(101)} + \bar{R}_{(105)} + \bar{R}_{(109)} + \bar{R}_{(132)} \end{aligned} \tag{138}$$

The values needed for the O_3 concentrations to evaluate some of the rates in Eq. (138) can be computed from the daylight steady-state expression on O_3:

$$[O_3] = \frac{k_{(22)}[O(^3P)][O_2][M]}{J_{(20)} + k_{(23)}[O(^3P)] + k_{(57)}[NO]} \tag{139}$$

where all the parameters in Eq. (139) are daylight values. In obtaining the average rates for Eq. (138), daylight values should be used for $[O_3]$, as well as [NO] and [NO_2], because either their reactions are unimportant at night or their concentrations are the same at night as they are during the day. The daylight NO_2 concentration can also be obtained from its daylight steady-state expression:

$$\bar{R}_{(55)} + \bar{R}_{(57)} + \bar{R}_{(86)} + \bar{R}_{(87)} + \bar{R}_{(100)} = \bar{R}_{(52)} + \bar{R}_{(54)} + \bar{R}_{(79)} + \bar{R}_{(85)} \quad (140)$$

so that Eq. (138) becomes

$$\begin{aligned} 2\bar{R}_{(19)} &+ \bar{R}_{(46)} + 2\bar{R}_{(51)} + \bar{R}_{(86)} + \bar{R}_{(87)} + \bar{R}_{(100)} + \bar{R}_{(104)} \\ &= 2\bar{R}_{(21)} + 2\bar{R}_{(23)} + 2\bar{R}_{(54)} + \bar{R}_{(79)} + \bar{R}_{(85)} + \bar{R}_{(96)} + \bar{R}_{(98)} \\ &\quad + \bar{R}_{(99)} + \bar{R}_{(101)} + \bar{R}_{(105)} + \bar{R}_{(109)} + \bar{R}_{(132)} \end{aligned} \quad (141)$$

Equation 141 can be solved for [$O(^3P)$] by using Eq. (139) to obtain [O_3] for which the ratio of the daylight average $O(^3P)$ concentration to that for the 24-hr average concentration $\overline{[O(^3P)]}$ can be taken as 2.0 for altitudes $\leq$80 km. Also it is necessary to know $2\bar{J}_{(19)}$, $\bar{J}_{(46)}$, $2\bar{I}_{(51)}$, $\bar{I}_{(52)}$, and a number of average species concentrations. The 24-hr average photodissociation coefficients are obtained from the parameters in Table II-1, and the average species concentrations are obtained by methods previously outlined. They are listed in Table II-17.

The values computed for [O_3] and $\overline{[O(^3P)]}$ at various altitudes are also given in Table II-17. They are somewhat sensitive to the average values used for the other species, so they may be no more accurate than a factor of two. First it can be seen that they are much lower than computed for a pure oxygen atmosphere (see Table II-4). At 20 km the computed value for odd oxygen ([O_3] + [$O(^3P)$]) is 6.2×10^{12} molecules/cm^3, which is slightly higher than the observed values of (2.4–4.9) $\times$ 10^{12} molecules/cm^3 for summertime (Table I-4). At 20 km the lifetime for odd oxygen can be computed from

$$\tau\{\text{odd}\} = \overline{[\text{odd O}]}/\textstyle\sum \bar{R}_r \quad (142)$$

where $\sum \bar{R}_r$ is the right-hand side of Eq. (141). The lifetime is so long that averaging for the seasons and diffusional loss can play some role, and the calculated concentrations are slightly high ($\leq$20%) because of these effects. With the correction, the calculated and observed concentrations are in reasonable agreement considering the large uncertainty in $\bar{I}_{(19)}$.

At 40 km, the calculated daylight value for [O_3] is 5.2×10^{11} molecules/sec, but the lifetime is of the order of 1 day. At night this concentration will be slightly reduced, and an average value close to the observed one of 5.1×10^{11} molecules/cm^3 (Table I-4) is obtained.

TABLE II-17

Calculation of Odd Oxygen versus Altitude for a Day with an Overhead Noonday Sun

Altitude (km)	Temp (°K)	$2\bar{I}_{(19)}$ [a] (molecules/cm^3-sec)	$\bar{J}_{(46)}$ [a] (sec^{-1})	$2\bar{J}_{(51)}$ [a] (sec^{-1})	$\bar{J}_{(87)}$ [a] (sec^{-1})	$\overline{[HO]}$ [a] (molecules/cm^3)	$\overline{[CH_2O]}$ [a] (molecules/cm^3)	[NO] [b] (molecules/cm^3)
20	219	1.93×10^5	1.04×10^{-9}	—	5.2×10^{-7}	9×10^5	3.0×10^6	1.11×10^9
40	268	5.79×10^6	1.60×10^{-7}	—	4.3×10^{-6}	7×10^6	2.0×10^6	1.3×10^9
60	253	2.41×10^6	4.0×10^{-7}	4.12×10^{-6}	—	3×10^6	—	6.0×10^7
80	177	7.50×10^5	4.0×10^{-7}	8.42×10^{-6}	—	1×10^6	—	5×10^7

Altitude (km)	$[NO_2]$ [b] (molecules/cm^3)	$[H_2O_2]$ [a] (molecules/cm^3)	$\overline{[H]}$ [a] (molecules/cm^3)	$\overline{[HO_2]}$ [a] (molecules/cm^3)	$[HNO_3]$ (molecules/cm^3)	$[N_2O]$ (molecules/cm^3)	$[O_3]$ [b] (molecules/cm^3)	$\overline{[O(^3P)]}$ [a] (molecules/cm^3)	τ{odd O} (sec)
20	1.5×10^9	1.0×10^9	—	1.7×10^7	7.1×10^9	4.1×10^{11}	6.2×10^{12}	2.5×10^6	2.9×10^7
40	3.6×10^8	3.6×10^7	—	9.0×10^6	1.62×10^8	1.6×10^9	5.2×10^{11}	5.2×10^8	9.0×10^4
60	4.6×10^4	—	1.4×10^6	3×10^6	$\sim 5.0 \times 10^3$	3.1×10^6	1.0×10^{10}	6.3×10^9	6.9×10^3
80	2.0×10^3	—	3×10^7	1.0×10^6	—	4.6×10^4	9.9×10^7	7.4×10^9	9.8×10^3

[a] 24-hr average.
[b] 12-hr daylight value.

TABLE II-18

Summary of Species Concentrations in the Upper Atmosphere for a Day with an Overhead Noonday Sun

Altitude (km)	Temperature (°K)	[M] (molecules/cm^3)	[N_2] (molecules/cm^3)	[O_2] (molecules/cm^3)	$\overline{\text{[odd O]}}$[a] (molecules/cm^3)	[NO_x][b] (molecules/cm^3)
20	219	1.7×10^{18}	1.4×10^{18}	3.6×10^{17}	2.9×10^{12}	9.8×10^{9}
40	268	8.1×10^{16}	6.4×10^{16}	1.7×10^{16}	5.1×10^{11}	1.86×10^{9}
60	253	7.2×10^{15}	5.7×10^{15}	1.5×10^{15}	$\sim 1.6 \times 10^{10}$	6×10^{7}
80	177	4.2×10^{14}	3.3×10^{14}	9.0×10^{13}	7.5×10^{9}	5.0×10^{7}
100	210	1.1×10^{13}	8.6×10^{13}	1.9×10^{12}	4.0×10^{11}	5.0×10^{7}

[a] Constant throughout the day at 20, 40, and 100 km; slightly variable at 60 and 80 km.
[b] $[NO_x] \equiv [NO] + [NO_2] + [NO_3] + 2[N_2O_5] + [HNO_2] + [HNO_3]$.
[c] $\sum H \equiv [H] + [HO] + [HO_2] + 2[H_2] + 2[H_2O] + 2[H_2O_2] + [HNO_2] + [HNO_3] + 2[CH_2O] +$ $4[CH_4]$.
[d] Constant throughout the day at 20 km, but somewhat variable at 40 km.

At 60 and 80 km, the lifetime of odd oxygen is considerably less than 1 day, and diurnal fluctuations will occur. However, our average values of about 1.6×10^{10} molecules/cm^3 at 60 km and 7.5×10^9 molecules/cm^3 at 80 km agree reasonably well with observations (see Chapter I).

Diurnal Cycle

With the information we have developed, we can now compute the diurnal chemical cycle. For those species whose concentrations are nearly invariant to the daily solar cycle, we have already obtained their concentrations and our results are summarized in Table II-18. With these values and the reactions leading to formation and removal of the other species the computation can be done. The rates of importance for each species are summarized in Table II-19. For chemical reactions, the rate coefficients are constant throughout the day. For the photochemical reactions, the photodissociation coefficients are computed as a function of the time of day from the approximate relationship

$$J_z = J_z^0 \exp\{-\zeta_z'(\sec\chi - 1)\} \tag{143}$$

where the parameters J_z^0 and ζ_z' are tabulated in Table II-1.

A step-by-step calculation using 2–100 sec time intervals gives the concentration profiles during a day for an overhead noonday sun. The results are summarized in Figs. II-13–II-17.

$\sum H^c$ (ppm)	$[N_2O]$ (ppm)	$\frac{[HNO_3]}{[NO_x]}$	$[H_2]$ (ppm)	$[H_2O]$ (ppm)	$\overline{[H_2O_2]}^d$ (molecules/cm^3)	$[CH_4]$ (ppm)	[CO] (ppm)
12.0	0.24	0.72	0.5	2.5	1.0×10^9	1.4	0.1
12.0	0.020	0.087	0.5	4.78	3.6×10^7	0.37	0.1
12.0	4.3×10^{-4}	—	0.5	4.9	—	0.044	0.1
12.0	1.1×10^{-4}	—	4.3	1.4	—	0.015	0.1
12.0	—	—	1.0	0.070	—	—	0.1

At 20 km (Fig. II-13) many of the species concentrations remain constant throughout the day. $[NO_x]$ is 9.8×10^9 molecules/cm^3, 72% of which is HNO_3, in good agreement with observed values (see Fig. I-12). The remaining NO_x is mainly NO and NO_2, with $[NO] = 1.17 \times 10^9$ molecules/cm^3 during daylight and disappearing at night. $[NO_2] = 1.5 \times 10^9$ molecules/cm^3 by day and 2.7×10^9 by night. Thus $[NO] + [NO_2] = 1.6$ ppb, which is slightly higher than the recent measured value of 1.3 ppb (Ackerman *et al.*, 1975). The HONO and CH_2O concentrations are nearly constant throughout the day, being $\sim(1.8 \pm 0.8) \times 10^7$ and $(3.0 \pm 0.3) \times 10^7$ molecules/cm^3, respectively, the changes in concentrations occurring near sunset or sunrise. The species H (not shown), HO, HO_2, and $O(^3P)$ vanish at night, but their daylight concentrations are 0.50 ± 0.25, $(1.3 \pm 0.4) \times 10^6$, $(3.5 \pm 1.0) \times 10^7$, and $(2.0 \pm 0.3) \times 10^6$ molecules/cm^3, respectively. The $O(^1D)$ and $O_2(^1\Delta)$ species disappear at night and are not important in the early morning and late afternoon. Their concentrations have reasonably sharp maxima at noon.

The computed concentrations at 40 km are shown in Fig. II-14. The NO concentration is 1.3×10^9 molecules/cm^3 during daylight in good agreement with observations (Fig. I-10) and very small at night, whereas $[NO_2] = 3.6 \times 10^8$ molecules/cm^3 during daylight and 1.6×10^9 molecules/cm^3 at night. The concentrations of HONO, CH_2O, H (not shown), HO, HO_2, and $O(^3P)$ show much greater daylight variations than at 20 km, but HO_2 decays more slowly and retains a significant concentration

TABLE II-19

Reactions Important in Formation and Removal of Atmospheric Species

Species	
O_3	$\sum R_f = R_{(22)}$
	$\sum R_r = R_{(20)} + R_{(23)} + R_{(41)} + R_{(57)} + R_{(98)} + R_{(105)}$
$O(^3P)$	$\sum R_f = R_{(19a)} + 2R_{(19b)} + R_{(20c)} + R_{(20d)} + R_{(24)} + R_{(41)} + R_{(104)}$
	$\sum R_r = 2R_{(21)} + R_{(22)} + R_{(23)} + R_{(96)} + R_{(101)}$
$O(^1D)$	$\sum R_f = R_{(19a)} + R_{(20a)}$
	$\sum R_r = R_{(24)}$
$O_2(^1\Delta)$	$\sum R_f = R_{(20a)}$
	$\sum R_r = R_{(35)} + R_{(40)} + R_{(41)}$
NO_2	$\sum R_f = R_{(55)} + R_{(57)} + R_{(86)} + R_{(87)} + R_{(100)}$
	$\sum R_r = R_{(52)} + R_{(54)} + R_{(79)} + R_{(85)}$
HNO_2	$\sum R_f = R_{(78)} + R_{(79)}$
	$\sum R_r = R_{(80)} + R_{(81)}$
H	$\sum R_f = R_{(93)} + R_{(94)} + R_{(95)} + R_{(96)} + R_{(118)} + R_{(137)}$
	$\sum R_r = R_{(97)} + R_{(98)} + R_{(107)}$
HO	$\sum R_f = R_{(80)} + R_{(87)} + R_{(93)} + R_{(94)} + R_{(98)} + R_{(99)} + R_{(100)} + R_{(101)}$
	$+ 2R_{(102)} + 2R_{(107b)} + 2R_{(108)} + R_{(120a)} + R_{(132)}$
	$\sum R_r = R_{(78)} + R_{(81)} + R_{(85)} + R_{(86)} + R_{(95)} + R_{(96)} + R_{(103)} + 2R_{(104)}$
	$+ R_{(105)} + R_{(119)} + R_{(137)}$
HO_2	$\sum R_f = R_{(97)} + R_{(105)} + R_{(109)} + R_{(118)} + R_{(119)} + 2R_{(131b)}$
	$\sum R_r = R_{(79)} + R_{(99)} + R_{(100)} + R_{(101)} + R_{(103)} + 2R_{(106)} + R_{(107)}$
H_2O_2	$\sum R_f = R_{(106)}$
	$\sum R_r = R_{(108)} + R_{(109)} + R_{(110)} + R_{(111)}$
CH_2O	$\sum R_f = R_{(118)} + R_{(119)} + R_{(120)}$
	$\sum R_r = R_{(131)} + R_{(132)} + R_{(133)}$

at night. On the other hand [O(^{1}D)] and [$O_2(^1\Delta)$] show less variation during daylight than at 20 km.

In the mesosphere, the fewest number of species retain a constant concentration during the day. Most of the species show significant changes with the sun's intensity and also at night. The species concentrations are very mutually dependent on each other. Consequently the calculations are the most difficult and least reliable at these altitudes. Furthermore,

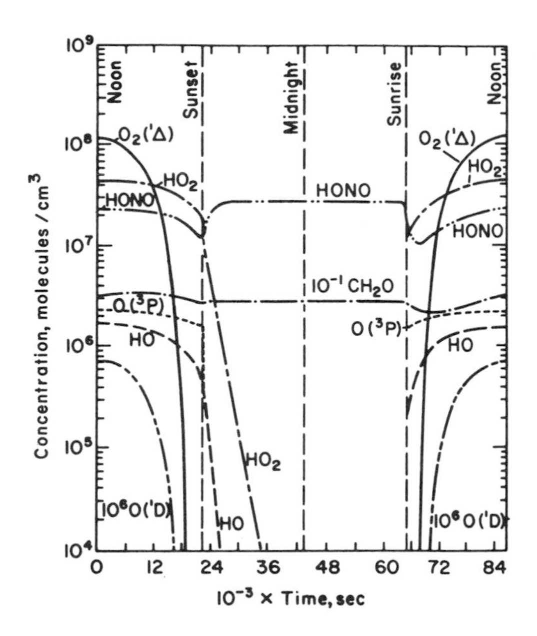

Fig. II-13. Calculated concentration profiles for a day with an overhead noonday sun as a function of the time of day at 20 km, 219°K. Concentrations (molecules/cm³):

$[M] = 1.7 \times 10^{18}$	$[HNO_3] = 7.1 \times 10^9$	$[H_2] = 8.5 \times 10^{11}$
$[N_2] = 1.4 \times 10^{18}$	$[NO] = 1.17 \times 10^9$ (day)	$[H_2O] = 4.25 \times 10^{12}$
$[O_2] = 3.6 \times 10^{17}$	$= 0$ (night)	$[H_2O_2] = 1.0 \times 10^9$
$[O_3] = 2.9 \times 10^{12}$	$[NO_2] = 1.5 \times 10^9$ (day)	$[CH_4] = 2.38 \times 10^{12}$
$[N_2O] = 4.08 \times 10^{11}$	$= 2.7 \times 10^9$ (night)	$[CO] = 1.7 \times 10^{11}$
$[NO_x] = 9.8 \times 10^9$		

many of the species concentrations are varying very sharply with altitude, as well as time of day. Thus experimental observations are difficult to make (an error in 1 km of altitude can give a factor of 2 or 3 change in some species concentrations), and they show large deviations (e.g., see the O_3 measurements in Fig. I-6). Our computed results at 60 km are shown in Fig. II-15. $[O_3]$ varies from a maximum of 1.0×10^{10} molecules/cm³ at noon to a minimum of 5×10^9 molecules/cm³ just after sunrise. At night the value is 1.15×10^{10} molecules/cm³. These values fall in the range of measured values (Fig. I-6). The daylight value computed for $[O_2(^1\Delta)]$ is $(2.5 \pm 0.5) \times 10^{10}$ molecules/cm³, in good agreement with observations (Fig. I-7). For $[O(^3P)]$ we compute a peak value of 1.25×10^{10} molecules/cm³ at noon, which stays relatively constant throughout the day and drops markedly at night. The midday values for [H], [HO],

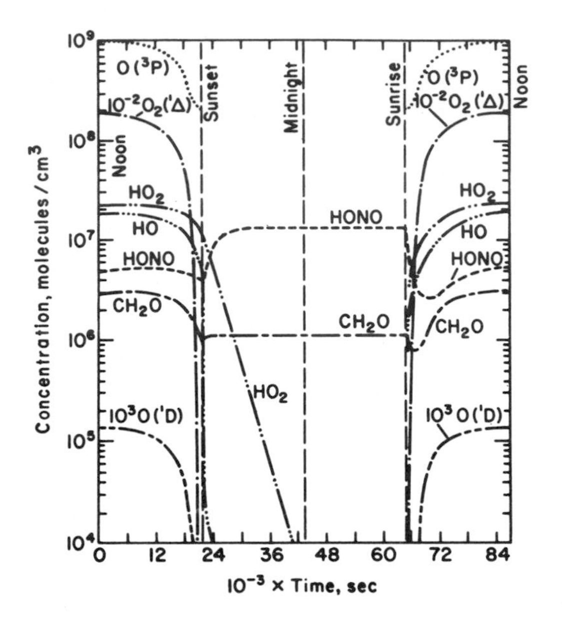

Fig. II-14. Calculated concentration profiles for a day with an overhead noonday sun as a function of the time of day at 40 km, 268°K. Concentrations (molecules/cm^3):

$[M] = 8.1 \times 10^{16}$	$[HNO_3] = 1.62 \times 10^8$	$[H_2] = 4.05 \times 10^{10}$
$[N_2] = 6.4 \times 10^{16}$	$[NO] = 1.3 \times 10^9$ (day)	$[H_2O] = 3.87 \times 10^{11}$
$[O_2] = 1.7 \times 10^{16}$	$= 0$ (night)	$[H_2O_2] = 3.8 \times 10^7$
$[O_3] = 5.1 \times 10^{11}$	$[NO_2] = 3.6 \times 10^8$ (day)	$[CH_4] = 3.00 \times 10^{10}$
$[N_2O] = 1.62 \times 10^9$	$= 1.6 \times 10^9$ (night)	$[CO] = 8.1 \times 10^9$
$[NO_x] = 1.86 \times 10^9$		

and $[HO_2]$ are 2.7×10^6, 5×10^6, and 3×10^6 molecules/cm^3, respectively. The twilight values for $[O(^3P)]$ and $[H]$ in Fig. II-15 are 7×10^9 and 2.0×10^6 molecules/cm^3, respectively, each of which is about an order of magnitude greater than deduced by Evans and Llewellyn (1973) from emission observations (see Fig. I-13). Their values should be lower than those calculated here, because for their measurements the noonday sun was not overhead—but a factor of 10 discrepancy is not reassuring.

At 80 km, the calculated concentrations (Fig. II-16) are essentially constant during the afternoon, but show significant variations at night and during the morning. Thus the daytime $O(^3P)$ concentration of 3×10^{10} molecules/cm^3 drops at night and O_3 is produced, its peak concentration reaching 5.8×10^8 molecules/cm^3. These values agree reasonably well with

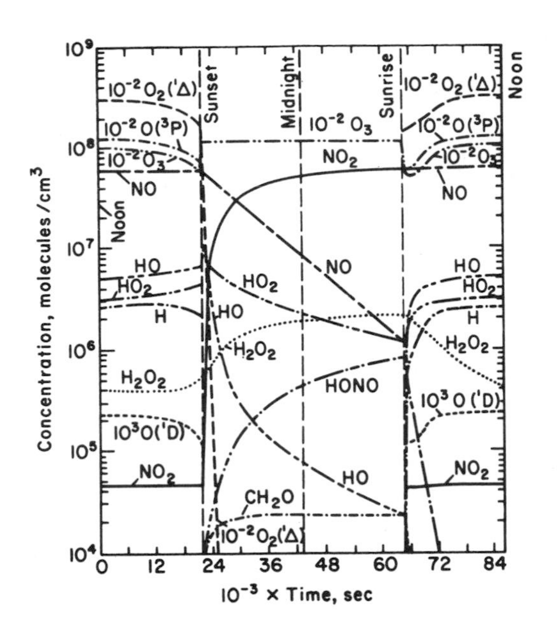

Fig. II-15. Calculated concentration profiles for a day with an overhead noonday sun as a function of the time of day at 60 km, 253°K. Concentrations (molecules/cm^3):

[M] = 7.2×10^{15}	[NO_x] = 6.0×10^7	[H_2O] = 3.53×10^9
[N_2] = 5.7×10^{15}	[HNO_3] = $\sim 5.0 \times 10^3$	[CH_4] = 3.17×10^8
[O_2] = 1.5×10^{15}	[H_2] = 3.6×10^9	[CO] = 7.2×10^8
[N_2O] = 3.1×10^6		

the observations reported in Chapter I, as does the daylight value of $\sim 3 \times 10^9$ molecules/cm^3 computed for [$O_2(^1\Delta)$]. The peak computed value for [H] is 1×10^8 molecules/cm^3. It is about three times larger than the twilight value observed by Evans and Llewellyn (1973).

At 100 km, the calculated nighttime O_3 concentrations of 3.4×10^6 molecules/cm^3 are somewhat lower than the observed values of $(1–2) \times 10^7$ molecules/cm^3 (Fig. I-6). The computed daylight concentration of O_3 is 3.3×10^5 molecules/cm^3. The computed daylight concentrations of O(^{1}D) and $O_2(^1\Delta)$ of 890 and 1.9×10^7 molecules/cm^3 agree with observations (Chapter I).

Other calculations have been made using the exact diffusion equations. Bowman *et al.* (1970) computed the concentration profiles of the minor species at 45° north latitude for values of the photodissociation coefficients fixed at $\frac{1}{4}$, $\frac{1}{2}$, or $\frac{3}{4}$ of their maximum values (i.e., for an overhead sun).

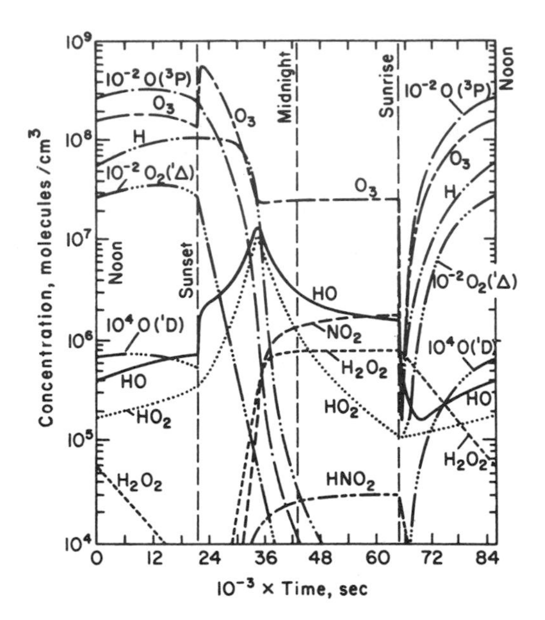

Fig. II-16. Calculated concentration profiles for a day with an overhead noonday sun as a function of the time of day at 80 km, 177°K. Concentrations (molecules/cm³):

$$[M] = 4.2 \times 10^{14} \quad [N_2O] = 4.62 \times 10^4 \quad [H_2O] = 5.88 \times 10^8$$
$$[N_2] = 3.3 \times 10^{14} \quad [NO] = 5.0 \times 10^7 \quad [CH_4] = 6.3 \times 10^6$$
$$[O_2] = 9.0 \times 10^{13} \quad [H_2] = 1.8 \times 10^9 \quad [CO] = 4.2 \times 10^7$$

Their results are compared with those shown in Figs. II-15, II-16, and II-17 in Table II-20.

Fukayama (1974) also included an exact diffusional treatment and computed the latitudinal distributions of the minor neutral species in the winter mesosphere and thermosphere. He assumed photodissociation coefficients fixed at their 24-hr daily average values. His results for 60° north latitude are summarized in Fig. II-18. His results at other latitudes show the same trends with altitude.

A calculation for the stratosphere was made by Shimazaki and Ogawa (1974), which showed the diurnal variation of the species including NO_3 and N_2O_5. Their results for 45° north latitude are summarized in Figs. II-19 and II-20 for the equinoxes. Their results for [HNO_3] and [NO_2]/[NO] are the same as those in Fig. II-13, but their values of [NO] + [NO_2] are about a factor of six lower. The values for the concentrations of HO, HO_2, and H_2O_2 at 20 km are markedly different in Figs. II-13 and

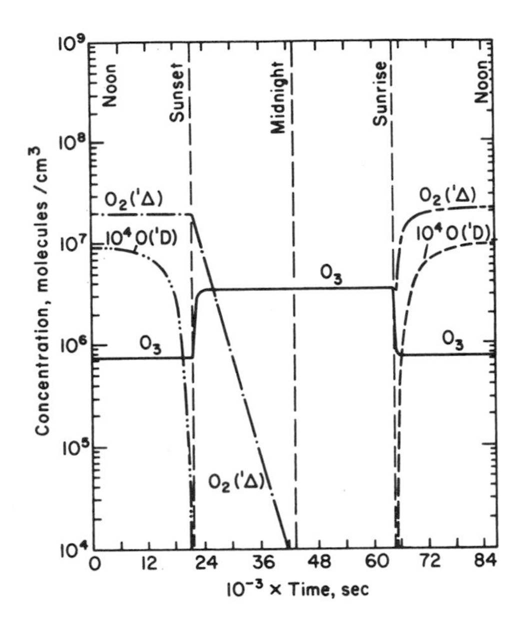

Fig. II-17. Calculated concentration profiles for a day with an overhead noonday sun as a function of the time of day at 100 km, 210°K. Concentrations (molecules/cm³):

$[M] = 1.1 \times 10^{13}$ $[O(^3P)] = 4.0 \times 10^{11}$ $[H_2] = 1.10 \times 10^7$
$[N_2] = 8.6 \times 10^{12}$ $[NO] = 5.0 \times 10^7$ $[H_2O] = 7.7 \times 10^5$
$[O_2] = 1.9 \times 10^{12}$ $[H] = 1.08 \times 10^8$ $[CO] = 1.1 \times 10^6$

II-20. This difference is due, at least in part, to the fact that for Fig. II-13 the noonday sun is overhead, whereas in Fig. II-20 it is not.

Elemental Balance

Now let us examine the overall processes occurring in each part of the atmosphere, i.e., look at overall effects resulting from the detailed chemistry considered above.

The thermosphere is rich in $O(^3P)$ atoms, the second most powerful oxidizing agent known (after fluorine atoms), and the oxidation processes are all energetically favorable. Yet the tremendous energy input from the sun drives the reactions in the endothermic direction and the daytime thermosphere in a reducing atmosphere! From the diffusive flow patterns deduced from the concentration profiles and the fact that the elemental composition of the atmosphere is fixed, i.e., total O, N, H, and C are

TABLE II-20

Comparison of Minor Constituent Concentrations with Those of Bowman et al. (1970)[a]

Altitude (km)	Source	$[O(^3P)]$ (molecules/cm³)	$[O_3]$ (molecules/cm³)	[H] (molecules/cm³)	[HO] (molecules/cm³)	$[HO_2]$ (molecules/cm³)	$[H_2]$ (molecules/cm³)	$[H_2O]$ (molecules/cm³)	$[H_2O_2]$ (molecules/cm³)
60	Fig. II-15	1.0×10^{10}	8×10^{9}	2.2×10^{6}	5×10^{6}	3×10^{6}	3.6×10^{9}	3.5×10^{10}	6×10^{6}
	Bowman *et al.* (1970)	8×10^{9}	4×10^{9}	1×10^{7}	8×10^{6}	3×10^{7}	8×10^{7}	5×10^{10}	1×10^{7}
80	Fig. II-16	2×10^{10}	1.0×10^{8}	6×10^{7}	4×10^{5}	2×10^{5}	1.8×10^{9}	5.9×10^{8}	1×10^{5}
	Bowman *et al.* (1970)	6×10^{10}	6×10^{7}	7×10^{8}	7×10^{5}	2×10^{6}	1×10^{8}	3×10^{9}	1×10^{6}
100	Fig. II-17	4.0×10^{11}	7.3×10^{5}	1.1×10^{8}	—	—	1.1×10^{7}	7.7×10^{5}	—
	Bowman *et al.* (1970)	1×10^{11}	2×10^{5}	1×10^{8}	70	1	2×10^{6}	3×10^{7}	20

[a] Values from Figs. II-15–II-17 are 12-hr average daylight values. Values from Bowman *et al.* (1970) were calculated assuming fixed photodissociation coefficients at one-half their values for an overhead sun.

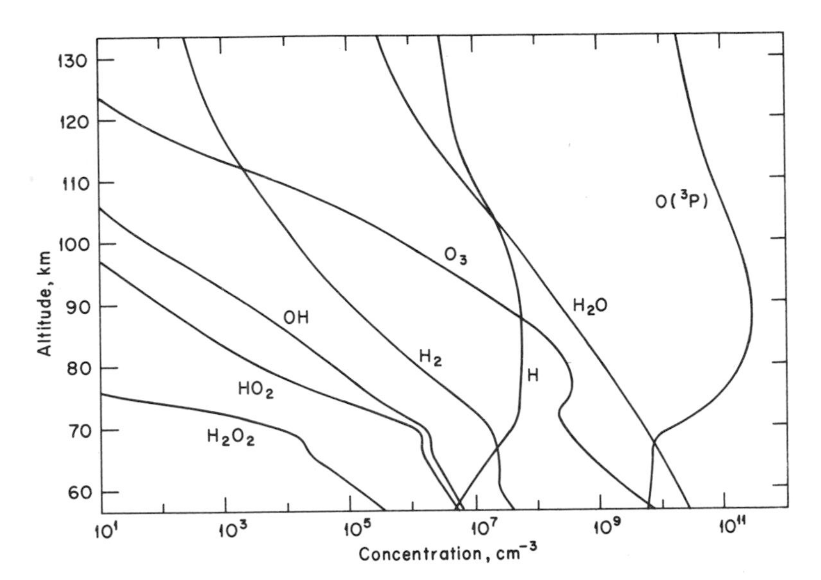

Fig. II-18. Calculated concentration profiles of minor neutral constituents for 60° north latitude in winter assuming the photodissociation constants are fixed at their 24-hr daily average. From Fukayama (1974) with permission of Pergamon Press.

mixed in constant proportions throughout the atmosphere, the overall chemical reactions, based on each element, can be deduced to be

O	$O_2 \rightarrow 2O(^3P)$	
N	$2NO \rightarrow N_2 + 2O(^3P)$	(lower thermosphere)
H	$H_2O, H_2 \rightarrow 2H$	
C	$CO_2 \rightarrow CO + O(^3P)$	

Each of the reactions is a reduction in that a less powerful oxidation agent is converted to a more powerful oxidation agent. The species on the left-hand sides of the reactions are flowing into the lower thermosphere, while those on the right-hand side are flowing out.

Likewise the daytime mesosphere also tends to be more reducing than oxidizing:

O	$O_2 \rightarrow 2O(^3P)$
N	$HNO_3 \rightarrow NO_3 \rightarrow NO_2 \rightarrow NO \rightarrow \frac{1}{2}N_2$
H	$CH_4, H_2O \rightarrow H_2, H$
C	$CH_4 \rightarrow CH_2O \rightarrow CO \rightarrow CO_2$

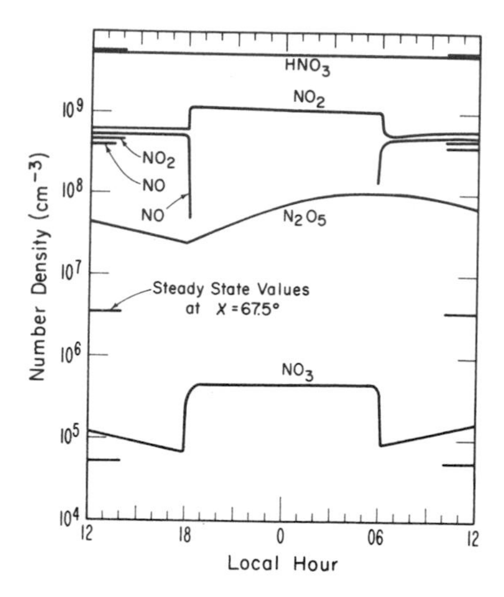

Fig. II-19. Diurnal variations in various nitrogen–oxygen compounds and [HNO_3] at 20 km and 45° north latitude for the equinoxes. The steady-state values for a solar zenith angle of 67.5° are also indicated. From Shimazaki and Ogawa (1974) with permission of the American Geophysical Union.

The first three cycles are reducing, whereas the minor carbon cycle is oxidizing.

On the other hand, the stratosphere is a strongly oxidizing atmosphere. The four cycles are

O	$O(^3P) + O_2 \rightarrow O_3$
N	$N_2O \rightarrow NO \rightarrow NO_2 \rightarrow HNO_3$
H	$CH_4, H_2O \rightarrow H_2O_2$
C	$CH_4 \rightarrow CO \rightarrow CO_2$

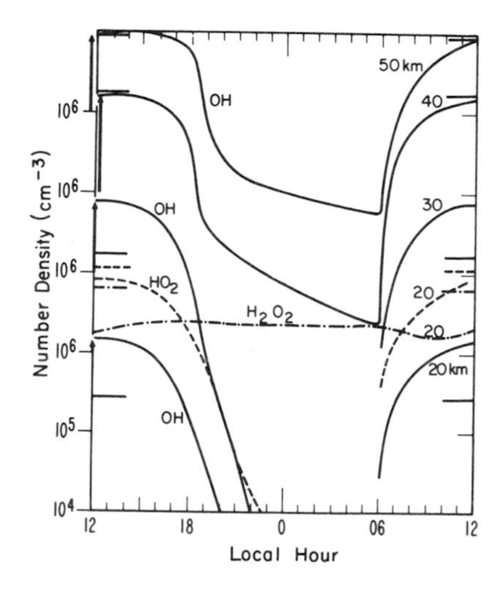

Fig. II-20. Diurnal variations in [OH] at various heights and in [HO_2] and [H_2O_2] at 20 km and 45° north latitude for the equinoxes. From Shimazaki and Ogawa (1974) with permission of the American Geophysical Union.

The oxygen cycle involves both oxidation and reduction, but from the overall energetics, we conclude that the net effect is oxidation. The other three cycles are strongly oxidizing. However, the nitrogen cycle has many facets. In addition to the one involving oxidation of N_2O to ultimately give HNO_3, the major cycle is one of reduction, reaction (46). The N_2 flows into the troposphere and to the surface of the earth, where it is removed by vegetation. Any NH_3 that reaches the stratosphere is also oxidized or reacts with HNO_3 to produce NH_4NO_3, which diffuses, along with unreacted HNO_3, through the troposphere to the surface of the earth, where it, along with N_2, becomes the fuel for plant life at the surface of the earth.

The O_3 produced in the stratosphere acts as a radiation shield for the earth, and the HNO_3 produced provides an essential food for plant life. Ironically we will see how these two absolutely necessary ingredients of life can prove to be so harmful to life when present at the surface of the earth at about the same mole fractions as in the stratosphere. Indeed, nature requires a very delicate balance, and an excess or deficiency of any compound may be equally disastorous.

The troposphere is a relatively quiescent region where relatively little chemistry occurs. However, in the surface of the earth, reduction processes dominate. For the four elements of interest, the dominating processes are

$$\text{O} \quad CO_2 \xrightarrow{h\nu} O_2$$

$$\text{N} \quad N_2, NH_4NO_3 \rightarrow N_2O, NH_3, NO, NO_2$$

$$\text{H} \quad H_2O \rightarrow H_2, NH_3, CH_4$$

$$\text{C} \quad CO_2 \xrightarrow{h\nu} CH_4, CO$$

Reduction of CO_2 occurs via photosynthesis ultimately to produce O_2 and CH_4. The nitrogen cycle is complex and involves both oxidation and reduction, the dominant process being the oxidation of N_2 to N_2O. However, HNO_3, which diffuses down from the stratosphere and combines with NH_3 in the troposphere to give NH_4NO_3, which is absorbed by the earth, ultimately yields NH_3, NO, and NO_2 in reducing processes.

The global atmosphere chemical cycles can be summarized as mainly oxidation occurring in the stratosphere, the oxidized products flowing upward into the mesosphere and thermosphere and downward to the surface of the earth, where reduction occurs. The reduced products then diffuse back toward the stratosphere, where they are again oxidized. The only exception is for the nitrogen cycle for which oxidation dominates on the surface of the earth, and reduction dominates everywhere in the atmosphere.

PERTURBATIONS

There are several ways in which the earth's atmosphere can be perturbed. Of concern in recent years is the possibility that man may be significantly perturbing the stratosphere and thus altering the ozone layer. One type of perturbation is the increase of NO_x and the oxides of hydrogen through atmospheric nuclear blasts or the exhaust emissions of supersonic transports (SST). Another perturbation of concern is the introduction of chlorofluoromethanes into the troposphere, which then diffuse into the stratosphere where they can photodissociate or react with $O(^1D)$ to produce radicals that attack O_3.

Supersonic Transports (SST)

The perturbations that could be introduced by the SST have been the subject of research from 1971 to 1975 by the Climatic Impact Assessment Program of the U.S. Dept. of Transportation. It is estimated that 2000 French Concordes or 500 American type SST's (Boeing 2707) would introduce 1.5×10^{12} gm/year of NO into the stratosphere (Grobecker *et al.*, 1974). This corresponds to a flux of 1.87×10^8 molecules/cm^2-sec averaged over the whole earth, and is more than twice the flux of 0.84×10^6 molecules/cm^2-sec of NO_x currently leaving the stratosphere (Table II-10). Furthermore, this NO would be introduced almost entirely in the 15–25 km region and mainly in the Northern Hemisphere, thus increasing the effect in the Northern Hemisphere lower stratosphere.

The adverse effect of excess NO_x on stratospheric O_3 was first pointed out by Crutzen (1971) and by Johnston (1971), who showed that O_3 would be depleted by the catalytic reactions (57) and (54). No NO is removed in this cycle, and thus it catalytically destroys the O_3 layer. However, it should be realized that most of the time NO_2 photodissociates rather than reacting with $O(^3P)$, in which case no odd oxygen is lost.

Several investigators computed the effect of increased NO on the stratospheric O_3 using one- and two-dimensional diffusion models. The results have been discussed in a National Academy of Sciences (U.S.) report (1975). For 20-km injection these results are summarized in Fig. II-21. It can be seen that the percent decrease in total O_3 rises approximately as the square root of the total increase in NO.

The first three-dimensional model for diffusion that included chemistry was reported by Alyea *et al.* (1975). They performed a three-year integration for 500 SST's assuming the injection occurred at 20 km in mid-latitudes of the Northern Hemisphere. They computed a 12% depletion of total stratospheric O_3 (16% in the Northern Hemisphere and 8% in

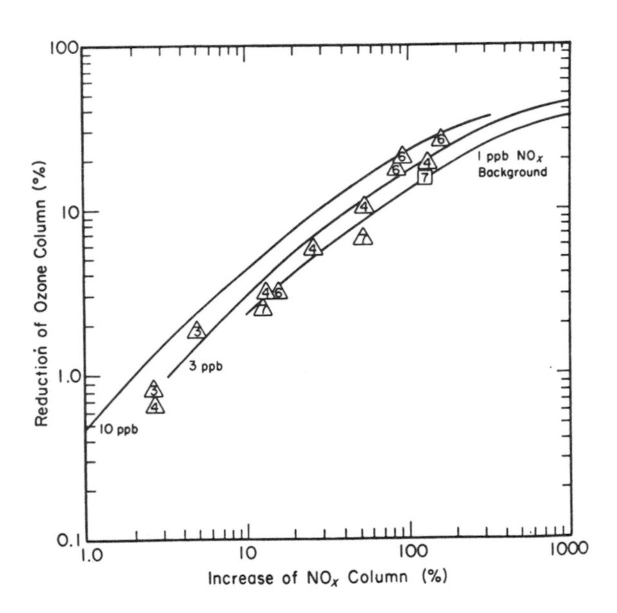

Fig. II-21. Calculated percentage reduction of vertical ozone column as a function of percentage increase of nitrogen oxide for injection of excess NO at 20 km. From National Academy of Sciences Report (1975) with permission.

the Southern Hemisphere), which is similar to the results of the one- and two-dimensional models.

The increase in damaging ultraviolet radiation (<3200 Å) is about twice the percent decrease in total ozone for small changes ($\leq 10\%$ changes in total O_3). (National Academy of Sciences Report, 1975). One of the most serious consequences of increased radiation below 3200 Å reaching the earth's surface is the increase it will cause in skin cancer incidence. The correlation between skin cancer incidence and exposure to ultraviolet radiation is given in Fig. II-22. For both types of skin cancer (melanoma and nonmelanoma) the logarithm of incidence increases linearly with the photon flux. In addition to the effect of increased ultraviolet radiation on human health, plant life and the climate may also be affected.

Chlorofluoromethanes

The possibility that chlorofluoromethanes could threaten the stratospheric ozone layer was first recognized by Molina and Rowland (1974a) and quickly supported by Cicerone *et al.* (1974) and Crutzen (1974). The problem results from the degradation of chlorofluoromethanes in the

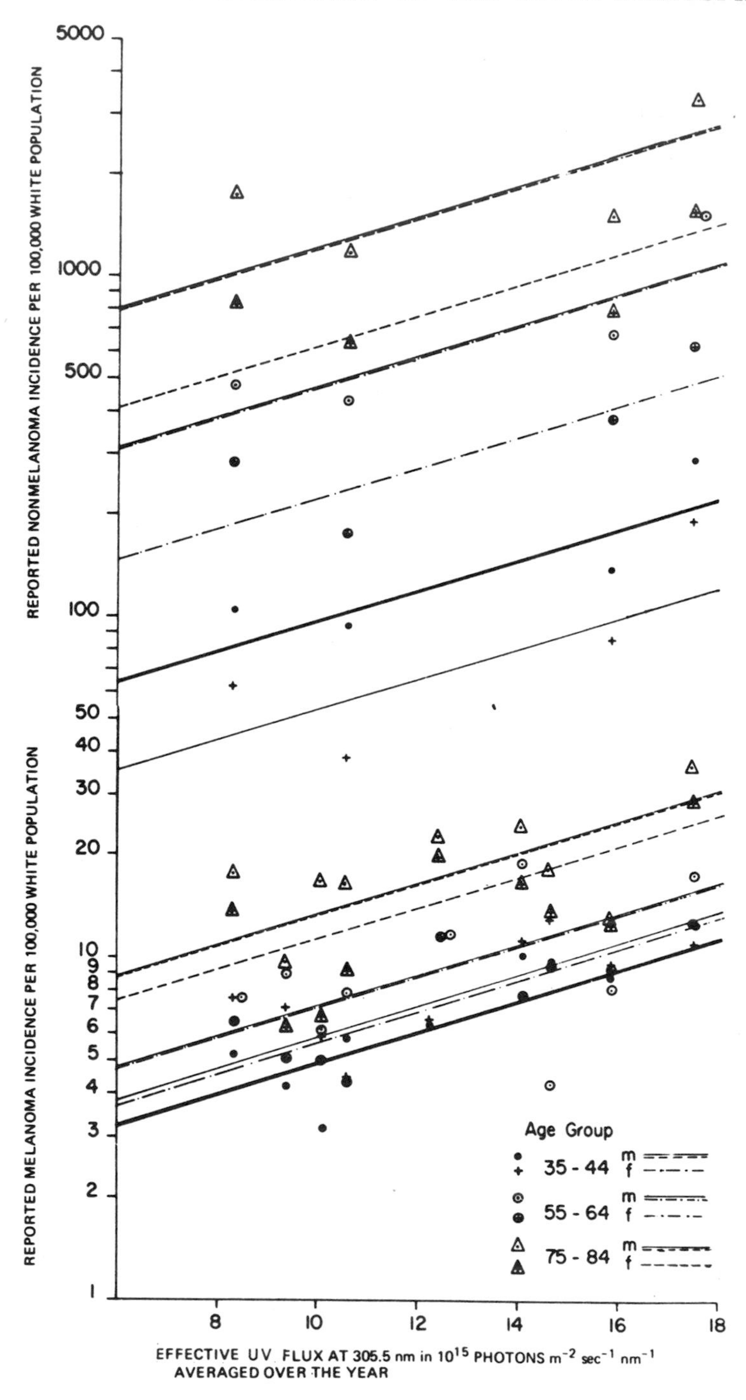
REPORTED NONMELANOMA INCIDENCE PER 100,000 WHITE POPULATION
5000
1000
500
100
50
40
30
20
10
9
8
7
6
5
4
3
2
1
REPORTED MELANOMA INCIDENCE PER 100,000 WHITE POPULATION
Age Group
35 - 44
55 - 64
75 - 84
m
f
m
f
m
f
8
10
12
14
16
18
EFFECTIVE UV FLUX AT 305.5 nm in 10^15 PHOTONS m^-2 sec^-1 nm^-1
AVERAGED OVER THE YEAR

stratosphere to produce Cl atoms, which remove O_3 via

$$Cl + O_3 \rightarrow ClO + O_2 \qquad (144)$$

This reaction is followed by

$$ClO + O(^3P) \rightarrow Cl + O_2 \qquad (145)$$

which regenerates Cl atoms so that a long chain process is involved, which conserves Cl atoms. An alternate chain involving the reaction

$$ClO + NO \rightarrow Cl + NO_2 \qquad (146)$$

does not lead to ozone removal since most of the time the NO_2 photodissociates to produce odd oxygen $(O(^3P))$. Termination is by

$$Cl + CH_4 \rightarrow HCl + CH_3 \qquad (147)$$

$$Cl + H_2 \rightarrow HCl + H \qquad (148)$$

The chain can be regenerated again by reaction of HCl with HO.

$$HO + HCl \rightarrow H_2O + Cl \qquad (149)$$

Eventually, however, the HCl diffuses to the troposphere.

The chlorofluoromethanes are used principally in aerosol spray cans and as refrigerants. They are chemically unreactive and accumulate in the troposphere. The mole fractions in the atmosphere have reached: CF_2Cl_2, $>1 \times 10^{-10}$ (Rowland and Molina, 1975); $CFCl_3$, $>8 \times 10^{-11}$ (Rowland and Molina, 1975); and CCl_4, $>7 \times 10^{-11}$ (Molina and Rowland, 1974b). Wilkniss *et al.* (1975) have found that tropospheric $CFCl_3$ has increased between 1971 and 1974, and that the increase is proportional to the increase of industrially produced amounts of $CFCl_3$ in the same time period.

Table II-21 lists the world production of CF_2Cl_2 and $CFCl_3$ by year. The chemical profile for fluorocarbons in the U.S. has a history of 8.7% growth per year from 1961 to 1971, the largest volume chlorofluoromethane being CF_2Cl_2; the world production rate has increased even faster. Based on the world production and the tropospheric concentrations, Rowland and Molina (1975) have calculated that each has a lifetime of >20 years.

The chlorofluoromethanes are removed in the stratosphere mainly by photodissociation and to a minor extent by reaction with $O(^1D)$. Rowland and Molina (1975) calculated the stratospheric photodissociation coefficients for CF_2Cl_2 and $CFCl_3$ assuming unit photodissociation efficiency, an

---

Fig. II-22. Reported skin cancer incidence rates per 100,000 white population among three age groups as a function of ultraviolet flux. The computations of ultraviolet flux were not adjusted for altitude. From National Academy of Sciences Report (1975) with permission.

TABLE II-21

Estimated World Production of CF_2Cl_2 *and* $CFCl_3$ *in kilotons* (10^9 gm)[a]

	$CFCl_3$		CF_2Cl_2	
Year	U.S.	World	U.S.	World
1950–1955	54	54	—	—
1956–1957	32	32	—	—
1958	23	23	59	59
1959	27	27	71	71
1960	33	40	75	87
1961	41	52	78	94
1962	56	72	94	117
1963	64	83	98	129
1964	67	93	103	143
1965	77	112	123	175
1966	77	122	130	196
1967	83	139	141	225
1968	93	165	148	256
1969	108	197	167	300
1970	111	217	170	329
1971	117	241	177	363
1972	136	285	191	422
1973	147	313	221	469
Cumulative to November 1, 1974		2576		3830

[a] Also during 1948–1972, 50% of the total fluorocarbons were used as aerosol propellants, 28% as refrigerants, 10% as plastics, 5% as solvents, and 7% as blowing agents, exports, and other miscellany. For 1971 the demand was 340×10^6 kg (750×10^6 lb); for 1972, 374×10^6 kg (825×10^6 lb); and for 1976 the estimated demand is 454×10^6 kg (1000×10^6 lb). From Rowland and Molina (1975) with permission of the American Geophysical Union.

assumption confirmed by Jayanty *et al.* (1975). The photodissociation coefficients as a function of altitude are displayed in Figs. II-23 and II-24, respectively, for CF_2Cl_2 and $CFCl_3$. For CCl_4 the photodissociation coefficients were computed by Molina and Rowland (1974b). Accurate rate coefficients for removal of the chlorofluoromethanes by $O(^1D)$ were obtained by Jayanty *et al.* (1975), who also established that the major, if not exclusive removal process was

$$O(^1D) + CF_nCl_{4-n} \rightarrow CF_nCl_{3-n} + ClO \qquad (150)$$

Thus the effective lifetimes for removal of the chlorofluoromethanes both

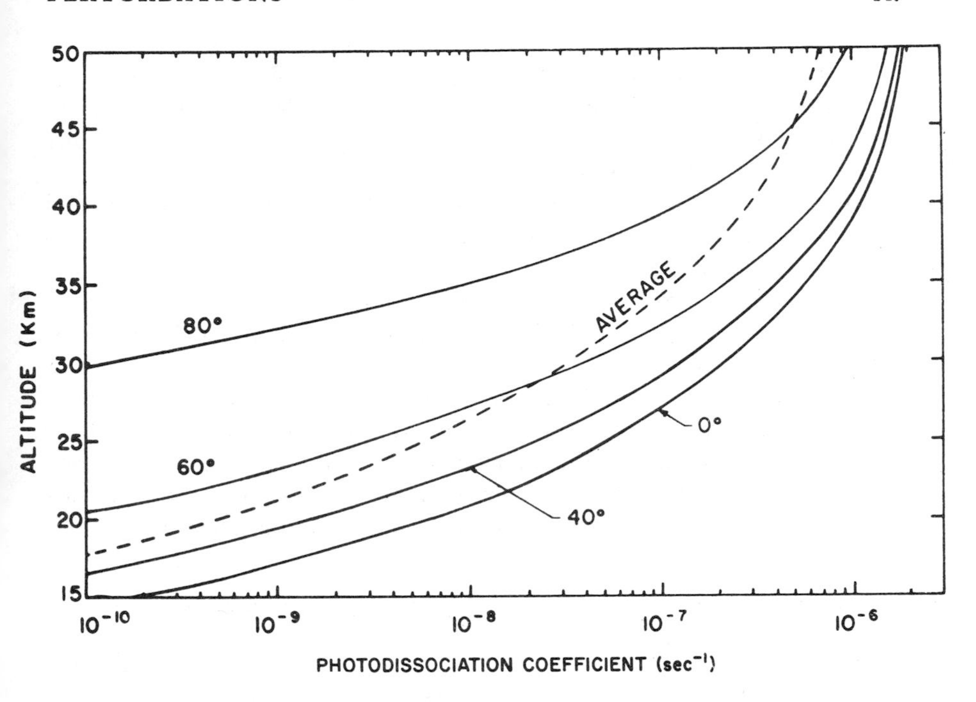

Fig. II-23. Photodissociation coefficient of CF_2Cl_2 in the stratosphere. From Rowland and Molina (1975) with permission of the American Geophysical Union.

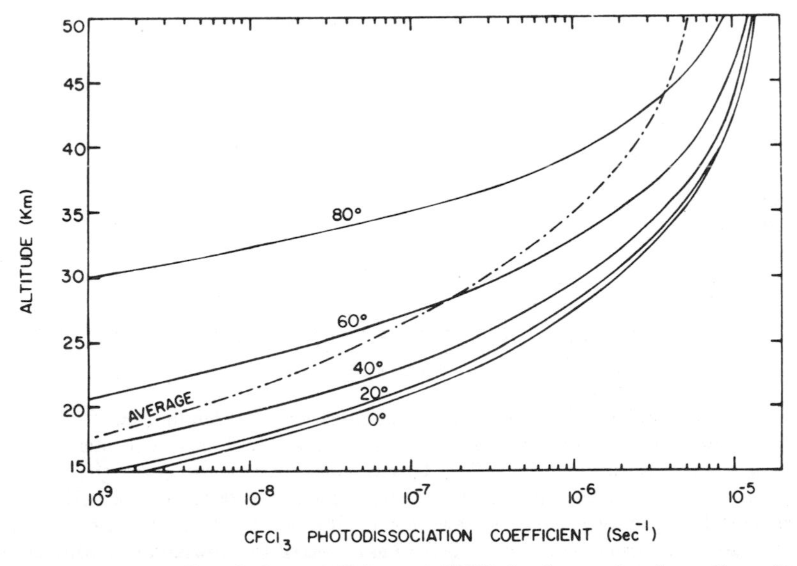

Fig. II-24. Photodissociation coefficient of $CFCl_3$ in the stratosphere. From Rowland and Molina (1975) with permission of the American Geophysical Union.

by photodissociation and O(^{1}D) attack can be computed and they are tabulated in Table II-22. Since most of the chlorofluoromethanes reside in the lower stratosphere, it can be years before they are removed, and any adverse effects will be delayed a generation.

The extent of the O_3 depletion has been the subject of a number of calculations. One of the earliest was that of Crutzen (1974), who predicted a 7% depletion in O_3, which was made for a steady production of chlorofluoromethanes at 1972 levels. A more detailed calculation has been made by Wofsy *et al.* (1975), and their results are shown in Fig. II-25 for several different assumptions. However, with any of the assumed conditions, a pronounced effect will not occur for many years. Nevertheless, a program of control cannot be delayed, since alleviating the perturbation will take at least as long. While more accurate recent measurements of various rate coefficients and concentrations have now been made, the perturbations suggested in Fig. II-25 are still reasonably accurate if the model is complete.

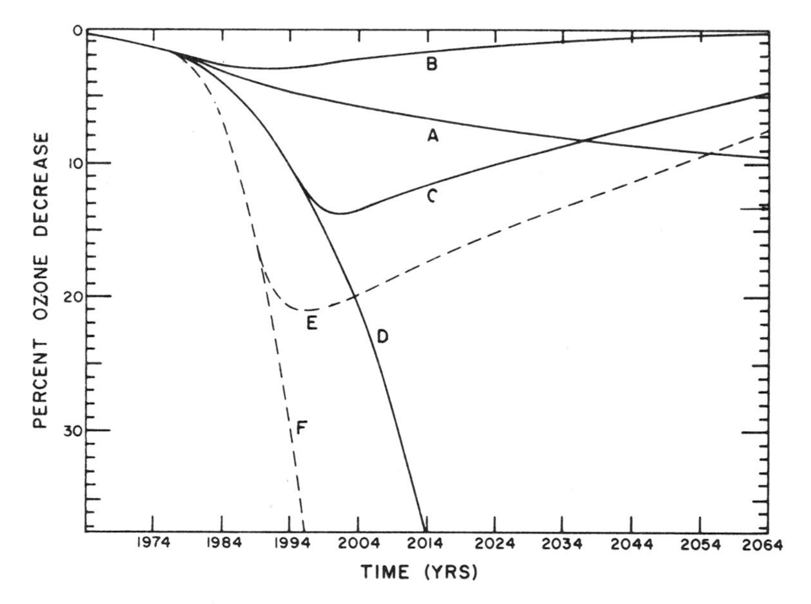

Fig. II-25. Calculated ozone destruction rates versus calendar year for six models of fluorocarbon use. A, constant production at 1972 levels; B, emissions increase at 10% per year and cease abruptly in 1978; C, emissions increase 10% per year and cease abruptly in 1995; D, emissions increase continuously at 10% per year; E, emissions increase at 22% per year and cease abruptly in 1987; F, emissions increase continuously at 22% per year. From Wofsy *et al.* (1975) as reported by Rowland and Molina (1975) with permission of the American Geophysical Union and the American Association for the Advancement of Science.

TABLE II-22

Lifetimes of Chlorofluoromethanes in the Stratosphere

Altitude (km)	$[O(^1D)]$[a] (molecules/cm^3)	CCl_4		$CFCl_3$		CF_2Cl_2		CF_3Cl,
		$\tau\{phot\}$[b] (yr)	$\tau\{O(^1D)\}$[c] (yr)	$\tau\{phot\}$[d] (yr)	$\tau\{O(^1D)\}$[c] (yr)	$\tau\{phot\}$[d] (yr)	$\tau\{O(^1D)\}$[c] (yr)	$\tau\{O(^1D)\}$[d] (yr)
20	0.25	3.5	274	6.60	384	63.4	480	1108
25	1.5	0.35	45.7	0.647	64.0	5.28	80.0	185
30	7.5	0.063	9.15	0.117	12.8	0.99	16.0	36.9
35	25	—	2.74	0.032	3.84	0.264	4.80	11.1
40	87.5	0.0063	0.78	0.014	1.10	0.106	1.37	3.17
45	175	—	0.39	0.008	0.55	0.063	0.69	1.58
50	150	—	0.46	0.006	0.64	0.045	0.80	1.85

[a] Average global yearly value.
[b] Reciprocal of global average photodissociation coefficients given by Molina and Rowland (1974b).
[c] Lifetime for removal by $O(^1D)$ from rate coefficients of Jayanty *et al.* (1975).
[d] Reciprocal of the global average photodissociation coefficients given by Rowland and Molina (1975).

REFERENCES

Ackerman, M., Fontanella, J. C., Frimout, D., Girard, A., Louisnard, N., and Muller, C. (1975). *Planet. Space Sci.* **23,** 651, "Simultaneous Measurements of NO and NO_2 in the Stratosphere."

Alyea, F. N., Cunnold, D. M., and Prinn, R. G. (1975). *Science* **188,** 117, "Stratospheric Ozone Destruction by Aircraft-Induced Nitrogen Oxides."

Arnold, S. J., and Ogryzlo, E. A. (1967). *Can. J. Phys.* **45,** 2053, "Some Reactions forming $O_2(^1\Sigma_g{}^+)$ in the Upper Atmosphere."

Bowman, M. R., Thomas, L., and Geisler, J. E. (1970). *J. Atmos. Terrest. Phys.* **32,** 1661, "The Effect of Diffusion Processes on the Hydrogen and Oxygen Constituents in the Mesosphere and Lower Thermosphere."

Brasseur, G., and Nicolet, M. (1973). *Aeronom. Acta* No. 113, "Chemospheric Processes of Nitric Oxide in the Mesosphere and Stratosphere."

Calvert, J. G., Kerr, J. A., Demerjian, K. L., and McQuigg, R. D. (1972). *Science* **175,** 751, "Photolysis of Formaldehyde as a Hydrogen Atom Source in the Lower Atmosphere."

Castellano, E., and Schumacher, H. J. (1972). *Chem. Phys. Lett.* **13,** 625, "The Kinetics and the Mechanism of the Photochemical Decomposition of Ozone with Light of 3340 Å Wavelength."

Cicerone, R. J., Stolarski, R. S., and Walters, S. (1974). *Science* **185,** 1165, "Stratospheric Ozone Destruction by Man-Made Chlorofluoromethanes."

Cieslik, S., and Nicolet, M. (1973). *Aeronom. Acta* No. 112, "The Aeronomic Dissociation of Nitric Oxide."

Crutzen, P. J. (1971). *J. Geophys. Res.* **76,** 7311, "Ozone Production Rates in an Oxygen-Hydrogen-Nitrogen Oxide Atmosphere."

Crutzen, P. J. (1974). *Geophys. Res. Lett.* **1,** 205, "Estimates of Possible Future Ozone Reductions from Continued Use of Fluoro-Chloro-Methanes (CF_2Cl_2, $CFCl_3$)."

Evans, W. F. J., and Llewellyn, E. J. (1973). *J. Geophys. Res.* **78,** 323, "Atomic Hydrogen Concentrations in the Mesosphere and the Hydroxyl Emissions."

Fukayama, K. (1974). *J. Atmos. Terrest. Phys.* **36,** 1297, "Latitudinal Distributions of Minor Neutral Hydrogen–Oxygen Constituents in the Winter Mesosphere and Lower Thermosphere."

Garvin, D., and Hampson, R. F. (1974). Nat. Bur. Std. Rep. NBSIR 74-430, "Chemical Kinetics Data Survey. VII. Tables of Rate and Photochemical Data for Modelling of the Stratosphere" (Revised).

Gilpin, R., Schiff, H. I., and Welge, K. H. (1971). *J. Chem. Phys.* **55,** 1087, "Photodissociation of O_3 in the Hartley Band. Reactions of $O(^1D)$ and $O_2(^1\Sigma_g{}^+)$ with O_3 and O_2."

Grobecker, A. J., Coroniti, S. C., and Cannon, R. H., Jr. (1974). U.S. Dept. of Transportation Climatic Impact Assessment Program Rep. DOT-TST-75-50, "The Effects of Stratospheric Pollution by Aircraft."

Hale, L. (1974). "Cospar Methods of Measurements and Results of Lower Ionosphere Structure," p. 219. Akademie Verlag, Berlin, "Positive Ions in the Mesosphere."

Izod, T. P. J., and Wayne, R. P. (1968). *Proc. Roy. Soc.* **A308,** 81, "The Formation, Reaction and Deactivation of $O_2(^1\Sigma_g{}^+)$."

Jayanty, R. K. M., Simonaitis, R., and Heicklen, J. (1975). *J. Photochem.* **4,** 381, "The Photolysis of Chlorofluoromethanes in the Presence of O_2 or O_3 at 213.9 nm and Their Reactions with $O(^1D)$."

Jayanty, R. K. M., Simonaitis, R., and Heicklen, J. (1976). *J. Phys. Chem.* **80,** 433, "The Reaction of NH_2 with NO and O_2."

Johnston, H. S. (1971). *Science* **173,** 517, "Reduction of Stratospheric Ozone by Nitrogen Oxide Catalysts from Supersonic Transport Exhaust."
Jones, I. T. N., and Wayne, R. P. (1969). *J. Chem. Phys.* **51,** 3617, "Photolysis of Ozone by 254-, 313-, and 334-nm Radiation."
Jones, I. T. N., and Wayne, R. P. (1970). *Proc. Roy. Soc.* **A319,** 273, "The Photolysis of Ozone by Ultraviolet Radiation. IV. Effect of Photolysis Wavelength on Primary Step."
Kearns, D. R. (1971). *Chem. Rev.* **71,** 395, "Physical and Chemical Properties of Singlet Molecular Oxygen."
Kockarts, G. (1971). *In* "Mesospheric Models and Related Experiments (G. Fiocco, ed.), p. 160. Reidel Publ., Dordrecht, Holland, "Penetration of Solar Radiation in the Schumann–Runge Bands of Molecular Oxygen."
Kuis, S., Simonaitis, R., and Heicklen, J. (1975). *J. Geophys. Res.* **80,** 1328, "Temperature Dependence of the Photolysis of Ozone at 3130 Å.
Lissi, E., and Heicklen, J. (1972). *J. Photochem.* **1,** 39, "Photolysis of O_3."
Molina, M. J., and Rowland, F. S. (1974a). *Nature (London)* **249,** 810, "Stratospheric Sink for Chlorofluoromethanes: Chlorine Atom-Catalyzed Destruction of Ozone."
Molina, M. J., and Rowland, F. S. (1974b). *Geophys. Res. Lett.* **1,** 309, "Predicted Present Stratospheric Abundances of Chlorine Species from Photodissociation of Carbon Tetrachloride."
National Academy of Sciences Report (1975). "Environmental Impact of Stratospheric Flight."
Nicolet, M. (1970). Ionosphere Res. Lab. Sci. Rep. No. 350, Penn State Univ., "Ozone and Hydrogen Reactions."
Nicolet, M. (1971). *In* "Mesospheric Models and Related Experiments" (G. Fiocco, ed.), p. 1. Reidel Publ., Dordrecht, Holland, "Aeronomic Reactions of Hydrogen and Ozone."
Nicolet, M., and Peetermans, W. (1972a). *Ann. Geophys.* **28,** 751, "The Production of Nitric Oxide in the Stratosphere by Oxidation of Nitrous Oxide."
Nicolet, M., and Peetermans, W. (1972b). *Aeronom. Acta* No. 103, "On the Vertical Distribution of Carbon Monoxide and Methane in the Stratosphere."
Ogryzlo, E. A. (1967). *Stanford Res. Inst. Int. Oxidat. Symp., San Francisco* **2,** 367, "Relaxation and Reactivity of Singlet Oxygen."
Osif, T. (1975). Unpublished Work at Penn State Univ., "Photooxidation of CH_2O."
Rowland, F. S., and Molina, M. J. (1975). *Rev. Geophys. Space Sci.* **13,** 1, "Chlorofluoromethanes in the Environment."
Shimazaki, T., and Ogawa, T. (1974). *J. Geophys. Res.* **79,** 3411, "A Theoretical Model of Minor Constituent Distributions in the Stratosphere Including Diurnal Variations."
Simonaitis, R., and Heicklen, J. (1972). *J. Chem. Phys.* **56,** 2004, "Kinetics and Mechanism of the Reaction of $O(^3P)$ with Carbon Monoxide."
Simonaitis, R., and Heicklen, J. (1976). *J. Phys. Chem.* **80,** 1, "Reactions of HO_2 with NO and NO_2 and of OH with NO."
Watanabe, K. (1954). *J. Chem. Phys.* **22,** 1564, "Photoionization and Total Absorption Cross Section of Gases. I. Ionization Potentials of Several Molecules. Cross Sections of NH_3 and NO.
Wilkniss, P. E., Swinnerton, J. W., Lamontagne, R. A., and Bressan, D. J. (1975). *Science* **187,** 832, "Trichlorofluoromethane in the Troposphere, Distribution and Increase, 1971–1974."
Wofsy, S. C., McElroy, M. B., and Sze, N. D. (1975). *Science* **187,** 535, "Freon Consumption: Implications for Atmospheric Ozone."

Chapter III

THE IONOSPHERE

The ionosphere is that region of space containing electrically charged species. It starts at about 60 km and extends upward, although the charged particle density is not zero below 60 km. However, below 60 km, ionization is not due to the sun's ultraviolet and X radiation primarily, but to cosmic radiation. This problem has been considered by Brasseur and Nicolet (1973) and will not be considered here.

The daytime electron concentration is about $10^2/cm^3$ at 60 km and rises to $\sim 10^5/cm^3$ at 100 km. There then follows a plateau until about 160 km, where the electron density then rises again, reaching a maximum of $\sim 10^6/cm^3$ at 220–250 km. Because of this plateau, the region above 160 km is referred to as the F2 region, whereas that part of the F region below 160 km (i.e., 120–160 km) is referred to as the F1 region. In addition, in the F1 region diffusion is not an important process, but in the F2 region diffusional separation is important. Since not much chemistry occurs above 160 km and physics dominates, we will not concern ourselves with the F2 region. Below 160 km the ion chemistry dominates the characterization of the ionosphere, and this will be the region of interest to us.

The positive-ion concentrations for O^+, NO^+, and O_2^+ at various solar zenith angles at altitudes from 100 to 200 km are listed in Tables III-1, III-2, and III-3, respectively. N_2^+ is also present (Table III-4), but at much lower concentrations. Typical daytime concentration profiles of these ions, as well as electrons, are shown in Fig. III-1. NO^+ and O_2^+ are the dominant ions, and they are about equally important. Their concentrations are $\sim 5 \times 10^4/cm^3$ from 100 to 160 km throughout most of the day. In fact, the experimental observations listed in Table III-2 for NO^+ indicate that $[NO^+]$ is slightly higher for a solar zenith angle of 75° than

TABLE III-1

O^+ Concentrations (particles/cm³) for Various Solar Zenith Angles

Altitude (km)	0° [a]	Typical daytime [b]	75° [c]	90° [c]
100	0.9	5	—	—
110	50	82	—	—
120	794	640	—	—
130	3,600	1,800	—	—
140	10,300	5,500	1,000	—
150	25,800	17,000	4,000	—
160	44,800	47,000	11,000	1,000
170	—	—	25,000	2,000
180	—	—	50,000	4,200
190	—	—	70,000	7,000
200	—	—	150,000	10,000

[a] Values at 0° are computed as described in text using photoionization rates of Hinteregger *et al.* (1964).

[b] Typical daytime values are a summary of experimental observations as reported in Norton and Barth (1970).

[c] 75 and 90° values are from experimental observations reported in Danilov (1970, p. 104).

for 0° at 100–130 km. This is probably an artifact due to experimental observations by different investigators at different times. At twilight the ion concentrations fall, those of O_2^+ falling more seriously. However, even at night these ions are present, the concentration of NO^+ being as high as $3.2 \times 10^3/cm^3$ at 130 km.

The O^+ ion is at a lower concentration than either O_2^+ or NO^+. However, its concentration rises rapidly with altitude and becomes as large as $[O_2^+]$ at 160 km for an overhead sun. It falls off rapidly with solar zenith angle and is essentially zero at night.

Since there are no negative ions present in the E and F regions, the electron concentration equals the sum of the positive-ion concentrations. In the D region, the situation is quite different, the electron density falling way below that of the total positive-ion concentration, and negative ions are important. The principal negative ions appear to be NO_3^-, CO_3^-, and CO_4^-, possibly hydrated, although thorough measurements are not available.

D region electron profiles were measured by Ferraro *et al.* (1974), who also summarized the results of some earlier investigations. Some of the

TABLE III-2

NO^+ Concentrations (particles/cm³) for Various Solar Zenith Angles

Altitude (km)	0° [a]	Typical daytime [b]	75° [c]	90° [c]	Night [d]
90	14,700	3,600	—	—	—
95	8,670	6,200	—	—	—
100	25,300	32,000	39,000	28,000	—
110	45,700	53,000	59,000	19,000	—
120	49,300	45,000	58,000	26,000	3,000
130	46,800(60,800)	43,000	49,000	34,000	3,200
140	37,500(59,900)	58,000	49,000	30,000	2,100
150	35,900(63,100)	74,000	57,000	30,000	700
160	31,800(66,100)	78,000	65,000	30,000	420
170	—	—	88,000	30,000	400
180	—	—	94,000	32,000	450
190	—	—	94,000	36,000	600
200	—	—	92,000	39,000	1,000

[a] Values at 0° are computed as described in text using photoionization rates of Hinteregger *et al.* (1964). Values in parentheses are based on computed, rather than observed, values of $[N_2^+]$.

[b] Typical daytime values are a summary of experimental observations as reported in Norton and Barth (1970).

[c] 75 and 90° values are from experimental observations reported in Danilov (1970, p. 105).

[d] Nighttime values are experimental observations reported by Donahue (1968).

results indicate a slight minimum at about 100/cm³ in the electron concentration at 65–70 km. Another group of experiments summarized by Bennett *et al.* (1972) does not show the minimum but indicates a plateau of 1000/cm³ between 70 and 85 km. A smooth compromise of these results is depicted in Fig. III-1. Below 60 km, the electron density drops sharply with decreasing altitude.

The positive-ion concentrations in the D region were measured by Narcisi, and his results are shown in Fig. III-2. Throughout most of this region O_2^+ is not important. The important lighter ions are NO^+ ($m/e = 30$), H_3O^+ ($m/e = 19$), and $H(H_2O)_2^+$ ($m/e = 37$). However, most of the ions are at $m/e > 45$ and presumably correspond to hydrates of NO^+ and H^+. Hale (1974) has shown that the positive-ion concentration goes through a minimum at ~62 km and rises to $\sim 4 \times 10^3$/cm³ at 30 km, even at night.

TABLE III-3

O_2^+ Concentrations (particles/cm^3) for Various Solar Zenith Angles

Altitude (km)	0° [a]	Typical daytime [b]	75° [c]	90° [c]	Night [d]
90	>470	2,400	—	—	—
95	>5,700	3,800	—	—	—
100	30,200	20,000	13,000	—	—
110	47,800	50,000	20,000	—	—
120	60,800	59,000	30,000	4,000	—
130	50,200	60,000	34,000	8,300	90
140	57,700	61,000	40,000	10,000	85
150	55,000	59,000	41,000	10,800	80
160	60,400	52,000	43,000	12,300	80
170	—	—	50,000	13,600	80
180	—	—	47,000	15,800	90
190	—	—	47,000	17,700	110
200	—	—	48,000	20,000	330

[a] Values at 0° are computed as described in text using photoionization rates of Hinteregger *et al.* (1964).

[b] Typical daytime values are a summary of experimental observations as reported in Norton and Barth (1970).

[c] 75 and 90° values are from experimental observations reported in Danilov (1970, p. 107).

[d] Nighttime values are experimental observations reported by Donahue (1968).

Ion production occurs through photoionization of the species present, i.e. N_2, O_2, O, and NO. The adiabatic (i.e., 0–0) ionization potentials for these molecules correspond to wavelengths at 796, 1027, 911, and 1340 Å, respectively. Therefore, NO is the only species photoionized by Lyman-α (1215.7 Å) and its photoionization dominates in the D region. The photoionization rates for NO at various altitudes are listed in Table III-5. However, in the E and F regions where short-wavelength radiation is present, NO photoionization is unimportant because NO is a minor constituent. Here the photoionization of the major constitutents dominates. Their photoionization rates have been computed by Hinteregger *et al.* (1964) for zero solar zenith angle, and they are tabulated in Table III-5. For other solar zenith angles, the total photoionization rates were computed by Ivanov-Kholodnyy and are shown in Fig. III-3.

Below 65 km, Lyman-α is effectively absent, and the photoionization

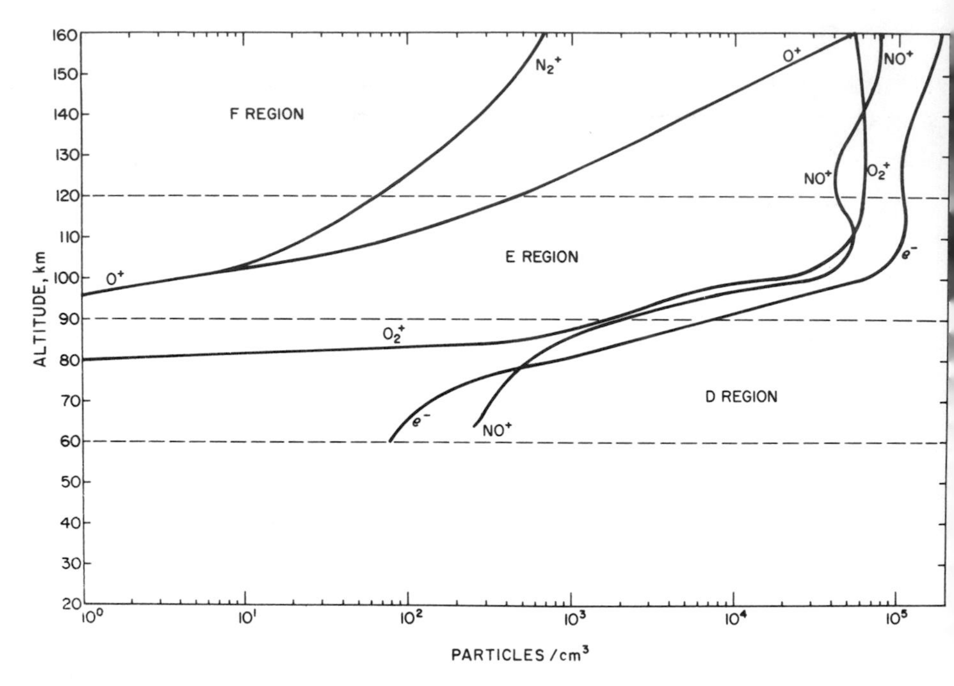

Fig. III-1. Typical daytime positive ion and electron concentrations in the ionosphere.

TABLE III-4

N_2^+ Concentrations (particles/cm^3) for an Overhead Sun (0°) and Typical Daytime Values

Altitude (km)	0° [a]	Typical daytime [b]
100	1.8	5
110	22	27
120	108	70
130	260	140
140	585	290
150	986	470
160	1610	660

[a] Values at 0° are computed as described in text using photoionization rates of Hinteregger *et al.* (1964).

[b] Typical daytime values are a summary of experimental observations as reported in Norton and Barth (1970).

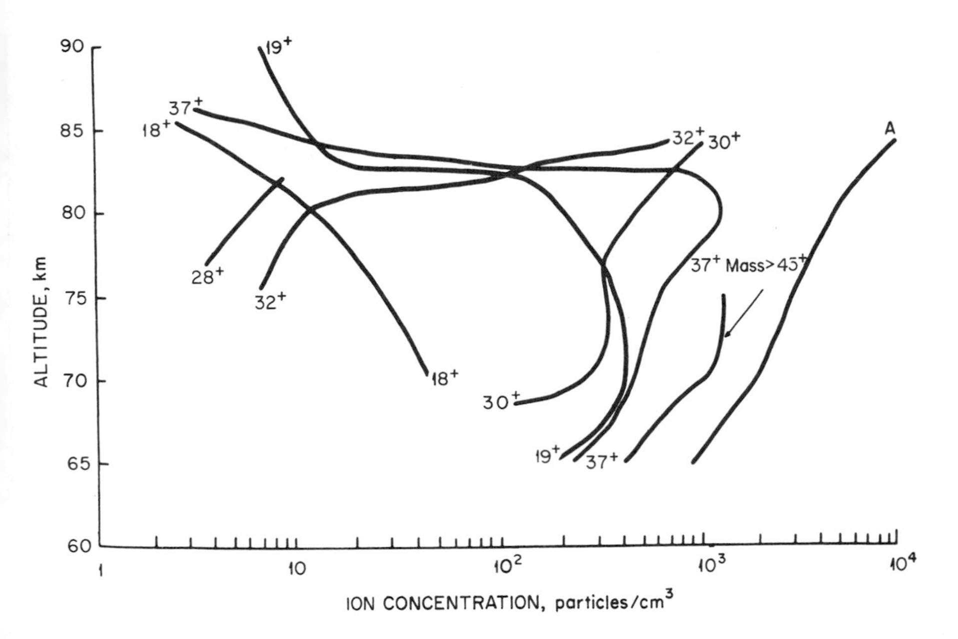

Fig. III-2. Distribution of positive ion concentrations in the D region based on the experimental data of R. S. Narcisi. Curve A represents total positive ions. From Danilov (1970, p. 19) with permission of Plenum Press.

rates become negligibly small, yet the ion concentrations increase with decreasing altitude. This phenomenon can be attributed to ion production by cosmic radiation, which overtakes photoionization rates at 65 km. These rates and consequently the ion concentrations at low altitudes are not altered by the diurnal cycle.

The galactic cosmic radiation, which is essentially isotropic as observed on the earth, is modulated by the interplanetary magnetic field. Thus the primary cosmic radiation that reaches the earth is minimum at periods of maximum solar activity. It consists of about 83% protons, 12% alpha particles, and 1% other elements.

This subject has been treated recently by Brasseur and Nicolet (1973). They have deduced the ionization rates from the data of Neher. The results are shown in Table III-6 for various geomagnetic latitudes between 35 and 85 km altitude. The importance of cosmic ray ionization increases with latitude and as the altitude drops. Thus minimum ionization (during the day) occurs at the equator near 60 km.

Ionization increases as the altitude decreases and becomes significant in the stratosphere. The ion pair production rate at 36 km is shown graphically

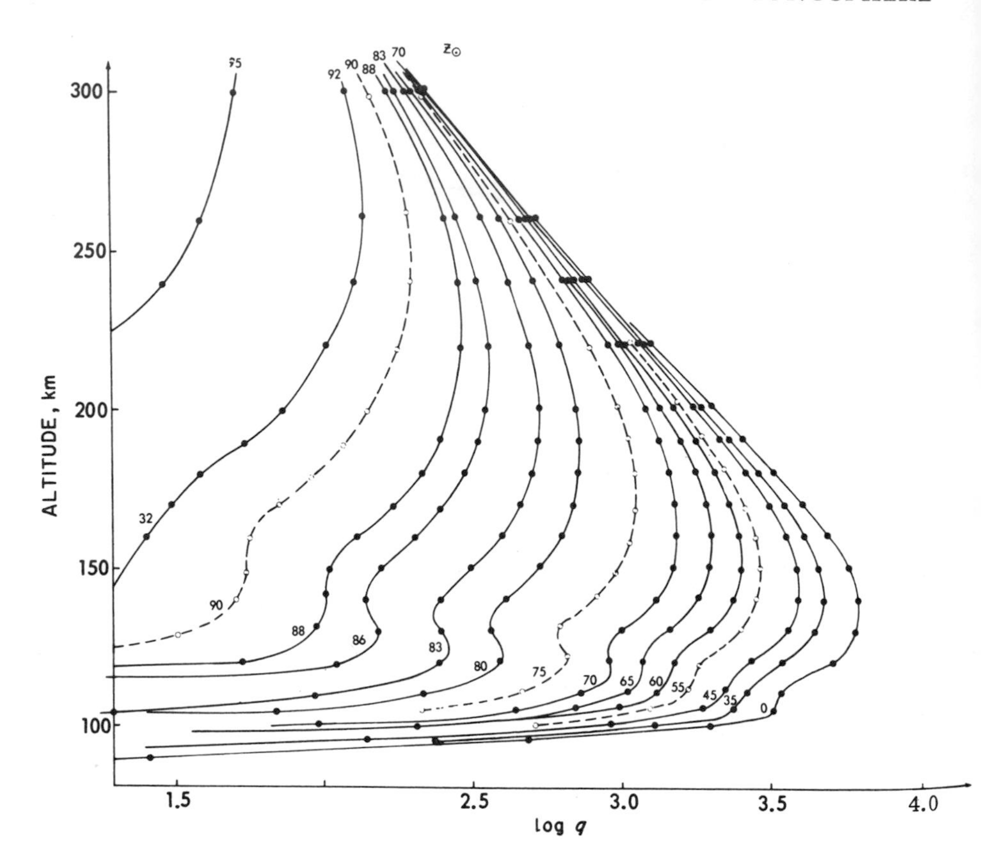

Fig. III-3. Variation of total photoionization rate with altitude at solar activity minimum for different solar zenith angles. From Ivanov-Kholodnyy (1966) as reported by Danilov (1970, p. 226) with permission of Plenum Press.

in Fig. III-4. It increases about a factor of 10 with latitude, and at the high latitudes is more important at minimum rather than at maximum solar activity. At even lower altitudes, ion pair production is further enhanced, passing through a maximum at about 13 km as shown in Fig. III-5. Here ion pair production can reach 40/cm³-sec, which is equivalent to the peak Lyman-α ionization rate of NO at 80 km for an overhead sun. Thus the ionosphere really extends all the way to the earth's surface. It is only the "electronosphere" that vanishes below 60 km.

In this chapter we shall examine the chemistry that is involved in the ionization cycle and see how it relates to the observations discussed above. In particular, there are four problems that immediately present them-

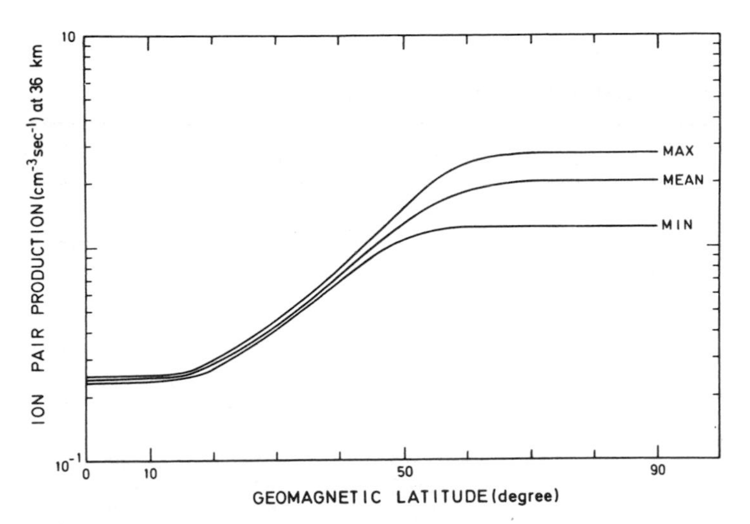

Fig. III-4. Ion pair production versus geomagnetic latitude by cosmic rays at different solar activities. Max is for minimum solar activity. From Brasseur and Nicolet (1973) with permission.

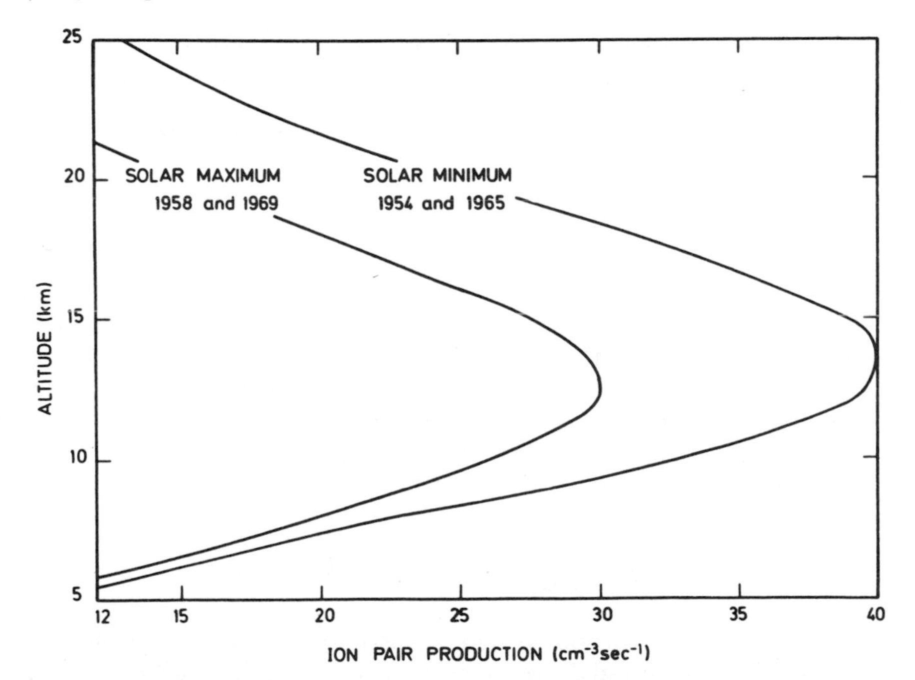

Fig. III-5. Ion pair production in the troposphere and stratosphere at high latitudes (Thule, Greenland) for minimum and maximum solar conditions. From Brasseur and Nicolet (1973) with permission.

TABLE III-5

Photoionization Rates for an Overhead Sun

Altitude (km)	Photoionization rate (particles/cm³-sec)			
	NO[a]	O_2[b]	O[b]	N_2[b]
65	0.1	—	—	—
70	2.0	—	—	—
75	8.1	—	—	—
80	15.8	0.009	—	0.02
85	17.3	0.3	—	0.3
90	21	20	0.03	23
95	20	300	7	100
100	18.5	1500	23	300
110	16.1	2800	330	1100
120	16.1	3000	1250	1600
130	14.5	2300	2000	1600
140	6.5	1900	2200	2000
150	—	1300	2300	2000
160	—	1000	2100	2000

[a] Computed from a photoionization coefficient of 5×10^{-7} sec^{-1} (Nicolet, 1970) at zero optical depth, the NO distribution given in Fig. I-11, and the O_2 photodissociation coefficient by Lyman-α given by Nicolet (1971).

[b] From Hinteregger *et al.* (1964).

selves and should be resolved in any discussion:

1. The minor role of N_2^+, despite the fact that it is produced at about the same rate as the other ions at 120–160 km.
2. Both $[O^+]$ and $[e]$ reach their maxima at ~250 km, even though their peak production rates are at ~140 km.
3. The appearance of NO^+ as a major ion, even though NO is a minor atmospheric constituent, and its production rate is negligible in the E and F regions.
4. NO is the major species that photoionizes in the D region, yet it is not the major positive ion in this region.

In examining the chemistry, we shall first examine the ion chemistry in the E and F1 regions, and then examine the neutral N atom and NO chemistry in the F1 region. After that we shall discuss the positive-ion cycle in the D region, followed by the negative-ion cycle.

TABLE III-6

Ion Production by Cosmic Rays (Mean Value; cm^{-3} sec^{-1})[a]

Altitude (km)	Geomagnetic latitude 0°	30°	50°	70°
85	3.4×10^{-4}	6.0×10^{-4}	1.7×10^{-3}	2.8×10^{-3}
80	8.2×10^{-4}	1.5×10^{-3}	4.2×10^{-3}	6.8×10^{-3}
75	1.8×10^{-3}	3.3×10^{-3}	9.3×10^{-3}	1.5×10^{-2}
70	3.8×10^{-3}	6.9×10^{-3}	2.0×10^{-2}	3.2×10^{-2}
65	7.4×10^{-3}	1.3×10^{-2}	3.8×10^{-2}	6.2×10^{-2}
60	1.4×10^{-2}	2.4×10^{-2}	7.0×10^{-2}	1.1×10^{-1}
55	2.5×10^{-2}	4.3×10^{-2}	1.2×10^{-1}	2.0×10^{-1}
50	4.4×10^{-2}	7.9×10^{-2}	2.2×10^{-1}	3.7×10^{-1}
45	8.1×10^{-2}	1.5×10^{-1}	4.1×10^{-1}	6.7×10^{-1}
40	1.5×10^{-1}	2.8×10^{-1}	7.9×10^{-1}	1.3
35	3.2×10^{-1}	5.6×10^{-1}	1.6	2.6

[a] From Brasseur and Nicolet (1973).

E AND F1 POSITIVE-ION CYCLE

In the E and F regions there are no negative ions. Therefore, to satisfy charge neutrality, the sum of the positive-ion concentrations must equal the electron concentration. As a result, the electron concentration can be computed from the overall ion production and loss rates.

Ion reactions are rapid and chemical processes always dominate over diffusion below 160 km. Also in this region the ion lifetimes during the day are sufficiently short so that the steady-state approximation is valid. Thus computations are relatively simple.

Electron Concentration

The overall ion production rate q is given as the sum of the individual photoionization rates:

$$q \equiv \phi\{N_2^+\}I\{N_2\} + \phi\{O^+\}I\{O\} + \phi\{O_2^+\}I\{O_2\} + \phi\{NO^+\}I\{NO\} \quad (1)$$

The ion loss reactions of importance are all the same type, that is, dissociative recombination,

$$N_2^+ + e \rightarrow 2N \quad (2)$$

$$O_2^+ + e \rightarrow 2O \quad (3)$$

$$NO^+ + e \rightarrow N + O \quad (4)$$

It should be noticed that O^+ does not have an analogous reaction and thus is very stable to electron removal. The rate coefficients for reactions (2) and (3) are not accurately known, but they must be similar to $k_{(4)}$, which is listed in Table III-7. With the assumption that $k_{(2)} \sim k_{(3)} \sim k_{(4)}$, we find that the steady-state approximation for total ions gives

$$q = \alpha[e]([N_2^+] + [O_2^+] + [NO^+]) \tag{5}$$

where the more common term α has been used in place of $k_{(2)}$ to connote the effective average positive-ion–electron annihilation coefficient.

Below 160 km, O^+ is a minor ion so that $[N_2^+] + [O_2^+] + [NO^+] \approx [e]$. Then $[e]$ can be computed readily as

$$[e] = (q/\alpha)^{1/2} \tag{6}$$

Because O^+ becomes more dominant with altitude, $[e]$ computed in this way is about 3% too small at 150 km and about 12% too small at 160 km.

Positive Ions

The reactions of importance pertaining to N_2^+ are

$$N_2 + h\nu \rightarrow N_2^+ \tag{7}$$

$$N_2^+ + O_2 \rightarrow N_2 + O_2^+ \tag{8}$$

$$N_2^+ + O \rightarrow NO^+ + N \tag{9a}$$

$$\rightarrow O^+ + N_2 \tag{9b}$$

$$N_2^+ + NO \rightarrow N_2 + NO^+ \tag{10}$$

$$N_2^+ + e \rightarrow 2N \tag{11}$$

The interaction of N_2^+ and O does not involve charge transfer but leads to rearrangement. From the above reactions, $[N_2^+]$ can be computed to be

$$[N_2^+] = \frac{\phi\{N_2^+\}I\{N_2\}}{k_{(8)}[O_2] + k_{(9)}[O] + k_{(10)}[NO] + (\alpha q)^{1/2}} \tag{12}$$

where $\alpha \equiv k_{(11)}$, $\phi\{N_2^+\}I\{N_2\}$ is the photoionization rate, and the right-hand side of Eq. (6) has been used for $[e]$. Values of the photoionization rate for an overhead sun for various altitudes have been calculated by Hinteregger *et al.* (1964) and Ivanov-Kholodnyy (1966). The two sets of calculations are listed in Table III-8 and they agree reasonably well between 120 and 160 km. There is some disagreement at lower altitudes; at 90 km, the discrepancy exceeds a factor of 2. The rate coefficients for reactions (8)–(10) are listed in Table III-7. The values of $[N_2^+]$ have been calculated for an overhead sun and are listed in Table III-4. For comparison purposes, observed typical daytime values are also shown in the table. As might be expected, the overhead sun values are somewhat greater than typical values except at 100 and 110 km. Since there is no known way to produce N_2^+ other than by direct photoionization at these altitudes, the observations must be atypically high at 100 and 110 km (i.e., correspond to a higher photoionization rate than used in the calculation).

The minor role of N_2^+ in the atmosphere (the first of the problems raised above) can be attributed to the fact that N_2 has a higher ionization potential than O_2, O, or NO, and thus readily transfers its charge via reactions (8)–(10).

For O^+, the chemistry is very simple:

$$O + h\nu \rightarrow O^+ \tag{13}$$

$$O^+ + O_2 \rightarrow O_2^+ + O \tag{14}$$

$$O^+ + N_2 \rightarrow NO^+ + N \tag{15}$$

The reaction of O^+ with NO is apparently very slow and need not be considered, even though it is exothermic. The rate coefficients are given in Table III-7.

The steady-state expression becomes

$$[O^+] = \frac{\phi\{O^+\}I\{O\} + k_{(9b)}[O][N_2^+]}{k_{(14)}[O_2] + k_{(15)}[N_2]} \tag{16}$$

From this expression the second problem raised above is answered, i.e., why $[O^+]$ continues to increase to $\sim$250 km when $\phi\{O^+\}I\{O\}$ goes through a maximum at 150 km (see Table III-5). The reason is that $[O_2]$ and $[N_2]$ drop much faster than $\phi\{O^+\}I\{O\}$ in the region from 150 to 250 km, so that $[O^+]$ continues to rise.

The photoionization rates for an overhead sun for O^+ are listed in Table III-5. The computed values for $[O^+]$ for an overhead sun using Eq. (16) are tabulated in Table III-1. Comparisons are made in the table with experimental values for typical daytime conditions as well as for other solar zenith angles. The trend is that $[O^+]$ rises rapidly with altitude and at any altitude falls rapidly as the solar zenith angle increases. The computed values are reasonably close to the typical daytime values at 120 and 160 km. They are somewhat larger between these altitudes, suggesting that other removal processes might play some role at these altitudes. However, at 110 km, and particularly at 100 km, the calculated values lie considerably below the observed values indicating either that some source term other than photoionization is principally responsible for the O^+ concentration at 100 km and below, or that the observed values are incorrect.

TABLE III-7

Reactions of the Ionosphere[a]

Reaction	Reaction number	k (cm^{3n}/sec)
	Reactions for E and F1 regions	
$NO^+ + e \rightarrow N + O$	(4)	$2.6 \times 10^{-3} T^{-1.5}$ [b]
$N_2^+ + O_2 \rightarrow N_2 + O_2^+$	(8)	6×10^{-11}
$N_2^+ + O \rightarrow NO^+ + N$	(9a)	1.3×10^{-10} [c]
$\rightarrow O^+ + N_2$	(9b)	1.0×10^{-11} [c]
$N_2^+ + NO \rightarrow N_2 + NO^+$	(10)	4×10^{-10}
$O^+ + O_2 \rightarrow O_2^+ + O$	(14)	2.0×10^{-11}
$O^+ + N_2 \rightarrow NO^+ + N$	(15)	1.2×10^{-12}
$O^+ + NO \rightarrow NO^+ + O$	—	$<1 \times 10^{-12}$
$O_2^+ + NO \rightarrow O_2 + NO^+$	(18)	7×10^{-10}
$O_2^+ + N \rightarrow NO^+ + O$	(19)	1.8×10^{-10} [c]
$N + NO \rightarrow N_2 + O$	(25)	2.7×10^{-11}
$N(^4S) + O_2 \rightarrow NO + O(^3P)$	(26)	$1.1 \times 10^{-14} \exp(-3150/T)$
$NO_2^+ + NO_2^- \rightarrow 2NO_2$	(33)	2.2×10^{-7} [d]
	Positive-ion neutral reactions in the D region	
$NO^+ + H_2O + M \rightarrow NO(H_2O)^+ + M$	(34)	1.6×10^{-28} [e]
$NO(H_2O)^+ + H_2O + M \rightarrow NO(H_2O)_2^+ + M$	(35)	1.3×10^{-27} [e]
$NO(H_2O)_2^+ + M \rightarrow NO(H_2O)^+ + H_2O + M$	(−35)	$5.2 \exp(-18500/RT) \times 10^{-1}$ [f]
$NO(H_2O)_2^+ + H_2O + M \rightarrow NO(H_2O)_3^+ + M$	(36)	1.9×10^{-27} [e]
$NO(H_2O)_3^+ + M \rightarrow NO(H_2O)_2^+ + H_2O + M$	(−36)	$8.0 \exp(-14700/RT) \times 10^{-2}$ [f]
$NO(H_2O)_3^+ + H_2O \rightarrow H(H_2O)_3^+ + HNO_2$	(37)	7.0×10^{-11} [g]
$NO^+ + N_2 + M \rightarrow NO(N_2)^+ + M$	(39)	3.5×10^{-31} [g]

Reaction	No.	Rate coefficient
$NO^+ + CO_2 + M \rightarrow NO(CO_2)^+ + M$	(40)	2.5×10^{-29} [g]
$NO(N_2)^+ + CO_2 \rightarrow NO(CO_2)^+ + N_2$	(41)	1.0×10^{-9} [g]
$NO(N_2)^+ + H_2O \rightarrow NO(H_2O)^+ + N_2$	(42)	1.0×10^{-9} [g]
$NO(CO_2)^+ + H_2O \rightarrow NO(H_2O)^+ + CO_2$	(43)	1.0×10^{-9} [g]
$NO(H_2O)^+ + H \rightarrow NO + H(H_2O)^+$	(44)	$<7 \times 10^{-12}$ [c,h]
Negative-species neutral reactions		
$e + O_2 + O_2 \rightarrow O_2^- + O_2$	(47)	$1.4 \times 10^{-29} (300/T) \exp\{-600/T\}$ [g]
$e + O_2 + N_2 \rightarrow O_2^- + N_2$	(48)	1.0×10^{-31} [g]
$O_2^- + O \rightarrow O_3 + e$	(49)	3.3×10^{-10}
$O_2^- + O_2(^1\Delta) \rightarrow 2O_2 + e$	(50)	2.0×10^{-10} [g]
$O_2^- + O_3 \rightarrow O_2 + O_3^-$	(51)	3.0×10^{-10}
$O_2^- + O_2 + M \rightarrow O_4^- + M$	(52)	1.0×10^{-30} [g]
$O_2^- + CO_2 + O_2 \rightarrow CO_4^- + O_2$	(53)	2.0×10^{-29} [g]
$O_3^- + CO_2 \rightarrow CO_3^- + O_2$	(54)	5.5×10^{-10}
$O_3^- + NO \rightarrow NO_2^- + O_2$	(55)	1.0×10^{-11} [c]
$O_3^- + O \rightarrow O_2^- + O_2$	(56)	1.4×10^{-10} [g]
$O_4^- + CO_2 \rightarrow CO_4^- + O_2$	(57)	4.3×10^{-10}
$O_4^- + O \rightarrow O_3^- + O_2$	(58a)	4.0×10^{-10}
$O_4^- + NO \rightarrow O_2 + NO_3^- \xrightarrow{NO} NO_2^- + NO_2$	(59)	2.5×10^{-10}
$CO_3^- + O \rightarrow O_2^- + CO_2$	(60)	8.0×10^{-11}
$CO_3^- + NO \rightarrow NO_2^- + CO_2$	(61)	1.8×10^{-11}
$CO_4^- + NO \rightarrow NO_3^- + CO_2$	(62)	4.8×10^{-11}
$CO_4^- + O \rightarrow CO_3^- + O_2$	(63a)	1.5×10^{-10}
$CO_4^- + O_3 \rightarrow O_3^- + CO_2 + O_2$	(64)	1.3×10^{-10} [c]
$NO_2^- + O_3 \rightarrow NO_3^- + O_2$	(65)	1.8×10^{-11}
$NO_2^- + H \rightarrow OH^- + NO$	(66)	4.0×10^{-10} [c]
$OH^- + O \rightarrow HO_2 + e$	(67)	2.0×10^{-10}
$OH^- + H \rightarrow H_2O + e$	(68)	1.8×10^{-9}
$OH^- + CO_2 + M \rightarrow HCO_3^- + M$	(69)	1.0×10^{-29} [g]

TABLE III-7—Continued

Reaction	Reaction number	k (cm^{3n}/sec)
	Photodetachment reactions	
$O_2^- + h\nu \rightarrow O_2 + e$	(70)	0.33 [g]
$O_3^- + h\nu \rightarrow O_3 + e$	(71)	1.3 [g]
$NO_2^- + h\nu \rightarrow NO_2 + e$	(72)	0.04 [g]
$NO_3^- + h\nu \rightarrow NO_3 + e$	(73)	1.5×10^{-3} [g]

[a] All rate coefficients from Garvin and Hampson (1974) unless otherwise specified.

[b] From Heicklen and Cohen (1968) for the temperature range 196–7500°K. This rate coefficient is taken to be the effective positive–ion electron annihilation coefficient α_e.

[c] From Ferguson (1974).

[d] From Mahan and Person (1964) at 300°K.

[e] From Howard *et al.* (1971).

[f] From room temperature rate coefficient (Howard *et al.*, 1971) and assuming that ΔH is the same as for the corresponding protonated reaction: $H(H_2O)_n^+ + H_2O + M \rightleftarrows H(H_2O)_{n+1}^+ + M$. The latter values of ΔH are given by Kebarle *et al.* (1967).

[g] From Rowe *et al.* (1974).

[h] For calculational purposes, we adopt 2×10^{-12}.

TABLE III-8

Comparison of Total Photoionization Rates for an Overhead Sun

Altitude (km)	$q\{0°\}$[a] $(\text{cm}^3\text{-sec})^{-1}$	$q\{0°\}$[b] $(\text{cm}^3\text{-sec})^{-1}$
90	46	108
95	321	630
100	1840	2150
110	4240	3300
120	4855	4700
130	5900	6600
140	6100	6600
150	5600	5300
160	5100	4500

[a] From Hinteregger *et al.* (1964).
[b] From Ivanov-Kholodnyy (1966).

For O_2^+, the only exothermic reactions are charge transfer to NO and electron capture. Thus the pertinent reactions are (8), (14), and

$$O_2 + h\nu \rightarrow O_2^+ \tag{17}$$

$$O_2^+ + NO \rightarrow O_2 + NO^+ \tag{18}$$

$$O_2^+ + N \rightarrow NO^+ + O \tag{19}$$

$$O_2^+ + e \rightarrow 2O \tag{20}$$

The steady-state expression for O_2^+ becomes

$$[O_2^+] = \frac{\phi\{O_2^+\}I\{O_2\} + (k_{(8)}[N_2^+] + k_{(14)}[O^+])[O_2]}{k_{(18)}[NO] + k_{(19)}[N] + (\alpha q)^{1/2}} \tag{21}$$

$[O_2^+]$ can be calculated for an overhead sun using the photoionization rates, $\phi\{O_2^+\}I\{O_2\}$, listed in Table III-5, the rate coefficients in Table III-7 (with $\alpha \equiv k_{(4)}$), the anticipated values of $[N]$, and the values of $[N_2^+]$ and $[O^+]$ in Tables III-4 and III-1, respectively. These computed rates are listed in Table III-3, along with observations at various solar zenith angles and at night.

At any altitude, $[O_2^+]$ decreases as the solar zenith angle increases, as expected. The only exceptions are at 130 and 140 km, where the calculated values are slightly low. The computations are approximate, because $[NO]$ is not known accurately, so that the discrepancy may not be meaningful.

TABLE III-9

Computed Positive-Ion–Electron Recombination Coefficients at Various Solar Zenith Angles

Altitude (km)	T (°K)	$q\{90°\}$[a] $(cm^3\text{-}sec)^{-1}$	$q\{75°\}$[a] $(cm^3\text{-}sec)^{-1}$	$q\{54°\}$[a] $(cm^3\text{-}sec)^{-1}$	$[e\{90°\}]$[b] (particles/ cm^3)	$[e\{75°\}]$[b] (particles/ cm^3)
100	210	—	—	710	—	—
110	265	—	500	1600	—	39,000
120	355	—	630	1850	—	63,000
130	420	35	590	2400	42,300	66,300
140	480	50	850	2800	40,000	60,000
150	530	53	920	2800	40,800	71,800
160	575	56	1000	2650	43,300	88,300

[a] From Ivanov-Kholodnyy (1966).
[b] Experimental results as reported by Danilov (1970, p. 109).
[c] Computed from $\alpha\{x\} = q\{x\}/[e\{x\}]^2$.

Another feature of interest is that $[O_2^+]$ goes through a maximum with altitude, the maximum occurring at about 140, 170, >200 km for typical values, 75, and 90°, respectively. The reason for the maximum and its shifting height with solar zenith angle comes from the changes in the terms in Eq. (21). Both the numerator and denominator of the right-hand side of Eq. (16) increase with height, but at different rates, leading to the observed behavior.

At night, both $[N_2^+]$ and $[O^+]$ fall essentially to zero, so that Eq. (21) would predict that $[O_2^+]$ should also fall to zero. However, observations indicate the persistence of O_2^+ at 130 km and above, although at much lower concentration, at night. The reason for the persistence of O_2^+ at night is not known. Presumably there is some minor production term that exists at night.

NO, having the lowest ionization potential of the atmospheric gases, is formed by charge exchange, but removed only by reaction with electrons in the E and F layers. This is the reason that NO^+ is such an important ion even though NO is a relatively minor atmospheric constituent. The reactions of pertinence are (4), (9), (10), (18), (19), and

$$NO + h\nu \rightarrow NO^+ + e \tag{22a}$$

$[e\{54°\}]^b$ (particles/ cm^3)	$\alpha\{90°\}^c$ (cm^3/sec)	$\alpha\{75°\}^c$ (cm^3/sec)	$\alpha\{54°\}^c$ (cm^3/sec)	$k_{(4)}$ (cm^3/sec)
1.6×10^5	—	—	2.8×10^{-8}	8.5×10^{-7}
1.7×10^5	—	3.3×10^{-7}	5.5×10^{-8}	6.0×10^{-7}
1.7×10^5	—	1.57×10^{-7}	6.4×10^{-8}	3.9×10^{-7}
1.8×10^5	1.95×10^{-8}	1.34×10^{-7}	7.4×10^{-8}	3.0×10^{-7}
2.1×10^5	3.1×10^{-8}	2.4×10^{-7}	6.6×10^{-8}	2.5×10^{-7}
2.4×10^5	3.2×10^{-8}	1.8×10^{-7}	4.8×10^{-8}	2.1×10^{-7}
3×10^5	3.0×10^{-8}	1.28×10^{-7}	3.0×10^{-8}	1.89×10^{-7}

The steady-state expression is

$$[NO^+] = \frac{\phi\{NO^+\}I\{NO\} + (k_{(9)}[O] + k_{(10)}[NO])[N_2^+] + (k_{(18)}[NO] + k_{(19)}[N])[O_2^+]}{(\alpha q)^{1/2}} \tag{23}$$

where again $k_{(4)}$ has been replaced by its equivalent α for generality. The rate coefficients are listed in Table III-7; the photoionization rates for an overhead sun in Table III-5; and $[N_2^+]$ and $[O_2^+]$ in Tables III-4 and III-3, respectively.

The computed values of $[NO^+]$ for 0° solar zenith angle are in Table III-2, as well as the observed values at various solar zenith angles. The NO^+ concentrations tend to increase with altitude and fall as the solar zenith angle increases, as expected from Eq. (23). The dramatic exception is at 100–130 km where $[NO^+]$ is largest at 75° solar zenith angle and is not much reduced even at 90°. The reason for this anomaly is not known. Even when the computed values for $[N_2^+]$ are used, the calculated values of $[NO^+]$ are less than the observed values at 150 and 160 km. It appears that we still have more to learn about the F1 region.

Significant NO^+ concentrations have been observed at night. At night, Eq. (23) is not applicable (both numerator and denominator tend to zero). As in the case of O_2^+, since NO^+ is being removed by reaction (4), there must be some other production term to compensate for the higher than expected observed concentrations.

Overall Production and Removal

If we now return to Eq. (6), we see that the electron concentration is a direct measure of the ratio between the overall ionization rate q and the effective recombination coefficient α. Equation (6) can then be used to compute effective values for α

$$\alpha = q/[e]^2 \tag{24}$$

The values for q for an overhead sun according to Hinteregger *et al.* (1964) are listed in Table III-8. Ivanov-Kholodnyy (1966) has compiled q values for many solar zenith angles, and these are shown graphically in Fig. III-3. For zero solar zenith angle the values obtained from this plot are also listed in Table III-8. The agreement between the two sets of calculations is satisfactory above 100 km. However, between 90 and 100 km, the values of Ivanov-Kholodnyy are noticeably higher, the discrepancy being a factor of 2.4 at 90 km. Presumably the values of Hinteregger are too low, giving rise to the low calculated ion concentrations at these altitudes, as discussed above.

From Fig. III-3, it is possible to obtain values for q at any zenith angle. This has been done for angles of 54, 75, and 90°, where experimental observations exist for electron densities. These values are listed in Table III-9, along with the values of α computed from them using Eq. (24). For comparison the dissociative recombination coefficient for NO, $k_{(4)}$, is also listed.

The values computed for $\alpha\{75°\}$ compare well with those for $k_{(4)}$, they being slightly on the lower side, as expected since O^+ does not undergo dissociative recombination. However, at 90 and 54° the values are very much lower, in fact, too low to be acceptable. If the observed electron densities are correct, then the photoionization rates during the observations must have been measurably higher than given in Fig. III-3.

N AND NO CHEMISTRY

Now that the ionic chemistry of the E and F regions has been introduced, we can complete the discussion of neutral chemistry neglected in the previous chapter. The reactions of concern in producing or removing N and NO are reaction (19) and

$$NO + h\nu \rightarrow NO^+ + e \quad (22a)$$

$$\rightarrow N(^4S) + O(^3P) \quad (22b)$$

$$N + NO \rightarrow N_2 + O \quad (25)$$

$$N(^4S) + O_2 \rightarrow NO + O(^3P) \quad (26)$$

$$NO^+ + e \rightarrow O + (1 - \gamma)N(^4S) + \gamma N(^2D) \quad (4)$$

$$N_2^+ + O(^3P) \rightarrow NO^+ + (1 - \delta)N(^4S) + \delta N(^2D) \quad (9a)$$

$$N_2^+ + e \rightarrow (2 - \beta)N(^4S) + \beta N(^2D) \quad (11)$$

$$N(^2D) + O_2 \rightarrow NO + O(^3P) \quad (27)$$

where allowance has been made for the possibility that excited $N(^2D)$ atoms are produced in reactions (4), (9a), and (11). $N(^2D)$ is known to be produced in the thermosphere since its emission has been observed. Reactions (4), (9a), and (11) are the only possibilities for its production.

In addition to the chemical reactions, diffusion of both NO and N should be considered. Fortunately a number of the above reactions are unimportant in the N-NO cycle in the thermosphere. These include reactions (22a), (11), and (26), and the diffusion of N atoms.

If we assume the steady-state hypothesis for odd N (i.e., $[NO] + [N(^4S)] + [N(^2D)]$), then

$$[N] = \frac{k_{(9a)}[N_2^+][O(^3P)] + k_{(4)}[NO^+][e] + D[NO](\xi_{NO} - \xi_n)(\dot{\xi}_{NO} + \xi_D)}{2k_{(25)}[NO] + k_{(19)}[O_2^+]} \quad (28)$$

$$\tau\{\text{odd N}\} = ([N] + [NO])/(2k_{(25)}[NO] + k_{(19)}[O_2^+])[N] \quad (29)$$

The values for the ion concentrations for overhead sun conditions can be obtained from Fig. III-1. The rate coefficients are listed in Table III-7. Values for $[O(^3P)]$ are in Table I-1. The diffusion parameters have been estimated and are listed in Table III-10 along with the computed values for $[N]$ and $\tau\{\text{odd N}\}$. The computed values of $[N]$ range from 1×10^7 to $4 \times 10^7/\text{cm}^3$ between 100 and 160 km. Experimental observations suggest $[N] = 10^8 - 10^{11}/\text{cm}^3$ at these altitudes (Balabonova *et al.*, 1974 and references therein). The calculated values are only approximate, since the steady-state assumption is not entirely justified, as can be seen from the values for $\tau\{\text{odd N}\}$. These values go from about 5×10^3 to 1.8×10^4 sec, which is comparable to the time of the diurnal cycle.

TABLE III-10

Calculation of N Atom Concentrations for an Overhead Sun

Altitude (km)	Temp (°K)	D (cm^2/sec)	[NO] ($particles/cm^3$)	$10^7\xi_D$ (cm^{-1})	$10^7\xi_n$ (cm^{-1})	$10^7\xi_{NO}$ (cm^{-1})	[N][a] ($particles/cm^3$)	τ {odd N}[a] (sec^{-1})	τ {NO}[a] (sec^{-1})	[NO][a] ($particles/cm^3$)
100	210	2×10^6	5.0×10^7	−9	16	1.2	8.3×10^5	2.25×10^4	3.07×10^4	4.6×10^7
110	265	7×10^6	4.0×10^7	−9	16	1.2	2.10×10^6	9.26×10^3	1.50×10^4	4.5×10^7
120	355	1.2×10^7	3.2×10^7	−9	16	1.2	1.72×10^6	1.13×10^4	1.78×10^4	4.4×10^7
130	420	3.5×10^7	2.9×10^7	−10	9	5	1.39×10^6	1.39×10^4	2.10×10^4	3.7×10^7
140	480	7.8×10^7	1.3×10^7	−7	7	8.6	3.65×10^6	6.40×10^3	9.22×10^3	1.8×10^7
150	530	1.4×10^8	5×10^6	−7	5.8	6	9.20×10^6	4.94×10^3	3.87×10^3	6.5×10^6
160	575	3.3×10^8	5×10^6	−7	~5	1	1.33×10^7	4.93×10^3	2.71×10^3	7.7×10^6

[a] Calculated.

Further information can be obtained by considering the steady-state expression for NO

$$R\{(22b)\} + R\{(25)\} = R\{(26)\} + \delta R\{(9a)\} + \gamma R\{(4)\} + R_D\{NO\} \quad (30)$$

$$\tau\{NO\} = (J\{(22b)\} + k_{(25)}[N])^{-1} \quad (31)$$

where $R\{x\}$ is the rate of reaction x. $J\{(22b)\}$ is 1.0×10^{-5} sec^{-1} in the thermosphere. The values for $\tau\{NO\}$ have been computed and are listed in Table III-10. They are comparable to those for $\tau\{odd\ N\}$ indicating that Eq. (30) is only approximately correct.

The value of γ is 0.77 ± 0.05 (Kley *et al.*, 1975). If the likelihood of $N(^4S)$ or $N(^2D)$ formation in reaction (4) is purely statistical, then γ should be 0.714 [^{4}S has 4 spin states, ^{2}D has five angular momentum states each with two spin states for a statistical weight of 10; the possibility of producing $N(^4S) + O(^1D)$ is not considered because it is spin forbidden]. Since the observed value for γ is close to that predicted on statistical grounds, we calculate δ on statistical grounds and obtain the same value as for γ, i.e., $\delta = 0.714$. With $\gamma = 0.77$ and $\delta = 0.714$, [NO] can be calculated and these values are listed in Table III-10. They are in good agreement with the accepted values.

D REGION

Positive Ions

The major source of ionization in the D region is photoionization of NO by the Lyman-α 1215.7 Å line, which penetrates to 65 km via reaction (22a). The NO^+ ions may undergo reactions to form other positive ions, but in order to conserve charge, the total concentration of positive ions must equal the total concentration of negatively charged species. Thus from steady-state considerations on the total charged species

$$q = [M^+](\alpha_e[e] + \alpha_{neg}([M^+] - [e])) \quad (32)$$

where $[M^+]$ is the total concentration of positive ions, q now becomes the rate of photoionization of NO, α_e is the effective positive-ion–electron annihilation coefficient, and α_{neg} is the effective positive-ion–negative-ion annihilation coefficient. Photoionization rates for an overhead sun are listed in Table III-5. The average observed electron concentrations (from Fig. III-1) are listed in Table III-11. For α_e we take $k_{(4)}$ (see Table III-7). The rate coefficient for the reaction

$$NO_2^+ + NO_2^- \rightarrow 2NO_2 \quad (33)$$

TABLE III-11

Computed and Observed Electron Concentrations in the D Region for an Overhead Sun

Altitude (km)	Temperature (°K)	[e][a] (particles/cm³)	[M⁺][b] (particles/cm³)	[M⁺][c] (particles/cm³)
65	232	100	496	900
70	211	140	2,230	1,900
75	194	260	4,220	2,700
80	177	900	5,170	4,100
85	160	2,500	4,170	12,000

[a] Average observed values (From Fig. III-1).
[b] Computed from Eq. (32).
[c] Total positive ion concentration found by Narcisi and reported in Danilov (1970, p. 19).

has been measured by Mahan and Person (1964) to be 2.1×10^{-7} cm³/sec at room temperature. If we consider it to have the same temperature dependence as $k_{(4)}$, then $k_{(33)} = 1.14 \times 10^{-3} T^{-1.5}$ cm³/sec. We take this value for α_{neg} and compute $[M^+]$. The computed and observed values of $[M^+]$ are listed in Table III-11. Except at 85 km, they are in reasonable agreement. They are much higher than the electron concentrations and thus imply that most of the negative charge in the D region is present as negative ions.

Narcisi's experiments gave ions of various m/e ratios and his results are shown in Fig. III-2. It can be seen that in the region 65–85 km the major ions are at $m/e = 30$, 19, 37, and >45, corresponding to NO^+, $H(H_2O)^+$, $H(H_2O)_2{}^+$, and heavier ions. The primary ion formed, NO^+, comprises only a minor fraction of all the positive ions.

The positive-ion reactions known to occur in the D region include

$$NO^+ + H_2O + M \rightarrow NO(H_2O)^+ + M \tag{34}$$

$$NO(H_2O)^+ + H_2O + M \rightleftarrows NO(H_2O)_2{}^+ + M \tag{35,−35}$$

$$NO(H_2O)_2{}^+ + H_2O + M \rightleftarrows NO(H_2O)_3{}^+ + M \tag{36,−36}$$

$$NO(H_2O)_3{}^+ + H_2O \rightarrow H(H_2O)_3{}^+ + HNO_2 \tag{37}$$

All the hydrated proton species can then be formed through the condensation reactions

$$H(H_2O)_n{}^+ + H_2O \rightleftarrows H(H_2O)_{n+1}^+ \tag{38,−38}$$

In addition to these reactions, rate coefficients have been measured for

$$NO^+ + N_2 + M \rightarrow NO(N_2)^+ + M \tag{39}$$

$$NO^+ + CO_2 + M \rightarrow NO(CO_2)^+ + M \tag{40}$$

Presumably these reactions can be followed by

$$NO(N_2)^+ + CO_2 \rightarrow NO(CO_2)^+ + N_2 \tag{41}$$

$$NO(N_2)^+ + H_2O \rightarrow NO(H_2O)^+ + N_2 \tag{42}$$

$$NO(CO_2)^+ + H_2O \rightarrow NO(H_2O)^+ + CO_2 \tag{43}$$

Also, we might expect the following reaction to occur to some extent, though its rate coefficient is $<7 \times 10^{-12}$ cm^3/sec (Ferguson, 1974).

$$NO(H_2O)^+ + H \rightarrow H(H_2O)^+ + NO \tag{44}$$

Rate coefficients for reactions (34)–(37) and (39)–(44) are listed in Table III-7.

In order to calculate the concentrations of the ions in the D region, we assume that all the positive ions react with negative species with the same effective annihilation rate coefficient α_{eff}. The value of α_{eff} can be deduced from

$$\alpha_{eff} = q/[M^+]^2 \tag{45}$$

Values for q are listed in Table III-5 for NO photoionization. For $[M^+]$ we use the calculated values listed in Table III-11. Then α_{eff} can be computed and these values, along with the concentrations of H and H_2O are listed in Table III-12. The overall particle densities are given in Table III-1. Furthermore, the steady-state applies to each ion and its loss or production by diffusion is unimportant. Thus, the ion concentrations can be computed based on a mechanism consisting of reactions (22a), (34)–(44), and the positive-ion–negative-species annihilation reaction.

In any calculation the reverse of reactions (39) and (40) must also be included. These should be rapid reactions, because since neither N_2 nor CO_2 have dipole moments, the heats of association for $NO(N_2)^+$ and $NO(CO_2)^+$ should be small. Unfortunately these rates are unknown. However, if they are fast enough, they will negate the forward reactions. Thus in our computation we neglect reactions (39) and (40) [and as a consequence, reactions (41)–(43)].

The results of the computation are shown in Table III-12. Qualitatively the results are satisfactory, i.e., NO^+ is the dominant ion near the top of the D region, and the hydrated proton is the dominant ion near the bottom of the D region. The values computed for $[NO^+]$ are in reasonable agreement with the observed values at 65 and 70 km. However, at higher alti-

TABLE III-12

Computed Positive-Ion Concentrations in the D Region for an Overhead Sun

Altitude (km)	Temperature (°K)	[H][a] (particles/cm³)	[H_2O][a] (particles/cm³)	α_{eff}[b] (cm³/sec)	[NO^+][c] (particles/cm³)
65	232	6×10^6	1.95×10^{10}	4.1×10^{-7}	8.2 (<100)
70	211	1×10^7	1.0×10^{10}	4.0×10^{-7}	489 (250)
75	194	1×10^8	3.5×10^9	4.5×10^{-7}	3302 (300)
80	177	5×10^8	1.0×10^9	6.4×10^{-7}	5084 (420)
85	160	3×10^8	4.0×10^8	9.1×10^{-7}	4150 (1300)

[a] From Bowman *et al.* (1970).

[b] Assuming all positive ions and all negative species annihilate each other at the same rate.

[c] Values in parenthesis are observed values in Fig. III-2.

tudes, the computed values are too large. The only apparent explanation for this discrepancy is that reaction (39) is not negligible. In spite of the fact that the atmosphere becomes more rarefied as the elevation increases, which tends to shift the equilibrium in reaction (39) to the left, reaction (39) becomes more important. Presumably this is due to the falling temperature and positive temperature coefficient of reaction (−39).

The hydrated proton has been included in the calculations as one ion. Actually a more detailed calculation can be made in which each $H(H_2O)_n^+$ ion is treated separately using the rate coefficient data for the set of reactions (38), (−38) [see Kebarle *et al.* (1967), and Rowe *et al.* (1974) for rate coefficient and equilibrium constant data]. If this is done, the values for [$H(H_2O)^+$] are too low compared to the concentrations of the more protonated ions (see Fig. III-2). The reason for this discrepancy is not known, and the implication is the presence of another, as yet unmeasured, reaction to produce $H(H_2O)^+$ efficiently.

It is interesting to see what the daily ion cycle looks like. We have calculated this cycle for a day with an overhead noonday sun by using the above mechanism. However, now the steady-state assumption does not apply so a step-by-step (using 5–100-sec steps) calculation was done. An altitude of 70 km was chosen since the noonday steady-state calculation indicated that the mechanism fit best at this altitude. The photoionization rate was calculated using Eq. II-18 with a photoionization rate for an overhead sun, q_0, of 2.0 (cm³-sec)$^{-1}$ and $\zeta_{70}' = 9.29$. The results are shown in Fig. III-6. The ions appear late in the morning and reach their maximum values at noon. The concentrations are not symmetric about the noonday

$[NO(H_2O)^+]$ (particles/cm³)	$[NO(H_2O)_2^+]$ (particles/cm³)	$[NO(H_2O)_3^+]$ (particles/cm³)	$[H(H_2O)_n^+]$ (particles/cm³)	$[M^+]$ (particles/cm³)
1.2	2.9	0.072	483	496
58	44	2.1	1630	2224
274	145	3.7	507	4233
70	9.3	0.010	23	5187
11	0.28	0.0013	1.8	4163

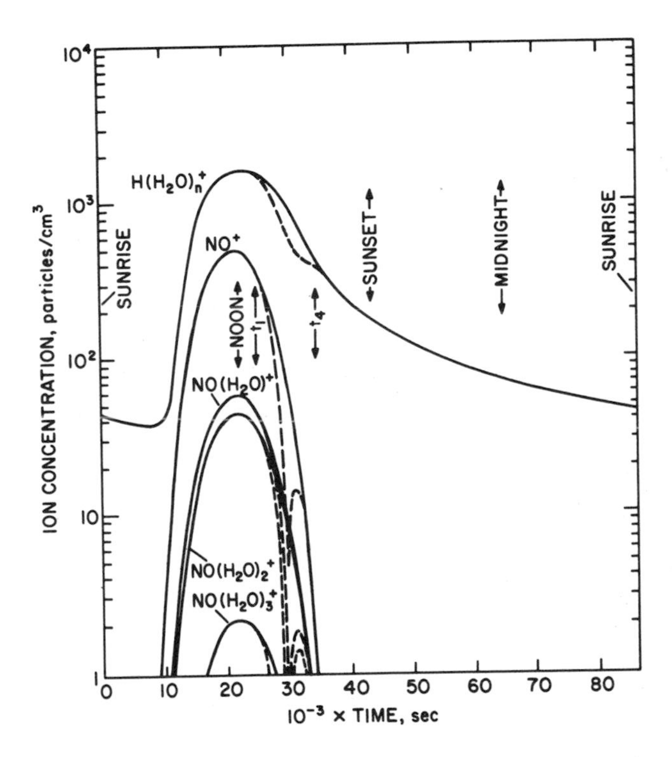

Fig. III-6. Calculated positive-ion concentrations at 70 km for a day in which the noonday sun is overhead ($q_0 = 2.0$ photons/cm³-sec; $\zeta_{70}' = 9.29$): ——, ordinary day; – – –, total eclipse of Nov. 12, 1966 ($t_1 = 24{,}600$, $t_2 = 29{,}376$, $t_3 = 29{,}500$, $t_4 = 34{,}440$ sec).

value. In fact the ions which are removed only by annihilation with negative species, i.e., $H(H_2O)_n^+$, do not vanish at sunset, but persist throughout the night. Even at sunrise they have maintained about 3% of their peak noonday concentration.

Let us see what happens to the positive ions during a total eclipse of the sun. During an eclipse, the photoionization rate $q\{t\}$ has the time dependence (Baker, 1969)

$$q\{t\} = q_0\{t\}(2/\pi)[\sin^{-1}\mathfrak{D} + \mathfrak{D}(1 - \mathfrak{D}^2)^{1/2}] \tag{46}$$

where $q_0\{t\}$ is the photoionization rate at time t for no eclipse,

$$\mathfrak{D} \equiv (t_2 - t)/(t_2 - t_1) \qquad \text{before totality}$$

$$\mathfrak{D} \equiv (t - t_3)/(t_4 - t_3) \qquad \text{after totality}$$

and t_1 is the time at the onset of the eclipse, t_2 the time at the onset of totality, t_3 the time at the end of totality, t_4 the time at the end of the eclipse.

For the total eclipse of November 12, 1966, over Brazil, the local times were (Baker, 1969)

$$t_1 = 12{:}50, \qquad t_2 = 14{:}09, \qquad t_3 = 14{:}11, \qquad t_4 = 15{:}34$$

In order to compare the results with an ordinary day for an overhead noonday sun at 70 km, we consider that the noonday sun was overhead on November 12, 1966, in Brazil. Actually this is not correct, since the noonday solar zenith angle was about 15–20°. Thus our calculated concentrations are about 3% too high for the actual eclipse.

The results are depicted in Fig. III-6. As the eclipse starts, the NO^+ ions fall rapidly and follow the photoionization rate. During totality they continue to fall rapidly and this fall continues to 29,700 sec (200 sec past the end of totality). Then these ion concentrations continue to rise, reaching their unperturbed values at the end of the eclipse. On the other hand, the hydrated protons, which are the terminating ions, show a considerable time lag. The fall in concentration is not nearly as dramatic as for NO^+ and its hydrated ions, being about 30%. Because of the time lag, the hydrated protons do not go through a concentration minimum. Furthermore the unperturbed concentration is not restored until about 37,000 sec, i.e., more than half an hour after the end of the eclipse. The rapid fall of NO^+ but the relatively minor drop in the hydrated proton concentrations agrees with experimental observations (Narcisi *et al.*, 1972).

Negative Ions

It is now well known that the electron concentrations are considerably less than the positive-ion concentrations in the D region. Therefore, the major negative species must be ions, and the difference between the positive-ion and electron concentrations is taken to be the total negative-ion concentration. Measurements have been made of the negative-ion composition above 70 km at night and during an eclipse (Narcisi *et al.*, 1971, 1972; Arnold and Krankowsky, 1971). The ions tentatively identified are O^-, O_2^-, Cl^-, NO_2^-, NO_3^-, CO_3^-, HCO_3^-, CO_4^-, $NO_3(H_2O)^-$, $CO_3(H_2O)^-$, and the higher hydrates of NO_3^-. Results are not quantitative and there is disagreement among the various flights as to which ions are dominant. No full-sun daytime measurements are available. In addition to the ions observed, O_3^- must also be present as a minor, but important ion.

We will present here a simple negative-ion scheme, based on the model of Rowe *et al.* (1974), which should explain the major features of negative-ion chemistry in the D region. For simplicity we ignore the hydrated ions, i.e., we assume the hydrated ions react indistinguishably from the non-hydrated ions. Initiation of the ion cycle is by photoionization of NO, reaction (22a), and termination is by positive-ion–negative-species annihilation. All positive-ion–electron reactions are assumed to proceed with a rate coefficient equal to that for reaction (4), i.e., $\alpha_e = 2.6 \times 10^{-3}T^{-1.5}$ cm^3/sec; all positive-ion–negative-ion reactions are assumed to have a rate coefficient $\alpha_{neg} = 1.14 \times 10^{-3}T^{-1.5}$ cm^3/sec. The ion-neutral reactions of importance are (Garvin and Hampson, 1973; Rowe *et al.*, 1974)

Electron reactions:

$$e + O_2 + O_2 \rightarrow O_2^- + O_2 \quad (47)$$

$$e + O_2 + N_2 \rightarrow O_2^- + N_2 \quad (48)$$

O_2^- reactions:

$$O_2^- + O \rightarrow O_3 + e \quad (49)$$

$$O_2^- + O_2(^1\Delta) \rightarrow 2O_2 + e \quad (50)$$

$$O_2^- + O_3 \rightarrow O_2 + O_3^- \quad (51)$$

$$O_2^- + O_2 + M \rightarrow O_4^- + M \quad (52)$$

$$O_2^- + CO_2 + M \rightarrow CO_4^- + M \quad (53)$$

O_3^- reactions:

$$O_3^- + CO_2 \rightarrow CO_3^- + O_2 \quad (54)$$

$$O_3^- + NO \rightarrow NO_3^- + O \quad (55)$$

$$O_3^- + O \rightarrow O_2^- + O_2 \quad (56)$$

O_4^- reactions:

$$O_4^- + CO_2 \rightarrow CO_4^- + O_2 \quad (57)$$

$$O_4^- + O \rightarrow O_3^- + O_2 \quad (58a)$$

$$O_4^- + NO \rightarrow O_2 + NO_3^- \xrightarrow{NO} NO_2^- + NO_2 \quad (59)$$

CO_3^- reactions:

$$CO_3^- + O \rightarrow O_2^- + CO_2 \quad (60)$$

$$CO_3^- + NO \rightarrow NO_2^- + CO_2 \quad (61)$$

CO_4^- reactions:

$$CO_4^- + NO \rightarrow NO_3^- + CO_2 \quad (62)$$

$$CO_4^- + O \rightarrow CO_3^- + O_2 \quad (63a)$$

$$CO_4^- + O_3 \rightarrow O_3^- + CO_2 + O_2 \quad (64)$$

NO_2^- reactions:

$$NO_2^- + O_3 \rightarrow NO_3^- + O_2 \quad (65)$$

$$NO_2^- + H \rightarrow OH^- + NO \quad (66)$$

OH^- reactions:

$$OH^- + O \rightarrow HO_2 + e \quad (67)$$

$$OH^- + H \rightarrow H_2O + e \quad (68)$$

$$OH^- + CO_2 + M \rightarrow HCO_3^- + M \quad (69)$$

Photodetachment:

$$O_2^- + h\nu \rightarrow O_2 + e \quad (70)$$

$$O_3^- + h\nu \rightarrow O_3 + e \quad (71)$$

$$NO_2^- + h\nu \rightarrow NO_2 + e \quad (72)$$

$$NO_3^- + h\nu \rightarrow NO_3 + e \quad (73)$$

The rate coefficients for all these reactions are listed in Table III-7.

In the above scheme, we have ignored, in addition to hydration reactions, the reverse of the association reactions. Actually these might play some role, especially the reverse of reaction (52).

It should be noticed that O^- is not included in the scheme. The existence of this ion in laboratory experiments is well known, but it is not clear how it is produced in the atmosphere. For many years it was thought to have been produced via

$$e + O_3 \rightarrow O_2 + O^- \quad (74)$$

but laboratory experiments have shown this reaction to be unimportant.

The step currently invoked to produce the O^- seen in the atmosphere is

$$O_4^- + O \rightarrow O^- + 2O_2 \tag{58b}$$

but we have ignored this reaction for simplicity. There is also another possible path for reaction (63) that has also been omitted from our consideration

$$CO_4^- + O \rightarrow O_3^- + CO_2 \tag{63b}$$

A pictoral representation of the negative-ion cycle is shown in Fig. III-7. The general flow is for electrons to pass, in turn, to O_2^-, O_4^-, and O_3^-, CO_4^- and CO_3^-, NO_2^-, and finally to NO_3^-. In the presence of H, NO_2^- also passes through OH^- to HCO_3^-. The HCO_3^- and NO_3^- ions are stable in that the only way they can be removed is by reaction with positive ions or photodetachment. Thus these are the terminating ions.

A steady-state computation has been carried out neglecting diffusion, using the above negative-ion cycle for overhead sun conditions. The results are given in Table III-13 at 65 and 70 km. At 65 km, the calculated values for concentrations of electrons agrees well with the observed values. However the calculated concentration of total negative species is lower than the corresponding observed value, but the discrepancy is less than a factor of two. The major negative ions in order of importance are CO_3^-, NO_3^-, CO_4^-, and O_2^-. At 70 km, the computed concentration for total negative species agrees well with the observed value, but the computed electron concentration is too high. The computed relative importance of the ions is CO_3^-, CO_4^-, O_2^-, and NO_3^-, in decreasing order. Thus, the major ion

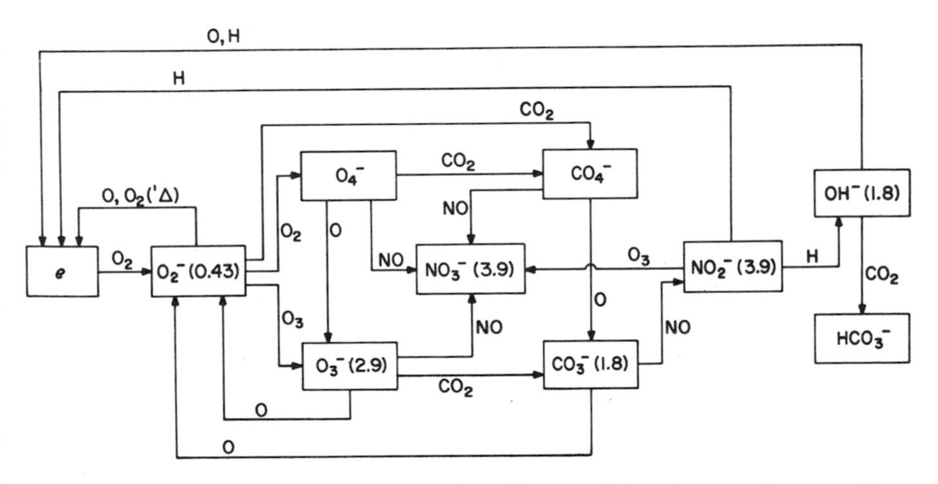

Fig. III-7. D region negative-species flow diagram. Values in parentheses are electron affinities in electron volts from Franklin and Harland (1974).

TABLE III-13

Computed Negative-Ion Concentrations in the D Region for an Overhead Sun[a]

Altitude (km)	65	70
Temperature (°K)	232	211
[M] (particles/cm^3)	3.9×10^{15}	2.0×10^{15}
[O] (particles/cm^3)	5.0×10^{9}	4.0×10^{9}
[$O_2(^1\Delta)$] (particles/cm^3)	1.5×10^{10}	8.0×10^{9}
[O_3] (particles/cm^3)	3.2×10^{9}	1.0×10^{9}
[O_2] (particles/cm^3)	8.2×10^{14}	4.2×10^{14}
[CO_2] (particles/cm^3)	1.27×10^{12}	6.5×10^{11}
[NO] (particles/cm^3)	2.0×10^{7}	2.0×10^{7}
[H] (particles/cm^3)	6.0×10^{6}	1.0×10^{7}
α_e (cm^3/sec)	7.4×10^{-7}	8.4×10^{-7}
α_{neg} (cm^3/sec)	3.2×10^{-7}	3.7×10^{-7}
q (photons/cm^3-sec)	0.1	2.0
[e] (particles/cm^3)	97 (100)	1120 (140)
[O_2^-] (particles/cm^3)	23	92
[O_3^-] (particles/cm^3)	0.070	0.117
[O_4^-] (particles/cm^3)	0.13	0.27
[CO_3^-] (particles/cm^3)	244	332
[CO_4^-] (particles/cm^3)	65	108
[NO_2^-] (particles/cm^3)	0.88	1.91
[NO_3^-] (particles/cm^3)	68	65
[OH^-] (particles/cm^3)	0.0020	0.0092
[HCO_3^-] (particles/cm^3)	0.62	0.21
[e] + [M^-] (particles/cm^3)	499 (900)	1720 (1900)

[a] Values in parenthesis are observed values (see Table III-11).

in the lower D region is CO_3^- during the day. However, at night NO_3^- is the only important negative ion. This agrees with the observations of Narcisi *et al.* (1971) that m/e 88 ($NO_3 \cdot H_2O^-$) was the dominant nighttime ion, but not with the observations of Arnold and Krankowsky (1971), who found several nighttime ions of which CO_3^- was dominant.

At 75–85 km, the negative-ion scheme outlined above fails, in that it predicts that almost all the negative charge is electrons, contrary to the observations. Possibly hydration may play a very important role in stabilizing the negative ions at these altitudes. The calculated electron (total negative species) concentrations for an overhead sun at 75, 80, and 85 km are, respectively, 2900, 3780, and 3640 electrons/cm^3, which can be compared with the corresponding observed values for total negative species of 2700, 4100, and 12,000. The agreement at 75 and 80 km is good, but not at 85 km. There is some scatter in the measurements of electron concentrations in the D region (Rowe *et al.*, 1974) but the upper limits

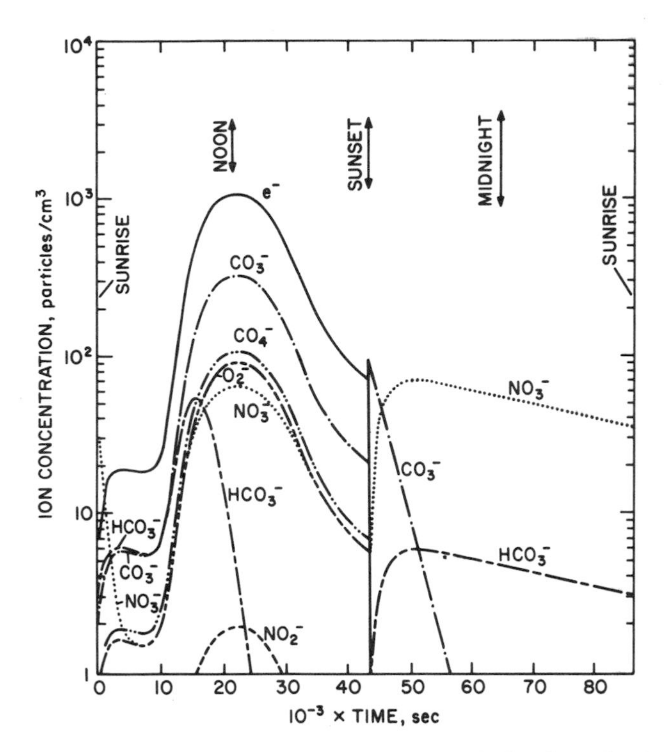

Fig. III-8. Calculated negative-ion concentrations at 70 km for a day in which the noonday sun is overhead (q_0 = 2.0 photons/cm³-sec; ζ_{70}' = 9.29).

appear to be 1000 electrons/cm³ at 75–80 km and about 500/cm³ at 70 km. The computed value of 3640 electrons/cm³ at 85 km is consistent with the observations.

A computed daily negative-ion profile at 70 km utilizing the above negative ion scheme is presented in Fig. III-8 for a step-by-step calculation using 1–100 sec time intervals with $q_0 = 2.0$ photons/cm³-sec and $\zeta_{70}' = 9.29$. It is assumed that the neutral species are constant throughout the day with the values listed in Table III-13, except for [O(^{3}P)] and [O_2($^1\Delta$)], which vanish at night. Diffusion has been neglected. It can be seen that in midday, electrons should be the dominant negative species and CO_3^- the dominant negative ion. However, after dark, [O(^{3}P)] and [O_2($^1\Delta$)] decrease in importance, and so do their reactions with O_2^- to induce electron detachment, as well as photodetachment. The electron and many ion concentrations are markedly reduced; but [HCO_3^-], [CO_3^-], and [NO_3^-] increase after dark, and the relative importance of CO_3^- and NO_3^- is interchanged. Later at night, the only ions of significance are NO_3^- and

HCO_3^-, with NO_3^- being dominant. HCO_3^- has an interesting behavior in that it is an important ion in the morning but vanishes in the afternoon to reappear again after sunset.

REFERENCES

Arnold, F., and Krankowsky, D. (1971). *J. Atmos. Terrest. Phys.* **33,** 1693, "Negative Ions in the Lower Ionosphere: A Comparison of a Model Computation and a Mass Spectrometric Measurement."

Baker, D. C. (1969). Ionospheric Res. Lab. Sci. Rep. No. 334, Penn State Univ., "Ionospheric D-Region Parameters from Blunt Probe Measurements during a Solar Eclipse."

Balabanova, V. N., Bychkova, K. D., and Martynenko, V. P. (1974). *J. Atmos. Terrest. Phys.* **36,** 1785, "The Atomic Nitrogen Amount in the Upper Atmosphere According to Measurement of the Ethylene Luminous Cloud Brightness."

Bennett, F. D. G., Hall, J. E., and Dickinson, P. H. G. (1972). *J. Atmos. Terrest. Phys.* **34,** 1321, "D-Region Electron Densities and Collision Frequencies from Faraday Rotation and Differential Absorption Measurements."

Bowman, M. R., Thomas, L., and Geisler, J. E. (1970). *J. Atmos. Terrest. Phys.* **32,** 1661, "The Effect of Diffusion Processes on the Hydrogen and Oxygen Constituents in the Mesosphere and Lower Thermosphere."

Brasseur, G., and Nicolet, M. (1973). *Aeronom. Acta* No. 113, "Chemospheric Processes of Nitric Oxide in the Mesosphere and Stratosphere."

Danilov, A. D. (1970). "Chemistry of the Ionosphere" (Engl. Transl.). Plenum Press, New York.

Donahue, T. M. (1968). *Science* **159,** 489, "Ionospheric Composition and Reactions."

Ferguson, E. E. (1974). *Rev. Geophys. Space Phys.* **12,** 703, "Laboratory Measurements of Ionospheric Ion–Molecule Reaction Rates."

Ferraro, A. J., Lee, H. S., Rowe, J. N., and Mitra, A. P. (1974). *J. Atmos. Terrest. Phys.* **36,** 741, "An Experimental and Theoretical Study of the D-Region. I. Mid Latitude D-Region Electron Density Profiles from the Radio Wave Interaction Experiment."

Franklin, J. L., and Harland, P. W. (1974). *Ann. Rev. Phys. Chem.* **25,** 485, "Gaseous Negative Ions."

Garvin, D., and Hampson, R. F. (1974). Nat. Bur. Std. Internal Rep. 74-430, "Chemical Kinetics Data Survey. VII. Tables of Rate and Photochemical Data for Modelling of the Stratosphere" (Revised).

Hale, L. (1974). "Cospar Methods of Measurements and Results of Lower Ionosphere Structure," p. 219. Akademie-Verlag, Berlin, "Positive Ions in the Mesosphere."

Heicklen, J., and Cohen, N. (1968). *Advan. Photochem.* **5,** 157, "The Role of Nitric Oxide in Photochemistry."

Hinteregger, H. E., Hall, L. A., and Schmidtke, G. (1964). *Space Res.* **5,** 1175, "Solar XUV Radiation and Neutral Particle Distribution in July, 1963 Thermosphere."

Howard, C. J., Rundle, H. W., and Kaufman, F. (1971). *J. Chem. Phys.* **55,** 4772, "Water Cluster Formation Rates of NO^+ in He, Ar, N_2, and O_2 at 296°K."

Ivanov-Kholodnyy, G. S. (1966). *Geomagn. Aeronom.* **6,** 382, "Ion Formation Intensity at Heights of 100–300 km."

Kebarle, P., Searles, S. K., Zolla, A., Scarborough, J., and Arshadi, M. (1967). *J. Amer. Chem. Soc.* **89,** 6393, "The Solvation of Hydrogen Ion by Water Molecules in the Gas

Phase. Heats and Entropies of Solvation of Individual Reactions: $H^+(H_2O)_{n-1}$ + $H_2O \rightarrow H^+(H_2O)_n$."

Kley, D., Lawrence, G. M., and Stone, E. J. (1975). 8th International Conference on Photochemistry, Edmonton, Canada, Paper v7, "Experimental Study of the NO^+ + $e \rightarrow N + O$ Recombination in the Vacuum U.V. Flash Photolysis of Nitric Oxide."

Mahan, B. H., and Person, J. C. (1964). *J. Chem. Phys.* **40,** 392, "Gaseous Ion Recombination Rates."

Narcisi, R. S., Bailey, A. D., Della Lucca, L., Sherman, C., and Thomas, D. M. (1971). *J. Atmos. Terrest. Phys.* **33,** 1147, "Mass Spectrometric Measurements of Negative Ions in the D- and Lower E-Regions."

Narcisi, R. S., Bailey, A. D., Wlodyka, L. E., and Philbrick, C. R. (1972). *J. Atmos. Terrest. Phys.* **34,** 647, "Ion Composition Measurements in the Lower Ionosophere during the November 1966 and March 1970 Solar Eclipses."

Nicolet, M. (1970). *Aeronom. Acta.* No. 71, "The Origin of Nitric Oxide in the Terrestrial Atmosphere."

Nicolet, M. (1971). *In* "Mesospheric Models and Related Experiments" G. Fiocco (ed.), D. Reidel Publishing Co., Dordrecht Holland, p. 1, "Aeronomic Reactions of Hydrogen and Ozone."

Norton, R. B., and Barth, C. A. (1970). *J. Geophys. Res.* **75,** 3903, "Theory of Nitric Oxide in the Earth's Atmosphere."

Rowe, J. N., Mitra, A. P., Ferraro, A. J., and Lee, H. S. (1974). *J. Atmos. Terrest. Phys.* **36,** 755, "An Experimental and Theoretical Study of the D-Region. II. A Semi-Empirical Model for Mid-Latitude D-Region."

Chapter IV

ATMOSPHERIC POLLUTANTS

In addition to the gases naturally present in the atmosphere, other gases can also be present, due to their introduction as man-made wastes. These usually occur in large urban areas, where very much higher concentrations of some of the natural gases can occur also for the same reason.

Pollution has long been a problem of man, and Wagner (1971) has pointed out that the earliest reference to this problem probably appears in Deuteronomy 23:12,13: "and thou shalt have a paddle upon thy weapon; and it shall be when thou wilt ease thyself abroad, thou shalt dig therewith and shall turn back and cover that which cometh from thee." Air pollution as a specific problem was well known in the Middle Ages, where as Wagner (1971) states: "The people of medieval towns and cities, being unable to return waste to the soil, dumped it from their windows into the street, requiring well-bred gentlemen of the day to defend their sensitivities by carrying an orange stuck with cloves to mask the foulness of the streets."

The first well-studied air pollution episode occurred in the Meuse Valley, Belgium, December 1–5, 1930, when a heavy fog covered the valley. Sixty-three people died and many others were ill with respiratory problems. In October 1948 in Donora, Pennsylvania, an inversion occurred, which resulted in a severe episode affecting 43% of the population. Twenty deaths occurred. The world's most disastrous episode occurred in London in December 1952 and accounted for 3900 excess deaths over that normally expected. Almost all the excess deaths occurred in people over 70 years of age, and persons with pre-existing pulmonary and cardiac diseases were most susceptible. One aspect of excess deaths that must be considered is the possibility that people slated to die in the next few weeks die a little early, and that after the episode there is a reduction in the death rate to compensate for this effect. After many, but not all, episodes such a reduction has been noted, but the reduction is not sufficient to compensate for all the excess deaths. The net effect is that such episodes really lead to an increase in the long-term death rate.

Many pollutants can occur in the atmosphere. It is convenient to discuss these pollutants in the following groups:

1. CO and CO_2
2. H–C–O compounds, usually referred to as hydrocarbons, although aldehydes, ketones, and other H–C–O compounds are also present.
3. Oxides of nitrogen, which when N_2O is excluded, are referred to as NO_x.
4. Oxidants that include O_3, singlet O_2, peroxyacetylnitrate (PAN), and peroxides.
5. Halogen-containing compounds.
6. Sulfur-containing compounds.
7. Particulate matter.

In this chapter we shall discuss the source, abundance, and effects of these pollutants. A few years ago this material was comprehensively discussed in a report by Robinson and Robbins (1968, 1969). We shall draw heavily but not exclusively on their work. More definitive sources are the Air Quality Criteria Reports issued by the U.S. Public Health Service in 1969–1971 on CO, hydrocarbons, oxides of nitrogen, oxidants, SO_2, and particulate matter.

CO AND CO_2

CO

The sources of CO have been estimated by Robinson and Robbins, and their estimates are shown in Table IV-1. The total worldwide annual CO emissions are about 340×10^6 tons/year, of which 274×10^6 (or 78.5%) are from man-made sources. Of these 193×10^6 tons (55.3% of the total) come from the combustion of gasoline (mostly automobiles) and this is the major source of CO in the troposphere.

In New York City alone, it was estimated that motor vehicles discharged 1.7×10^6 tons of CO in 1967 (Eisenbud, 1970). For Los Angeles County, the value was 3.46×10^6 tons in 1969 from motor vehicles, which was $>97\%$ of the total input of CO into Los Angeles County.

The natural sources of CO include forest fires, oxidation of terpenes emitted in forests, and the production of CO by marine and plant life.

TABLE IV-1

Estimated World-Wide Annual CO Emissions[a]

Source	Source consumption (tons × 10^6)	CO emission factor (%)	CO emission (tons × 10^6)
Gasoline	379	51.0	193
Coal			
Power	1219	0.025	0.30
Industry	781	0.15	1.17
Residential	404	2.50	10.1
Coke and gas	615	0.0055	0.03
Wood and noncommercial fuel	1260	3.5	44.0
Incineration	500	5.0	25.0
Forest fires	315	3.5	11.0
Marine siphonophores	4	100	4[b]
Terpene reactions	60	100	60
Total			349

[a] From Robinson and Robbins (1969).
[b] Considerable uncertainty in this number. It could be as high as 360 × 10^6 tons.

Estimates of this last source are rather uncertain, but they are much higher than the 4 × 10^6 tons/yr estimated by Robinson and Robbins (1969). However, it is known that CO is produced in this manner, an interesting marine source being *Nereocyctis,* a kelp or seaweed with large bladders that have been found to contain CO at concentrations up to 800 ppm (8000 times background values) (Robinson and Robbins, 1969).

Jaffe (1973) has made more recent estimates for CO emissions. His estimates for global man-made emissions for 1970 are listed in Table IV-2. Of the 359 × 10^6 metric tons (396 × 10^6 tons) emitted globally in 1970, about 133 × 10^6 metric tons (147 × 10^6 tons) were from the U.S., of which >75% came from fossil fuel combustion. The natural sources of CO are much larger than previously thought. The ocean sources are much greater than estimated by Robbins and Robinson. Linnenbom *et al.* (1973) estimate them as 240 × 10^6 tons/yr. An additional source of CO is the decay of chlorophyl in plants, which could be as high as 54 × 10^6 metric tons/yr (61 × 10^6 tons/yr) (Jaffe, 1973). Additional vegetation sources could raise this estimate to 90 × 10^6 metric tons/yr (101 × 10^6 tons/yr) (Jaffe, 1973).

TABLE IV-2

Estimated Global Anthropogenic CO *Sources for 1970*[a]

Sources	World fuel consumption[b] (10^6 metric tons/yr[d])	World CO emissions[c] (10^6 metric tons/yr[d])
Mobile		
Motor vehicles	439	199
Gasoline	—	197
Diesel	—	2
Aircraft (aviation gasoline, jet fuel)	84	5
Watercraft	—	18
Railroads	—	2
Other (nonhighway) motor vehicles (construction equipment, farm tractors, utility engines, etc.)	—	26
Stationary		
Coal and lignite	2983	4
Residual fuel oil	682	<1
Kerosene	69	<1
Distillate fuel oil	411	<1
Liquefied petroleum gas	34	<1
Industrial processes (petroleum refineries, steel mills, etc.)	—	41
Solid waste disposal (urban and industrial)	1130	23
Miscellaneous (agricultural burning, coal bank refuse, structural fires)	—	41
Total anthropogenic CO	—	359

[a] From Jaffe (1973).
[b] Based on production data. Assumption is made that consumption equals production.
[c] CO emission obtained by multiplying weight of fuel by appropriate CO factor.
[d] To convert metric tons to short tons, multiply by 1.1023.

The CO that is produced diffuses into the upper atmosphere where it is consumed by reaction with HO radicals in the stratosphere. From the emission rate and the surface area of the earth, the flux is computed to be 4.21×10^{10} molecules/cm^2 sec. In the upper atmosphere, CO is also produced by CH_4 oxidation and CO_2 photodecomposition. Since all the CH_4 ultimately oxidizes to CO, and since the CH_4 emissions from the earth are at least 4 times those of CO, it can be seen that the CO produced at the earth is minor in the overall atmospheric balance.

The expected lifetimes of CO in the troposphere can be computed, since the ambient CO concentration is known to be 0.1 ppm. Utilizing the diffusion equations

$$(\tau_{\text{diff}}\{CO\})^{-1} = D(\xi_{CO} + \xi_D)(\xi_{CO} - \xi_n) \tag{1}$$

and

$$F\{CO\} = D[CO](\xi_{CO} - \xi_n) \tag{2}$$

we find

$$\tau_{\text{diff}}\{CO\} = [CO]/F\{CO\}(\xi_{CO} + \xi_D) \tag{3}$$

The values for ξ_{CO} and ξ_D are $\sim 14 \times 10^{-7}$ and ~ 0 cm^{-1}, respectively. Thus $\tau_{\text{diff}}\{CO\}$ is computed to be about 1.3 yr at the earth's surface.

Measurements of ambient (nonurban) CO concentrations have been made, and they also have been summarized by Robinson and Robbins (1968, 1969). Individual measurements vary widely from 0.03 to 0.90 ppm, a factor of 30. However, if daily average measurements are made over the North Pacific Ocean, then much better consistency is obtained. Such measurements are shown as a function of longitude in Fig. IV-1. The values range from 0.08 to 0.39 ppm, but there is no apparent trend with longitude, the average value being 0.15 ppm. Figure IV-2 shows the longitudinal average of the daily average CO concentrations as a function of latitude. There is clearly a change with latitude, ranging from a maximum value of 0.19 ppm at 50° N to 0.04 ppm at 50° S. There is no known explanation for this phenomenon. The overall ground average value is 0.1 ppm. Robbins *et al.* (1973) have shown that this value has been the same for several centuries, indicating that the industrial revolution has had little effect on background CO levels.

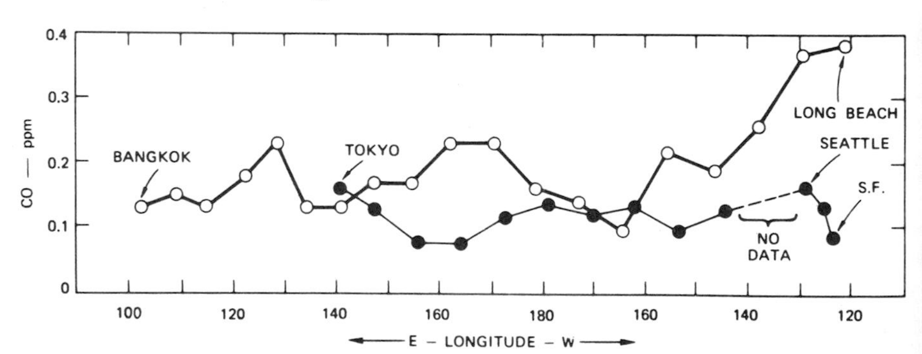

Fig. IV-1. Average daily CO concentrations as a function of longitude across the North Pacific, USNS Perseus voyages 41-AB and 41-H, June–July (1968). (Open circles, Perseus 41-AB; solid circles, Perseus 41-H.) From Robinson and Robbins (1969) with permission of the authors.

Of course, since CO is produced mainly from man-made sources, urban concentrations must be very much greater than 0.1 ppm. The U.S. National Air Pollution Control Administration (now the Environmental Protection Agency) data show a three-year average of 7.3 ppm for 1964 to 1966 for off-street sites in five major cities (NAPCA Publication AP-62, 1970). Instantaneous concentrations of 200 ppm have been obtained at tunnel entrances. Maximum hourly average values in Los Angeles of 35 ppm in 1967 were typical. Similar values have also been reported for New York City.

The city with the worst CO problem in the United States is Chicago (Tokyo may be worse). In the period 1964–1967 yearly maximum values of >60 ppm were observed in Chicago. Since then, traffic but not traffic density has increased. Yet, because of improved combustion in more modern cars, the CO levels in Chicago have dropped in recent years and 1972 daily maximum values were more commonly 20–30 ppm.

The CO concentration can be correlated directly with traffic density. The average diurnal variation of CO in Chicago for 1962 to 1964 is shown in Fig. IV-3. The peaks at 7–8 A.M. and about 4 P.M. during the weekdays correspond to the rush hours. During the weekends, the morning peak is absent, but levels rise toward evening.

Some of the highest levels of CO reported have been 235 ppm measured in a traffic island and 360 ppm at another street-level location in London,

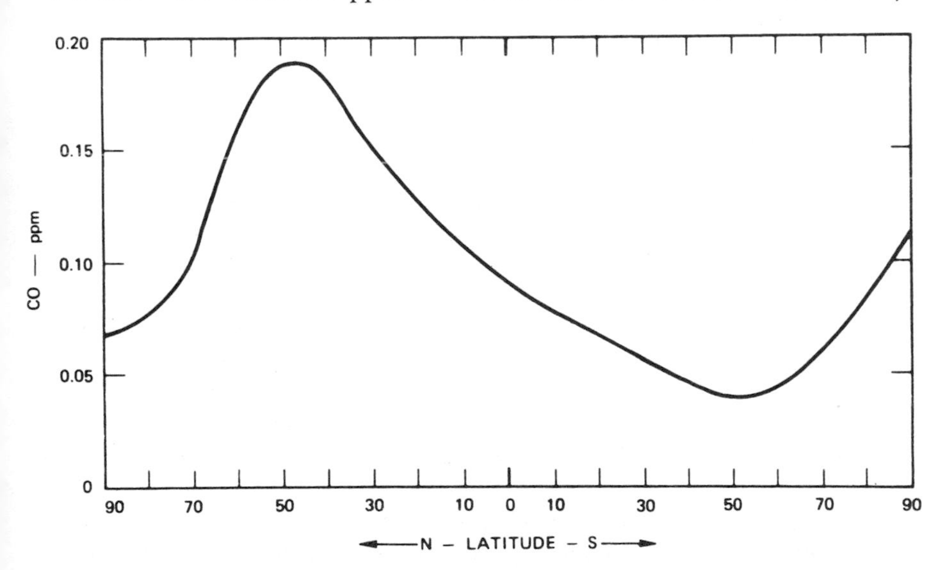

Fig. IV-2. Average latitude distribution of CO in the Pacific Area. USNS Perseus voyages 1967–1968. From Robinson and Robbins (1969) with permission of the authors.

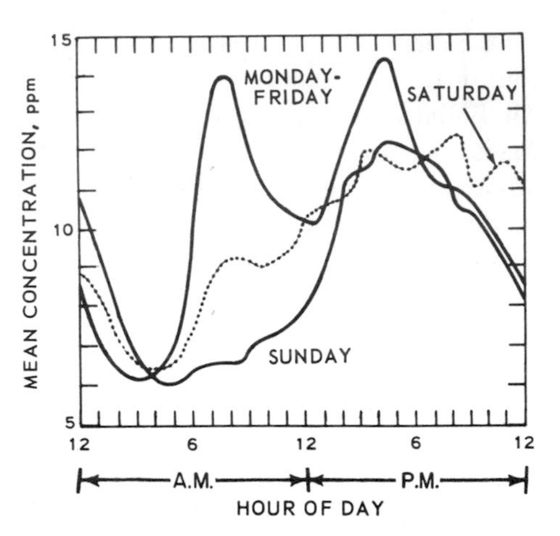

Fig. IV-3. Average diurnal variation of CO levels in Chicago, 1962–1964. From NAPCA Publication AP-62 (1970) with permission of the Environmental Protection Agency.

1957. In Paris in 1967, 1.5% of the readings exceeded 100 ppm, and 0.07% exceeded 300 ppm (NAPCA Publication AP-62, 1970).

Denver also has high CO concentrations, the yearly maximum values exceeded 70 ppm in 1965 and 1966. This is a particularly high level, since Denver, which is located at an altitude of 5280 ft, already has thin air.

CO is an odorless, colorless gas of low chemical reactivity. In fact, its only known chemical reactions in the atmosphere are with HO and $O(^3P)$, and these give negligible removal of CO in urban atmospheres. Its only known physiological activity is to combine with hemoglobin in the blood and thus reduce oxygen uptake. However, this effect is completely reversible, and inhalation of a CO-free atmosphere for just 1–2 hr removes most of the CO; thus CO does not accumulate in the body.

The activity of CO is that at 30 ppm in the air, 5% of the sites in hemoglobin are occupied, and O_2 uptake is reduced by the same amount. Thus the deleterious effect of CO is due to suffocation and not to any direct chemical change. Physiological effects are first noticeable in some individuals at 10 ppm, where judgment may become impaired due to effects on the central nervous system. At 30 ppm prolonged exposure, cardiovascular changes are detectable in people with coronary heart disease or emphysema. Above 40 ppm, physical effects are detectable in regard to cardiovascular activity in most people over 40. Prolonged exposure to 200 ppm CO leads to headaches, although many individuals can function normally at these levels. At long exposures to 600 ppm, there can be loss of consciousness; and finally, at higher levels, death.

As a consequence of the above facts, safety levels have been developed. The threshold limits values accepted in 1972 by the American Conference of Governmental Industrial Hygenists is 50 ppm for an 8-hr day, 40-hr week exposure. The U.S. Environmental Protection Agency has recommended more stringent standards for ambient urban air of 35 ppm for 1 hr and 9 ppm for 8 hr exposure. Needless to say, many cities do not yet meet these standards. In fact, the California standards in the 1960s were the more lenient 120 ppm/hr or 30 ppm for 8 hr, but these levels were exceeded in Los Angeles County on at least 19 days in each of the years 1960–1966 (Lemke *et al.*, 1967).

Perhaps a better perspective of the CO pollution problem can be obtained from other considerations. The major cause of CO pollution is cigarette smoking. Eisenbud (1970), who was the first administrator of the Environmental Protection Administration of the City of New York, has reported that smoking one pack of cigarettes per day is "equivalent to continuous exposure to 50 ppm of CO in ambient air." The CO content of the mainstream smoke from a cigarette is 40,000 ppm! Yet it is not the CO from cigarettes smoking, but the tars and nicotine, that is considered harmful to health!

Possibly the danger of CO has been greatly overemphasized. A dramatic argument for this point of view comes from the realization that the reduction in blood oxygen uptake by the lungs is the same for absolutely CO-free air at 11,200 ft elevation (because of the pressure drop) as for continuous exposure to air with >90 ppm of CO at sea level (NAPCA Publication AP-62, 1970)! Most people do not consider mountain air hazardous. In fact, quite the reverse, they go to the mountains, "because it is healthy."

CO_2

The CO_2 emission sources estimates for 1965 are shown in Table IV-3. All the CO_2 comes from combustion, essentially all of it man-made. The total 1965 worldwide rate of 14.08×10^9 tons/yr corresponds to a flux of 1.08×10^{12} molecules/cm^2 sec, which is very much larger than the CO_2 that can be absorbed into the earth (primarily by solution into the oceans and photosynthesis), and the atmospheric CO_2 concentration is continuing to increase. In regard to the CO_2 balance, man-made changes are having a significant effect.

The man-made CO_2 production has increased by almost an order of magnitude since 1890, and it is estimated that this increase will be 18-fold from 1890 to 2000. As a result the CO_2 concentration in the atmosphere has risen from 296 ppm in 1896 to 325 ppm at the present time. During 1958 to 1963, the annual increase was 0.25%. This corresponds to 5.51×10^9 tons/yr in the whole atmosphere. The amount returning to the earth is 8.57×10^9 tons/yr, which indicates that the earth can accommodate about 61% of the perturbation at the present time.

TABLE IV-3

Estimated Annual Global CO_2 Emissions Based on 1965 Fuel Usage[a]

Fuel	Fuel consumption (tons × 10^9)	CO_2 emission factor (%)	CO_2 emission (tons × 10^9)
Coal, lignite, etc.	3.074	248	7.33
Petroleum	1.273	317	4.03
Gasoline	0.379	—	—
Kerosene	0.100	—	—
Fuel oil	0.287	—	—
Residual oil	0.507	—	—
Natural gas	0.432	275	1.19
Waste incineration	0.500	92	0.46
Wood fuel	0.466	146	0.68
Forest fires	0.324	120	0.39
Total			14.08

[a] From Robinson and Robbins (1969).

One must consider the global effects of depositing so much CO_2 into the atmosphere. It is estimated that a 10% increase in CO_2 could raise the earth's temperature by 0.5°C. In fact, the annual average temperature rose 0.4°C from 1880 to 1940. However, from 1940 to 1967 the average annual temperature dropped 0.2°C, indicating that other factors must be compensating (Robinson and Robbins, 1969). It has been suggested that the other great pollutant, particulate matter, has had more than a compensating effect because of its turbidity, although this has not been proven. However, Rasool and Schneider (1971) compute that a factor of four increase in global background aerosol concentration could reduce the surface temperature by as much as 3.5°K and trigger another ice age.

Another effect of increasing atmospheric CO_2 concentrations would be to increase the dissolved CO_2 content in the oceans. However, as CO_2 dissolves it makes the water more acid, which because of the ocean's buffering effect dissolves the limestone to produce HCO_3^- and thus reduce the CO_2 content. As a result, the 10% increase in atmospheric CO_2 (from 296 in 1896 to 325 ppm at present) caused only a 1% increase in oceanic CO_2 concentration (Robinson and Robbins, 1969).

Fortunately, CO_2 is a nontoxic gas, having no known physiologically harmful activity. In 1972 the threshold safety limit set by the American

Conference of Governmental Industrial Hygienists was 5000 ppm for a 40-hr week exposure, well above atmospheric levels.

There appears to be no harmful short-term effect yet known as a result of CO_2 accumulation. However, if CO_2 continues to increase at an accelerating rate, it is certain that ecological effects must occur at some time, probably in the next hundred years. As stated by Robinson and Robbins (1968):

> It seems ironic that, in air pollution technology, we are so seriously concerned with small scale events, such as the photochemical reactions of trace concentrations of hydrocarbons and the effect on vegetation of a fraction of a part per million of SO_2, whereas the abundant pollutants—CO_2 and submicron particles—which we generally ignore because they have little local effect, may be the cause of serious worldwide environmental changes.

HYDROCARBONS

Emissions

Estimates of the annual worldwide emissions of hydrocarbons are given in Table IV-4. The overall emission rate is 1858.3×10^6 tons/yr, of which the overwhelming amount is CH_4 from natural sources. Additional CH_4 comes from some of the man-made emissions, and this brings the total CH_4 emission to $>1600 \times 10^6$ tons/yr ($>86\%$ of the total). This amount corresponds to a flux of 8.13×10^{11} molecules/cm^2-sec. The amount of CH_4 in the atmosphere is 2.4×10^{19} molecules/cm^3, so that its lifetime is 0.94 yr.

The second most important group of emissions is the terpene group from vegetation, which accounts for another 170×10^6 tons/yr (9.15%). A number of hydrocarbons in these emissions have been identified by Rasmussen (1964): α-pinene, β-pinene, myrcene, *d*-limonene, santene, camphene, *n*-heptane, isoprene, α-ionene, β-ionene, and α-irone.

Less than 5% of the hydrocarbon emissions are from man-made sources, although these include the vast majority of the non-CH_4- and nonterpene-type compounds. According to Robinson and Robbins (1968), the five major sources of man-made pollution in decreasing order of importance are gasoline combustion (38.5%), incineration (28.3%), solvent evaporation (11.3%), petroleum evaporation and transfer losses (8.8%), and refinery wastes (7.1%). These five categories comprise 95.8% of all man-made hydrocarbon emissions.

Table IV-5 shows the estimate of the man-made annual emissions in the U.S. Of these emissions it can be seen that the U.S. contributes about 40% of the total world contribution. Lcs Angeles County alone emits

TABLE IV-4

Estimate of Annual Worldwide Hydrocarbon Emission (Late 1960s)[a]

Source	Source consumption (tons $\times 10^6$)	Hydrocarbon emission factor (%)	Hydrocarbon emissions (tons $\times 10^6$)
Coal	—	—	2.9
Power	1219	0.016	0.2
Industrial	1369	0.051	0.7
Domestic and commercial	404	0.50	2.0
Petroleum	—	—	48.5
Refineries	11,317 bbl	56 tons/10^4 bbl	6.3
Gasoline	379	9.0	34
Kerosene	100	<0.1	<0.1
Fuel oil	287	0.035	0.1
Residual oil	507	0.039	0.2
Evaporation and transfer loss	379	2.06	7.8
CH_4 from natural sources	—	—	1600
Paddy fields	—	—	210
Swamps	—	—	630
Humid tropics	—	—	672
Other (mines, etc.)	—	—	88
Terpenes from vegetation	—	—	170
Coniferous forest	—	—	50
Hardwood forest, cultivated land, and steppes	—	—	50
Carotene decomposition of organic matter	—	—	70
Other	—	—	36.9
Solvent use	—	—	10
Incinerators	500	5.0	25
Wood fuel	466	0.15	0.7
Forest fires	324	0.37	1.2
Total	—	—	1858.3

[a] From Robinson and Robbins (1968).

1.27×10^6 tons/yr, which is 1.44% of the total world rate. Of the U.S. contribution, 52% is from transportation; and in Los Angeles County, about 70% is from transportation.

The natural hydrocarbons in the U.S. have been estimated by Tuesday (1972) to be 50×10^6, 22×10^6, and 0.04×10^6 tons/yr, respectively, for CH_4, terpenes, and C_2H_4. Thus in the U.S. about 34% of the hydrocarbon

TABLE IV-5
Estimate of U.S. Annual Hydrocarbon Emissions by Source Category

	1968[a]		1969[b]		1971[b,c]
Source	Hydrocarbon emission (tons × 10⁶)	(%)	Hydrocarbon emission (tons × 10⁶)	(%)	Hydrocarbon emission (tons × 10⁶)
Transportation	**16.6**	**51.9**	**19.8**	**53.0**	—
Motor vehicles (gasoline)	15.2	47.5	16.9	45.2	
Motor vehicles (diesel)	0.4	1.2	0.2	0.5	—
Aircraft	0.3	1.0	0.4	1.1	—
Railroads	0.3	1.0	0.1	0.3	—
Vessels	0.1	0.2	0.3	0.8	—
Nonhighway use, motor fuels	0.3	1.0	1.9	5.1	—
Fuel combustion—stationary	**0.7**	**2.2**	**0.9**	**2.4**	—
Coal	0.2	0.7	0.1	0.3	0.2
Fuel oil	0.1	0.3	0.1	0.3	0.4
Natural gas	0.0	0.0	0.3	0.8	—
Wood	0.4	1.2	0.4	1.1	—
Industrial processes	**4.6**	**14.4**	**5.5**	**14.7**	—
Primary metals	—	—	0.3	0.8	—
Petroleum refining	—	—	2.3	6.2	2.0
Chemical	—	—	0.8	2.1	1.4
Others	—	—	2.1	5.6	1
Solid waste disposal	**1.6**	**5.0**	**2.0**	**5.3**	**4.5**
On-site incineration	—	—	0.5	1.3	—
Open dumps, burned	—	—	1.2	3.2	—
Wigwam burners	—	—	0.3	0.8	—
Miscellaneous	**8.5**	**26.5**	**9.2**	**24.6**	—
Forest fires	2.2	6.9	2.9	7.8	2.4–3.0
Structural fires	0.1	0.2	0.1	0.3	—
Coal refuse	0.2	0.6	0.1	0.3	—
Organic solvent evaporation	3.1	9.7	3.1	8.3	7.1
Gasoline marketing	1.2	3.8	1.3	3.5	3.2
Agricultural burning	1.7	5.3	1.7	4.5	4.2
Total	32.0	100.0	37.4	100.0	—

[a] From NAPCA Publication AP-64 (1970).
[b] From Tuesday (1972). [c] From Garner (1972).

emissions are man-made. However with the introduction of pollution control devices on motor vehicles, it appears that 1969 was the peak output year. It has been estimated that annual transportation emissions in the U.S. increased from 12.1×10^6 tons/yr in 1965 to the maximum of 19.8×10^6 in 1969, and will decrease to 8.4×10^6 tons/yr by 1974 and 2.9×10^6 tons/yr by 1985.

Typical composition of the auto exhausts is given in Table IV-6. The major hydrocarbons are CH_4, C_2H_4, C_2H_2, C_3H_6, and C_4 hydrocarbons. Of the aldehydes, by far the most dominant is CH_2O, but CH_3CHO, propionaldehyde, acetone, acrolein, and benzaldehyde are also important. The amount of aldehydes in auto exhaust may be comparable to that of the hydrocarbons (Pitts, 1970).

If unburned gasoline is lost, through evaporation and transfer, the vapor composition typically has the following molar composition

Hydrocarbon:	C_3H_8	i-C_4H_{10}	C_4H_8	n-C_4H_{10} i-C_5H_{12}	C_5H_{10}	n-C_5H_{12}	All C_6H_{14}	Aromatics
Mole %:	1.4	0–7	1–7	60	3–12	2–10	2–7	1–4

Incinerator emissions have variable compositions, but gases identified in the effluent from municipal and commercial incinerators are (Garner, 1972)

Methane	CH_3CHO	Fluoranthene	CCl_2O
Benzene	HCOOH	Oleic acid	Acetone
Toluene	CH_3COOH	Palmitoleic acid	Benzo[*a*]pyrene
C_2H_4	Palmitic acid	Ethyl acetate	Pyrene
CH_2O	Coronene	Ethyl stearate	Benzo[*e*]pyrene
			Benzo[*a*]anthracene

Industrial incinerators emit the same type of pollutants.

An increasing source of man-made pollution is solvent evaporation. Table IV-7 shows the annual production of organic solvents and their estimated usage in the U.S. for 1968. By far the greatest used are naphthas, but aromatics, alcohols, ketones, and halocarbon compounds are also important. It is interesting to note that the total usage estimate is 7.1×10^6 tons/yr, more than twice that estimated in Table IV-5 for 1968 and 1969.

Concentrations

Many hydrocarbons have been identified in urban air and they are listed in Table IV-8. They include almost all the paraffins and olefins of

TABLE IV-6

Typical Auto Exhaust Compositions[a]

Volume % of total hydrocarbons		Volume % of total aldehyde	
Compound	Volume %	Compound	Volume %
Methane[b]	14–18	Formaldehyde	60–73
Ethylene[b]	15–19	Acetaldehyde	7–14
Acetylene[b]	8–14	$C_2H_5CHO + CH_3COCH_3$	0.4–16
Propylene[b]	6–9	Acrolein	2.6–9.8
n-Butane	2–5	Butyraldehydes	1–4
Isopentane	2–4	Crotonaldehyde	0.4–1.4
Toluene	3–8	Valeraldehydes	0.4
Benzene[b]	2.4	Benzaldehyde	3.2–8.5
n-Pentane	2.5	Tolualdehydes	2–7
m- and *p*-Xylene	1.9–2.5	Others	0–10
Butenes[b]	2–6		
Ethane[b]	1.8–2.3		
2-Methylpentane	1.5		
n-Hexane	1.2		
Isooctane	1.0		
Other	22–30		

[a] From Tuesday (1972).
[b] Combustion products.

six or less carbon atoms as well as isooctane (2,2,4-trimethylpentane), C_2H_2, propyne, benzene, a number of simple alkyl benzenes, and a number of aldehydes and ketones.

The city where sampling has been most extensive is Los Angeles, since it has had the most severe problems. Average concentrations as well as high values for 1965 to 1967 are listed in Table IV-8. By far the most dominant hydrocarbon is methane. However, there is a large natural background value for this hydrocarbon, and it is also much less reactive than the others. The other alkanes, of which ethane, propane, and the butanes are most important, have average values of 0.34 ppm, although 10% of the time the values exceed 0.50 ppm. These are very much larger than the background values, which are of the order of 0.003 ppm. The alkenes are next in importance, having average values of 0.10 ppm compared to background values of $\sim 1 \times 10^{-3}$ ppm. Of the alkenes, C_2H_4, C_3H_6, and the butenes are the most important.

The aromatics are the next most important group, having average values of 0.11 and peak values of 0.33 ppm. Of these toluene is the most important, but benzene and the xylenes are also significant. The alkynes, mostly C_2H_2, are next, their average values being 0.04 ppm. Finally there are the aldehydes and ketones, mostly CH_2O and CH_3CHO, but their concentrations are quite variable.

Lonneman *et al.* (1974) reported hydrocarbon concentrations for various locations in the metropolitan New York area from June to August 1969. In Manhattan, the overall average concentrations for this period were: total paraffins, 0.37; total olefins, 0.096; total aromatics, 0.30 ppm. The aromatics were about one-third toluene, about one-third C_8 species, and about one-third C_9 and C_{10} species. The olefin breakdown gave: C_2H_4, 0.047; C_3H_6, 0.014; C_4 olefins, 0.017 ppm.

Apparently there are no reported urban air measurements on alcohol concentrations, although Robinson and Robbins (1968) report significant background values at Point Barrow, Alaska. There the sum of CH_3OH and C_2H_5OH is $(0.1–0.3) \times 10^{-3}$ ppm, and *n*-butanol encompasses the incredible range $(4–12) \times 10^{-3}$ ppm. Thus *n*-butanol is comparable to CH_2O, CH_3CHO, CH_3COCH_3, *n*-hexane, and *n*-pentane, and is exceeded significantly only by CH_4.

Plate I. Dry sepal in orchid caused by C_2H_4 in ambient air at about 0.1 ppm for 6 hr. Flower on left shows severe necrosis of sepals after C_2H_4 exposure. Flower on right was control grown in charcoal-filtered air. From Jacobson and Hill (1970) with permission of the Air Pollution Control Association. Picture courtesy of Professor Ellis F. Darley.

Plate II. Intercostal collapse of leaf tissue on periwinkle (*Vinca* sp.) injured by short-term, high-concentration NO_2 exposure. From Jacobson and Hill (1970) with permission of the Air Pollution Control Association. Picture courtesy of Professor O. Clifton Taylor.

Plate III. Primary leaves from pinto bean (*Phaseolus vulgaris* L) plants. Bronzing and glazing over entire lower surface of leaf injured by PAN. Noninjured leaf from plant grown in activated-charcoal filtered atmosphere. Picture courtesy of Professor O. Clifton Taylor.

Plate IV. Damage to leaves of scarlet oak exposed to 0.25 ppm of O_3 in air for 4 hr. The leaf on the left is the control. Picture courtesy of Professor F. A. Wood.

Plate I

Plate II

Plate III

Plate IV

Plate V

Plate VI

Plate VII

Plate VIII

Effects

Hydrocarbons are relatively nontoxic materials. However, they do have some effects that can be unpleasant, if not harmful. Plants are sensitive to the simple olefins, particularly C_2H_4. At levels of 0.1 ppm C_2H_4 for 8 hr, many plants undergo damage (Jacobson and Hill, 1970). In fact, the African marigold, which is the most sensitive to ethylene, was found to develop epinasty after a 24-hr exposure to only 0.001 ppm. The orchid is also sensitive and dry sepal has been reported after 6-hr exposure to 0.1 ppm as shown in Plate I. Slight dry sepal injury has been observed at exposure to 0.01 ppm for 24 hr. The other olefins are much less toxic and ordinarily would not cause plant damage at their ambient urban concentrations. Propylene and acetylene, the next most toxic olefins, require 60–100 times the concentrations of C_2H_4 to cause the same effects.

Ethylene is a growth hormone in plants; it causes a general reduction in growth, stimulates lateral growth, and decreases apical dominance. Plant leaves or modified leaves may develop epinasty or show chlorosis, necrosis, or abscission (Jacobson and Hill, 1970).

Plate V. Chlorine injury to soybean (*Glycine max* Merr.) produced in the laboratory with 6 ppm for 10 min. Injury occurs as a severe edge necrosis combined with upper-surface necrotic lesions occasionally extending through the leaf. Lesions show heavy pigmentation. The pattern is somewhat dissimilar to that normally associated with ozone in that the lesions tend to run together and do not remain as small individual lesions. From Jacobson and Hill (1970) with permission of the Air Pollution Control Association. Picture courtesy of Professor O. Clifton Taylor.

Plate VI. Hydrogen chloride injury to maple (*Acer* sp.) observed in the laboratory with 7 ppm for 4 hr. Injury is observed primarily as a severe edge necrosis showing dark reddish-tan necrotic areas. From Jacobson and Hill (1970) with permission of the Air Pollution Control Association. Picture courtesy of Dr. Norman L. Lacasse.

Plate VII. Typical accute SO_2 injury to yellow summer squash (*Cuburbita pepo* L). All squash and pumpkin plants bleach to a white or ivory color in the acutely injured areas. Note that the new top leaves not fully expanded at the time of fumigation were not injured. Picture courtesy of Professor Thomas W. Barrett.

Plate VIII. Acute SO_2 injury to the leaves of quince (*Cydonia Oblonga* Mill.). Picture courtesy of Professor Thomas W. Barrett.

TABLE IV-7

Industrial Solvents[a]

Solvent	1968 production (10^8 lb)	Estimated solvent usage (10^8 lb)	Solvent	1968 production (10^8 lb)	Estimated solvent usage (10^8 lb)
Special naphthas	86.5	86.5	Turpentine	2.4	0.4
Perchloroethylene	6.4	5.7	Isopropyl acetate	0.4	0.4
Ethanol	21.3	5.3	Ethyl ether	1.0	0.4
Trichloroethylene	5.2	4.9	Monochlorobenzene	5.8	0.3
Toluene	49.1	4.8	Isopropanol	20.7	0.2
Acetone	13.6	4.1	Diethyleneglycol	0.3	0.1
Xylene	38.3	3.6	Methyl acetate	0.1	0.1
Fluorocarbons[b]	8.2	3.5	Cresols	1.2	0.05
Methyl ethyl ketone	4.5	3.2	Phenol	15.1	—
1,1,1-Trichloroethane	3.0	2.8	Chloroethylene	14.6	—
Methylene chloride	3.0	2.8	1,2-Dichloroethane	0.5	—
Methanol	38.2	2.7	Carbon tetrachloride	7.6	—
Ethylene dichloride	48.0	2.4	Pinene	1.2	—
Ethyl acetate	1.8	1.7	Cyclohexanol	7.2	—
Cyclohexane	20.4	1.6	Cyclohexanone	4.8	—
Methyl isobutyl ketone	1.8	1.5	Ethyl benzene	40.3	—
Hexanes	2.3	1.3	Isobutyl alcohol	1.1	—
Benzene	67.0	~1.0	Chloromethane	1.4	—
n-Butanol	4.3	0.9	*n*-Butyl acetate	0.6	—
Nitrobenzene	4.0	0.5			
			Total	—	142.7

[a] From Garner (1972).
[b] Fluorocarbons used as aerosols; 1969 production figures.

TABLE IV-8

Concentrations (ppm) *in Los Angeles for 1965–1967 of Hydrocarbon Pollutants Identified in Urban Air*

Pollutant	1965		1966		1967		
	Av[a]	High	Av[a,b]	Highest[a,b]	Av[c]	High[c,d]	Background[g]
CH_4	3.22	—	—	—	2.4	3.5	1.5
Other Alkanes	0.3412	—	—	—	—	—	—
Ethane	0.098	—	—	—	0.102[e]	0.18[e]	4×10^{-4}
Propane	0.049	—	—	—	0.076[f]	0.12[f]	—
n-Butane	0.064	—	—	—	0.046	0.080	5×10^{-4}
Isobutane	0.013	—	—	—	0.012	0.020	—
n-Petane	0.035	—	—	—	0.021	0.035	$4\text{–}22 \times 10^{-4}$
Isopentane	0.043	—	—	—	0.035	0.056	—
Cyclopentane	0.004	—	—	—	—	—	—
n-Hexane	0.012	—	—	—	—	—	8×10^{-4}
2-Methylpentane	—	—	—	—	—	—	—
3-Methylpentane	0.008	—	—	—	—	—	—
2,3-Dimethylbutane	0.0012	—	—	—	—	—	—
2,3-Dimethylbutane	0.014	—	—	—	—	—	—
Cyclohexane	—	—	—	—	—	—	—
Methylcyclopentane	—	—	—	—	—	—	—
2-Methylhexane	—	—	—	—	—	—	—
3-Methylhexane	—	—	—	—	—	—	—
2,3-Dimethylpentane	—	—	—	—	—	—	—
2,4-Dimethylpentane	—	—	—	—	—	—	—
2,2,4-Trimethylpentane	—	—	—	—	—	—	—
Nonane and decane	—	—	—	—	—	—	—

TABLE IV-8—Continued

Pollutant	1965		1966		1967		
	Av[a]	High	Av[a,b]	Highest[a,b]	Av[c]	High[c,d]	Background[g]
Alkenes	**0.1020**	—	—	—	—	—	—
Ethylene	0.060	—	—	—	—	—	4×10^{-4}
Propylene	0.018	—	—	—	0.0105	0.021	—
Allene	0.0001	—	—	—	—	—	—
1-Butene	0.007	—	—	—	0.0055	0.010	—
Isobutene							
trans-2-Butene	0.0014	—	—	—	0.002	0.005	—
cis-2-Butene	0.0012	—	—	—			
1,3-Butadiene	0.002	—	—	—	0.002	0.005	—
trans-2-Pentene	0.003	—	—	—	—	—	—
cis-2-Pentene	0.0013	—	—	—	—	—	—
1-Pentene	0.002	—	—	—	0.003	—	—
2-Methyl-1-butene	0.002	—	—	—			
2-Methyl-2-butene	0.004	—	—	—	—	—	—
3-Methyl-1-butene	—	—	—	—	—	—	—
2-Methyl-1,3-butadiene	—	—	—	—	—	—	—
Cyclopentene	—	—	—	—	—	—	—
1-Hexene	—	—	—	—	—	—	—
cis-2-hexene	—	—	—	—	—	—	—
trans-2-Hexene	—	—	—	—	—	—	—

cis-3-Hexene	—	—	—	—	—	—	—
trans-3-Hexene	—	—	—	—	—	—	—
2-Methyl-1-pentene	—	—	—	—	—	—	—
4-Methyl-1-pentene	—	—	—	—	—	—	—
4-Methyl-2-pentene	—	—	—	—	—	—	—
Alkynes	**0.0404**	—	—	—	—	—	—
Acetylene	0.039	—	—	—	—	—	—
Propyne	0.0014	—	—	—	—	—	—
Aromatics	**0.0850**	—	**0.106**	**0.330**	—	—	—
Benzene	0.032	—	0.015	0.057	—	—	1×10^{-4}
Toluene	0.053	—	0.037	0.129	0.030	0.050	—
o-Xylene	—	—	0.008	0.033	0.065	0.011	—
m-Xylene	—	—	0.016	0.061	0.012	0.021	—
p-Xylene	—	—	0.006	0.025	0.005	0.010	—
Ethylbenzene	—	—	0.006	0.022	0.005	0.009	—
m-Ethyltoluene	—	—	0.008	0.027	0.0075	0.015	—
p-Ethyltoluene	—	—					
n-Propylbenzene	—	—	0.002	0.006	0.0045	0.008	—
Isopropylbenzene	—	—	0.003	0.012			
1,2,4-Trimethylbenzene	—	—	—	—	0.012	0.023	—
sec-Butylbenzene	—	—	0.009	0.030			
Isobutylbenzene	—	—					
tert-Butylbenzene	—	—	0.002	0.006			
1,3,5-Trimethylbenzene	—	—	0.003	0.011			

TABLE IV-8—Continued

Pollutant	1965		1966		1967		
	Av[a]	High	Av[a,b]	Highest[a,b]	Av[c]	High[c,d]	Background[g]
Aldehydes and ketones							
Formaldehyde	—	—	—	—	—	—	—
Acetaldehyde	—	—	—	—	—	—	$1\text{–}11 \times 10^{-4}$
Propionaldehyde	—	—	—	—	—	—	—
Acetone	—	—	—	—	—	—	7×10^{-4}
Acrolein	—	—	—	—	—	—	—
Benzaldehyde	—	—	—	—	—	—	—
Butyraldehydes	—	—	—	—	—	—	—
Crotonaldehyde	—	—	—	—	—	—	—
Valeraldehydes	—	—	—	—	—	—	—
Tolualdehydes	—	—	—	—	—	—	—

[a] From Lonneman *et al.* (1968).
[b] Values for 26-day period in Sept.–Nov. 1966.
[c] From Altshuller *et al.* (1971).
[d] Value exceeded by 10% of measurements.
[e] C_2H_6 includes C_2H_4.
[f] C_3H_8 includes C_2H_2.
[g] Measured at Point Barrow, Alaska. From Robinson and Robbins (1968).

Other effects harmful to humans, such as odor threshold, eye irritation, respiratory tract irritation (coughing or sneezing), or drowsiness, dizziness, or drunkeness are not a problem with hydrocarbons in urban air. Table IV-9 lists the concentrations at which these effects occur, and recommended industrial and air standards. In all cases the harmful concentrations are so high that there is no danger in ambient air from any of these compounds.

The only possible exceptions to the above generality appears to be with CH_2O and acrolein. Both of these aldehydes are lachrymators and can irritate the respiratory tract. For acrolein, the industrial threshold value limit has been set at 0.1 ppm. The ambient air quality standard is 0.03 ppm for 24 hr in Russia and the very stringent value of 0.003 ppm for 1 hr in West Germany. Los Angeles air would not be able to meet the last standard, but it is probably unreasonably low.

Formaldehyde begins to cause lachrymation in some individuals at levels of 0.15 ppm for prolonged exposures. Even so, the industrial threshold value limit has been set at 2 ppm, which may be dangerously high. In 1969, the American Industrial Hygiene Association recommended that ambient levels of CH_2O should not exceed 0.1 ppm. European air quality standards are set at 0.03 ppm in West Germany and 0.01 ppm in Russia (NAPCA Publication AP-64, 1970).

It should be realized that ambient air standards must be lower than industrial standards for two reasons. In industries often only one or two gases are impurities in the air, but in cities many gases are present at the same time. As a result they can have an additive effect, and the levels must be set to account for the sum of all the individual levels. Also, in industries only healthy people or those insensitive to the particular pollutant are employed. Cities have much larger, unscreened populations that include the sick, sensitive, and allergic.

Carcinogens

Of particular concern are the compounds that may induce cancer. Several compounds that are carcinogenic to animals have been identified in urban air in all large cities in which surveys have been conducted. It is true, however, that in no case has a suspected organic carcinogenic been demonstrated to produce lung cancer in humans. These carcinogens include the polynuclear aromatic hydrocarbons of which benzo[*a*]pyrene (BaP) is strongly active and benzo[*b* and *j*]fluoranthenes are moderately active (Olsen and Haynes, 1969). It should be pointed out that there is no known correlation of BaP levels in air with lung cancer. There are also polynuclear aza and imino heterocyclic compounds; large quantities of the former are found in some air pollution source effluents. In addition, it is

TABLE IV-9

Concentrations (ppm) *of Hydrocarbons for Various Symptoms*

Compound	Odor threshold[a]	Eye irritation	Respiration tract affected	Drunkenness, drowsiness, or dizziness (10-min exp.)	Industrial standard[b] (8-hr day, 40-hr wk)	Plant damage (8-hr exposure)	Air standard (30 min)	Air standard (24 hr)
CH_4	None	None	None	$>10^5$	None	None	—	—
Other alkanes								
Ethane	None	None	None	$>5 \times 10^4$	None	—	—	—
Propane	$>2 \times 10^4$	$>10^5$	$>10^5$	1×10^5	None	—	—	—
n-Butane	>5000	—	—	1×10^4	—	—	—	—
Isobutane	—	—	—	—	—	—	—	—
n-Pentane	—	>5000	>5000	>5000	500	—	—	—
Isopentane	—	—	—	—	—	—	—	—
Cyclopentane	—	—	—	—	—	—	—	—
n-Hexane	—	—	—	5000	500	—	—	—
2-Methylpentane	—	—	—	—	—	—	—	—
3-Methylpentane	—	—	—	—	—	—	—	—
2,2-Dimethylbutane	—	—	—	—	—	—	—	—
2,3-Dimethylbutane	—	—	—	—	—	—	—	—
Cyclohexane	—	—	—	—	300	—	—	—
Methylcyclopentane	—	—	—	—	—	—	—	—
2-Methylhexane	—	—	—	—	—	—	—	—
3-Methylhexane	—	—	—	—	—	—	—	—
2,3-Dimethylpentane	—	—	—	—	—	—	—	—
2,4-Dimethylpentane	—	—	—	—	—	—	—	—

2,2,4-Trimethylpentane	—	—	—	—	—	—	—	—
Nonane and decane	—	—	—	—	—	—	—	—
Alkenes								
Ethylene	—	—	—	—	None	0.1[c]	—	—
Propylene	—	—	—	—	—	6	—	—
Allene	—	—	—	—	—	—	—	—
1-Butene	—	—	—	—	—	200	—	—
Isobutene	—	—	—	—	—	—	—	—
trans-2-Butene	—	—	—	—	—	—	—	—
cis-2-Butene	—	—	—	—	—	—	—	—
1,3-Butadiene	—	—	—	—	1000	10,000	—	—
trans-2-Pentene	—	—	—	—	—	—	—	—
cis-2-Pentene	—	—	—	—	—	—	—	—
1-Pentene	—	—	—	—	—	—	—	—
2-Methyl-1-butene	—	—	—	—	—	—	—	—
2-Methyl-2-butene	—	—	—	—	—	—	—	—
3-Methyl-1-butene	—	—	—	—	—	—	—	—
2-Methyl-1,3-butadiene	—	8000	8000	—	—	—	—	—
Cyclopentene	—	—	—	—	—	—	—	—
1-Hexene	—	—	—	—	—	—	—	—
cis-2-Hexene	—	—	—	—	—	—	—	—
trans-2-Hexene	—	—	—	—	—	—	—	—
cis-3-Hexene	—	—	—	—	—	—	—	—
trans-3-Hexene	—	—	—	—	—	—	—	—
2-Methyl-1-pentene	—	—	—	—	—	—	—	—
4-Methyl-1-pentene	—	—	—	—	—	—	—	—
4-Methyl-2-pentene	—	—	—	—	—	—	—	—
Alkynes								
Acetylene	—	—	—	1×10^5	—	10	—	—
Propyne	—	—	—	—	1000	—	—	—

TABLE IV-9—Continued

Compound	Odor threshold[a]	Eye irritation	Respiration tract affected	Drunkenness, drowsiness, or dizziness (10-min exp.)	Industrial standard[b] (8-hr day, 40-hr wk)	Plant damage (8-hr exposure)	Air standard (30 min)	Air standard (24 hr)
Aromatics								
Benzene	4.68	—	100	—	25	—	—	—
Toluene	2.14	—	—	>200	100	—	—	—
o-Xylene	—	—	—	—	—	—	—	—
m-Xylene	—	—	—	—	100	—	—	—
p-Xylene	0.47	—	—	—	—	—	—	—
Ethylbenzene	—	—	—	—	100	—	—	—
m-Ethyltoluene	—	—	—	—	—	—	—	—
p-Ethyltoluene	—	—	—	—	—	—	—	—
n-Propylbenzene	—	—	—	—	—	—	—	—
Isopropylbenzene	—	—	—	—	—	—	—	—
1,2,4-Trimethylbenzene	—	—	—	—	25[d]	—	—	—
sec-Butylbenzene	—	—	—	—	—	—	—	—
Isobutylbenzene	—	—	—	—	—	—	—	—
tert-Butylbenzene	—	—	—	—	—	—	—	—
1,3,5-Trimethylbenzene	—	—	—	—	25[d]	—	—	—
Aldehydes and ketones								
Formaldehyde	1.0	0.15	0.4	—	2	—	0.03[e]	0.01[f,g]
Acetaldehyde	0.21	—	—	—	100	—	—	—
Propionaldehyde	—	—	134	—	—	—	—	—
Acetone	100	—	—	—	1000	—	—	—
Acrolein	0.21	1.0	0.60	—	0.1	—	0.003[e]	0.03[f]

Benzaldehyde	—	—	—	—	—	—	—	—
Butyraldehydes	—	>200	>200	—	—	—	—	—
Crotonaldehyde	—	4.0	—	—	—	2	—	—
Valeraldehydes	—	—	—	—	—	—	—	—
Tolualdehydes	—	—	—	—	—	—	—	—

[a] From Leonardos *et al.* (1969).

[b] Threshold limit values, 8-hr day, 40-hr week exposure recommended by American Conference of Governmental Industrial Hygienists (1972).

[c] African marigolds and orchids have lower thresholds.

[d] Sum of trimethylbenzenes =25.

[e] In West Germany.

[f] In Russia.

[g] In Czechoslovakia.

suspected that some airborne polynuclear carbonyl compounds and alkylating agents (epoxides, peroxides, and lactones) may also be carcinogenic. The airborne carcinogens are in the atmosphere as adsorbed matter on soot particles.

There are three possible natural sources of carcinogens (Olsen and Haynes, 1969). Bituminous coal contains BaP, but studies in England reveal a lower incidence of lung cancer among coal miners than in the population at large. Industrial asbestos has adhered on it natural oils containing BaP, and asbestos miners do have a greater potential for developing malignancy than the population at large. However, this potential may just be due to particulate asbestos and not to the BaP. It is also possible that some molds in the environment may produce airborne carcinogenic compounds, although this has not been proven.

Incomplete combustion is the principal source of man-made carcinogens. The sources and estimated emissions are shown in Table IV-10. The

TABLE IV-10

Estimated Annual Benzo[a]pyrene (BaP) Emissions for the United States[a]

Source	Estimated BaP emission rate	Estimated annual consumption or production	Estimated annual BaP emission (tons)
Heat generation	($\mu g/10^6$ Btu)	(10^{15} Btu)	
Coal			
Residential			
Hand-stoked	1,400,000	0.26	400
Underfeed	44,000	0.20	9.7
Commercial	5,000	0.51	2.8
Industrial	2,700	1.95	5.8
Electric generation	90	6.19	0.6
Oil	200	6.79	1.5
Gas	100	10.57	1.2
Total	—	—	421.6
Refuse burning	(μg/ton)	(10^6 tons)	
Incineration			
Municipal	5,300	18	0.1
Commercial	310,000	14	4.8
Open burning			
Municipal refuse	310,000	14	4.8
Grass, leaves	310,000	14	4.8
Auto components	26,000,000	0.20	5.7
Total	—	—	20.2

TABLE IV-10—Continued

Source	Estimated BaP emission rate	Estimated Annual consumption or production	Estimated annual BaP emission (tons)
Industries	(μg/bbl)	(10^6 bbl)	
Petroleum catalytic cracking (catalyst regeneration)			
FCC[b]			
No CO boiler[c]	240	790	0.21
With CO boiler	14	790	0.012
HCC[d]			
No CO boiler	218,000	23.3	5.6
With CO boiler	45	43.3	0.0024
TCC[e] (air lift)			
No CO boiler	90,000	131	13.0
With CO boiler	<45	59	<0.0029
CC (bucket lift)			
No CO boiler	—	119	0.0041
With CO boiler	<31	0	0
Asphalt road mix	50 μg/ton	187,000 tons	0.000010
Asphalt air blowing	<10,000 μg/ton	4,400 tons	<0.000048
Carbon-black manufacturing	Atmospheric samples indicate that BaP emissions from these processes are not extremely high		
Steel and coke manufacturing			
Chemical complex			
Total	—	—	18.8
Motor vehicles	(μg/gal)	(10^{10} gal)	
Gasoline			
Automobiles	170	4.61	8.6
Trucks	>460	2.01	>10
Diesel	690	0.257	2.0
Total	—	—	>20.6
Total (all sources tested)	—	—	481

[a] From Hangebrauck *et al.* (1967) as reported in Olsen and Haynes (1969) with permission.

[b] FCC: fluid catalytic cracker.

[c] CO boiler: carbon monoxide waste heat boiler.

[d] HCC: houdriflow catalytic cracker.

[e] TCC: thermofor catalytic cracker.

TABLE IV-11

Concentrations of Benzo[a]pyrene ($\mu g/m^3$) in Ambient Air for Several U.S. Cities with High Levels[a]

City	Summer 1958	Winter 1958	Winter 1959	Summer 1960	Winter 1960	1st quarter 1966	2nd quarter 1966	3rd quarter 1966	4th quarter 1966
Montgomery, Alabama	—	—	0.024	—	—	—	—	—	—
Indianapolis, Indiana	—	—	0.026	—	—	0.011	0.011	0.00707	0.0124
Hammond, Indiana	—	—	0.039	—	—	0.00572	0.00572	0.00248	0.00149
St. Louis, Missouri	—	—	0.054	—	—	0.00698	0.00698	0.00167	0.00563
Charlotte, North Carolina	—	—	0.039	—	—	0.00509	0.00509	0.00077	0.0119
Cleveland, Ohio	—	—	0.024	—	—	0.00356	0.00356	0.00207	0.00338
Youngstown, Ohio	—	—	0.028	—	—	0.00567	0.00567	0.00599	0.0118
Altoona, Pennsylvania	—	—	0.061	—	—	—	—	—	—
Columbia, South Carolina	—	—	0.024	—	—	0.00239	0.00239	0.00045	—
Chattanooga, Tennessee	—	—	0.031	—	—	0.00621	0.00621	0.00167	0.0193
Knoxville, Tennessee	—	—	0.024	—	—	—	—	—	—
Richmond, Virginia	—	—	0.045	—	—	—	—	—	—
Birmingham, Alabama	0.006	0.074	—	0.003	0.062	0.0163	0.0163	0.00536	0.036
Cincinnati, Ohio	0.002	0.026	—	0.0012	0.018	—	—	—	—
Detroit, Michigan	0.0034	0.031	—	—	—	—	—	—	—

[a] From Olsen and Haynes (1969).

sources include heat generation inceration, industrial processes (mainly petroleum cracking), and motor vehicle emissions. Of the 481 tons/yr estimated BaP emissions in the United States, 400 tons come from hand-stoked residential coal heating.

The polyaromatics are produced by pyrolysis of hydrocarbons followed by pyrosynthesis at 400–500°C. During incomplete combustion some of the hydrocarbon fuel is cracked. These cracked fractions, particularly C_2H_2 and 1,3-C_4H_6 can combine to form polyaromatics and ultimately soot particles (see Chapter V).

Polyaromatic hydrocarbons have been found to be present in all atmospheric studies. Of the nine polyaromatic hydrocarbons analyzed, BaP was found to comprise 5–20% of the total. The concentrations of BaP found in several U.S. urban cities with the highest levels are shown in Table IV-11. The levels reported in the study of 1958–1960 are higher than those reported in 1966. It is not clear whether this represents an improvement in air quality or in analytical procedures. However, it is interesting to note that while urban areas have higher concentrations than rural areas, the largest cities do not necessarily have the worst problems, and Boston, New York, Philadelphia, Chicago, Baltimore, Washington, D.C., Los Angeles, and San Francisco all have lower levels than the highest reported for the cities in Table IV-11. In fact, the highest levels reported were 0.061 $\mu g/m^3$ in Altoona, Pennsylvania, in the winter of 1959. However, even this value is very much lower than the threshold limit value of 200 $\mu g/m^3$ (0.019 ppm based on BaP) for an 8-hr day, 40-hr week exposure adopted by the American Conference of Governmental Industrial Hygienists (1972) for coal tar pitch volatiles that contain polyaromatic molecules.

Of course, the winter months have the highest levels, since that is when heating is at a peak and reflects the dominance of heating in affecting concentration levels. The contribution from automobiles was shown in a study in Detroit, Michigan (Colucci and Begeman, 1965) to be 18% in the air over freeways, 5% in downtown air, and 42% from the suburbs in winter.

OXIDES OF NITROGEN

Of the oxides of nitrogen, the only important man-made pollutants are NO and NO_2, and we shall confine our discussion to them.

Emissions

It is estimated that 500×10^6 tons of NO are emitted worldwide each year from natural sources, principally bacterial action. In addition, man-

TABLE IV-12

Worldwide Urban Emissions of Nitrogen Oxides (as NO_2)[a]

Fuel and source type	Fuel usage	Emission factor	NO_2 emission (tons)
Coal			
Power generation	$1,219 \times 10^6$ tons[b]	20 lb/ton[e]	12.2×10^6
Industrial	$1,369 \times 10^6$ tons[b]	20 lb/ton[e]	13.7×10^6
Domestic and commercial	404×10^6 tons[b]	5 lb/ton[e]	1.0×10^6
Petroleum			
Refinery production	$11,317 \times 10^6$ bbl[b]	6 ton/10^5 bbl[f]	0.7×10^6
Gasoline	379×10^6 tons[b]	0.113 lb/gal[e]	7.5×10^6
Kerosene	100×10^6 tons[b]	0.072 lb/gal[e]	1.3×10^6
Fuel oil	287×10^6 tons[b]	0.072 lb/gal[e]	3.6×10^6
Residual oil	507×10^6 tons[b]	0.104 lb/gal[e]	9.2×10^6
Natural gas			
Power generation	2.98×10^{12} ft^3 [c]	390 lb/10^6 ft^3 [e]	0.6×10^6
Industrial	10.72×10^{12} ft^3 [c]	214 lb/10^6 ft^3 [e]	1.1×10^6
Domestic and commercial	6.86×10^{12} ft^3 [c]	116 lb/10^6 ft^3 [e]	0.4×10^6
Others			
Incineration	500×10^6 tons	2 lb/ton[d]	0.5×10^6
Wood fuel	466×10^6 tons[d]	1.5 lb/ton[d]	0.3×10^6
Forest fire	324×10^6 tons[b]	5 lb/ton[g]	0.8×10^6
		Total	52.9×10^6

[a] From Robinson and Robbins (1969) with permission.
[b] 1967 U.S. Statistical Abstracts.
[c] Figure is $1.28 \times$ U.S. usage as per 1967 U.S. Statistical Abstracts.
[d] World Forest Inventory, U.N. 1963.
[e] From Mayer (1965).
[f] From Elkin (1962).
[g] From Gerstle and Kemnitz (1967).

made sources of NO and NO_2 account for another 53×10^6 tons/yr of NO_x (computed as NO_2). From this flux and the worldwide average ambient concentration of 7.5×10^{10} molecules/cm³ (3×10^{-3} ppm) of NO_x,* the lifetime for NO_x is found to be about 9 days from the formula

$$\tau = [NO_x]/F\{NO_x\}\,(\xi_{NO_x} + \xi_D) \qquad (4)$$

where $F\{NO_x\} = 4.1 \times 10^{10}$ molecules/cm²-sec and ξ_{NO_x} and ξ_D are, re-

* NO_x here refers to only NO + NO_2. Concentrations of NO_3 and N_2O_5 are negligible. N_2O is not included.

spectively, $\sim 21 \times 10^{-7}$ and ~ 0 cm^{-1}. On a global scale, the man-made contribution to NO_x production is significant but not dominant. The flux of 4.1×10^{10} molecules/cm^2-sec for $NO + NO_2$ calculated from the worldwide emissions is offset essentially by the downward flow of HNO_3 from the stratosphere.

The worldwide man-made urban emissions of nitrogen oxides are shown in Table IV-12. Almost half of these emissions are from coal combustion, with petroleum product combustion being almost as large. The U.S. accounts for almost 40% of the world emissions; the data for the late 1960s are shown in Table IV-13. Los Angeles County, which accounts for over ½% of the world emissions, shows a dramatic difference in that >68% of the NO_x came from motor vehicles alone in 1969. It should also be noticed

TABLE IV-13

Summary of Emissions (10^6 tons/yr) of Nitrogen Oxides

Source	U.S. 1966[a]	U.S. 1967[a]	U.S. 1968[a]	Los Angeles 1967[b]	Los Angeles 1969[c]
Transportation	**7.6**	**7.6**	**8.1**	**0.202**	**0.242**
Motor vehicles	6.6	6.7	7.2	0.197	0.236
Other	1.0	0.9	0.9	0.0047	0.0055
Fuel combustion	**6.7**	**9.5**	**10.0**	**0.102**	
Coal	4.0	3.8	4.0	—	
Fuel oil	0.9	1.0	1.0	—	
Natural gas	1.6	4.2	4.5	—	
Wood	0.2	0.2	0.2	—	
LPG and kerosene	—	0.3	0.3	—	
Industrial processes	**0.2**	**0.2**	**0.2**	**0.0197**	
Petroleum refining, marketing, products	—	—	—	0.0164	**0.113**
Ferrous metals	—	—	—	0.0011	
Glass, asphalt other minerals	—	—	—	0.0022	
Solid waste disposal	**0.5**	**0.6**	**0.6**	**0.0036**	
Miscellaneous	**1.7**	**1.7**	**1.7**	—	
Man-made	0.5	0.5	0.5	—	
Forest fires	1.2	1.2	1.2	—	
Total	16.7	19.6	20.6	0.327	0.345

[a] From APCO Publication No. AP-84 (1971).
[b] From Lemke *et al.* (1967).
[c] From Mosher *et al.* (1969).

that from 1967 to 1969 motor vehicle emissions of NO_x rose 20% in Los Angeles County and all sources had an increase of 10%. In the U.S. as a whole, emissions also increased in the late 1960s. From 1966 to 1968, the total NO_x emissions rose over 23%, but principally from fuel combustion.

Concentration

The worldwide background concentrations of NO and NO_2 have been estimated by Robinson and Robbins (1968, 1969). They show marked differences with geographical location, ranging from $<0.6 \times 10^{-3}$ ppm in Antarctica up to 10×10^{-3} ppm in the Panama forests. From all the world data Robinson and Robbins conclude that background values can be considered to be 2×10^{-3} ppm NO and 4×10^{-3} ppm NO_2 over land areas between 65° north and 65° south latitude, and ten times less for each constituent elsewhere in the world. The world grand average is about 1×10^{-3} ppm NO (2.5×10^{10} molecules/cm^3) and 2×10^{-3} ppm NO_2 (5.0×10^{10} molecules/cm^3).

The concentrations of NO_x in Los Angeles are a very strong seasonal function as shown in Fig. IV-4. They peak in the winter and are minimal

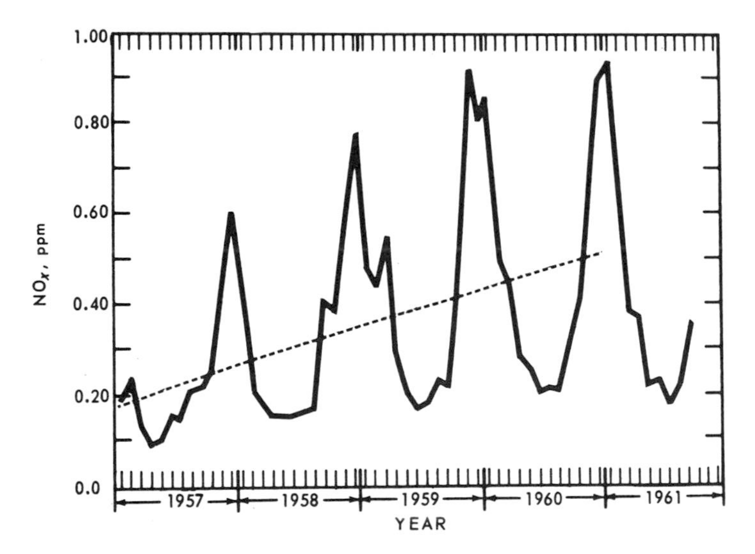

Fig. IV-4. Monthly means of daily maximum NO_x concentrations at Los Angeles Civic Center, 1957–1961. From APCO Publication No. AP-84 (1971) with permission of the Environmental Protection Agency.

in the summer, mainly reflecting the meteorological conditions in the Los Angeles Basin. However, this effect is also observed in Chicago (1966), Denver (1966), and Bayonne (1967), although to a slighter degree. Undoubtedly some other seasonal factor is also at work. It should also be noticed in Fig. IV-4 that superimposed on the seasonal variation is an upward drift in NO_x concentrations from <0.2 ppm at the beginning of 1957 to over 0.5 ppm in 1961. On two occasions, Los Angeles County recorded total NO_x concentrations above 3 ppm, but $[NO_2]$ alone has rarely exceeded 1 ppm.

More detailed information on NO_x concentrations in eight major U.S. cities is given in Tables IV-14 and IV-15. These values have been measured by the Continuous Air Monitoring Program (CAMP) and are not accurate for values <0.10 ppm. (Up to 200% errors may be involved in individual readings.) Even average values above 0.10 ppm may be in error because they are averages of values containing readings <0.10 ppm. If we examine yearly averages for NO_x, Chicago, Cincinnati, and St. Louis passed through maxima in 1966, but Philadelphia and Washington, D.C., increased in average levels throughout the 1960s. Insufficient data are available for Denver, Los Angeles, and San Francisco.

Effects and Standards

The oxides of nitrogen can have deleterious effects on materials, vegetation, and animal and human life (APCO Publication No. AP-84, 1971). Some textile dye fading has been shown to be due to NO_x. In a Berkeley, California, test cotton fiber showed loss of strength when exposed to ambient air, although this may have been due to oxidant rather than NO_x exposure. Nitric acid aerosols attack nylon and have caused "runs" in nylon stockings. High nitrate content in airborne dust led to premature (<2 yr) cracking of nickel–brass wire springs in some relays located in the Los Angeles area central office of Pacific Telegraph Co. in 1959. In several U.S. cities the nickel bases of palladium-topped electrical contacts showed corrosion due to anions, principally nitrates, in accumulated dust.

Ordinarily, ambient air NO_x is not harmful to plants. However, plant damage has been seen from NO_2 near a HNO_3 factory where the ambient NO_2 levels are high. The plant injury that can be caused in plants is summarized in Table IV-16. At concentrations of 0.15 ppm or greater for long exposures, chronic effects begin to appear in tomato plants. Other plants

TABLE IV-14

Nitric Oxide Concentration[a] at Camp Sites, by Averaging Time and Frequency, 1962–1968[b]

City and averaging time	Maximum concentration (ppm) for year							Data available (%)	% of time concentration (ppm) is exceeded							
	1962	1963	1964	1965	1966	1967	1968		0.01	0.1	1	10	30	50	70	90
Chicago																
5 min	0.66	0.63	0.97	0.67	0.74	0.69	0.66	79	0.67	0.52	0.33	0.18	0.11	0.08	0.05	0.02
1 hr	0.61	0.60	0.91	0.62	0.64	0.63	0.61	79	0.64	0.50	0.32	0.18	0.11	0.08	0.05	0.02
8 hr	0.41	0.39	0.49	0.34	0.42	0.41	0.40	83	—	0.40	0.27	0.17	0.11	0.08	0.06	0.03
1 day	0.36	0.35	0.35	0.28	0.34	0.26	0.21	86	—	—	0.23	0.15	0.11	0.09	0.07	0.04
1 month	0.17	0.15	0.16	0.13	0.14	0.15	0.11	94	—	—	—	0.13	0.11	0.08	0.07	0.06
1 yr	0.10	0.10	0.10	0.10	0.10	0.08	0.07	100	—	—	—	—	—	0.10	—	—
Cincinnati																
5 min	0.60	0.54	0.68	0.63	1.18	1.46	1.26	70	1.08	0.57	0.29	0.09	0.03	0.02	0.01	0.00
1 hr	0.58	0.50	0.65	0.62	1.00	1.38	1.02	71	0.98	0.55	0.29	0.09	0.03	0.02	0.01	0.00
8 hr	0.38	0.22	0.34	0.40	0.46	0.86	0.58	74	—	0.40	0.24	0.08	0.04	0.02	0.01	0.01
1 day	0.29	0.15	0.31	0.31	0.37	0.36	0.34	77	—	—	0.19	0.08	0.04	0.03	0.02	0.01
1 month	0.06	0.06	0.10	0.07	0.07	0.06	0.08	80	—	—	—	0.06	0.04	0.03	0.02	0.02
1 yr	0.03	0.03	0.04	0.03	0.04	0.03	—	86	—	—	—	—	—	0.03	—	—
Denver																
5 min	—	—	—	0.46	0.59	0.48	0.68	68	0.55	0.42	0.23	0.09	0.04	0.02	0.01	0.00
1 hr	—	—	—	0.40	0.54	0.36	0.61	68	0.52	0.39	0.22	0.09	0.04	0.02	0.01	0.00
8 hr	—	—	—	0.16	0.30	0.19	0.33	72	—	0.24	0.17	0.08	0.04	0.02	0.02	0.01
1 day	—	—	—	0.14	0.23	0.11	0.21	74	—	—	0.13	0.08	0.04	0.03	0.02	0.01
1 month	—	—	—	0.05	0.08	0.07	0.07	79	—	—	—	0.07	0.05	0.03	0.02	0.01
1 yr	—	—	—	0.03	0.04	0.04	0.04	100	—	—	—	—	—	0.04	—	—
Los Angeles																
5 min	2.11	0.97	1.32	—	—	—	—	72	1.23	0.85	0.54	0.23	0.08	0.03	0.01	0.00
1 hr	1.42	0.79	1.24	—	—	—	—	72	1.12	0.80	0.53	0.23	0.08	0.03	0.01	0.00
8 hr	0.60	0.49	0.60	—	—	—	—	78	—	0.55	0.41	0.21	0.09	0.04	0.02	0.01
1 day	0.46	0.38	0.44	—	—	—	—	81	—	—	0.36	0.19	0.10	0.05	0.03	0.01

Philadelphia																
5 min	1.58	1.72	1.35	1.03	1.98	1.74	1.68	75	1.60	0.95	0.37	0.12	0.05	0.03	0.01	0.00
1 hr	1.57	1.51	1.15	0.89	1.87	1.49	1.44	76	1.49	0.93	0.36	0.12	0.05	0.03	0.01	0.00
8 hr	0.93	0.91	0.52	0.53	1.00	0.72	0.87	80	—	0.72	0.31	0.11	0.06	0.03	0.02	0.01
1 day	0.46	0.58	0.25	0.31	0.48	0.34	0.37	82	—	—	0.25	0.10	0.06	0.04	0.02	0.01
1 month	0.08	0.08	0.09	0.07	0.13	0.10	0.12	87	—	—	—	0.09	0.06	0.05	0.03	0.02
1 yr	0.04	0.05	0.04	0.04	0.06	0.06	0.06	100	—	—	—	—	—	0.04	—	—
St. Louis																
5 min	—	—	0.84	0.52	0.61	0.45	0.44	76	0.51	0.33	0.19	0.08	0.04	0.02	0.01	0.00
1 hr	—	—	0.75	0.37	0.57	0.41	0.41	76	0.41	0.31	0.18	0.08	0.04	0.02	0.01	0.00
8 hr	—	—	0.22	0.17	0.26	0.23	0.18	81	—	0.22	0.14	0.07	0.04	0.02	0.01	0.01
1 day	—	—	0.14	0.12	0.25	0.17	0.13	84	—	—	0.10	0.07	0.04	0.03	0.02	0.01
1 month	—	—	0.05	0.06	0.05	0.07	0.05	87	—	—	—	0.05	0.04	0.03	0.03	0.02
1 yr	—	—	0.03	0.03	0.03	0.04	0.03	100	—	—	—	—	—	0.03	—	—
San Francisco																
5 min	0.50	1.64	0.81	—	—	—	—	72	1.25	0.64	0.43	0.18	0.09	0.05	0.03	0.01
1 hr	0.42	1.30	0.63	—	—	—	—	72	0.87	0.60	0.42	0.18	0.09	0.05	0.03	0.01
8 hr	0.26	0.52	0.41	—	—	—	—	77	—	0.43	0.31	0.18	0.09	0.06	0.03	0.01
1 day	0.15	0.32	0.23	—	—	—	—	77	—	—	0.22	0.15	0.10	0.07	0.05	0.02
1 month	0.06	0.11	0.13	—	—	—	—	83	—	—	—	0.11	0.09	0.08	0.06	0.05
1 yr	—	0.09	0.09	—	—	—	—	67	—	—	—	—	—	0.09	—	—
Washington																
5 min	0.68	0.73	1.03	0.74	1.15	1.26	0.89	79	—	—	0.42	0.10	0.03	0.01	0.01	0.00
1 hr	0.65	0.68	0.88	0.62	1.02	1.14	0.69	80	0.88	0.62	0.33	0.09	0.03	0.01	0.01	0.00
8 hr	0.39	0.51	0.54	0.39	0.55	0.55	0.43	84	—	0.48	0.27	0.09	0.04	0.02	0.01	0.00
1 day	0.22	0.26	0.29	0.21	0.41	0.36	0.31	86	—	—	0.20	0.08	0.04	0.02	0.01	0.01
1 month	0.07	0.08	0.07	0.06	0.09	0.08	0.10	93	—	—	—	0.07	0.04	0.03	0.02	0.01
1 yr	0.03	0.04	0.03	0.03	0.04	0.04	0.04	100	—	—	—	—	—	0.03	—	—

[a] Determined by continuous Griess–Saltzman method.
[b] From APCO Publication No. AP-84 (1971) with permission of the Environmental Protection Agency.

TABLE IV-15

Nitrogen Dioxide Concentration[a] at Camp Sites, by Averaging Time and Frequency, 1962–1968[b]

City and averaging time	Maximum concentration (ppm) for year							Data available (%)	% of time concentration (ppm) is exceeded							
	1962	1963	1964	1965	1966	1967	1968		0.01	0.1	1	10	30	50	70	90
Chicago																
5 min	0.26	0.23	0.79	0.21	0.35	0.29	0.29	79	0.28	0.18	0.13	0.07	0.05	0.04	0.03	0.02
1 hr	0.22	0.21	0.47	0.18	0.31	0.25	0.18	80	0.25	0.18	0.12	0.07	0.05	0.04	0.03	0.02
8 hr	0.16	0.17	0.22	0.15	0.20	0.15	0.12	84	—	0.16	0.12	0.07	0.05	0.04	0.04	0.03
1 day	0.12	0.13	0.15	0.09	0.15	0.11	0.10	87	—	—	0.10	0.07	0.05	0.04	0.04	0.03
1 month	0.06	0.06	0.07	0.05	0.08	0.07	0.06	93	—	—	—	0.06	0.05	0.04	0.04	0.03
1 yr	0.04	0.04	0.05	0.04	0.06	0.05	0.05	100	—	—	—	—	—	0.04	—	—
Cincinnati																
5 min	0.26	0.30	1.00	0.21	0.30	0.70	1.19	79	0.26	0.16	0.09	0.05	0.04	0.03	0.02	0.02
1 hr	0.23	0.25	0.34	0.17	0.24	0.24	0.56	79	0.24	0.15	0.09	0.05	0.04	0.03	0.02	0.02
8 hr	0.12	0.14	0.14	0.10	0.13	0.13	0.16	83	—	0.12	0.08	0.05	0.04	0.03	0.02	0.02
1 day	0.07	0.09	0.10	0.08	0.09	0.07	0.10	86	—	—	0.07	0.05	0.04	0.03	0.03	0.02
1 month	0.04	0.04	0.05	0.04	0.05	0.04	0.05	94	—	—	—	0.04	0.04	0.03	0.03	0.02
1 yr	0.03	0.03	0.03	0.03	0.04	0.03	0.03	100	—	—	—	—	—	0.03	—	—
Denver																
5 min	—	—	—	0.31	0.30	0.37	0.38	71	0.31	0.18	0.11	0.06	0.04	0.03	0.03	0.02
1 hr	—	—	—	0.28	0.27	0.33	0.25	72	0.27	0.17	0.10	0.06	0.04	0.03	0.03	0.02
8 hr	—	—	—	0.14	0.18	0.17	0.12	76	—	0.13	0.09	0.05	0.04	0.03	0.03	0.02
1 day	—	—	—	0.09	0.09	0.08	0.12	79	—	—	0.08	0.05	0.04	0.03	0.03	0.02
1 month	—	—	—	0.04	0.04	0.04	0.05	81	—	—	—	0.04	0.04	0.03	0.03	0.03
1 yr	—	—	—	0.04	0.03	0.04	0.04	100	—	—	—	—	—	0.04	—	—
Los Angeles																
5 min	—	0.57	1.27	—	—	—	—	63	0.66	0.44	0.25	0.12	0.06	0.04	0.03	0.01
1 hr	—	0.52	0.68	—	—	—	—	63	0.55	0.40	0.24	0.11	0.06	0.04	0.03	0.01
8 hr	—	0.37	0.27	—	—	—	—	67	—	0.27	0.23	0.11	0.06	0.05	0.03	0.02
1 day	—	0.20	0.26	—	—	—	—	71	—	—	0.18	0.10	0.07	0.05	0.04	0.02

Philadelphia															
5 min	0.24	0.36	0.37	0.24	0.29	0.44	0.22	79	0.25	0.18	0.11	0.07	0.04	0.03	0.03	0.02
1 hr	0.22	0.32	0.26	0.20	0.23	0.23	0.20	79	0.23	0.18	0.11	0.07	0.04	0.04	0.03	0.02
8 hr	0.14	0.19	0.14	0.10	0.15	0.13	0.12	84	—	0.13	0.10	0.06	0.05	0.04	0.03	0.02
1 day	0.10	0.14	0.10	0.07	0.10	0.10	0.09	86	—	—	0.08	0,06	0.05	0.04	0.03	0.02
1 month	0.05	0.06	0.05	0.04	0.05	0.05	0.05	90	—	—	—	0.05	0.04	0.04	0.03	0.03
1 yr	0.04	0.04	0.04	0.04	0.04	0.04	0.04	100	—	—	—	—	—	0.04	—	—
St. Louis																
5 min	—	—	0.31	0.28	0.21	0.17	0.20	81	0.21	0.13	0.08	0.05	0.03	0.03	0.02	0.01
1 hr	—	—	0.22	0.13	0.20	0.16	0.18	81	0.20	0.13	0.08	0.05	0.03	0.03	0.02	0.01
8 hr	—	—	0.14	0.08	0.14	0.09	0.09	87	—	0.11	0.07	0.05	0.03	0.03	0.02	0.01
1 day	—	—	0.12	0.05	0.10	0.06	0.05	89	—	—	0.07	0.04	0.03	0.03	0.02	0.01
1 month	—	—	0.04	0.03	0.04	0.03	0.04	92	—	—	—	0.04	0.03	0.03	0.02	0.02
1 yr	—	—	0.04	0.03	0.03	0.03	0.02	100	—	—	—	—	—	0.03	—	—
San Francisco																
5 min	0.50	0.38	0.52	—	—	—	—	70	0.43	0.31	0.18	0.08	0.06	0.04	0.03	0.01
1 hr	0.27	0.33	0.41	—	—	—	—	70	0.38	0.29	0.18	0.08	0.06	0.04	0.03	0.01
8 hr	0.16	0.23	0.31	—	—	—	—	75	—	0.23	0.16	0.08	0.06	0.04	0.03	0.01
1 day	0.10	0.13	0.18	—	—	—	—	75	—	—	0.12	0.07	0.06	0.04	0.03	0.02
1 month	0.04	0.06	0.08	—	—	—	—	83	—	—	—	0.07	0.05	0.05	0.04	0.02
1 yr	0.03	0.05	0.06	—	—	—	—	100	—	—	—	—	—	0.03	—	—
Washington																
5 min	0.37	0.24	0.24	0.25	0.19	0.24	0.26	79	0.22	0.16	0.10	0.06	0.04	0.03	0.03	0.02
1 hr	0.30	0.22	0.23	0.23	0.18	0.21	0.24	79	0.21	0.15	0.10	0.06	0.04	0.03	0.03	0.02
8 hr	0.10	0.11	0.14	0.10	0.09	0.14	0.13	84	—	0.11	0.09	0.06	0.04	0.03	0.03	0.02
1 day	0.07	0.09	0.10	0.07	0.07	0.09	0.08	86	—	—	0.08	0.06	0.04	0.04	0.03	0.02
1 month	0.04	0.04	0.04	0.04	0.04	0.05	0.05	95	—	—	—	0.05	0.04	0.04	0.03	0.03
1 yr	0.03	0.03	0.04	0.03	0.03	0.04	0.05	100	—	—	—	—	—	0.03	—	—

[a] Determined by continuous Griess–Saltzman method.

[b] From APCO Publication No. AP-84 (1971) with permission of the Environmental Protection Agency.

susceptible at progressively higher levels are Navel oranges, pinto beans, alfalfa, and oats. With 3 ppm of NO, there is an immediate reduction in photosynthesis in pinto bean and tomato plants, but they show complete recovery when the NO is removed. The leaf damage caused by high-level NO_2 poisoning of periwinkle is shown in Plate II.

There is no known case of human poisoning by NO. The toxicological findings have been reviewed (APCO Publication No. AP-84, 1971). Guinea pigs exposed to 16–50 ppm for 4 hr showed no effect, but mice exposed at 2500 ppm for 6–7 min showed physiological effects. However, they recovered rapidly when the NO was removed (continued exposure led to death in 12 min).

Continued exposure to 30 ppm NO_2 caused no fatalities in animals, but higher levels did. NO_2 exposure affects the respiratory tract, but there was no noticeable effect in animals exposed to levels <5 ppm. However, if the animals were killed and autopsied, changes in lung tissues were noticed at lower levels, e.g., rabbits exposed to 0.25 ppm for 4 hr/day for 6 days showed structural changes in lung collogen.

In man NO_2 affects the respiratory tract. The odor threshold is 0.12 ppm, and NO_2 may be one of the principal odors in diesel exhaust fuel (APCO Publication No. AP-84, 1971). In addition, NO_2 is a brown gas, and on smoggy days about 20% of the brown haze is due to this gas. Chronic respiratory effects begin at 10–40 ppm exposure. At 25–75 ppm, the effects are more severe, but recovery is complete after exposure is terminated.

An epidemiological study in Chattanooga, Tennessee, was done in the late 1960s on the effects of NO_2 in four residential areas (Shy *et al.*, 1970a,b). One area, near a large TNT plant, had high NO_2 and low particulate levels. Another had high suspended particulate and low NO_2 exposure. The two other areas were "clean" controls. The results indicated that school children exposed for six months to NO_2 levels >0.06 ppm had a higher incidence of respiratory tract illness than students in the "clean" areas. The National Air Sampling Network (NASN) has shown that the yearly average NO_2 levels in the 1960s exceeded 0.06 ppm in 10% of U.S. cities with populations $<50{,}000$; in 54% of cities of 50,000 to 500,000; and 85% of cities $>500{,}000$.

The threshold limit values in 1972 approved by the American Conference of Governmental Industrial Hygienists were 25, 5, and 2 ppm, respectively, for NO, NO_2, and HNO_3 for an 8-hr day, 40-hr week exposure. The U.S. ambient air standards for NO_2 are 0.25 ppm for 1 hr and 0.05 ppm for the yearly average. The former limit was exceeded in Los Angeles County on at least 78 days in each of the years 1960–1966.

TABLE IV-16

Chronic Injury and Physiological Effects of NO *and* NO_2[a]

Plant	Exposure			Effects
	Concentration		Time[b] (days)	
	ppm	mg/m³		
		NO		
Gramineae				
Oat (*Avena sativa* L.)	0.6	0.7	V	Reduced rate of apparent photosynthesis
Leguminosae				
Alfalfa (*Medicago sativa* L.)	0.6	0.7	V	Reduced rate of apparent photosynthesis
Pinto bean (*Phaseolus vulgaris* L.)	3–10	3.7–12.3	V	Immediate reduction up to 70% in apparent photosynthesis
Solanaceae				
Tomato (*Lycopersicon esculentum* Mill)	3–10	3.7–12.3	V	Immediate reduction up to 70% in apparent photosynthesis
		NO_2		
Gramineae				
Oat (*Avena sativa* L.)	0.6	1.1	V	Reduced rate of apparent photosynthesis
Leguminosae				
Alfalfa (*Medicago sativa* L.)	0.6	1.1	V	Reduced rate of apparent photosynthesis
Pinto bean (*Phaseolus vulgaris* L.)	0.3	0.6	10–19	Decrease in dry weight, increase in chlorophyll content per unit weight
Rutaceae				
Navel orange (*Citrus sinensis* Osbeck)	0.5–1.0	0.9–1.9	35	Chlorosis of leaves, extensive leaf-drop
	0.25	0.47	240	Increased leaf-drop, reduced yield
Solanaceae				
Tomato (*Lycopersicon esculentum* Mill)	0.15–0.26	0.3–0.5	10–22	Decrease in dry weight and leaf area, increase in chlorophyll, downward curvature of leaves

[a] From APCO Publication No. AP-84 (1971) with permission of the Environmental Protection Agency.

[b] V = variable.

OXIDANT

Oxidant refers to that group of compounds in the air of highly oxidative nature. The principal compound of this group is ozone, but also included are peroxides, hydroperoxides, and peroxy nitrates, of which peroxyacetyl nitrate (PAN), the lowest homolog, is most prevalent. However, peroxypropionyl, peroxybutyryl, peroxyisobutyryl, and peroxybenzoyl nitrates have also been identified.

The oxidants are the most annoying of all the air pollutants, although strictly speaking they are not really air pollutants since they are not emitted into the air. They are produced *in situ* through photochemical oxidation, and this subject will be discussed in detail in Chapter VI.

Concentration

Background levels of O_3 usually vary between 0.01 and 0.03 ppm with an average value of 0.025 ppm (6×10^{11} molecules/cm^3-sec). Possibly the most interesting background measurement was made by Steinberger (1974) in Jerusalem, Israel, on the Days of Atonement, 1970–72. Jerusalem has no industry, and vehicular traffic is prohibited on the Day of Atonement in Jerusalem. She found values of 0.025 ppm (6.3×10^{11} molecules/cm^3) on each of those days. However, urban atmospheres have very much higher concentrations, the highest recorded value being 0.90 ppm on September 13, 1955, in Vernon, California (in Los Angeles County) (Lemke *et al.*, 1967). This is a very high level indeed when it is realized that the odor threshold is 0.015 ppm. Although O_3 is not a lachrymator, eye irritation starts at $\sim$0.10 ppm of oxidant due to the peroxy nitrate compounds, CH_2O, and acrolein. Peroxybenzoyl nitrate is the most effective lachrymator known to be present in the atmosphere and is 200 times more eflective than CH_2O (Heuss and Glasson, 1968).

The threshold limit values for 1972 adopted by the American Conference on Governmental and Industrial Hygienists for O_3 for 8-hr/day, 40-hr/week exposure is 0.1 ppm. The California standard for the 1960s was 0.15 ppm for 1 hr, and the U.S. federal standard in the early 1970s was 0.08 ppm of oxidant for 1 hr, not to be exceeded more than once a year.

The actual situation is considerably worse than the standards would permit, as shown in Table IV-17, which summarizes the data in several U.S. cities for 1964–67. The maximum hourly average was an incredible 0.58 ppm in Los Angeles and 0.46 ppm in Pasadena during this period. On one out of every three days in these cities, the oxidant maximum hourly average exceeded the 0.15 ppm level. In at least 10 of the 12 cities

TABLE IV-17

Summary of Maximum Oxidant Concentrations Recorded in Selected Cities, 1964–1967[a]

Station	Total days of available data	Number and percent of total days with maximum hourly average equal to or greater than concentration specified: 0.15 ppm		0.10 ppm		0.05 ppm		Maximum hourly average (ppm)	Peak concentration (ppm)
		Days	Percent of days	Days	Percent of days	Days	Percent of days		
Pasadena	728	299	41.1	401	55.1	546	75.0	0.46	0.67
Los Angeles	730	220	30.1	354	48.5	540	74.0	0.58	0.65
San Diego	623	35	5.6	130	20.9	440	70.6	0.38	0.46
Denver[b]	285	14	4.9	51	17.9	226	79.3	0.25	0.31
St. Louis	582	14	2.4	59	10.1	362	62.2	0.35	0.85
Philadelphia	556	13	2.3	60	10.9	233	41.9	0.21	0.25
Sacramento	711	16	2.3	104	14.6	443	62.3	0.26	0.45
Cincinnati	613	10	1.6	55	9.0	319	52.0	0.26	0.32
Santa Barbara	723	11	1.5	76	10.5	510	70.5	0.25	0.28
Washington, D.C.	577	7	1.2	65	11.3	313	54.2	0.21	0.24
San Francisco	647	6	0.9	29	4.5	185	28.6	0.18	0.22
Chicago	530	0	0	24	4.5	269	50.8	0.13	0.19

[a] From NAPCA Publication No. AP-63 (1970) with permission of the Environmental Protection Agency.

[b] Eleven months of data beginning February 1965.

listed, the federal standard of 0.08 ppm/hr was exceeded at least 3 days a month.

The days of highest oxidant concentrations show a seasonal variation as depicted in Fig. IV-5. In Los Angeles these peak oxidant days occurred in the summer, and apparently every day of the summers of 1964 and 1965 the 0.15 ppm/hr level was exceeded. In Denver the peak is in July; and in Phoenix, in May. Most of the oxidant is O_3, but an important second component is PAN. Its levels are shown in Fig. IV-6 for the 1966–1967 year at Riverside, California. The PAN levels more or less follow the total oxidant levels and account for 2–6% of the oxidant. As in Los Angeles, the Riverside levels are at a peak in summer and a minimum in December.

Fortunately, with the introduction of more stringent pollution control devices on automobiles in the early 1970s, oxidant levels have fallen in the central cities. Altshuller (1975) summarized the results of the CAMP site measurements from 1964 to 1973 in six central cities. The results by month are listed in Table IV-18 for the number of days the federal oxidant

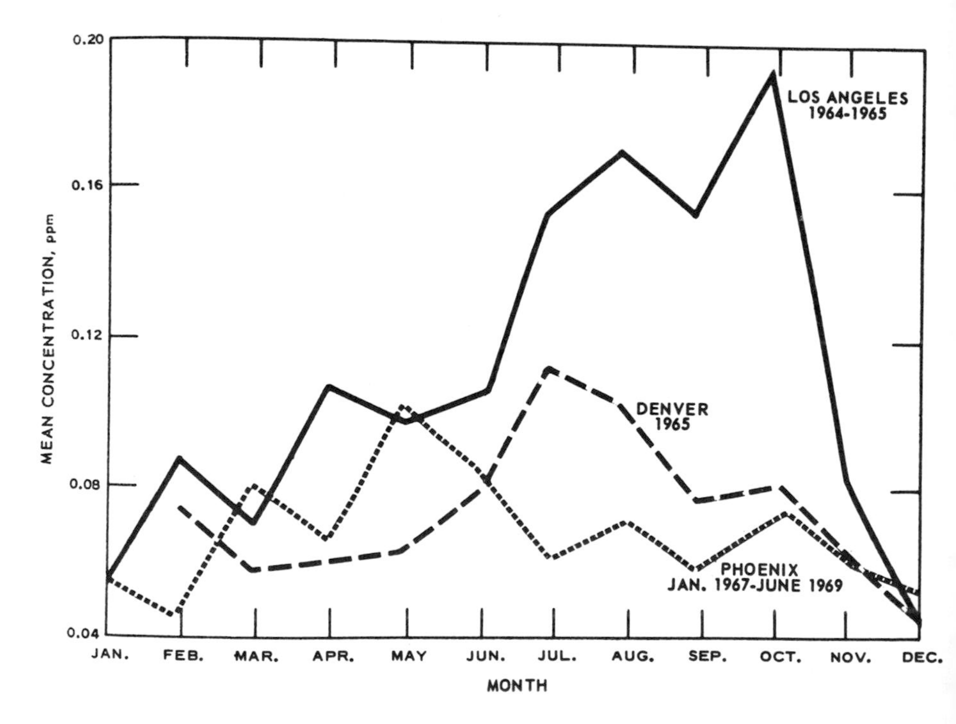

Fig. IV-5. Monthly variation of mean daily maximum 1-hr average oxidant concentrations for three selected cities. From NAPCA Publication No. AP-63 (1970) with permission of the Environmental Protection Agency.

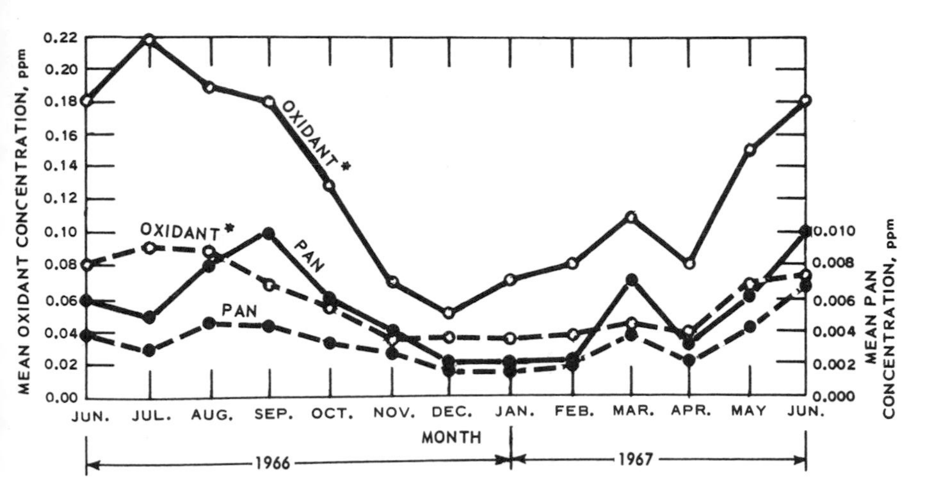

Fig. IV-6. Monthly variation of oxidant and PAN concentrations at the Air Pollution Research Center, Riverside, California June 1966–June 1967. —, Monthly means of daily maximum 1-hr average concentrations; – –, monthly means of 1-hr average concentrations; *, oxidant by MAST, continuous 24 hr, PAN by PANALYZER, sequential 6 A.M. to 4 or 5 P.M. only. From NAPCA Publication No. AP-63 (1970) with permission of the Environmental Protection Agency.

standard was exceeded. It is clear that the summer months are the worst because of the enhanced photochemical activity during this season. On a significant fraction of days during the summer, the federal standard of 0.08 ppm has been exceeded. However, the trend with advancing years is encouraging as seen by the data in Fig. IV-7. With the possible exception of Chicago, the air quality in the central cities as regards oxidant levels was considerably improved in the early 1970s over that in the 1960s. Interestingly, there does not appear to be much, if any, difference between weekdays and Sundays.

On the negative side, the outlying suburbs of the cities are not showing improvement. In the Los Angeles Basin, Azusa, Riverside, and San Bernardino are not showing the same downward trend, and in the perimeter of Washington, D.C., there was a higher frequency of days of elevated O_3 and higher maximum O_3 concentrations in 1972 and 1973 than earlier (Altshuller, 1975).

Recent measurements also indicate large rural O_3 levels. Thus in Appalachia there were several days of hourly average O_3 concentrations of 0.05–0.07 ppm and peak values of 0.12–0.17 ppm (Corn *et al.*, 1975). In rural New York, the average daily O_3 concentration for the first 17 days

TABLE IV-18

Distribution of Days by Month of Year for 1964–1973 with Oxidant Concentrations Exceeding 0.08 ppm[a]

Month	Chicago	Cincinnati	Denver	Philadelphia	St. Louis	Washington, D.C.	Total
January	0	0	0	0	0	0	0
February	0	0	2 (0)	0	1 (0)	0	3 (0)
March	0	0	6 (1)	0	4 (1)	1 (1)	11 (3)
April	0	5 (1)	11 (1)	3 (0)	7 (1)	3 (0)	29 (3)
May	2 (1)	21 (3)	14 (4)	10 (2)	20 (4)	19 (5)	86 (19)
June	7 (2)	33 (3)	14 (1)	35 (8)	44 (7)	47 (8)	180 (29)
July	15 (3)	43 (3)	49 (11)	52 (12)	41 (8)	43 (5)	243 (42)
August	21 (5)	34 (6)	44 (4)	50 (8)	42 (10)	55 (8)	246 (41)
September	5 (0)	27 (1)	29 (3)	18 (3)	9 (3)	44 (5)	132 (15)
October	0	3 (0)	15 (1)	8 (2)	5 (2)	1 (0)	32 (5)
November	0	1 (0)	6 (2)	1 (0)	3	1 (0)	12 (2)
December	0	0	1 (0)	1 (0)	0	0	2 (0)
Total	50 (11)	168 (17)	191 (28)	178 (35)	176 (36)	214 (32)	977 (159)
June–August (% of total)	86 (91)	65 (71)	56 (57)	77 (80)	72 (70)	68 (66)	68 (70)
May–October (% of total)	100 (100)	96 (94)	86 (86)	97 (100)	91 (95)	98 (97)	94 (95)

[a] From Altshuller (1975). Numbers in parentheses represent Sundays in the test period.

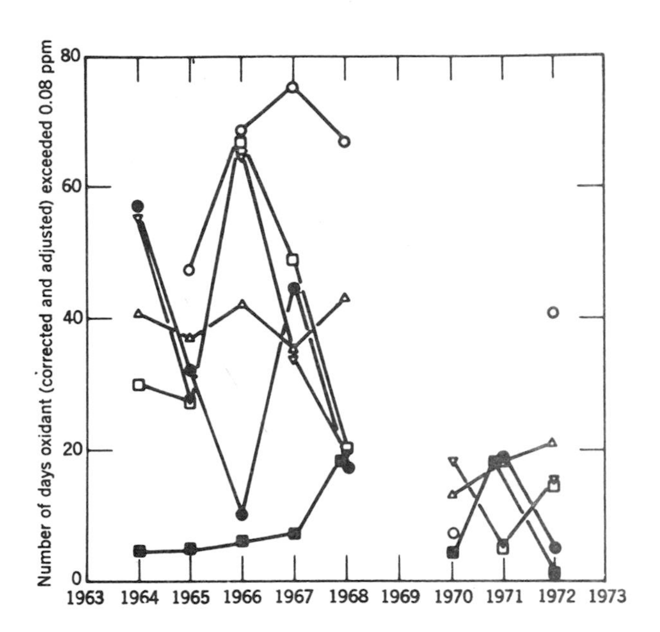

Fig. IV-7. Trends in oxidant results by year at CAMP sites. ○, Denver; □, St. Louis; △, Washington, D.C.; ▽, Philadelphia; ●, Cincinnati; ■, Chicago. From Altshuller (1975) with permission of the Air Pollution Control Association.

of August 1973 ranged between 0.030 and 0.102 ppm, with about one-half those days over 0.060 ppm (Stasiuk and Coffey, 1974). High rural values correlate with high urban values, and it has been suggested that the high rural values could be caused by transport of urban pollutants. However, Stasiuk and Coffey (1974) showed that for rural New York, the levels are too high for transport to account for them, and that background terpenes and NO produced in the woods are probably responsible for the high O_3 levels.

Effects

Oxidant is usually considered to be the most unpleasant and harmful of the air pollutants. However, O_3 may also have one beneficial effect, since it can kill some types of bacteria and protozoa. For example, streptococcus salivarius shows 90% mortality at 0.025 ppm O_3 and 60–80% relative humidity for 30 min, although for other bacteria and protozoa considerably higher concentrations are needed (>1 ppm) for reasonable mortality rates

TABLE IV-19

Relative Phytotoxicity of Four Members of the Peroxyacyl Nitrates, Indicated by Preliminary Fumigation on Two Species of Plants[a]

Species	Fumigation time (hr)	Toxicant[b] PAN Conc (ppb)	PAN Injury (%)	PPN Conc (ppb)	PPN Injury (%)	PBN Conc (ppb)	PBN Injury (%)	P *iso*-BN Conc (ppb)	P *iso*-BN Injury (%)
Pinto bean	0.5	—	—	100	90	100	80	100	80
	1.0	140	55	24	7	30	59	—	—
	4.0	40	90	10	86	—	—	—	—
	8.0	20	44	5	100	—	—	—	—
Rosy morn petunia	0.5	—	—	50	7	100	90	25	18
	1.0	140	33	24	35	12	45	—	—

[a] From Taylor (1969) as reported in NAPCA Publication No. AP-63 (1970) with permission of the Environmental Protection Agency.

[b] PAN = peroxyacetyl nitrate; PPN = peroxypropionyl nitrate; PBN = peroxybutyryl nitrate; P *iso*-BN = peroxyisobutyryl nitrate.

(NAPCA Publication No. AP-63, 1970). Because of this property of O_3, germicidal lamps were popular in the 1940s and 1950s to kill bacteria by producing O_3 with ultraviolet radiation. However, because of the deleterious properties of O_3, this practice has been discontinued.

Plants are susceptible to damage, both by O_3 and peroxyacetylnitrate (PAN). In many species of plants PAN causes glazing or bronzing of the under surface of leaves, an example of which is shown in Plate III. Table IV-19 summarizes the concentrations of the various peroxyacylnitrates found in air needed to cause damage to pinto beans and petunia.

Ozone causes leaf damage and causes reduction in yield and growth. The damage of 0.25 ppm exposure for 4 hr to scarlet oak is shown in Plate IV. The levels at which damage to various species occur are shown in Table IV-20. The peanut even has slight damage at 0.02 ppm of O_3, which is the natural background value.

TABLE IV-20

Threshold Susceptibility of Plants to Acute Injury from Ozone[a]

Species	Concentration[b] (ppm)	Time (hr)	Effects[c]
Tobacco (*Nicotiana tabacum* L.)			
Bel-W3	0.15	0.5	1
	0.75	0.2	1
	0.15	1.0	1
	0.08	2.0	2
	0.03	8.0	1
	0.15	2.0	2
White Gold	0.05	3.0	1
Bean (*Phaseolus vulgaris* L.)	0.15	0.5	1
	0.40	0.33	1
	0.08	4.0	1
Oat (*Avena sativa*)	0.12	2.0	1
White pine (*Pinus strobus* L.)	0.10	4.0	1
Alfalfa (*Medicago sativa* L.)	0.20	2.0	2
Tomato (*Lycopersicon esculentum* Mill.)	0.08	1.0	1
Radish (*Raphanus sativus* L.)	0.08	2.0	1
Onion (*Allium cepa* L.)	0.40	2.0	3
Peanut (*Arachis hypogaea*)	0.02	24–48	1

[a] From NAPCA Publication No. AP-63 (1970) with permission of the Environmental Protection Agency.

[b] Concentrations have been corrected to neutral KI values. Reported Mast oxidant values were multiplied by a factor of 1.5.

[c] Severity of injury: 1 = slight, 2 = moderate, 3 = severe.

TABLE IV-21

LD_{50} *of Ozone for Various Species after 3-hr Exposure*[a]

Species	Ozone (ppm)
Guinea pigs	51.7
Rabbits	36.0
Mice	21.0
Rats	21.8
Cats	34.5

[a] From Mittler *et al.* (1956).

O_3 also causes serious damage to textiles and elastomers. With textiles, both the fabric and the dye can be attacked. In elastomers, cracking occurs, but only if the elastomer is under stress. Since the most common use of elastomers is rubber in rubber tires, a stress situation does occur. Experiments have shown that at 2–3% stress, 0.01–0.02 ppm of O_3 (background values) can cause cracks perpendicular to the stress axis. At 100% stress, an O_3 level of 0.02 ppm causes cracking in 65 min. In fact, crack development in rubber is so easy that it has been used as an analytical method to detect O_3, both qualitatively and quantitatively.

In animals and humans, O_3 principally attacks the lungs and respiratory tract, but changes have been observed in the heart, liver, and brains of animals (NAPCA Publication AP-63, 1970). Peroxy nitrates are not known to cause effects (at atmospheric levels) other than eye irritation. In mice the peroxyacylnitrates have the same toxicity as NO_2.

In animals O_3 poisoning can be fatal and the LD_{50} concentrations (dosage needed for 50% mortality) are shown in Table IV-21. It can be seen that levels >20 ppm are needed for 50% mortality in the animals tested. At levels >6 ppm hemorrhage and edema occur; at 4–6 ppm, hemorrhagic change; but at 3.2 ppm for 18–22 hr, no pathological changes were observed in the lungs of rats. However, prolonged exposure (>1 yr) at 1.0 ppm can lead to (1) long-term pulmonary effects, (2) lung tumor acceleration, and (3) aging. In addition, the animals develop a tolerance for O_3.

As far as human health is concerned, several studies have been made (NAPCA Publication AP-63, 1970). There is no demonstrated correlation of ambient oxidation concentrations to lung cancer mortality, general mortality, or hospital admissions. One study of asthma patients has shown an increased incidence of attacks on days when the maximum oxidant level exceeds 0.25 ppm or the maximum hourly average exceeds 0.15 ppm. The athletic performance of high school cross-country runners has been shown to decrease at 0.10 ppm oxidant.

The effect of short-term exposure of O_3 on humans is summarized in Table IV-22. The odor is present at background values and gives air its "fresh" smell. In most healthy individuals the first serious irritation starts at 0.3 ppm, and the symptoms continue to increase in severity as the concentration grows. However, all persons exposed to 3 ppm of O_3 or less have shown complete and rapid recovery when the O_3 is removed. One man accidentally exposed to 9 ppm for an unspecified time recovered after two weeks, but still complained of fatigue and shortness of breath after 9 months.

TABLE IV-22

Effects of O_3 Exposure on Humans

Effect	$[O_3]$ (ppm)	Duration of exposure (hr)
Odor threshold	0.015	—
Background	0.025	—
Federal standard	0.08	1[a]
Athletic performance affected	0.10	—
Eye irritation[b]	—	—
California standard	0.15	1
Asthmatics have increased attacks	0.15	1
Nasal and throat irritation	0.30	—
Changes in pulmonary functions	0.5–1.0	1–2
Coughing, exhaustion	0.94	1.5
Lack of coordination, inability to express thoughts, chest pains, cough for 2 days, tired for 2 weeks	1.5–2	2
Sleepiness	3.0	1
Headache, shortness of breath, increased pulse	4.0	$\frac{1}{2}$

[a] Not more than once a year.

[b] O_3 is not an eye irritant, but at oxidant levels of $\sim$0.10 ppm eye irritation usually began because of presence of PAN, peroxybenzoyl nitrate, CH_2O, and acrolein.

HALOGENATED COMPOUNDS

A number of halogenated hydrocarbons are used as solvents and therefore evaporate into the air. Their production and estimated usage are listed in Table IV-23, along with their odor thresholds and recommended industrial threshold limit values (TLV). All of the TLVs are way above any possible atmospheric concentrations, even though these are not known. The physiological effect of chlorinated compounds is to dissolve fat, and thus they tend to attack the liver. It has recently been reported (1973) that C_2H_3Cl may be carcinogenic at 50 ppm.

Recently much attention has been given to the fluorochloromethanes because of the possible damage they may cause the stratospheric ozone layer when they photodissociate. The two important ones are $CFCl_3$, used primarily in aerosol spray cans, and CF_2Cl_2, used as a refrigerant as well as in aerosol spray cans. These compounds have no natural origin and are entirely man-made. Their total cumulative world production as of 1974

TABLE IV-23

Halocarbon Industrial Solvents

	1968 Production[a] ($\times 10^8$ lb)	Estimated usage[a] ($\times 10^8$ lb)	Odor threshold[b] (ppm)	TLV industrial[c] (ppm)
C_2Cl_4	6.4	5.7	4.68	100
C_2Cl_3H	5.2	4.9	21.4	100
Fluorocarbons[d]	8.2	3.5	—	—
CCl_3CH_3	3.0	2.8	—	350
CH_2Cl_2	3.0	2.8	214	250
$C_2H_2Cl_2$	48.0	2.4	—	200
C_6H_5Cl	5.8	0.3	0.21	75
C_2ClH_3	14.6	—	—	200[e]
$CClH_2CClH_2$	0.5	—	—	50
CCl_4	7.6	—	100	10
CH_3Cl	1.4	—	>10	100

[a] From Garner (1972).
[b] From Leonardos *et al.* (1969).
[c] From Am. Conf. Government Hygienists (1972).
[d] Fluorocarbons used as aerosols; 1969 production figures.
[e] Carcinogenic at 50 ppm.

TABLE IV-24

Other Halogenated Compounds in the Air

Compound	Odor threshold[a] (ppm)	TLV, industrial[b]
Cl_2	0.31	1 ppm
HCl	10	5 ppm
HF	—	3 ppm
HBr	—	—
CCl_2O	1.0	0.1 ppm
CF_2O	—	—
DDT	—	1 mg/m^3
Chlordane	—	0.5 mg/m^3
Dieldrin	—	0.25 mg/m^3
Heptachlor	—	0.5 mg/m^3
Lindane	—	0.5 mg/m^3

[a] From Leonardos *et al.* (1969).
[b] From American Conference of Government Hygienists (1972).

was 4390 million lb for $CFCl_3$ and about twice that much for CF_2Cl_2 (Rowland and Molina, 1974). Their production rates increased 8.7%/yr between 1961 and 1974. The sum of their average concentrations in the troposphere is about 1×10^{-4} ppm, with about $\frac{2}{3}$ CF_2Cl_2 and $\frac{1}{3}$ $CFCl_3$, the same as their fraction produced. However, urban concentrations can be very much higher, and Hester *et al.* (1974) report values of 7×10^{-4} ppm for CF_2Cl_2 and 5.6×10^{-4} ppm for $CFCl_3$ in Los Angeles in a moderately smoggy day in July 1970. Readings were about 25% lower when the air was clear in February 1973.

Because of the introduction of chlorofluoromethanes into the atmosphere it is expected that there will be at least a 5% depletion in the stratospheric ozone layer in the next 20–40 years even if the current production rate remains constant (Rowland and Molina, 1974). This could lead to a 10% increase in harmful ultraviolet radiation reaching the earth's surface, which could increase skin cancer incidence by at least 8000 cases a year in the white U.S. population.

Other halogenated compounds may also be in the atmosphere and these are listed in Table IV-24. A number of these are introduced as herbicide sprays. Two of the more important may be Cl_2 and HCl, both of which have significant concentrations in localized areas near industrial plants using them. HCl is also produced in the combustion and atmospheric oxidation of chlorinated hydrocarbons. Phosgene (CCl_2O) is a major oxidation product of C_2Cl_4.

Chlorine can cause plant damage, and alfalfa and radish show injury at 0.1 ppm for 2 hr. Chlorine injury to soybean exposed to 6 ppm for 10 min is shown in Plate V. Shriner and Lacasse (1969) found that Bonny Best tomato plants showed bronzing followed by necrosis after 72 hr when exposed to 5 ppm HCl for 2 hr. Plate VI shows HCl damage to maple at slightly higher dosages.

SULFUR COMPOUNDS

The sulfur compounds in the atmosphere consist of H_2S, mercaptans, SO_2, SO_3, H_2SO_4, sulfites, sulfates, and organic sulfur aerosols. The most important of these are H_2S, SO_2, and sea spray sulfites and sulfates.

Emissions

The only known natural source of SO_2 emission is from volcanic eruptions, which also produce H_2S. However, H_2S has major natural production sources from plant and animal decay. The SO_2 comes from various pollutant sources, these being the combustion of fossil fuels, the refining of petroleum, the smelting of ores containing sulfur, the manufacture of H_2SO_4, the burning of refuse, paper making, and the burning or smoldering of coal refuse banks. The trend has been away from SO_2 pollution by low-level disperse sources and toward large point sources, except for space heating and diesel trucks using fuels of high sulfur content. The large source emissions contain lower concentrations of polynuclear hydrocarbons and higher concentrations of nitrogen oxides and sulfuric acid (NAPCA Publication No. AP-50, 1970).

The worldwide emissions of sulfur compounds, as estimated by Robinson and Robbins (1969) are summarized in Table IV-25 for both the Northern and Southern Hemispheres. Almost all the man-made pollutants are in the Northern Hemisphere. The Northern Hemisphere accounts for 93% of the SO_2 and 69% of the total sulfur content emitted. It should be noticed that while there is more tonnage of SO_2 than H_2S emitted, more sulfur is deposited in the atmosphere in the form of H_2S. On the basis of sulfur content, H_2S, SO_2, and the sea spray sulfite and sulfate account for 101 (46.3%), 73 (33.5%), and 44 (20.2%) million tons/yr, respectively.

The long-term historical trend in SO_2 emissions is summarized in Table IV-26. From 1860 to 1965, SO_2 emissions have increased 28-fold. From 1860 to 1910, they doubled every 16 yr. Then the exponential rate slowed, but the absolute rate continued to increase except for the period from 1920 to 1930. The last doubling occurred from 1940 to 1965.

The atmospheric SO_2 emissions in the U.S. for 1963 and 1966 are shown in Table IV-27. The U.S. output is about one-fourth of the world's output, with over half coming from coal combustion. In regard to sulfuric acid manufacture, which is done either by the chamber or contact process, the emissions from uncontrolled factories are as follows (Public Health Service Publication No. 999-AP-13, 1965). For the chamber process, the H_2SO_4 in the exit gases come from the Gay Lussac tower and has 5–30 mg H_2SO_4/ft^3 of gas. The mist contains 10% dissolved nitrogen oxides, and over 90%

TABLE IV-25

Total Hemispheric Sulfur Emissions (10^6 tons/yr)[a]

Source	Total	Northern Hemisphere	Southern Hemisphere
SO_2	**146**	**136**	**10**
Coal	102	98	4
Petroleum combustion and refining	28.5	27.1	1.4
Copper smelting	12.9	8.6	4.3
Lead smelting	1.5	1.2	0.3
Zinc smelting	1.3	1.2	0.1
H_2S	**107**	**69**	**38**
Biological (land)	72	52	20
Biological (sea)	32	14	18
SO_2 pollutant sources	3	3	0
Kraft paper mills	0.06	0.06	0
Sea spray (sulfates, sulfites)	**44**[b]	**19**[b]	**25**[b]

[a] For the years 1963–66. From Robinson and Robbins (1969).

[b] Expressed as tons of sulfur. Actual tonnage is larger.

TABLE IV-26

Estimated Historical SO_2 *Emission* ($\times 10^6$ *tons*)[a]

	Coal	Petroleum	Copper	Lead	Zinc	Total
1860	5.0	0.0	0.22	—	—	5.22
1870	7.8	0.01	0.24	—	—	8.05
1880	12.2	0.07	0.26	—	—	12.53
1890	18.7	0.17	0.68	—	—	19.53
1900	28.1	0.33	1.60	0.47	0.15	30.65
1910	42.1	0.70	2.84	0.61	0.26	46.51
1920	49.5	1.79	3.26	0.54	0.22	55.31
1930	51.3	3.12	3.52	0.93	0.46	59.33
1940	61.0	4.62	5.46	0.97	0.53	72.58
1950	66.0	8.30	5.92	0.86	0.60	81.68
1960	95.7	19.9	10.0	1.28	1.0	127.88
1965	102.0	28.5	12.9	1.5	1.3	146.20

[a] From Robinson and Robbins (1969).

of the particles are $>3\ \mu m$ in diameter. For the contact process, the exit gas from the absorber contains SO_2, H_2SO_4, and unreacted SO_3. The stack gas is 0.1–0.5 volume % SO_2. The H_2SO_4 mist comprises 3–15, and the unreacted SO_3 comprises 0.5–48 mg per standard cubic foot of gas. Usually the SO_3 level is near the lower end of the range.

Man-made sources of H_2S are mainly from the same industrial sources that produce SO_2. Sheehy *et al.* (1963) estimate industrial sources of H_2S to be 2% of SO_2 emitted or 3×10^6 tons/yr on a worldwide basis. Of particular significance is the H_2S emitted from kraft paper mills. This amounts to only 0.06×10^6 tons/yr, so is no national or global pollution problem. However, H_2S has an odor threshold of 0.47 ppb (exactly 1000 times lower than for SO_2). Wherever it is emitted it is of considerable annoyance. Anywhere that SO_2 is emitted, the unpleasant odor is more likely to be due mainly to the H_2S present rather than the SO_2 itself.

Robinson and Robbins (1968, 1969), as well as others, list sea spray as a contribution to sulfur emissions. Actually the reverse must be occurring to balance the sulfur cycle, and sulfates and sulfites presumably are being absorbed by the earth in the biological cycle. Robinson and Robbins (1968) estimate that the 212×10^6 tons/yr of sulfur emitted return to the earth via 90×10^6 and 71×10^6 tons/yr precipitation and dry deposition into land and sea areas, respectively, 26×10^6 tons/yr uptake by vegetation no land, and 25×10^6 tons/yr gaseous absorption by the oceans.

TABLE IV-27

Atmospheric Sulfur Dioxide Emissions in 1963 and 1966 by Source[a]

Process	Sulfur dioxide[b]			
	1963		1966	
	Tons	% of total emissions	Tons	% of total emissions
Burning of coal				
Power generation (211,189,000 tons, 1963 data)	9,580,000	41.0	11,925,000	41.6
Other combustion (112,630,000 tons, 1963 data)	4,449,000	19.0	4,700,000	16.6
Subtotal	14,029,000	60.0	16,625,000	58.2
Combustion of petroleum products				
Residual oil	3,703,000	15.9	4,386,000	15.3
Other products	1,114,000	4.8	1,218,000	4.3
Subtotal	4,817,000	20.7	5,604,000	19.6
Refinery operations	1,583,000	6.8	1,583,000	5.5
Smelting of ores	1,735,000	7.4	3,500,000	12.2
Coke processing	462,000	2.0	500,000	1.8
Sulfuric acid manufacture	451,000	1.9	550,000	1.9
Coal refuse banks	183,000	0.8	100,000	0.4
Refuse incineration	100,000	0.4	100,000	0.4
Total emissions	23,360,000	100.0	28,562,000	100.0

[a] From NAPCA Publication No. AP-50 (1970) with permission of the Environmental Protection Agency.

[b] A small amount of this tonnage is converted to sulfuric acid mist before discharge to the atmosphere. The rest is eventually oxidized and/or washed out. Only under unusual meteorologic conditions does accumulation occur. The increasing output to sulfur oxides due to increasing power demand is evident.

Concentrations

Throughout the whole troposphere over the whole globe, mole fractions of the major sulfur compounds are relatively constant. The average values are 0.2 ppb each for SO_2 and H_2S, although the H_2S value is based on only one measurement in Wales. For sulfates the value is 2 $\mu g\ SO_4/m^3$ at the earth's surface. These values are summarized in Table IV-28.

TABLE IV-28

Average Tropospheric Concentrations of Sulfur Compounds[a]

Material	Average concentration	Average concentration as sulfur ($\mu g/m^3$)
SO_2	0.2 ppb	0.25
H_2S	0.2 ppb	0.25
SO_4^{2-}	2 $\mu g/m^3$	0.7

[a] From Robinson and Robbins (1968).

Very much higher concentrations of H_2S have been found over an oil field. Graedel *et al.* (1974) measured H_2S in Artesia Junction, New Mexico, over the Permian oil basin. They found peak concentrations (20-min averages) >1 ppm and long-term averages of 12 to 24 ppb, very much higher than global average ambient levels. The highest concentrations occurred at night for conditions of low wind turbulence.

The SO_2 measurements are of particular interest since 93% of the SO_2 is produced in the Northern Hemisphere. The measurements in this hemisphere generally vary between 0.1 and 0.3 ppb, although values as high as 1 ppb were found off the southeast coast of Florida by Junge (1963). Surprisingly, values of 0.3–1 ppb were found by Cadle *et al.* (1967) in Panama (near the equator) and in Antarctica. It is not known why there is such a uniform (or possibly higher in Antarctica) distribution of atmospheric SO_2, when the input is so nonuniform.

With the above concentrations and the flux of the various sulfur compounds into the atmosphere (98, 73, and 44 $\times 10^6$ tons/yr of sulfur, respectively, for H_2S, SO_2, and sulfate), the lifetimes become, respectively, 4.0, 6.5, and 22.7 days.

In the early 1960s a worldwide aerosol band was discovered in the lower stratosphere at 17–24 km (Robinson and Robbins, 1968). The relative aerosol concentration is 4 times that in the troposphere and consists of particles mainly 0.1–2 μm, 80% of which is sulfate (mainly H_2SO_4). There are also traces of Al, Cl, Ca, and Fe. The particles in the stratosphere presumably grow until they reach 2–4 μm, at which point they fall to the earth by gravitational settling. About 3 $\times 10^4$ tons/yr (as S) is cycled, and a lifetime of 6 months has been estimated. The aerosol layer scatters sunlight, preferentially favoring the shorter wavelengths, and is probably the cause of the purple coloration of the sky at twilight.

Although the average concentrations of SO_2 in the troposphere are 0.2 ppb, the concentrations in urban air are orders of magnitude greater, as shown in Table IV-29. This table summarizes maximum hourly, daily, and yearly averages for eight U.S. cities from 1962 to 1967 as determined from CAMP monitoring. Of these cities the severest SO_2 problem is in Chicago, which has had peak hourly, daily, and monthly averages of 1.69, 0.79, and 0.18 ppm, respectively, during 1963–1964. Philadelphia is second with

TABLE IV-29

CAMP Data on SO_2 Concentrations for Several U.S. Cities[a]

City	SO_2 concentration (ppm)					
	1962	1963	1964	1965	1966	1967
	Maximum hourly average					
Chicago	0.86	1.69	1.12	1.14	0.98	1.11
Cincinnati	0.46	0.48	0.55	0.57	0.41	0.42
Denver	—	—	—	0.36	0.26	0.17
Los Angeles	0.13	0.19	0.29	—	—	—
Philadelphia	1.03	0.85	0.84	0.94	0.66	0.77
St. Louis	—	—	0.73	0.96	0.84	0.55
San Francisco	0.11	0.26	0.16	—	—	—
Washington	0.38	0.48	0.62	0.35	0.45	0.37
	Maximum daily average					
Chicago	0.36	0.71	0.79	0.55	0.48	0.65
Cincinnati	0.11	0.11	0.14	0.18	0.10	0.13
Denver	—	—	—	0.06	0.05	0.02
Los Angeles	0.06	0.07	0.10	—	—	—
Philadelphia	0.35	0.46	0.43	0.35	0.36	0.35
St. Louis	—	—	0.26	0.19	0.18	0.21
San Francisco	0.05	0.05	0.08	—	—	—
Washington	0.18	0.25	0.22	0.20	0.25	0.15
	Yearly average					
Chicago	0.10	0.14	0.18	0.13	0.09	0.12
Cincinnati	0.03	0.03	0.04	0.03	0.03	0.02
Denver	—	—	—	0.02	0.01	—
Los Angeles	0.03	0.02	0.02	—	—	—
Philadelphia	0.09	0.06	0.09	0.08	0.09	0.10
St. Louis	—	—	0.06	0.05	0.04	0.03
San Francisco	—	0.01	0.02	—	—	—
Washington	0.05	0.05	0.04	0.05	0.04	—

[a] From NAPCA Publication No. AP-50 (1970).

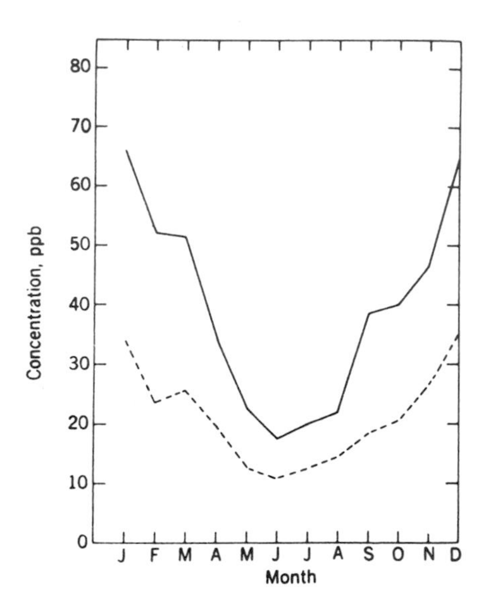

Fig. IV-8. Annual cycle of sulfur dioxide concentration in suburban Long Island from monthly means, 1968–1969. —, Elmont; – –, Centereach. From Raynor *et al.* (1974) with permission of the Air Pollution Control Association.

1.03, 0.46, and 0.10 ppm, respectively. By comparison, the highest daily and annual averages recorded by the NASN network were 0.38 and 0.17 ppm, both in New York City. These three cities have the most serious SO_2 problems in the U.S. The CAMP data show little, if any, trend with time for the period 1962–1967, although New York City air has been improving in regard to SO_2 content. The SO_2 concentration is strongly dependent on season, increasing in the winter when heating increases. The seasonal variation for suburban Long Island in 1968–1969 is shown in Fig. IV-8, and the pronounced summer minimum is apparent.

For six U.S. cities, the percentage of time any SO_2 concentration was exceeded in the 1962–1967 time period is shown in Fig. IV-9. From this graph the six cities rank, in order of decreasing SO_2 problem, Chicago, Philadelphia, St. Louis, Cincinnati, Los Angeles, and San Francisco. In Chicago almost 0.1% of the time (8 hr/yr) the SO_2 concentrations exceeded 1.0 ppm.

The values discussed so far refer to measurements at test stations representative of a city. In actuality, since SO_2 is emitted from point stationary sources, there is a large fluctuation depending on proximity to the emission source. Tests have shown that the maximum SO_2 concentrations in the vicinity of a point source can be 100–200 times the annual average for the area. Values up to 5 ppm have been recorded. This effect is nicely shown in a study by Mosher *et al.* (1969) in Los Angeles, where the SO_2 emissions tend to be concentrated in the southwest coastal region. Their data are shown in Table IV-30. They correlate annual average of

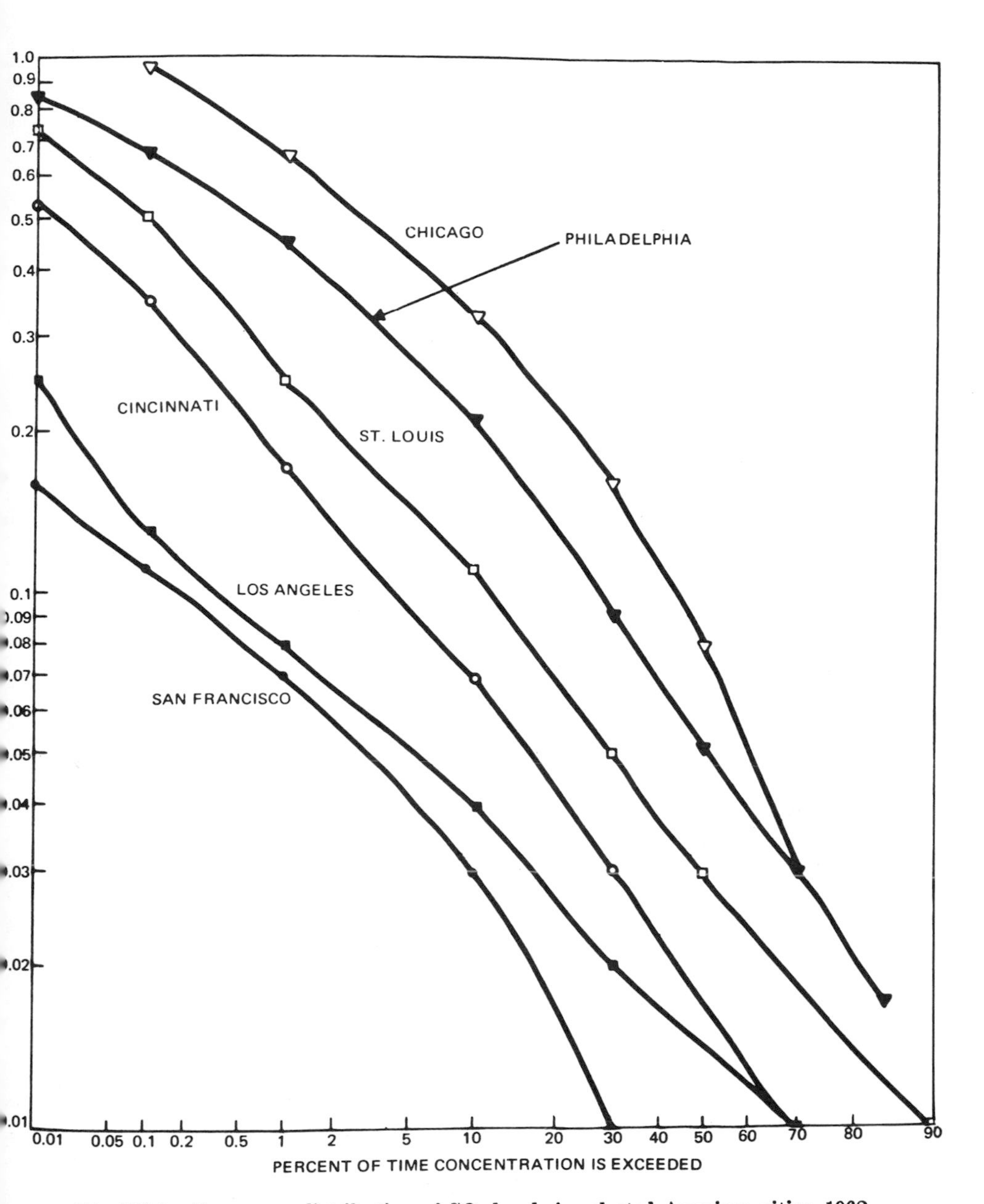

Fig. IV-9. Frequency distribution of SO_2 levels in selected American cities, 1962–1967. From NAPCA Publication No. AP-50 (1970) with permission of the Environmental Protection Agency.

TABLE IV-30

Annual Average of the Monthly Maximum Hourly Average SO_2 *Concentrations at Los Angeles Sampling Stations*[a]

Sampling station	Distance[b] (mi)	Annual average (ppm)												
		1956	1957	1958	1959	1960	1961	1962	1963	1964	1965	1966	1967	1968
Southwest coastal	2	0.48	0.42	0.34	0.20	0.24	0.28	0.20	0.24	0.28	0.28	0.28	0.28	—
Southwest coastal	4	0.26	0.22	0.26	0.16	0.16	0.17	0.13	0.20	0.20	0.19	0.20	0.23	0.19
South coastal	6	0.12	0.22	0.14	0.11	0.11	—	—	0.13	0.13	0.14	0.21	0.21	0.28
Northwest coastal	11	—	—	—	0.06	0.06	0.08	0.04	0.06	0.08	0.06	0.07	0.08	0.07
Central	13	0.12	0.12	0.10	0.07	0.06	0.09	0.08	0.10	0.06	0.05	0.06	0.07	0.11
West San Gabriel Valley	22	0.08	0.12	0.09	0.06	0.07	0.06	0.08	0.07	0.06	0.06	0.08	0.07	0.05
East San Fernando Valley	18	—	0.10	0.08	0.10	0.07	0.08	0.06	0.06	0.08	0.08	0.10	0.09	0.09

[a] From Mosher *et al.* (1969).
[b] Approximate distance from nearest coastal power plant.

monthly maximum hourly average SO_2 concentrations at distances of 2 to 18 mi from the nearest coastal power plant for the years 1956–1968. There is a steady decrease with distance. At 18 mi distance the value is 0.08 ppm, more or less invariant with the year. At 2 mi the values dropped from 0.48 in 1956 to 0.20 in 1959, and thereafter rose slightly to 0.28 in the middle 1960s.

Sulfuric acid is an important atmospheric impurity and its concentration can be correlated with SO_2, from which it is formed in the atmosphere by oxidation. In a Los Angeles study the weight ratio of H_2SO_4/SO_2 was found to vary from 0.037 to 3.0, the ratio increasing as the SO_2 concentration decreases. As $[SO_2]$ increases so does $[H_2SO_4]$, but at a lower rate, until some critical SO_2 concentration is reached, after which the H_2SO_4 concentration drops with further increases in $[SO_2]$. The critical SO_2 concentration varies with location; in central Los Angeles a maximum H_2SO_4 concentration of 30 $\mu g/m^3$ was found for $[SO_2] = 0.05$–0.10 ppm.

In a London study, the SO_2 concentrations were found to vary between 0.13 and 0.58 ppm, the weight ratio of H_2SO_4 to SO_2 being 0.011–0.013, generally. However, on one misty day the ratio rose to 0.023 (Coste and Courtier, 1936). During the episode of December 2–5, 1957, the weight ratio was 0.053 and the SO_2 concentration 1.47 ppm. This corresponds to a H_2SO_4 level of 0.051 ppm.

Effects

Many materials are affected by the oxides of sulfur. In paints, a particular effect is slowing of drying time, which can occur in the presence of 1–2 ppm SO_2. H_2SO_4 and sulfates are known to corrode metals. For a mild low-carbon steel tested in Chicago in 1963–1964, Upham (1967) deduced the formula

$$\%\ \text{wt. loss/yr} = 5.41[SO_2] + 9.5 \qquad (5)$$

where $[SO_2]$ is in ppm, and the equation is valid for $[SO_2]$ between 0.03 and 0.12 ppm as a mean annual average.

Carbonate stone, often used in building material, is converted by H_2SO_4 to the more soluble sulfates. These are leached away by rain water. Cleopatra's Needle, the large stone obelisk moved from Alexandria, Egypt, to London, has suffered more deterioration in the damp smoky acid atmosphere of London in 70–80 years than in the earlier 3000 or more years of its history (NAPCA Publication AP-50, 1970).

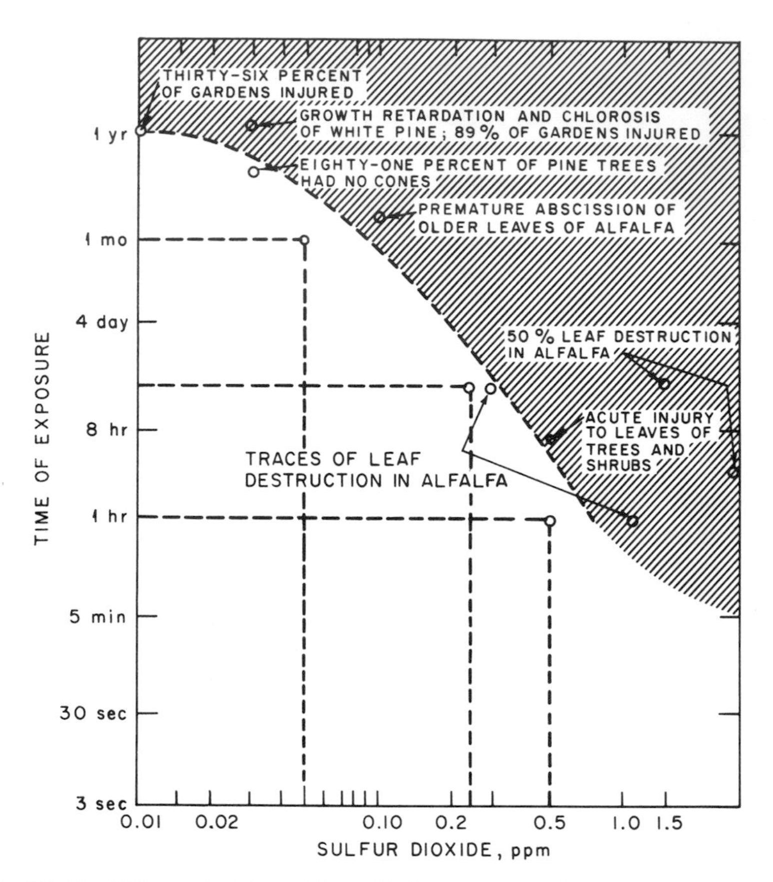

Fig. IV-10. **Effects of sulfur oxides pollution on vegetation. Shaded area: range of concentrations and exposure times in which injury to vegetation has been reported. Unshaded area: undetermined significance. Source unknown.**

Cellulose vegetable fibers lose tensile strength when exposed to H_2SO_4 and H_2SO_3. Dyes, leather, and paper are also harmed by SO_2, probably as a result of oxidation to its acids.

In vegetation, acute injury by SO_2 often results in an ivory coloration as shown in Plate VII, but sometimes the coloration is reddish brown. A dramatic example of this latter coloration in quince is shown in Plate VIII. Chronic injury over a period of days or weeks leads to gradual yellowing.

Figure IV-10 relates the onset of plant damage to SO_2 concentrations and duration of exposure. In the region of 0.02 to 0.5 ppm SO_2, corresponding to typical pollution levels, the onset of plant damage approximately follows the relationship

$$[SO_2]t^{0.39} = 1 \tag{6}$$

where $[SO_2]$ is in ppm and the exposure time t is in hours.

In animals no effects have been reported at SO_2 levels below 33 ppm. However, at higher levels the lung is attacked and ultimately death can be caused by asphyxia. The effects of H_2SO_4 are the same. Table IV-31 summarizes some test results in animals, showing the levels needed to kill 50% of the population. H_2SO_4 is at least 100 times as lethal as SO_2 in guinea pigs. Mice are more susceptible to SO_2 poisoning than guinea pigs. The H_2SO_4 study on guinea pigs is interesting in that it shows the larger aerosols are more toxic than the smaller ones. Also toxicity is increased as the temperature is reduced to 0°C.

For man also, the oxides of sulfur have unpleasant and serious effects (NAPCA Publication AP-50, 1970). The taste threshold for SO_2 is 0.3 ppm and the odor threshold is 0.5 ppm. For H_2SO_4 the odor threshold is 0.6 mg/m³ (0.15 ppm). Some subjects exposed to 5 ppm for 10–30 min showed increased lung resistance, and higher exposures aggrevate pulmo-

TABLE IV-31

SO_2 *and* H_2SO_4 *Exposure Required to Kill 50% of Test Animals* (LD_{50})[a]

Animal	Conc. (ppm)	Conc. (mg/m³)	Exposure time (hr)	Comments
		SO_2		
Mice	150	400	847	—
Mice	1000	2670	4	—
Guinea Pigs	130	347	154	—
Guinea Pigs	1000	2670	20	—
		H_2SO_4		
Guinea Pigs				
1–2 months old	—	18	8	—
18 months old	—	50	8	—
200–250 gm	—	27	8	2.7 μm aerosol
200–250 gm	—	47	8	6.8 μm aerosol (10°C)
200–250 gm	—	60	8	0.8 μm aerosol

[a] From NAPCA Publication No. AP-50 (1970).

nary problems. However, most subjects show adaptation and higher levels are required in a second exposure to produce the same effects. With H_2SO_4 at 0.35–0.5 mg/m^3 (0.088–0.125 ppm) in 1-μm aerosols, Amdur *et al.* (1952) found that the respiration rate increased after 3 min with no further change up to 15-min exposure. There was complete recovery after removal of the insult.

The effect of air pollution episodes in London is shown in Table IV-32. These air pollution episodes are usually attributed to SO_2 and soot, each of which is about equally important. The first and worst of these episodes occurred in December, 1952 and accounted for 3900 excess deaths. The SO_2 and soot levels each reached 4000 μg/m^3 (1.5 ppm for SO_2). In the last of these episodes in December 1962, the peak SO_2 level was 3300 μg/m^3, about 83% of that in 1952, but only for 1, rather than 2, days.

TABLE IV-32

Survey of Selected Acute Air Pollution Episodes in Greater London[a,b]

	Dec. 1952	Jan. 1956	Dec. 1962	Dec. 1957	Dec. 1956	Jan. 1955	Jan. 1959
Duration of the cumulation period in days	5	5	5	5	10	11	5
Number of days with maximum pollution	2	2	1	1	5	1 × 3[c]	1
SO_2 level preceding episode	500	300	400	300	300	300	300
SO_2 maximum	4000	1500	3300	1600	1100	1200	800
SO_2 increase per day	1200	500	1000	325	400	450	250
Soot level preceding episode	400	500	200	400	400	500	400
Soot maximum	4000	3250	2000	2300	1200	1750	1200
Soot increase per day	1200	1300	600	500	400	600	400
Number of excess deaths	3900	1000	850	800	400	240	200
Number of days with excess mortality	18	10	13	10	6	6	6
Daily mortality expected under circumstances	300	330	310	300	270	320	325
Average daily mortality in the period (excess mortality as a percent of normal)	170	130	120	125	125	112	110

[a] From NAPCA Publication No. AP-50 (1970) with permission of the Environmental Protection Agency.

[b] The SO_2 and soot concentrations mentioned are average values over 24 hr expressed in μg/m^3.

[c] Maximum pollution values of one day's duration occurred three times.

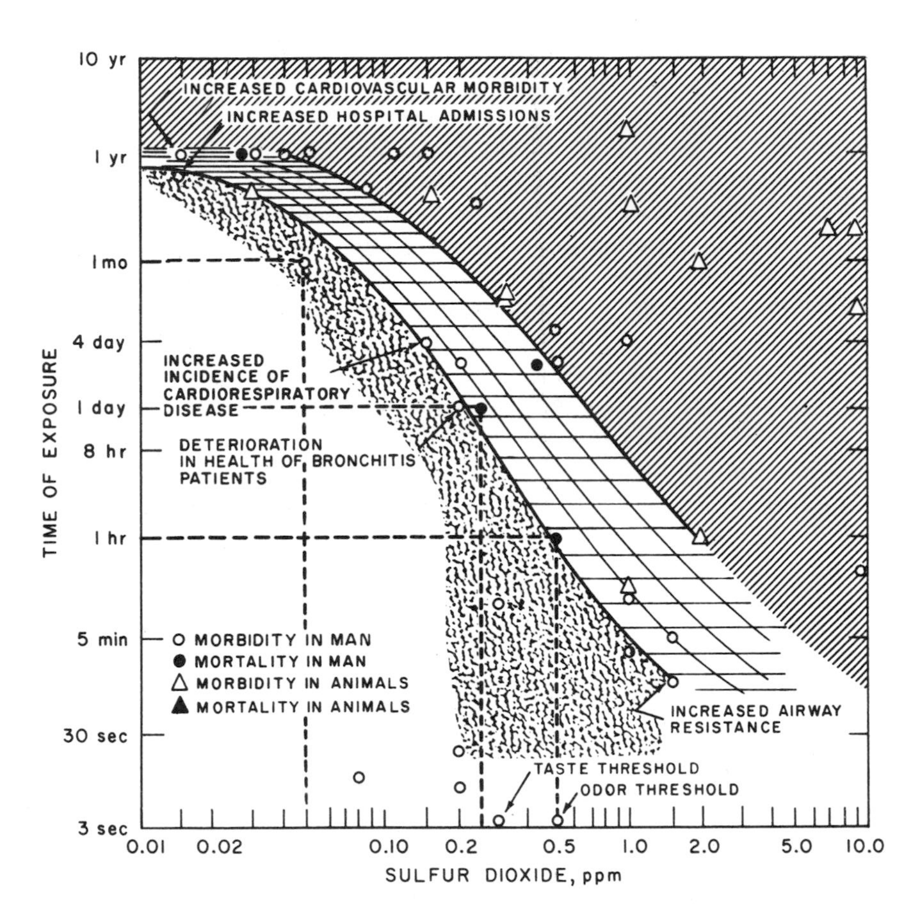

Fig. IV-11. Effects of SO_2 pollution on health. Shaded area: range of concentrations and exposure times in which deaths have been reported in excess of normal expectation. Cross-hatched area: significant health effects reported. Stippled area: health effects suspected. Source unknown.

The soot levels were half those in 1952, and excess deaths were considerably less, being 850 instead of 3900. In December 1957, the soot levels were higher than in December 1962, but mortality was less, as the SO_2 levels were only about one-half those of the 1962 episode. One interesting and anomolous observation is that in January 1955 both SO_2 and soot levels exceeded those in December 1956, but mortality was lower. Obviously some other variable parameter was also playing a role. That SO_2 may really be important was shown in a Rotterdam episode in which excess deaths were observed at daily average levels of SO_2 as low as 0.19 ppm for 3–4 days, even though particulate levels were low.

A survey of many studies suggests that excess deaths begin when $[SO_2] > 0.25$ ppm and soot exceeds 750 $\mu g/m^3$. The excess deaths occur in the aged and persons with pulmonary and cardiac disease.

In addition to affecting mortality, morbidity (disease) rates also rise with SO_2 pollution. A study in Chicago on patients with severe chronic bronchitis revealed that acute illness in persons over 55 years of age was 13.6% if the previous day's 24-hr SO_2 level was at 0.05 ppm, 17–19% at 0.10–0.25 ppm, and increased sharply above this level, reaching 26.5% at $[SO_2]$ in excess of 0.30 ppm. Patients with the same disease who were under 55 years of age, showed no increase until the previous days average exceeded 0.30 ppm (Carnow *et al.*, 1968).

Figure IV-11 summarizes the levels and duration of SO_2 exposure needed for the onset of increased morbidity and mortality. For the region of

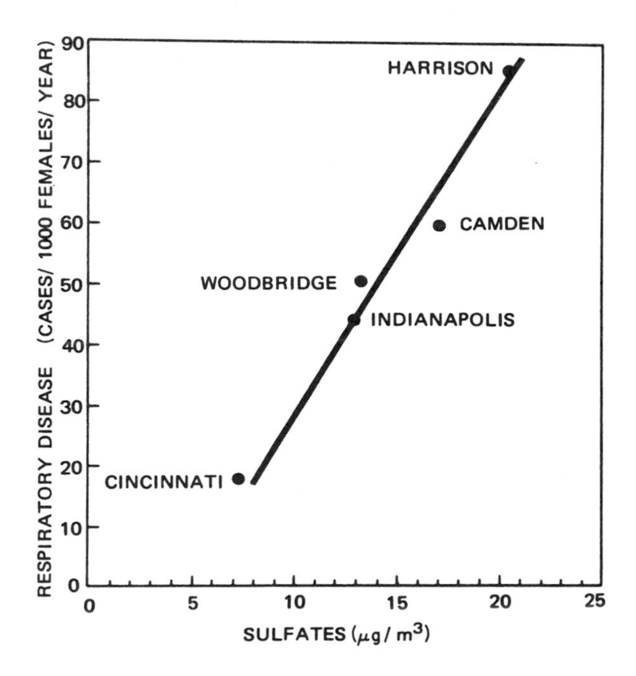

Fig. IV-12. Incidence of respiratory disease lasting more than 7 days in women versus concentration of sulfates in the city air at test sites. From Anderson and Ferris (1965) with permission of the American Medical Association.

0.05 to 1.0 ppm SO_2, these onsets can be estimated from the approximate relationship

$$[SO_2]t^{0.32} = 0.50 \qquad \text{for morbidity increase} \tag{7}$$

$$[SO_2]t^{0.38} = 2.0 \qquad \text{for mortality increase} \tag{8}$$

where $[SO_2]$ is in ppm, and the exposure time t is in hours.

Sulfates have also been shown to affect human health. A study of respiratory disease in women (Anderson and Ferris, 1965) showed an increase as a function of sulfate concentration in five cities. The results are shown graphically in Fig. IV-12.

As a result of the effect on human health, air quality standards have been set. The 8-hr day, 40-hr week standards recommended by the American Conference of Governmental Hygienists (1972) are 5 ppm for SO_2, 10 ppm for H_2S, and 1 mg/m^3 (0.24 ppm) for H_2SO_4. For ambient air the California standards are 1 ppm SO_2 for 1 hr or 0.3 ppm SO_2 for 8 hr. The U.S. federal standards for SO_2 are 0.5 ppm for 3 hr and 0.14 ppm for 24 hr, neither of which should be exceeded more than once a year. In addition the recommended annual average is $\leq$0.03 ppm.

PARTICULATE MATTER

Particulate matter includes particles from molecular size to greater than 10 μm in diameter. However, particles below 0.1 μm in diameter are generally not much of a problem, and particles $>$10 μm are removed from the atmosphere by gravitational settling. Therefore, it is the particles of 0.1 to 10 μm diameter that are of most concern.

Emissions

The particle emissions from various locations in the U.S. are tabulated in Table IV-33. Both New York City and St. Louis have serious particulate matter problems. In New York City in November 1969, the major source of particulate emission was from domestic heating, which accounted for about one-third of all emissions. Total incineration accounted for slightly more, divided about evenly between municipal and on-site incineration. Since domestic fuel consumption, on-site incineration, and motor vehicles are generally from private use, private use accounts for about 65% of all

TABLE IV-33

Sources of Particulate Matter (tons/yr)

Source	New York–New Jersey 1966[a]	New York City Nov. 1969[b]	Washington 1965–1966[a]
Fuel combustion	134,410 (58.1%)	33,200 (48.0%)	19,280 (55.4%)
Power generation	40,042 (17.3%)	6,400 (9.2%)	9,912 (28.5%)
Industrial	33,599 (14.5%)	4,500 (6.5%)	351 (1.0%)
Domestic	41,073 (17.8%)	22,300 (32.3%)	3,166 (9.1%)
Commercial and govt.	19,696 (8.5%)	—	5,851 (16.8%)
Incineration	41,734 (18.0%)	26,020 (37.7%)	8,155 (23.4%)
Municipal	—	13,330 (19.3%)	—
On site	—	12,690 (18.4%)	—
Transportation	35,245 (15.2%)	9,900 (14.3%)	6,245 (18.0%)
Motor vehicles	33,761 (14.6%)	—	5,678 (16.3%)
Other	1,484 (0.6%)	—	567 (1.7%)
Industrial processes	19,914 (8.5%)	—[f]	1,110 (3.2%)
Cement plants	—	—	—
Chemical plants	—	—	—
Grain industry	—	—	—
Metals	—	—	—
Other	19,914 (8.6%)	—	1,110 (3.2%)
Organic solvents	—	—	—
Total	231,303	69,120	34,790

[a] From NAPCA Publication No. AP-49 (1969).
[b] From Eisenbud (1970).
[c] From Lemke *et al.* (1967).
[d] From Whitby (1971).

St. Louis 1963[a]	Los Angeles 1965[a]	Los Angeles County Jan. 1967[c]	Los Angeles County 1968–1969[d]
86,800 (58.9%)	8,580 (19.3%)	18,250 (35.7%)	5,100 (13.0%)
22,400 (15.2%)	4,825 (10.9%)	—	—
39,000 (26.5%)	730 (1.6%)	—	—
19,900 (13.5%)	2,425 (5.4%)	—	—
5,500 (3.7%)	600 (1.4%)	—	—
15,800 (10.7%)[e]	365 (0.8%)	365 (0.7%)	—
—	—	—	—
—	—	—	—
7,100 (4.8%)	21,535 (48.6%)	20,400 (40.0%)	20,800 (52.8%)
4,700 (3.2%)	17,155 (38.7%)	16,400 (32.7%)	16,400 (41.6%)
2,400 (1.6%)	4,380 (9.9%)	4,000 (7.8%)	4,400 (11.2%)
37,556 (25.4%)	8,395 (19.0%)	9,850 (19.3%)	7,300 (18.5%)
3,600 (2.4%)	—	—	—
—	2,920 (6.6%)	2,920 (5.7%)	—
6,695 (4.5%)	—	—	—
12,433 (8.3%)[g]	2,920 (6.6%)[h]	2,190 (4.3%)[h]	—
14,828 (10.2%)	2,555 (5.8%)	4,385 (8.6%)	—
—	5,470 (12.3%)	2,550 (5.0%)	6,200 (15.7%)
147,256	44,345	51,050	39,400

[e] 14,100 tons/yr in open burning and 1700 tons/yr in incinerators.
[f] Included in industrial fuel consumption.
[g] All ferrous metals.
[h] Half ferrous, half nonferrous metals.

particulate emissions in New York City. The usual "whipping boys" of industry and power generation together account for only 15.7% of particulate emissions. St. Louis has the complication that industrial processes contribute a considerable fraction (25.4%) of the particulates. In addition, industrial fuel consumption contributes 26.5%. Since part of the city's power generation also goes into industry, the St. Louis problem (at least in 1963) is primarily from industrial sources.

In computing particulate emissions, it is necessary to know the emission factor (i.e., amount of particulates per unit consumption) for each source. Such estimates have been made and they are listed in Table IV-34.

Concentrations

Even clean air contains particulate matter, and even the purest air contains about 300 particles/cm^3, although these particles are less than 0.02 μm in diameter. Very polluted air can contain 10^5 particles/cm^3. In terms of mass, the background concentrations are about 10 μg/m^3. Thus if there are 10^3 particles/cm^3 and if they have a density of 1 gm/cm^3, then the average diameter corresponding to 10 μg/m^3 is 0.12 μm. In urban areas 60–220 μg/m^3 is typical, and in a heavily polluted area, levels may approach 2000 μg/m^3. The particle concentration usually is highest in midwinter, when fuel oil consumption is at a peak. A daily maximum may also occur between 6 and 8 A.M. in the morning because of rush hour traffic. In those cities in which photochemical pollution dominates (e.g., Los Angeles), the aerosol concentration rises sharply during the morning rush hour traffic, but continues to climb until about noon. The dustfall (i.e., particles >10 μm that settle by gravitation) runs 0.35–3.5 μg/cm^2-month in a typical city, but can go up to 70 μg/cm^2-month near offensive sources.

Table IV-35 shows the average concentrations of suspended particles in the most-bothered U.S. cities for the period 1961–1965. It can be seen that Chattanooga has the most serious problem. Then follow the large metropolitan cities interspersed with some smaller cities that are heavily industrialized. The Los Angeles–Long Beach area ranks 12th, having a mean concentration of 145.5 μg/m^3 for this period. Nevertheless during this 5-yr period, the California State Standards (sufficient to reduce visibility to <3 mi for relatively humidity $<70\%$) were exceeded on 831 (45.6%) of the days!

The overall U.S. average concentrations of suspended particles decreased $\sim$19% from 1957 to 1963 and then showed a 6% increase in 1964 as measured at "every year" stations of the NASN network (McMullen and Smith, 1965).

TABLE IV-34

Emission Factors for Selected Categories of Uncontrolled Sources of Particulates[a,b]

Emission source	Emission factor
Natural gas combustion	
Power plants	15 lb/million ft^3 of gas burned
Industrial boilers	18 lb/million ft^3 of gas burned
Domestic and commercial furnaces	19 lb/million ft^3 of gas burned
Distillate oil combustion	
Industrial and commercial furnaces	15 lb/thousand gal of oil burned
Domestic furnaces	8 lb/thousand gal of oil burned
Residual oil combustion	
Power plants	10 lb/thousand gal of oil burned
Industrial and commercial furnaces	23 lb/thousand gal of oil burned
Coal combustion	
Cyclone furnaces	2 × (ash %) lb/ton of coal burned
Other pulverized coal-fired furnaces	13–17 × (ash %) lb/ton of coal burned
Spreader stokers	13 × (ash %) lb/ton of coal burned
Other stokers	2–5 × (ash %) lb/ton of coal burned
Incineration	
Municipal incinerator (multiple chamber)	17 lb/ton of refuse burned
Commercial incinerator (multiple chamber)	3 lb/ton of refuse burned
Commercial incinerator (single chamber)	10 lb/ton of refuse burned
Flue-fed incinerator	28 lb/ton of refuse burned
Domestic incinerator (gas-fired)	15 lb/ton of refuse burned
Open burning of municipal refuse	16 lb/ton of refuse burned
Motor vehicles	
Gasoline-powered engines	12 lb/10^3 gal of gasoline burned
Diesel-powered engines	110 lb/10^3 gal of diesel fuel burned
Grey iron cupola furnaces	17.4 lb/ton of metal charged
Cement manufacturing	38 lb/bbl of cement produced
Kraft pulp mills	
Smelt tank	20 lb/ton of dried pulp produced
Lime kiln	94 lb/ton of dried pulp produced
Recovery furnaces[c]	150 lb/ton of dried pulp produced
Sulfuric acid manufacturing	0.3–7.5 lb. acid mist/ton of acid produced
Steel manufacturing:	
Open-hearth furnaces	1.5–20 lb/ton of steel produced
Electric arc furnaces	15 lb/ton of metal charged

[a] From NAPCA Publication No. AP-49 (1969) with permission of the Environmental Protection Agency.

[b] For more detailed data, consult "Control Techniques for Particulate Air Pollutants," U.S. Department of Health, Education, and Welfare, Dec. 1968.

[c] With primary stack gas scrubber.

TABLE IV-35

Suspended Particle Concentrations (Geometric Mean of Center City Station) in Urban Areas 1961–1965[a]

Standard metropolitan statistical area	Total suspended particles		Benzene-soluble organic particles	
	μg/m³	Rank	μg/m³	Rank
Chattanooga	180	1	14.5	2
Chicago–Gary–Hammond–E. Chicago	177	2	9.5	19.5
Philadelphia	170	3	10.7	12.5
St. Louis	168	4	12.8	4
Canton	165	5	12.7	5
Pittsburgh	163	6	10.7	12.5
Indianapolis	158	7	12.6	6
Wilmington	154	8	10.2	15
Louisville	152	9	9.6	18
Youngstown	148	10	10.5	14
Denver	147	11	11.7	8.5
Los Angeles–Long Beach	145.5	12	15.5	1
Detroit	143	13	8.4	28
Baltimore	141	14.5	11.0	10
Birmingham	141	14.5	10.9	11
Kansas City	140	16.5	8.9	23
York	140	16.5	8.1	34
New York–Jersey City–Newark–Passaic–Patterson–Clifton	135	18	10.1	16
Akron	134	20	8.3	30.5
Boston	134	20	11.7	8.5
Cleveland	134	20	8.3	30.5
Cincinnati	133	22.5	8.8	25
Milwaukee	133	22.5	7.4	42
Grand Rapids	131	24	7.2	44.5
Nashville	128	25	11.9	7
Syracuse	127	26	9.3	23
Buffalo	126	27.5	6.0	56
Reading	126	27.5	8.8	25
Dayton	123	29	7.5	40.5
Allentown–Bethlehem–Easton	120.5	30	6.8	50
Columbus	113	31.5	7.5	40.5
Memphis	113	31.5	7.6	39

[a] From NAPCA Publication No. AP-49 (1969) with permission of the Environmental Protection Agency.

TABLE IV-36

Arithmetic Mean and Maximum Urban Particulate Concentrations in the U.S., Biweekly Samplings, 1960–1965[a]

Pollutant	Number of stations	Concentrations ($\mu g/m^3$)	
		Arith. average[b]	Maximum
Suspended particulates	291	105	1254
Fractions			
Benzene-soluble organics	218	6.8	—[c]
Nitrates	96	2.6	39.7
Sulfates	96	10.6	101.2
Ammonium	56	1.3	75.5
Antimony	35	0.001	0.160
Arsenic	133	0.02	—[c]
Beryllium	100	<0.0005	0.010
Bismuth	35	<0.0005	0.064
Cadmium	35	0.002	0.420
Chromium	103	0.015	0.330
Cobalt	35	<0.0005	0.060
Copper	103	0.09	10.00
Iron	104	1.58	22.00
Lead	104	0.79	8.60
Manganese	103	0.10	9.98
Molybdenum	35	<0.005	0.78
Nickel	103	0.034	0.460
Tin	85	0.02	0.50
Titanium	104	0.04	1.10
Vanadium	99	0.050	2.200
Zinc	99	0.67	58.00
Gross beta radioactivity	323	(0.8 pCi/m^3)	(12.4 pCi/m^3)

[a] From NAPCA Publication No. AP-49 (1969) with permission of the Environmental Protection Agency.

[b] Arithmetic averages are presented to permit comparable expression of averages derived from quarterly composite samples; as such they are not directly comparable to geometric means calculated for previous years' data. The geometric mean for all urban stations during 1964–1965 was 90 $\mu g/m^3$, for the nonurban stations, 28 $\mu g/m^3$.

[c] No individual sample analyses performed.

The composition of particulate matter is shown in Table IV-36 for U.S. cities. It can be seen that, in order of importance, the composition is sulfates, benzene-soluble organics, nitrates, iron, and ammonium, with many other metals being present in smaller amounts. In fact, more metals are present than analyzed in the U.S. cities. A study of Ottawa, Canada, for 1968 showed considerable amounts of Na and Ca, and lesser amounts of K and Mg. These results are summarized in Table IV-37. It can be seen from these results that there are high daily variations, encompassing a factor of 62 for particulate matter and 100 for K. The average total soluble sulfate composition also differs some from the U.S. tests, being 18.4% in Ottawa rather than 10.0% in the U.S.

Of particular interest because of its toxicity is lead, which comprised 0.75% of the particulate matter in the U.S. studies, i.e., 0.79 $\mu g/m^3$. Lead is introduced into the air mainly from gasoline combustion, this source accounting for 5×10^8 lb of lead in the U.S. in 1968, 10% of which was emitted in California.

A study of lead levels in San Diego was made by Chow and Earl (1970). They report that background Pb levels over the mid-Pacific Ocean are $\sim 10^{-3}$ $\mu g/m^3$, but that in downtown San Diego these were 2.7 $\mu g/m^3$ in 1969. This exceeds the ambient air standard of 2 $\mu g/m^3$ recommended by the World Health Organization, but is less than the value of 10 $\mu g/m^3$ recommended in 1972 by the American Industrial Hygiene Association. (The industrial threshold value for an 8-hr day, 40-hr week recommended in 1972 by the American Conference of Governmental Industrial Hygienists was 150 $\mu g/m^3$ in inorganic compounds, fumes, and dusts.) The Pb aerosols account for about 3 to 4% of the suspended particulate matter in San

TABLE IV-37

Particulate Composition in Ottawa, Canada for 1968 ($\mu g/m^3$)[a]

Component	Mean of the monthly mean of daily average	Maximum monthly mean of daily average	Minimum monthly mean of daily average	Maximum daily average	Minimum daily average
Particulates	56.1	96.2	22.0	267	4.3
Ca	2.01	2.90	0.88	7.80	0.19
Mg[b]	0.18	0.27	0.12	1.23	0.02
K	0.38	0.58	0.11	2.92	0.03
Na	2.61	4.09	1.43	11.59	0.23
H_2SO_4	1.08	2.23	0.60	8.00	0.25
Fixed soluble sulfate[c]	9.24	15.3	4.21	37.7	2.23

[a] Courtesy of Monkman (1969).
[b] First 8 months only.
[c] Excluding H_2SO_4.

Diego. The San Diego dust is 0.84% lead and 950 $\mu g/m^3$ of lead settles per day in the dust. Chow and Earl (1970) concluded that in the late 1960s the Pb aerosol was increasing in San Diego at the rate of 5%/yr.

Tepper and Levin (1972) studied the ambient air Pb levels during 1968–1971 in Cincinnati, Philadelphia, Los Angeles, Los Alamos, Houston, Chicago, Washington, D.C., New York, and rural farms outside Cincinnati (Oheana). They found the annual mean Pb concentrations to be 0.14–0.20 in Los Alamos, 0.31–0.33 in Oheana, 1.0–1.3 in suburban areas, and 1.1–2.3 in the cities, with one station in Philadelphia reporting an average value of 3.75 and one in downtown Los Angeles reporting 4.55 $\mu g/m^3$. The lead levels at stations in three of the cities are shown for the period 1959–1971 in Fig. IV-13. Although there is considerable year-to-year fluctuation, the general trend is one of increasing lead concentrations in each city.

Tepper and Levin also measured the Pb levels in the blood of living humans in the cities tested during 1968–1971. They found these levels to be higher in smokers than nonsmokers, and higher in urban than nonurban residents. Yet there was no direct correlation with ambient air levels and blood Pb levels.

Meteorological Effects

The most obvious effect of particulate matter is that it scatters light, and thus reduces visibility. In Leningrad the yearly average reduction in sunlight due to particulates is estimated as 40%, 10% in summer and 70% in winter. An interesting correlation with economics was noted in 1932, when because of the depression and reduced industrial activity in the U.S.,

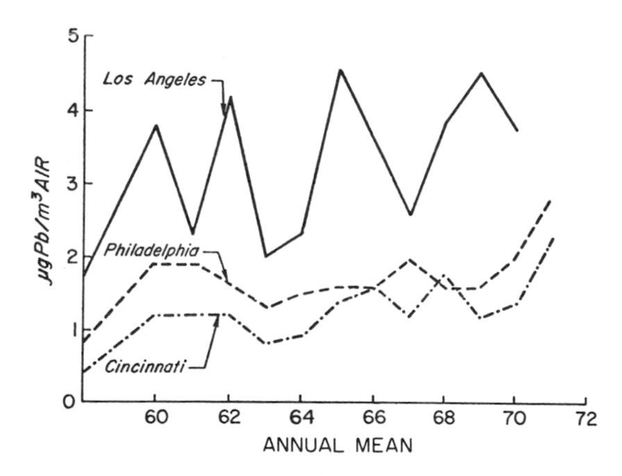

Fig. IV-13. Annual mean lead concentrations in three cities, 1959–1971, from the National Air Sampling Network. From Tepper and Levine (1972) with permission.

aerosol concentrations were significantly lower and solar radiation significantly higher than in 1931.

For low particulate concentrations, little total solar radiation is lost even though it is scattered, because most of the scattering is in the forward direction. However, as the particle concentrations increase, light losses increase due to absorption and back scattering. The relative solar radiation levels with particulate concentration are approximately (NAPCA Publication No. AP-49, 1969)

Particulates ($\mu g/m^3$):	50	100	200	400
Relative total radiation:	105	100	95	90
UV radiation:	104	100	92	77

Radiative scattering depends on the particle size in relation to the wavelength of light scattered. For particles much smaller than visible light, i.e., <0.1 μm diameter, Rayleigh scattering occurs. It is proportional to the sixth power of the diameter and the fourth power of the radiation frequency. This is the scattering done by molecules and small particles (Aitken nuclei). It accounts for the reddening sun toward dawn and twilight and the blue color of the sky (multiple scattering, first out of the sun's path and then back to earth). For particles similar in size to the wavelength of the scattered light, i.e., 0.1–1.0 μm diameter for visible light, the scattering law is complex, but has been worked out by Mie, and is known as Mie scattering. For larger particles, >1.0 μm in diameter, scattering goes as the square of the particle diameter.

The scattering properties of particles >0.1 μm in diameter depend on many things. However, some average characteristics have been observed for atmospheric hazes. If the relative humidity is $<70\%$, then the scattering coefficient b_{scat} varies as the 1.0–1.5 power of the radiation frequency. Consequently, as with Raleigh scattering, the violet light is preferentially scattered, and the sun appears redder through a haze.

The scattered radiation loss can be related to a Beer–Lambert law, just as absorbed radiation is,

$$\mathcal{I} = \mathcal{I}_0 \exp\{-\epsilon_{scat} C Z\} \tag{9}$$

where $\mathcal{I}$ and $\mathcal{I}_0$ are the transmitted and incident intensity, ϵ_{scat} the specific scattering coefficient, C the particle concentration, and Z the distance.

The scattering coefficient b_{scat} is defined as

$$b_{scat} \equiv \epsilon_{scat} C \tag{10}$$

and is usually measured because either ϵ_{scat} or C may not be known. However, a good representative value for ϵ_{scat} for atmospheric particulates (at $<70\%$ relative humidity) is 3.3×10^{-6} $m^2/\mu g$ at 5000 Å.

The visual range L_V is usually defined as the distance at which black and white contrast drops to 2%, the limit of visual discernment. It is related to b_{scat} by the simple relationship

$$L_V = 3.9/b_{scat} \tag{11}$$

Thus for $\epsilon_{scat} = 3.3 \times 10^{-6}$ m^2/μg, and particulates at background levels (i.e., 10 μg/m^3), the visual range is about 118 km. For a relatively high pollution level of 300 μg/m^3, the visual range drops to about 3.9 km. Radiation can further be reduced by light absorption by NO_2, which absorbs throughout the visible spectrum. At 5000 Å its absorption coefficient is 69 M^{-1} cm^{-1}, which corresponds to 2.9×10^{-4} (ppm-m)$^{-1}$. At a fairly high pollution level of 0.35 ppm, the radiation loss due to NO_2 at 5000 Å would be 10% of that due to particulates at 300 μg/m^3; visibility would be reduced to 3.6 km. In Los Angeles, where particulates and NO_2 are often associated with each other, the visibility reduction is generally attributed to be ~80% to particulates and ~20% to NO_2.

For relative humidities >70%, particulate properties may be considerably different, because of water condensation to produce fogs. At 70% relative humidity, deliquesence starts for many salts suspended in the air and concentrated water droplets form to produce fogs. In fact, particles act as centers for water condensation in clouds. Particles >1 μm act as centers for precipitation condensation nuclei in clouds, because they start falling due to gravitational settling, coagulating with smaller particles on the way down through the cloud, and growing to large raindrops. Because of excess particulates in urban atmospheres, there is a tendency for the urban factor to increase precipitation for five reasons:

1. water vapor addition from combustion sources and processes,
2. thermal updrafts from local heating,
3. updrafts from increased function turbulence,
4. added condensation nuclei leading to more cloud formation,
5. added nuclei, which may act as freezing nuclei for supercooled cloud particles.

Effects on Health

Suspended particles in the air are efficiently collected in the respiratory tract. As a result they can be toxic. Their toxicity increases as their diameters decrease below 1.0 μm because more pass through the nasal passages into the lungs. The computed collection efficiency as a function of particle radius is shown in Fig. IV-14. The minimum collection efficiency is for particles of 0.2-μm radius, but even these particles are collected with a

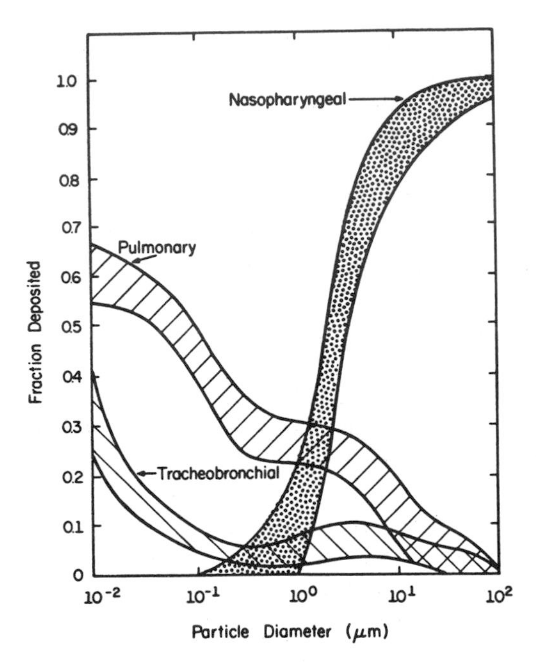

Fig. IV-14. Calculated fraction of particles deposited in the respiratory tract as a function of particle diameter. From Natusch and Wallace (1974) with permission of the American Association for the Advancement of Science.

computed 30% efficiency. (Experimental results suggest 50% efficiency; NAPCA Publication No. AP-49, 1969.) For particles >3 μm radius, the collection efficiency is 100%, but essentially all of these are collected in the nasal passages and do not reach the lungs. Experimental results in general confirm the predictions, except the observed collection efficiencies are somewhat higher.

Particles that enter the lung may be deleterious for three reasons. They may be toxic themselves, carry adsorbed toxic material, or interfere with the clearance of other airborne gases. The toxic group of particulates includes asbestos, H_2SO_4, BeO, and lead (as oxides or bromides). Soot is an especially good adsorbant, and often carries gases such as SO_2.

It has been observed that particulates given in conjunction with toxic gases tend to increase their toxicity, but if given before the toxic gas, may reduce its toxicity (NAPCA Publication No. AP-49, 1969). Thus mixtures of smoke and SO_2 increase mortality in animals over that for pure SO_2. However, if the smoke were given first, followed by SO_2, mortality was reduced. Wagner *et al.* (1961) found that oil mists would protect mice against later inhalation of NO_2 and O_3. However, if the oil was given together with the NO_2 and O_3, toxicity was enhanced. The prior beneficial effect of particulates has been attributed to the fact that they either coat or cause secretions that coat lung tissue, and thus protect it.

Particulate matter in urban atmospheres has been correlated with mortality, although it is difficult to sort out this effect from SO_2, since the two often occur together. In Detroit in September 1952 there was a rise in infant mortality and deaths to cancer patients for 3 days, when particulate levels exceeded 200 $\mu g/m^3$ and $[SO_2]$ was >1.0 ppm. In Osaka, Japan, in December 1962, there were 60 excess deaths during a period when the mean daily average particulate concentration exceeded 1000 $\mu g/m^3$ and $[SO_2] > 0.1$ ppm. A study in Buffalo, New York in 1959–1961 showed that the death rate rose with particulate levels (Winklestein *et al.*, 1967, 1968; Winklestein and Kantor, 1969).

Aerosols of soluble salts of ferrous, manganese, and vanadium ions increase the toxicity of SO_2, because they oxidize it to H_2SO_4.

The U.S. federal standards for particulates are 260 $\mu g/m^3$ for 24 hr, but not to be exceeded more than once a year, and 75 $\mu g/m^3$ as an annual average.

Other Effects

The most obvious effect of particles on materials is just to settle as dirt, thus disfiguring paints, finishes, works of art, etc. In addition, particulate matter can lead to corrosion of metals. Hygroscopic particles, such as sulfates and chlorides, act as corrosion nuclei. Also particles may act as a reservoir for adsorbed gases such as H_2S, SO_2, H_2SO_3, and H_2SO_4. The settling of dust on vegetation may also be harmful, both to the plant and to animals that eat the plant. In particular, it has been noticed that cement kiln dust settles on leaves and in moist air forms a relatively thick crust of gelatinous calcium silicate that interferes with photosynthesis and starch formation.

Odors have been associated with particulate matter. Some of this may be due to adsorbed gases, but this does not seem to be the sole reason. It has been noticed that if particulates are removed from diesel exhausts, the odor diminishes, even though there is no CH_2O or acrolein adsorbed on the particles. It is not clear what causes the diminution of odor intensity.

REFERENCES

Air Pollution Control Office Publ. No. AP-84 (1971). "Air Quality Criteria for Nitrogen Oxides."

Altshuller, A. P. (1975). *J. Air Pollut. Contr. Ass.* **25,** 19, "Evaluation of Oxidant Results at CAMP Sites in the United States."

Altshuller, A. P., Lonneman, W. A., Sutterfield, F. D., and Kopczynski, S. L. (1971). *Environ. Sci. Tech.* **5**, 1009, "Hydrocarbon Composition of the Atmosphere of the Los Angeles Basin.—1967."

Amdur, M. O., Silverman, L., and Drinker, P. (1952). *Amer. Med. Ass. Arch. Ind. Hyg. Occup. Med.* **6**, 306, "Inhalation of Sulfuric Acid Mist by Human Subjects."

Amer. Conf. of Govt. Ind. Hygienists (1972), "Threshold Limit Values for Chemical Substances and Physical Agents in the Workroom Environment with Intended Changes for 1972."

Anderson, D. O., and Ferris, B. G. Jr. (1965). *Arch. Environ. Health* **10**, 307, "Air Pollution Levels and Chronic Respiratory Disease."

Cadle, R. D., Pate, J., and Lodge, J. P. (1967). Amer. Chem. Soc. 154th Nat. Meeting, Chicago, Illinois.

Carnow, B. W., and Lepper, M. H. (1968). Presented at the Nat. Meeting Air Pollut. Contr. Ass., Paper No. 68-39, "Chicago Air Pollution Study—Dose Effects of Varying SO_2 Levels and Acute Illness in Patients with Chronic Bronchopulmonary Diseases."

Chow, T. J., and Earl, J. L. (1970). *Science* **169**, 577, "Lead Aerosols in the Atmosphere: Increasing Concentrations."

Colucci, J. M., and Begeman, C. R. (1965). *J. Air Pollut. Contr. Ass.* **15**, 113, "The Automotive Contribution to Air-Borne Polynuclear Aromatic Hydrocarbons in Detroit."

Corn, M., Dunlap, R. W., Goldmuntz, L. A., and Rogers, L. H. (1975). *J. Air Pollut. Contr. Ass.* **25**, 16, "Photochemical Oxidants: Sources, Sinks, and Strategies."

Coste, J. H., and Courtier, G. B. (1936). *Trans. Faraday Soc.* **32**, 1198, "Sulphuric Acid as a Disperse Phase in Town Air."

Eisenbud, M. (1970). *Science* **170**, 706, "Environmental Protection in the City of New York."

Elkin, H. F. (1962). *In* "Air Pollution" (A. C. Stern, ed.), Vol. III. Academic Press, New York. "Petroleum Refinery Emissions."

Garner, J. (1972). Environmental Protection Agency Rep., "Hydrocarbon Emission Effects Determination: A Problem Analysis."

Gerstle, R. W., and Kemnitz, D. A. (1967). *J. Air Pollut. Contr. Ass.* **17**, 324. "Atmospheric Emissions from Open Burning."

Graedel, T. E., Kleiner, B., and Patterson, C. C. (1974). *J. Geophys. Res.* **79**, 4467, "Measurements of Extreme Concentrations of Tropospheric Hydrogen Sulfide."

Hangebrauck, R. P., von Lehmden, D. J., and Meeker, J. E. (1967). U.S. Public Health Serv. Publ. 999-AP-33, "Sources of Polynuclear Hydrocarbons in the Atmosphere."

Hester, N. E., Stephens, E. R., and Taylor, O. C. (1974). *J. Air Pollut. Contr. Ass.* **24**, 591, "Fluorocarbons in the Los Angeles Basin."

Heuss, J. M., and Glasson, W. A. (1968). *Environ. Sci. Tech.* **2**, 1109, "Hydrocarbon Reactivity and Eye Irritation."

Jacobson, J. S., and Hill, A. C. (1970). Air Pollut. Contr. Ass., "Recognition of Air Pollution Injury to Vegetation: A Pictorial Atlas."

Jaffe, L. S. (1973). *J. Geophys. Res.* **78**, 5293, "Carbon Monoxide in the Biosphere: Sources, Distribution, and Concentrations."

Junge, C. E. (1963). "Air Chemistry and Radioactivity," p. 123. Academic Press, New York.

Lemke, E. E., Shaffer, N. R., Thomas, G., and Verssen, J. A. (1967). Air Pollut. Contr. District, County of Los Angeles, "Air Pollution Data for Los Angeles County January 1967."

Leonardos, G., Kendall, D., and Barnard, N. (1969). *J. Air Pollut. Contr. Ass.* **19,** 91, "Odor Threshold Determinations of 53 Odorant Chemicals."

Linnenbom, V. J., Swinnerton, J. W., and Lamontagne, R. A. (1973). *J. Geophys. Res.* **78,** 5333, "The Ocean as a Source for Atmospheric Carbon Monoxide."

Lonneman, W. A., Bellar, T. A., and Altshuller, A. P. (1968). *Environ. Sci. Tech.* **2,** 1017, "Aromatic Hydrocarbons in the Atmosphere of the Los Angeles Basin."

Lonneman, W. A., Kopczynski, S. L., Darley, P. E., and Sutterfield, F. D. (1974). *Environ. Sci. Tech.* **8,** 229, "Hydrocarbon Composition of Urban Air Pollution."

Mayer, M. (1965). U.S. Public Health Service, "Pollutant Emission Factors."

McMullen, T. B., and Smith R. (1965). Public Health Serv. Publ. No. 999-AP-19, "The Trend of Suspended Particulate in Urban Air: 1957–1964."

Mittler, S., Hedrick, D., King, M., and Gaynor, A. (1956). *Ind. Med. Surg.* **25,** 301, "Toxicity of Ozone. I. Acute Toxicity."

Monkman, J. L. (1969). Canadian Dept. of Health, Education, and Welfare, private communication.

Mosher, J. C., Macbeth, W. G., Leonard, M. J., Mullins, T. P., and Brunelle, M. F. (1969). Presented at Air Pollut. Contr. Ass. 62nd Ann. Meeting, New York, Paper No. 69-28, "The Distribution of Contaminants in the Los Angeles Basin Resulting from Atmospheric Reactions and Transport."

National Air Pollution Control Administration Publication No. AP-49 (1969). "Air Quality Criteria for Particulate Matter."

National Air Pollution Control Administration Publication No. AP-50 (1970). "Air Quality Criteria for Sulfur Oxides."

National Air Pollution Control Administration Publication No. AP-62 (1970). "Air Quality Criteria for Carbon Monoxide."

National Air Pollution Control Administration Publication No. AP-64 (1970). "Air Quality Criteria for Hydrocarbons."

Natusch, D. F. S., and Wallace, J. R. (1974). *Science* **186,** 695, "Urban Aerosol Toxicity: The Influence of Particle Size."

Olsen, D., and Haynes, J. L. (1969). Nat. Air Pollut. Contr. Administration Publ. No. APTD 69-43, "Preliminary Air Pollution Survey of Organic Carcinogens."

Pitts, J. N. Jr. (1970). *Ann. N. Y. Acad. Sci.* **171,** 239, "Singlet Molecular Oxygen and the Photochemistry of Urban Atmospheres."

Public Health Service Publication No. 999-AP-13 (1965). "Atmosphere Emissions from Sulfuric Acid Manufacturing Processes."

Rasmussen, R. A. (1964). Ph.D. Dissertation, Washington University, St. Louis, "Terpenes: Their Analysis and Fate in the Atmosphere."

Rasool, S. I., and Schneider, S. H. (1971). *Science* **173,** 138, "Atmospheric Carbon Dioxide and Aerosols: Effects of Large Increases on Global Climate."

Raynor, G. S., Smith, M. E., and Singer, I. A. (1974). *J. Air Pollut. Contr. Ass.* **24,** 586, "Temporal and Spatial Variation in Sulfur Dioxide Concentrations in Suburban Long Island, New York."

Robbins, R. C., Cavanagh, L. A., Salas, L. J., and Robinson, E. (1973). *J. Geophys. Res.* **78,** 5341, "Analysis of Ancient Atmospheres."

Robinson, E., and Robbins, R. C. (1968). Stanford Res. Inst. Project PR-6755 Final Rep., "Sources, Abundance, and Fate of Gaseous Atmospheric Pollutants."

Robinson, E., and Robbins, R. C. (1969). Stanford Res. Inst. Project PR-6755 Supple. Rep., "Sources, Abundance, and Fate of Gaseous Atmospheric Pollutants Supplement."

Rowland, F. S., and Molina, M. J. (1974). At. Energy Comm. Rep. No. 1974-1, "Chlorofluoromethanes in the Environment."

Sheehy, J. P., Henderson, J. J., Harding, C. I., and Danis, A. L. (1963). U.S. Public Health Serv. Publ. No. 999-AP-3. "A Pilot Study of Air Pollution in Jacksonville, Florida."

Shriner, D. S., and Lacasse, N. L. (1969). *Phytopathology* **59**, 402 (abstr.), "Distribution of Chlorides in Tomato following Exposure to Hydrogen Chloride Gas."

Shy, C. M., Creason, J. P., Pearlman, M. E., McClain, K. E., Benson, F. B., and Young, M. M. (1970a). *J. Air Pollut. Contr. Ass.* **20**, 539, "The Chattanooga School Children Study: Effects of Community Exposure to Nitrogen Dioxide. I. Methods, Description of Pollutant Exposure and Results of Ventilatory Function Testing."

Shy, C. M., Creason, J. P., Pearlman, M. E., McClain, K. E., Benson, F. B., and Young, M. M. (1970b). *J. Air Pollut. Contr. Ass.* **20**, 582, "The Chattanooga School Children Study: Effects of Community Exposure to Nitrogen Dioxide. II. Incidence of Acute Respiratory Illness."

Stasiuk, W. N., and Coffey, P. E. (1974). New York State Dept. of Environmental Conservation Rep. BTS-5, "Rural and Urban Ozone Relationships in New York State."

Steinberger, H. (1974). *Israel J. Earth Sci.* **23**, 19, "A Study of the Influence of Meteorological Parameters on Oxidant Concentration in Jerusalem."

Taylor, O. C. (1969). *J. Air Pollut. Contr. Ass.* **19**, 347, "Importance of Peroxyacetyl Nitrate (PAN) as a Phytotoxic Air Pollutant."

Tepper, L. B., and Levin, L. S. (1972). Dept. of Environmental Health, College of Medicine, Univ. of Cincinnati, "A Survey of Air and Population Lead Levels in Selected American Communities."

Tuesday, C. S. (1972). Private communication, "Sources of Hydrocarbon Emissions."

Upham, J. B. (1967). *J. Air Pollut. Contr. Ass.* **17**, 398, "Atmospheric Corrosion Studies in Two Metropolitan Areas."

Wagner, R. H. (1971). "Environment and Man." Norton, New York.

Wagner, W. D., Dobrogorski, O. J., and Stokinger, H. E. (1961). *Arch. Environ. Health* **2**, 523, "Antagonistic Action of Oil Mists on Air Pollutants."

Whitby, K. T. (1971). Environmental Protection Agency Publ. APTD-0630, "Aerosol Measurements in Los Angeles Smog," Vol. I.

Winkelstein, W., Jr., and Kantor, S. (1969). *Arch. Environ. Health* **18**, 544, "Stomach Cancer: Positive Association with Suspended Particulate Air Pollution."

Winkelstein, W. Jr., Kantor, S., Davis, E. W., Maneri, C. S., and Mosher, W. E. (1967). *Arch. Environ. Health* **14**, 162, "The Relationship of Air Pollution and Economic Status to Total Mortality and Selected Respiratory System Mortality in Men."

Winkelstein, W. Jr., Kantor, S., Davis, E. W., Maneri, C. S., and Mosher, W. E. (1968). *Arch. Environ. Health* **16**, 401, "The Relationship of Air Pollution and Economic Status to Total Mortality and Selected Respiratory System Mortality in Men. (II. Oxides of Sulfur)."

Chapter V

HYDROCARBON OXIDATION

Many atmospheric pollutants, including hydrocarbons, aldehydes, ketones, peroxides, epoxides, acids, CO, CO_2, polynuclear aromatic hydrocarbons, soot, and the oxides of nitrogen and sulfur, result from combustion processes. Therefore, in this chapter we shall examine combustion processes of hydrocarbons to illustrate the conditions that lead to various pollutants. As an example of a typical hydrocarbon, we shall consider butane, but it will become apparent that to understand butane combustion, we must also understand the oxidation of all the products of the reaction, i.e., alkenes, aldehydes, ketones, and CO.

Combustion processes lead to different results under different conditions. In particular the product distributions can be markedly different in oxygen-rich and oxygen-lean mixtures. At relatively low temperatures, i.e., $<\sim 450°C$, slow combustion occurs. At high temperatures, ~ 450–$1200°C$, the combustion becomes more rapid and ignition can occur, leading to explosions or flames. Finally, at the very high temperatures ($>2000°C$) in the internal combustion engine, explosive combustion occurs, which is extremely rapid.

The unique feature of an oxidation as opposed to other reactions is the chain branching. At high temperatures, chain branching can be effected through ground-state oxygen atoms, $O(^3P)$. This species is extremely reactive and is a diradical, i.e., it has two unpaired electrons. Thus it can react with a hydrocarbon by abstracting a hydrogen atom and give two highly reactive free radicals. In this way one reactive species is converted to two reactive species. If this process or other processes leading to free-radical generations are repeated more rapidly than radical removal, a runaway situation arises, which can lead to ignition. The ignition is accompanied by a large release of energy, since the reactions involved are highly exothermic.

SLOW COMBUSTION

In the low-temperature combustion of hydrocarbons, the reaction starts slowly, then accelerates until about one-half completed, and finally slows down until the reaction is completed. This behavior is generally attributed to the oxidation of an initial product of the reaction in such a manner that the radicals produced enhance the rate of oxidation of the hydrocarbon. Thus the initial product, generally accepted to be a hydroperoxide at lower temperatures, is really an intermediate that produces more and more active radicals. This process is known as delayed or degenerate chain branching. At lower temperatures the intermediate is an organic hydroperoxide. Above 450°C, H_2O_2 decomposition may be important.

The initial products of the reaction include organic hydroperoxides, H_2O, and H_2O_2. Aldehydes, principally CH_2O, and to some extent ketones are also produced. Above about 400°C, olefins, H_2, and CH_4 are important, and they rise in importance as the temperature or fuel-to-O_2 ratio is en-

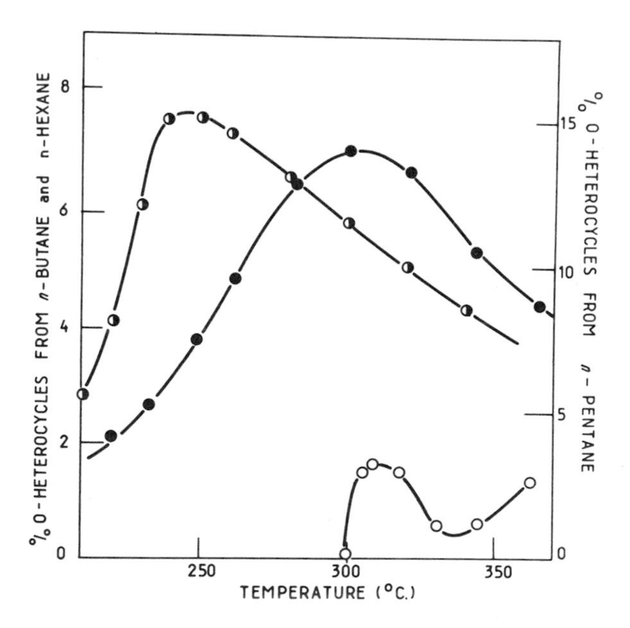

Fig. V-1. Yields of O-heterocycles as a function of temperature; ○ = 2,3-dimethyloxiran from *n*-butane, P = 240 Torr; ● = 2,4-dimethyloxetan from *n*-pentane, P = 141 Torr; ◑ = 2,5-dimethyltetrahydrofuran from *n*-hexane, P = 135 Torr. The hydrocarbon-to-O_2 concentration ratios are 0.29, 0.75, and 1.25 for *n*-C_4H_{10}, *n*-C_5H_{12}, and *n*-C_6H_{14}, respectively. From Berry *et al.* (1968) with permission of the American Chemical Society.

hanced. Organic acids are formed only in the low-temperature range. Alcohols are sometimes seen, usually for fuel-rich conditions. With the higher hydrocarbons, O-heterocycles are produced under some conditions. The main final products are CO, CO_2, and H_2O.

The production of O-heterocycles is of consequence, since there is some fear that these may be carcinogenic. The importance of these at reaction temperatures of 250 to 300°C was demonstrated in a classic paper by Berry *et al.* (1968). They showed that up to 30% yield of 2,4-dimethyloxetane could be produced in the oxidation of n-C_5H_{12}, and that up to 10% 2,3-dimethyloxirane and 2,5-dimethylhydrofuran could be produced in the oxidation of n-C_4H_{10} and n-C_6H_{14}, respectively. These products are formed early in the reaction. Typical results as a function of temperature are shown in Fig. V-1. For the three hydrocarbons, the respective hydrocarbon-to-O_2 concentration ratios were 0.29, 0.75, and 1.25 for n-C_4H_{10}, n-C_5H_{12}, and n-C_6H_{14}, respectively.

Surface effects have also been shown to be important in hydrocarbon oxidations. The pressure histories of CH_4 oxidation at ~500°C are shown in Fig. V-2 for various surfaces. The oxidation rates are fastest in freshly

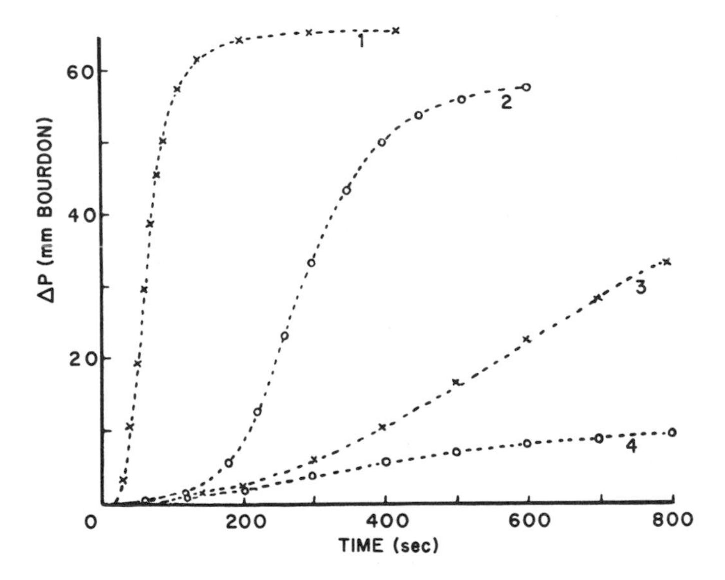

Fig. V-2. Comparison of rates of reaction with various surfaces for $2CH_4 + O_2$ in a silica vessel. 1.8 mm Bourdon = 1 Torr. Curve 1, 450 Torr in new or HF-treated vessel; curve 2, 500 Torr in aged vessel; curve 3, 500 Torr in heat-treated vessel; curve 4, 500 Torr in PbO-coated vessel. From Hoare (1965) with permission of Gordon and Breach.

cleaned silica vessels. Aging, heat treating, or coating with PbO diminishes the reaction rate, presumably by enhancing termination reactions or reducing initiation reactions that occur on the wall.

Initial Stages

The initiation step in combustion is not known but presumably consists of free-radical-forming processes occurring either homogeneously or heterogeneously. A step often invoked is

$$RH + O_2 \rightarrow R + HO_2 \tag{1}$$

The alkyl radical produced, R, reacts with O_2. Two reactions are possible, and for the example of $i\text{-}C_4H_9$, these are

$$i\text{-}C_4H_9 + O_2 \leftrightarrows i\text{-}C_4H_9O_2 \qquad (2a, -2a)$$

$$\rightarrow i\text{-}C_4H_8 + HO_2 \qquad (2b)$$

Benson (1968) has argued that reaction (2b) is truly a parallel path of reaction (2a), and does not occur exclusively through reaction (2a) followed by decomposition of the peroxy radical $i\text{-}C_4H_9O_2$:

$$i\text{-}C_4H_9O_2 \rightarrow i\text{-}C_4H_8 + HO_2 \tag{3}$$

Actually none of reactions (2b), (3), and (−2a) are important in the slow combustion, except possibly with C_2H_6, since the main fate of the peroxy radical for the higher hydrocarbons is internal abstraction. This process occurs easily if the radical contains a 3-hydrogen:

$$i\text{-}C_4H_9O_2 \longrightarrow \left[\text{cyclic: } H_2C\text{–}H\cdots O\text{–}O\text{–}C(H)(CH_3)\text{–}CH_2\right] \longrightarrow \cdot CH_2CH(CH_3)COOH \tag{4}$$

The peroxy radical can also abstract a hydrogen atom from the parent alkane

$$RO_2 + RH \rightarrow RO_2H + R \tag{5}$$

This can be an important reaction in the low-temperature (100–200°C) oxidation of the lower alkanes. Nalbandyan (1967) has shown that the corresponding hydroperoxides are the major products for several sensitized oxidations of CH_4 and C_2H_6 at 100–150°C.

The main course of the oxidation is determined by the fate of the

i-C_4H_8OOH radical produced in reaction (4). It can undergo the reactions

$$i\text{-}C_4H_8OOH + O_2 \rightleftharpoons HOOC_4H_8OO \quad (6a,-6a)$$

$$\rightarrow i\text{-}C_4H_7OOH + HO_2 \quad (6b)$$

$$i\text{-}C_4H_8OOH \rightarrow \text{O-heterocycle} + HO \quad (7a)$$

$$\xrightarrow{\text{wall}} CH_2O + C_3H_6 + HO \quad (7b)$$

$$\rightarrow i\text{-}C_4H_8 + HO_2 \quad (7c)$$

Reaction (6a) can be followed by abstraction of a hydrogen from the alkane via reaction (5) to give a dihydroperoxide:

$$HOOC_4H_8OO + RH \rightarrow HOOC_4H_8OOH + R \quad (8)$$

At low temperatures, the main products are hydroperoxides. Thus Allara *et al.* (1968) found C_4 hydroperoxides as the principal products in the peroxide-initiated oxidation of isobutane at 100°C. They considered the product to be *tert*-C_4H_9OOH, but their analytical scheme would not have distinguished among the various C_4 hydroperoxides.

From 250 to 350°C, the major reaction can be HO ejection to yield the O-heterocycle, which is the major initial hydrocarbon product at these temperatures (Berry *et al.*, 1968). A dominant reaction of the RO_2 radicals formed in 3-ethylpentane and 3-methylpentane oxidation at 250–410°C (Barat *et al.*, 1971) was isomerization via reaction (4), followed by HO ejection, reaction (7a). Possibly aldehydes and lower olefins may be formed in a similar process, reaction (7b), although apparently carbonyl compounds are only formed heterogeneously (Knox and Kinnear, 1971). For *n*-pentane oxidation at 250–400°C, Knox and Kinnear report that acetone was produced heterogeneously.

At temperatures above 350°C, the major hydrocarbon product of the oxidation of alkanes can be the corresponding olefins (Minkoff and Tipper, 1962), which are produced mainly via reaction (7c).

The HO and HO_2 radicals produced react rapidly with the parent hydrocarbon:

$$HO + i\text{-}C_4H_{10} \rightarrow H_2O + C_4H_9 \quad (9)$$

$$HO_2 + i\text{-}C_4H_{10} \rightarrow H_2O_2 + C_4H_9 \quad (10)$$

The HO_2 radicals are also the terminating radicals. They can terminate on the walls of the reaction vessel or by reaction with other peroxy radicals:

$$2HO_2 \rightarrow H_2O_2 + O_2 \quad (11)$$

$$HO_2 + C_4H_9O_2 \rightarrow C_4H_9O_2H + O_2 \quad (12)$$

$$HO_2 + HOOC_4H_8O_2 \rightarrow HOOC_4H_8OOH + O_2 \quad (13)$$

The wall termination of HO_2 leads to H_2O_2 or H_2O and accounts for an important part of the observed surface effects.

An interesting feature of the above mechanism is that no alkoxy radicals have been included, and thus no aldehydes or ketones [except by the wall reaction (7b)] are produced. The implication is that aldehydes are not initial products of the homogeneous oxidation and can appear only with an induction period. Thus they are not the intermediates responsible for degenerate chain branching. This role is reserved to the hydroperoxides.

There are additional minor reactions that may play a role under some conditions for slow combustion. There is the possibility that the alkyl radical will decompose unimolecularly. For C_4H_9 radicals, the possible routes are

$$C_4H_9 \rightarrow H + C_4H_8 \tag{14a}$$

$$\rightarrow CH_3 + C_3H_6 \tag{14b}$$

In this way H atoms and smaller free radicals, usually CH_3, as well as olefins with fewer carbons than the original alkane are formed. In fuel-rich mixtures, new radicals, as well as lower-molecular-weight alkanes, can also be produced:

$$R' + RH \rightarrow R'H + R \tag{15}$$

If R' is H or CH_3, then H_2 or CH_4 is produced, respectively. This explains the H_2 and CH_4 production at high temperatures and high fuel-to-oxygen ratios, since H and CH_3 are produced by the cracking reactions (14a) and (14b) in all paraffin hydrocarbon combustions.

The H atom plays a special role in that it can also react with O_2:

$$H + O_2 + M \rightarrow HO_2 + M \tag{16}$$

$$H + O_2 \rightarrow HO + O(^3P) \tag{17}$$

At low temperatures and high pressures reaction (16) dominates, but reaction (17) increases in importance as the temperature rises or the pressure is reduced. Reaction (17) is of special interest since it produces $O(^3P)$, which enters the chain branching reaction

$$O(^3P) + RH \rightarrow HO + R \tag{18}$$

Model computations have been made for the initial stages of the homogeneous slow combustion of an oxygen-rich mixture of O_2 and i-C_4H_{10}. The reactions used and their rate coefficients are listed in Table V-1. In order to simplify the computations, no distinction was made between i-C_4H_9 and *tert*-C_4H_9 radicals. Averages of their rate coefficients are used. There are 16 reactions of importance. However, in order to make the model computation more complete, an additional 11 unimportant reactions were included. They had almost no effect on the results.

The calculations were done for concentrations of $[O_2] = 2.5 \times 10^{18}$, $[i\text{-}C_4H_{10}] = 2.5 \times 10^{17}$, and $[M] = 3.5 \times 10^{18}$ molecules/cm³ at 50° intervals between 400 and 700°K. The resulting steady-state radical concentrations for the important radicals are shown in Fig. V-3. With the exception of $[HOOC_4H_8OO]$, all the radical concentrations rise sharply with temperature. The $HOOC_4H_8OO$ concentration goes through a maximum at 500°K.

The rates of the important reactions are tabulated in Table V-2. They all increase with temperature except for the reaction producing the dihydroperoxide, which goes through a maximum at 500–550°K. This result

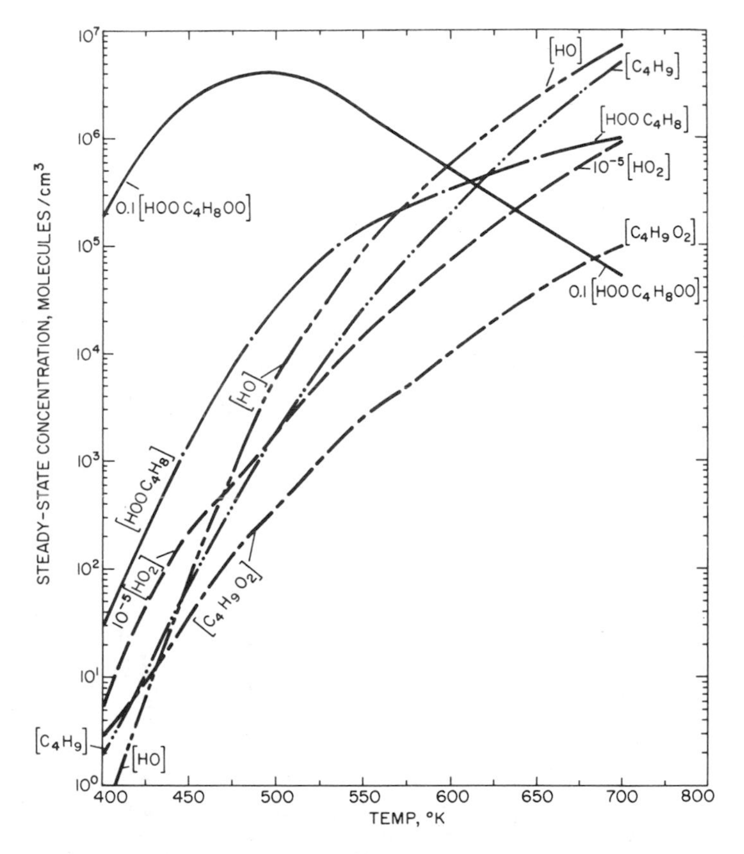

Fig. V-3. Semilog plot of the steady-state concentrations of the major radicals versus temperature in the initial stages of the slow combustion of $i\text{-}C_4H_{10}$ based on a model calculation for $[O_2] = 2.5 \times 10^{18}$, $[i\text{-}C_4H_{10}] = 2.5 \times 10^{17}$, and $[M] = 3.5 \times 10^{18}$ molecules/cm³.

TABLE V-1

Rate Coefficients for Reactions in the Slow Combustion of $i\text{-}C_4H_{10}$

Reaction	A (cm^{3n}/sec)[a]	E_a (kcal/mole)	Temp range (°K)	M	Reference
	Important reactions				
$i\text{-}C_4H_{10} + O_2 \rightarrow HO_2 + C_4H_9$	1.0×10^{-11}	44.0	—	—	Estimate
$HO_2 + HO_2 \rightarrow H_2O_2 + O_2$	3.0×10^{-11}	1.0	300–1000	—	Garvin and Hampson (1974)
$HO_2 + HOOC_4H_8OO \rightarrow HOOC_4H_8OOH + O_2$	5.0×10^{-12}	0.0	—	—	Estimate
$C_4H_9 + O_2 \rightarrow C_4H_9O_2$	2.0×10^{-12}	0.0	—	—	Estimated by Knox (1968)
$C_4H_9 + O_2 \rightarrow C_4H_8 + HO_2$	2.5×10^{-12}	5.0	—	—	Average of estimates by Knox (1968) and Heicklen (1968)
$C_4H_9O_2 \rightarrow C_4H_9 + O_2$	2.5×10^{14}	28.0	—	—	Estimate
$HO + i\text{-}C_4H_{10} \rightarrow H_2O + C_4H_9$	8.7×10^{-12}	0.8	—	—	Donovan and Husain (1972)
$C_4H_9O_2 + C_4H_{10} \rightarrow C_4H_9O_2H + C_4H_9$	2.5×10^{-13}	7.7	—	—	Estimated by Fish (1967)[c]
$C_4H_9O_2 \rightarrow HOOC_4H_8$	1.0×10^{11}	8.3	—	—	Estimated by Fish (1967)
$HO_2 + C_4H_{10} \rightarrow H_2O_2 + C_4H_9$	1.0×10^{-12}	10.0	—	—	Estimate
$HOOC_4H_8 + O_2 \rightarrow HOOC_4H_8OO$	1.0×10^{-13}	0.0	—	—	Estimated by Fish (1967)
$HOOC_4H_8 + O_2 \rightarrow C_4H_7OOH + HO_2$	2.5×10^{-12}	5.0	—	—	Average of estimates by Knox (1968) and Heicklen (1968)
$HOOC_4H_8OO \rightarrow HOOC_4H_8 + O_2$	2.5×10^{14}	28.0	—	—	Estimate
$HOOC_4H_8OO + i\text{-}C_4H_{10} \rightarrow HOOC_4H_8OOH + C_4H_9$	2.5×10^{-13}	7.7	—	—	Estimated by Fish (1967)[c]
$HOOC_4H_8 \rightarrow HO$ + O-heterocycle	1.0×10^{11}	13.0	—	—	Estimated by Fish (1967)[d]
$HOOC_4H_8 \rightarrow HO_2 + i\text{-}C_4H_8$	2.5×10^{14}	23.0	—	—	Estimate

Reaction	A	E	Temp. range	M	Source
	Unimportant reactions				
$C_4H_9 \rightarrow H + C_4H_8$	1.0×10^{14}	48.0	—	—	Estimate
$C_4H_9 \rightarrow CH_3 + C_3H_6$	1.0×10^{15}	39.0	—	—	Estimate
$H + O_2 + M \rightarrow HO_2 + M$	1.3×10^{-32}	0.0	400–700	N_2, O_2	Baulch *et al.* (1972)
$H + O_2 \rightarrow HO + O(^3P)$	3.7×10^{-10}	16.9	700–2500	—	Baulch *et al.* (1972)
$O(^3P) + i\text{-}C_4H_{10} \rightarrow HO + C_4H_9$	irrelevant				
$CH_3 + O_2 + M \rightarrow CH_3O_2 + M$	8.0×10^{-32}	0.0	300–500	—	Heicklen (1968)
$CH_3 + O_2 \rightarrow CH_2O + HO$	1.0×10^{-12}	9.0	300–1400	—	Clark *et al.* (1971b) and estimate[b]
$CH_3O_2 + M \rightarrow CH_3 + O_2 + M$	1.0×10^{-6}	26.0	300–500	—	Heicklen (1968)
$CH_3O_2 + i\text{-}C_4H_{10} \rightarrow CH_3O_2H + C_4H_9$	1.6×10^{-13}	7.7	—	—	Estimate[c]
$HO_2 + C_4H_9O_2 \rightarrow C_4H_9O_2H + O_2$	5.0×10^{-12}	0.0	—	—	Estimate
$HO_2 + CH_3O_2 \rightarrow CH_3O_2H + O_2$	5.0×10^{-12}	0.0	—	—	Estimated by Heicklen (1968)

[a] Units are cm^6/sec for third-order reactions, cm^3/sec for second-order reactions, and sec^{-1} for first-order reactions.

[b] Rate coefficient of Clark *et al.* (1971b) at 1100–1400°K and estimated Arrhenius parameters.

[c] For abstraction of a tertiary hydrogen.

[d] For furan formation.

is of the utmost consequence since it identifies the dihydroperoxide as the degenerate chain branching molecule and explains the negative temperature coefficient for the overall oxidation between 300 and 450°C.

The initial rates of i-C_4H_{10} consumption and the production of the significant hydroperoxides are also listed in Table V-2. Except for the dihydroperoxide they all increase with temperature. Note that the initial i-C_4H_{10} consumption is more than 6 orders of magnitude greater than the rate of the initiating reaction at 400°K and still over 2 orders of magnitude greater than the rate of the initiating reaction at 700°K. Yet these rates are too slow to be of any significance. Even at 700°K, the computed lifetime of i-C_4H_{10} consumption is almost 3 hr. Thus these rates correspond only to the initial induction period, and delayed chain-branching processes must follow.

The relative initial product yields, based on i-C_4H_{10} consumption, for the important products are shown in Fig. V-4. Below 470°K, the major initial carbon-containing product is the dihydroperoxide. Between 470 and 640°K, the major products are the O-heterocycles, and above 640°K, i-C_4H_8 becomes the dominant carbon-containing product. Notice also that H_2O production is coupled to O-heterocycle production, and that H_2O_2, which is produced only 5% of the time at 400°K, becomes the major product above 635°K.

Hydroperoxide Decomposition

The hydroperoxides are unstable compounds. They decompose readily on metal surfaces to give aldehydes and H_2O:

$$RCH_2OOH \rightarrow RCH(O) + H_2O \tag{19}$$

Thus any oxidation in a metal reactor will not give hydroperoxides, but rather aldehydes in their place. In the gas phase, decomposition occurs via a different route by cleavage of the O—O bond, which has a bond enthalpy of only ~43 kcal/mole:

$$ROOH \rightarrow RO + OH \tag{20}$$

Presumably the dihydroperoxides have even weaker O—O bonds, and their decomposition can become important above ~450°K. In this respect H_2O_2 is quite different, since its O—O bond enthalpy is 51 kcal/mole. Thus its decomposition

$$H_2O_2 \rightarrow 2HO \tag{21}$$

is not important below ~450°C.

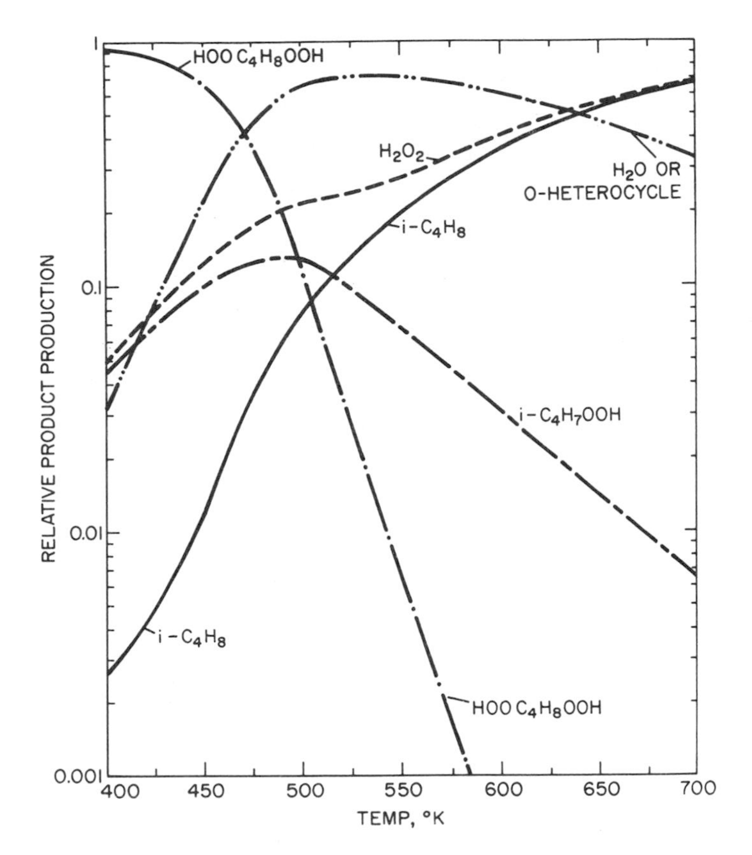

Fig. V-4. Semilog plot of the major product yields (relative to i-C_4H_{10} consumption) versus temperature in the initial stages of the slow combustion of i-C_4H_{10} based on a model calculation for $[O_2] = 2.5 \times 10^{18}$, $[i\text{-}C_4H_{10}] = 2.5 \times 10^{17}$, and $[M] = 3.5 \times 10^{18}$ molecules/cm^3.

The alkoxy radicals produced in reaction (20) can lead to aldehyde and alcohol formation via

$$R_1R_2CHO \rightarrow R_1 + R_2CH(O) \tag{22}$$

$$R_1R_2CHO + O_2 \rightarrow R_1R_2C(O) + HO_2 \tag{23}$$

$$R_1R_2CHO + RH \rightarrow R_1R_2CHOH + R \tag{24}$$

Since reaction (24) generally has a higher activation energy than reaction (23), alcohols will only tend to be formed at high fuel-to-O_2 ratios. Both the alcohols and aldehydes are secondary products of the oxidation.

TABLE V-2

Some Important Rates in a Model Calculation of the Initial Stages of the Slow Combustion of i-C_4H_{10}[a]

Species or reaction	Reaction rate (molecules/cm³-sec)						
	400°K	450°K	500°K	550°K	600°K	650°K	700°K
				Reactant and product rates			
$-d[i\text{-}C_4H_{10}]/dt$	7.87×10^6	3.54×10^8	8.72×10^9	1.30×10^{11}	9.91×10^{11}	5.61×10^{12}	2.59×10^{13}
$d[HOOC_4H_8OOH]/dt$	7.26×10^6	2.41×10^8	1.08×10^9	9.50×10^8	4.90×10^8	2.49×10^8	1.32×10^8
$d[C_4H_9O_2H]/dt$	10.4	4.33×10^2	9.88×10^3	1.38×10^5	1.00×10^6	5.41×10^6	2.40×10^7
$d[C_4H_7OOH]/dt$	3.49×10^5	3.52×10^7	1.13×10^9	8.74×10^9	3.05×10^{10}	7.84×10^{10}	1.68×10^{11}
				Rate of initiating reaction			
$C_4H_{10} + O_2 \rightarrow HO_2 + C_4H_9$	5.92	2.78×10^3	3.81×10^5	2.13×10^7	6.11×10^8	1.05×10^{10}	1.19×10^{11}
				Rates of important terminating reactions			
$HO_2 + HO_2 \rightarrow H_2O_2 + O_2$	1.72	1.47×10^3	3.47×10^5	2.12×10^7	6.10×10^8	1.05×10^{10}	1.19×10^{11}
$HO_2 + HOOC_4H_8O_2 \rightarrow O_2 + HOOC_4H_8O_2H$	4.20	1.29×10^3	3.55×10^4	1.16×10^5	1.71×10^5	2.12×10^5	2.41×10^5
				Rates of important reactions involving C_4H_9 and $C_4H_9O_2$			
$C_4H_9 + O_2 \rightarrow C_4H_9O_2$	7.87×10^6	3.56×10^8	8.69×10^9	1.28×10^{11}	9.73×10^{11}	5.47×10^{12}	2.51×10^{13}
$C_4H_9 + O_2 \rightarrow C_4H_8 + HO_2$	1.83×10^4	1.66×10^6	7.10×10^7	1.65×10^9	1.84×10^{10}	1.43×10^{11}	8.64×10^{11}
$C_4H_9O_2 \rightarrow C_4H_9 + O_2$	0.340	2.40×10^2	5.34×10^4	4.77×10^6	1.63×10^8	3.25×10^9	4.44×10^{10}
$HO + C_4H_{10} \rightarrow H_2O + C_4H_9$	2.39×10^5	7.37×10^7	5.78×10^9	9.33×10^{10}	5.96×10^{11}	2.58×10^{12}	8.62×10^{12}
$C_4H_9O_2 + C_4H_{10} \rightarrow C_4H_9O_2H + C_4H_9$	10.4	4.31×10^2	9.90×10^3	1.38×10^5	1.00×10^6	5.41×10^6	2.40×10^7
$C_4H_9O_2 \rightarrow HOOC_4H_8$	7.87×10^6	3.55×10^8	8.70×10^9	1.28×10^{11}	9.73×10^{11}	5.47×10^{12}	2.51×10^{13}
$HO_2 + i\text{-}C_4H_{10} \rightarrow H_2O_2 + C_4H_9$	3.90×10^5	4.30×10^7	1.19×10^9	3.56×10^{10}	3.94×10^{11}	3.01×10^{12}	1.72×10^{13}

	Rates of important reactions involving $HOOC_4H_8$ or $HOOC_4H_8OO$						
$HOOC_4H_8 + O_2 \rightarrow HOOC_4H_8OO$	7.50×10^{6}	3.75×10^{8}	6.88×10^{9}	3.38×10^{10}	8.00×10^{10}	1.50×10^{11}	2.44×10^{11}
$HOOC_4H_8 + O_2 \rightarrow C_4H_7OOH + HO_2$	3.48×10^{5}	3.50×10^{7}	1.12×10^{9}	8.71×10^{9}	3.02×10^{10}	7.84×10^{10}	1.68×10^{11}
$HOOC_4H_8OO \rightarrow HOOC_4H_8 + O_2$	2.37×10^{5}	1.34×10^{8}	5.80×10^{9}	3.28×10^{10}	7.95×10^{10}	1.50×10^{11}	2.44×10^{11}
$HOOC_4H_8OO + C_4H_{10} \rightarrow HOOC_4H_8OOH + C_4H_9$	7.26×10^{6}	2.41×10^{8}	1.08×10^{9}	9.49×10^{8}	4.89×10^{8}	2.49×10^{8}	1.32×10^{8}
$HOOC_4H_8 \rightarrow HO$ + O-heterocycle	2.39×10^{5}	7.37×10^{7}	5.78×10^{9}	9.33×10^{10}	5.96×10^{11}	2.59×10^{12}	8.62×10^{12}
$HOOC_4H_8 \rightarrow HO_2 + i\text{-}C_4H_8$	2.05×10^{3}	2.55×10^{6}	6.13×10^{8}	2.47×10^{10}	3.38×10^{11}	2.50×10^{12}	1.62×10^{13}

[a] $[C_4H_{10}] = 2.5 \times 10^{17}$ molecules/cm^3, $[O_2] = 2.5 \times 10^{18}$ molecules/cm^3, $[M] = 3.5 \times 10^{18}$ molecules/cm^3.

With the above mechanism the observation of cool flames and their negative activation energies can be explained in a manner analogous to that of Benson (1972). As the temperature of the oxidation is raised above 200°C, light tends to be emitted, and "cool flames" are observed. These increase in intensity until temperatures of about 300–350°C. Above these temperatures the flame is quenched, and the oxidation rate drops with increasing temperature until a temperature of about 450°C is reached, at which point oxidation is again enhanced with rising temperature.

The explanation follows from the model calculation of the initial stages of the oxidation. The delayed chain branching is carried mainly by the dihydroperoxide, whose production rate increases to about 550°K. Above this temperature, the rate of dihydroperoxide production decreases, and the delayed chain branching diminishes in importance. The reaction slows down as the temperature is raised. However as the temperature is raised, H_2O_2 becomes an important product, and it can act as the delayed chain branching agent above 450°C, where its decomposition is important.

Aldehyde Oxidation

Formaldehyde oxidation leads to a pressure increase as shown in Fig. V-5 for oxidations carried out at 337°C. The results are similar in a cleaned or aged reaction vessel. At 275°C, the main products have been shown to be (Hoare, 1965) CO and H_2O. Also found were CO_2, H_2, HCOOH, $HC(O)O_2H$, and $(HOCH_2O)_2$.

The oxidation of CH_2O is initiated by H atom abstraction, either by O_2 or a free radical:

$$CH_2O + O_2 \rightarrow HCO + HO_2 \tag{25}$$

$$R + CH_2O \rightarrow RH + HCO \tag{26}$$

The HCO radical is known to undergo three reactions:

$$HCO + M \rightarrow H + CO + M \tag{27}$$

$$HCO + O_2 + M \leftrightharpoons HCO_3 \tag{28,-28}$$

$$HCO + O_2 \rightarrow HO_2 + CO \tag{29}$$

The formation of CO_2 and OH in the HCO–O_2 interaction has been shown to be unimportant. Reaction (27) is of significance at high temperatures because the H atom can lead to chain branching via reaction (17). However, under most conditions, HCO is removed by reaction with O_2, or the H atom produced in reaction (27) adds to O_2 via reaction (16) to give HO_2. Reaction (28) must occur, because it is the only route that can lead to the observed reaction products HCO_3H and HCOOH. The possible fate

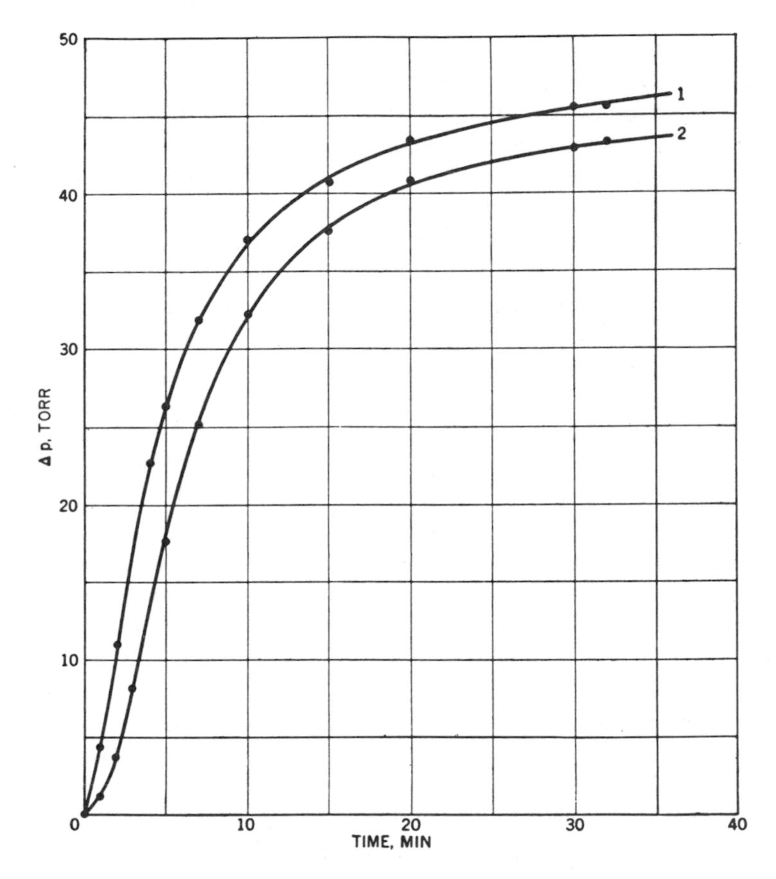

Fig. V-5. Thermal reaction rate of CH_2O oxidation in cleaned and aged silica vessels at 337°C. Curve 1, HNO_3-cleaned, $[CH_2O]$ = 127.4 Torr, $[O_2]$ = 79.3 Torr; curve 2, aged vessel, $[CH_2O]$ = 127.6 Torr, $[O_2]$ = 68.6 Torr. From Scheer (1955) with permission of the Combustion Institute.

of the HCO_3 radical, besides the reverse of reaction (28), is

$$HCO_3 + RH \rightarrow HCO_3H + R \tag{30}$$

The HCO_3H can then decompose to give the series of reactions

$$HCO_3H \rightarrow HCO_2 + OH \tag{31}$$

$$HCO_2 + RH \rightarrow HCOOH + R \tag{32}$$

$$HCO_2 + M \rightarrow H + CO_2 + M \tag{33}$$

$$HCO_2 + O_2 \rightarrow HO_2 + CO_2 \tag{34}$$

Thus HCO_3 can lead to the formation of HCO_3H, HCOOH, or CO_2, but not CO. Since CO and H_2O are the major products of the reaction, the most important fate of HCO is reaction (29) [or the equivalent sequence of reaction (28) followed by decomposition of HCO_3 to CO + HO_2].

The higher aldehydes undergo a slightly different oxidation than CH_2O, because radicals are produced that have no H atom on the α carbon atom. For both CH_3CHO and C_2H_5CHO, the main products are the corresponding peracids in both photochemical and thermal oxidations below 150°C. The reaction is accompanied by a pressure decrease. Above 150°C, CO, CO_2, CH_2O, alcohols, and acids are produced. Even at 230°C, where the peracids are not the major final product, the reaction first gives a pressure decrease and then an increase (Gál *et al.*, 1958). Thus peracid formation is indicated as an intermediate product, which then decomposes. The aldehyde oxidation also yields lower aldehydes and olefins. Baldwin and Walker (1969) studied the oxidation of C_2H_5CHO in boric acid-coated vessels at 440°C. The C_2H_5CHO oxidized to give about 80% C_2H_4 and about 10% CH_3CHO in a homogeneous reaction. Baldwin *et al.* (1971b) restudied the reaction at 440°C in a KCl-coated reaction vessel and found increased amounts of CH_3CHO. The high yield was attributed to a heterogeneous reaction forming both C_2H_4 and CH_3CHO in almost equal proportions. Acetaldehyde oxidation shows the negative activation energy between 350 and 400°C (Baldwin *et al.*, 1971a) similar to that in paraffin oxidation. Above this regime, the oxidation is accompanied by a pressure increase from the very beginning of the reaction, as reported at 440°C by Baldwin *et al.* (1971a).

The mechanism for CH_3CHO oxidation can be summarized as follows (the higher aldehydes presumably behave similarly):

$$CH_3CHO + O_2 \rightarrow CH_3CO + HO_2 \tag{35}$$

$$CH_3CHO + R \rightarrow CH_3CO + RH \tag{36}$$

$$CH_3CO + O_2 \rightarrow CH_3CO_3 \tag{37}$$

$$\rightarrow CH_3O + CO_2 \tag{38}$$

$$CH_3CO_3 + RH \rightarrow CH_3CO_3H + R \tag{39}$$

The evidence indicates that other possible routes for reaction (37), viz.,

$$CH_3CO + O_2 \rightarrow CH_3O_2 + CO \tag{40}$$

$$\rightarrow CH_2CO + HO_2 \tag{41}$$

do not occur. The peracetic acid can thermally decompose to give

$$CH_3CO_3H \rightarrow CH_3CO_2 + OH \tag{42}$$

$$CH_3CO_2 + RH \rightarrow CH_3CO_2H + R \tag{43}$$

$$CH_3CO_2 \rightarrow CH_3 + CO \tag{44}$$

The final products of the reaction, in addition to CH_3COOH, CO, CO_2 and H_2O, then come from CH_3 and CH_3O radical oxidation.

Since the aldehydes have an easily abstractable H atom, an important reaction is the removal of the chain terminating radical HO_2:

$$HO_2 + RCHO \rightarrow H_2O_2 + RCO \tag{45}$$

Thus the chain is propagated, and the oxidation is accelerated in the presence of aldehydes.

Olefin Oxidation

Above about 300°C, the olefins start to oxidize. With ethylene, major products are CO, CO_2, H_2O, CH_2O, and $\overline{CH_2CH_2O}$. Presumably the epoxides are produced via

$$RO_2 + \text{olefin} \rightarrow RO + \text{epoxide} \tag{46}$$

Supporting evidence for this reaction was given by Ray and Waddington (1971), who examined the cooxidation of acetaldehyde with 12 alkenes and found epoxides as major products. They proposed that the oxidation was carried through the CH_3CO_3 radical.

The oxidation to produce the lower-molecular-weight carbon products presumably occurs through HO and RO radical oxidation. Unfortunately the details of the reactions are not known.

The net reaction of HO with olefins may be summarized as (say for C_2H_4):

$$HO + C_2H_4 \rightarrow CH_2O + CH_3 \tag{47a}$$

$$\rightarrow HOC_2H_4 \underset{O_2}{\rightarrow} HO + 2CH_2O \tag{47b}$$

With RO, the situation is even more obscure, and we must settle for a reaction analogous to reaction (47b):

$$RO + C_2H_4 \rightarrow ROC_2H_4 \overset{O_2}{\rightarrow} RO + 2CH_2O \tag{48}$$

The oxidation of C_2H_2 starts at about 250°C. The product and reactant time histories for a reaction at 330°C are shown in Fig. V-6. The main initial products are CO, CH_2O, and glyoxal. Possibly the initial overall reaction can be summarized as

$$C_2H_2 + O_2 \rightarrow CO + CH_2O \tag{49a}$$

$$\rightarrow (CHO)_2 \tag{49b}$$

The overall reaction is second order in $[C_2H_2]$, zero order in $[O_2]$, and has $E_a = 33.3$ kcal/mole from 250 to 400°C. The detailed mechanism is un-

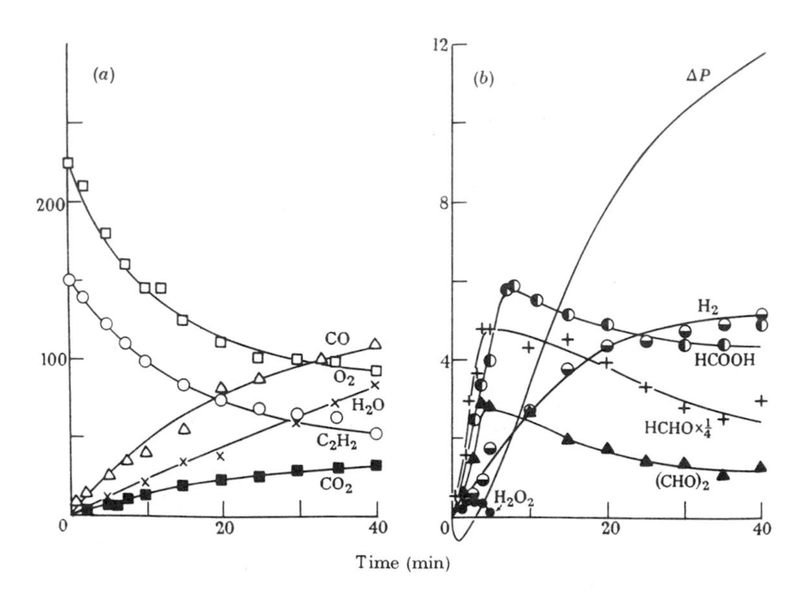

Fig. V-6. Variation of reactants and products during the reaction between C_2H_2 and O_2 at 330°C: $[C_2H_2]_0$ = 150 Torr, $[O_2]_0$ = 225 Torr. (a) Pressure of reactants and products, (b) pressure of products and ΔP. From Hay and Norrish (1965) with permission of the Royal Society.

known, but probably $(CHO)_2$ is the degenerate chain branching intermediate (Williams and Smith, 1970). Its oxidation leads to glyoxalic acid, which has been observed under some conditions (Minkoff and Tipper, 1962), as well as CO, CO_2, H_2O, and possibly HCOOH. Probably most of the HCOOH, as well as some of the CO, CO_2, and H_2O, comes from CH_2O oxidation. The origin of the small amount of hydrogen produced is not known.

For the sake of completeness we include the reactions of $O(^3P)$ with olefins. For small alkenes, the reaction gives an excited aldehyde that usually decomposes to radical fragments. Thus for C_2H_4

$$O(^3P) + C_2H_4 \rightarrow CH_3CHO^* \rightarrow CH_3 + HCO \tag{50}$$

with C_2H_2, the reaction is

$$O(^3P) + C_2H_2 \rightarrow CO + CH_2 \tag{51a}$$

CH_4 Oxidation

CH_4 oxidation is unique among the hydrocarbons, because CH_4 has only one carbon atom, and internal rearrangements are not easy. It does not oxidize readily below 450°C, and its oxidation is relatively slow above this temperature.

Methane oxidation exhibits the same histories for the slow combustion as higher hydrocarbons. Typical pressure histories for CH_4 are shown in Fig. V-7. First there is an induction period followed by acceleration, and finally a slowdown in the oxidation.

Typical results for the product distribution versus reaction time are shown for CH_4 in Fig. V-8. It can be seen that CH_4 removal is slow at first but accelerates as CH_2O and H_2O_2 grow in importance; the maximum removal rate for CH_4 corresponds to the maxima in CH_2O and H_2O_2 pressures. The final carbon-containing products are CO and CO_2.

The results of a model calculation corresponding to the experiment shown in Fig. V-8 are given in Fig. V-9. The reactions and rate coefficients used in the calculation are listed in Table V-3. In the mechanism the oxidation of CH_3 radicals is considered to be entirely

$$CH_3 + O_2 \rightarrow CH_2O + HO \qquad (52)$$

This reaction is the analog corresponding to O-heterocycle formation with the higher hydrocarbons. The omission of other reactions is, of course, an oversimplification to make the computations tractable. The reactions

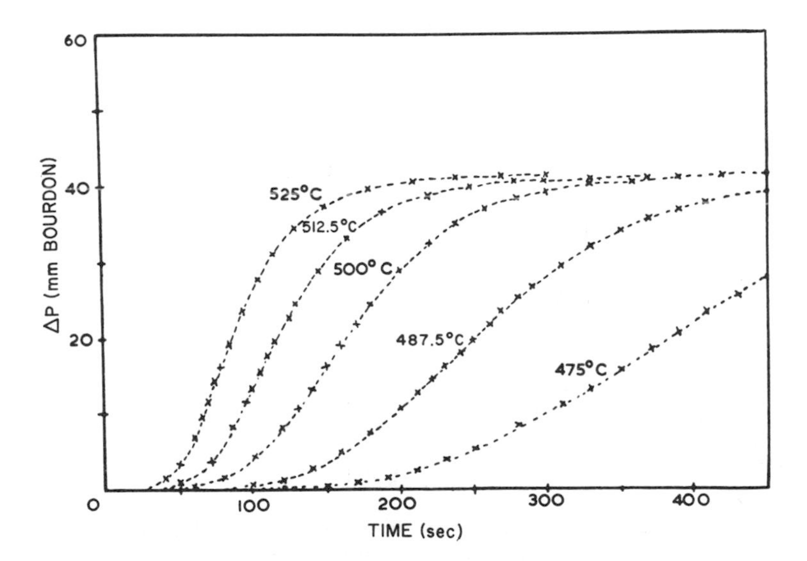

Fig. V-7. Typical pressure–time curves in the oxidation of CH_4 starting with 200 Torr CH_4 and 100 Torr O_2 in an HF-treated silica vessel. (1.8 mm Bourdon = 1 Torr.) From Hoare (1965) with permission of Gordon and Breach.

omitted are

$$CH_3 + O_2 + M \leftrightharpoons CH_3O_2 + M \qquad (53,-53)$$

$$CH_3O_2 + RH \rightarrow CH_3OOH \qquad (54)$$

$$CH_3OOH \rightarrow CH_3O + OH \qquad (55)$$

$$CH_3O \rightarrow CH_2O + H \qquad (56)$$

$$CH_3O + O_2 \rightarrow CH_2O + HO_2 \qquad (57)$$

The justification for omitting these reactions is that at 745°K, the rate of reaction (54) for RH=CH_4 is much slower than that of reaction (52). (The ratio of rate coefficients is $\sim 10^{-3}$, whereas the equilibrium ratio $[CH_3O_2]/[CH_3] = 7.2$.) Of course, omitting this sequence introduces some error into the computation, especially since reaction (55) is a chain branching step.

In spite of the simplification, the model computation reproduces the main features of the combustion history shown in Fig. V-8. Both $[CH_2O]$

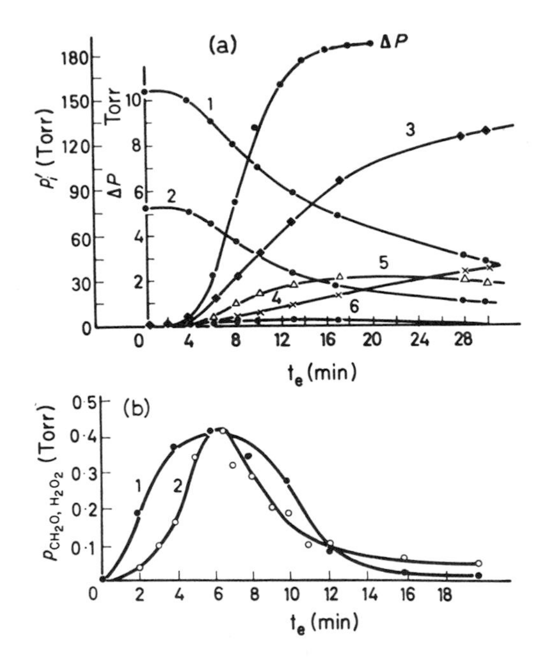

Fig. V-8. Reactant and product time histories in the oxidation of CH_4 ($[O_2]_0/[CH_4]_0 = 2$) at 472°C and an initial total pressure of 235 Torr in an HF-treated vessel. Curves in (a): 1, O_2; 2, CH_4; 3, H_2O; 4, CO_2; 5, CO; 6, H_2. Curves in (b): 1, CH_2O; 2, H_2O_2. From Karmilova *et al.* (1960a,b) as reported in Minkoff and Tipper (1962) with permission of Butterworths.

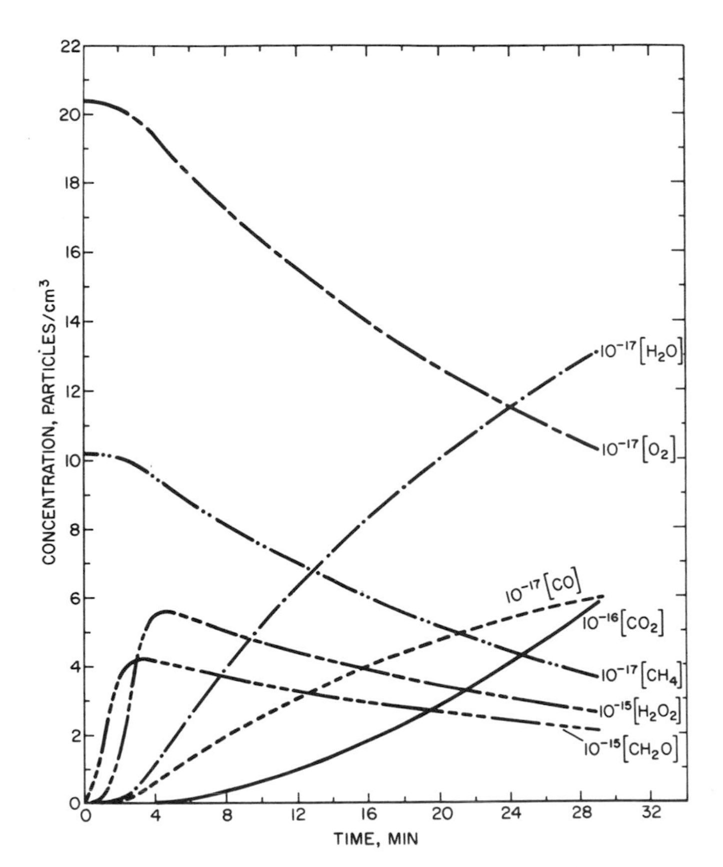

Fig. V-9. Plot of the reactant and product concentrations during the combustion of a 1 : 2 CH_4–O_2 mixture at 745°K based on a model calculation with $[M] \equiv [O_2] + 4[CH_4] + 3[CH_2O] + 2[H_2O_2] + [CO] + 3[CO_2] + 3[H_2O]$.

and $[H_2O_2]$ reach peak values between 4×10^{15} and 6×10^{15} molecules/cm³ in good agreement with observation (0.4 Torr = 5.2×10^{15} molecules/cm³). Experimentally the times of the maxima occur at 6 min, while the calculation indicates somewhat shorter times, i.e., 3 min for CH_2O and 5 min for H_2O_2. However, after the maximum is reached the computed oxidation is 2–3 times slower than observed.

Perhaps the fact that the computed oxidation is too slow in the later stages of the reaction can be explained by the omission of reactions (53)–(57). These reactions may not be too important in the early stages. However, as CH_2O and H_2O_2 accumulate they can be removed much more efficiently as a result of reaction (54), which produces the chain branching agent CH_3OOH. Thus in the later stages of the reaction, the computed

TABLE V-3

Rate Coefficients for Reactions in CH_4 *Oxidation*

Reaction	A^a (cm^{3n}/sec)	E_a (kcal/mole)	Temp. range (°K)	M	Reference
		Initiation reactions			
$O_2 + CH_4 \rightarrow HO_2 + CH_3$	3.0×10^{-11}	62.0	—	—	Estimate
$O_2 + CH_2O \rightarrow HO_2 + HCO$	3.4×10^{-11}	38.9	713–813	—	Baldwin *et al.* (1974)
$O_2 + H_2O_2 \rightarrow HO_2 + HO_2$	3.0×10^{-11}	41.0	—	—	Estimate
$H + O_2 \rightarrow HO + O(^3P)$	3.7×10^{-10}	16.9	700–2500	—	Baulch *et al.* (1972)
$H_2O_2 \rightarrow 2HO$	5.0×10^{13}	51.0	—	—	Estimate
$O(^3P) + CH_4 \rightarrow HO + CH_3$	3.5×10^{-11}	9.1	350–1000	—	Garvin and Hampson (1974)
$O(^3P) + CH_2O \rightarrow HO + HCO$	2.0×10^{-11}	2.9	—	—	Estimate
$O(^3P) + H_2O_2 \rightarrow HO + HO_2$	2.8×10^{-12}	4.2	283–373	—	Garvin and Hampson (1974)
		Termination reactions			
$2HO_2 \rightarrow H_2O_2 + O_2$	3.0×10^{-11}	1.0	300–1000	—	Garvin and Hampson (1974)
$HO + HO_2 \rightarrow H_2O + O_2$	8.3×10^{-11}	1.0	300–1000	—	Lloyd (1974)
$HO_2 + CH_3 \rightarrow CH_4 + O_2$	1.0×10^{-11}	0.0	—	—	Estimate
$2CH_3 \rightarrow C_2H_6$	2.2×10^{-11}	0.0	745	—	Clark *et al.* (1971a) and Bass and Laufer (1973)
		Chain reactions			
$HO_2 + CH_4 \rightarrow H_2O_2 + CH_3$	1.7×10^{-12}	19.0	—	—	Estimate
$HO_2 + CH_2O \rightarrow H_2O_2 + HCO$	1.7×10^{-12}	8.0	200–1000	—	Garvin and Hampson (1974)
$HCO + M \rightarrow H + CO + M$	1.9×10^{-9}	30.6	298–373	—	Hikida *et al.* (1971), Wang *et al.* (1973), and JANAF tables
$HCO + O_2 \rightarrow HO_2 + CO$	7.0×10^{-11}	1.3	300–1800	—	Garvin and Hampson (1974)

$H + O_2 + M \rightarrow HO_2 + M$	1.3×10^{-32}	0.0	745	O_2	Baulch *et al.* (1972)
$HO_2 + M \rightarrow H + O_2 + M$	3.5×10^{-9}	46.0	300–2000	Ar	Baulch *et al.* (1972)
$HO + CO \rightarrow H + CO_2$	1.4×10^{-13}	0.0	200–400	—	Garvin and Hampson (1974)
$HO + CH_4 \rightarrow H_2O + CH_3$	4.7×10^{-11}	5.0	300–2000	—	Garvin and Hampson (1974)
$HO + CH_2O \rightarrow H_2O + HCO$	4.0×10^{-11}	0.5	300–1600	—	Garvin and Hampson (1974)
$HO + H_2O_2 \rightarrow H_2O + HO_2$	1.7×10^{-11}	1.8	300–800	—	Garvin and Hampson (1974)
$CH_3 + O_2 \rightarrow CH_2O + HO$	1.0×10^{-12}	9.0	300–1400	—	Clark *et al.* (1971b) and estimate[b]
$CH_3 + CH_2O \rightarrow CH_4 + HCO$	1.0×10^{-11}	4.0	—	—	Estimate
$CH_3 + H_2O_2 \rightarrow CH_4 + HO_2$	1.0×10^{-11}	6.0	—	—	Estimate

[a] Units are cm^6/sec for third-order reactions; cm^3/sec for second-order reactions; and sec^{-1} for first-order reactions.

[b] Rate coefficient of Clark *et al.* at 1100–1400°K and estimated Arrhenius parameters.

concentrations of CH_2O and H_2O_2 are too large, while the oxidation rate of CH_4 is too small.

RAPID COMBUSTION

Above about 450°C, combustion processes become rapid, the heat can no longer be dissipated, and the process becomes adiabatic; the temperature rises as the reaction proceeds. In addition to combustion, pyrolysis of the hydrocarbons occurs at these temperatures, leading to cracking, pyrosynthesis, and soot formation in fuel-rich mixtures.

Oxidation

Above 450°C there are three major changes in the reaction mechanism that leads to the accelerated rate of oxidation.

First is the fact that H_2O_2 begins to decompose thermally [reaction (21)]. Thus the self-removal of HO_2 to produce H_2O_2 is no longer a chain termination step, but only converts the HO_2 radicals to HO radicals; previously minor terminating reactions must now be invoked:

$$H + HO_2 \rightarrow H_2 + O_2 \tag{58}$$

$$HO + HO_2 \rightarrow H_2O + O_2 \tag{59}$$

$$2H + M \rightarrow H_2 + M \tag{60}$$

$$H + O(^3P) + M \rightarrow HO + M \tag{61}$$

$$H + OH + M \rightarrow H_2O + M \tag{62}$$

$$2O(^3P) + M \rightarrow O_2 + M \tag{63}$$

$$R + HO_2 \rightarrow RH + O_2 \tag{64}$$

$$2R \rightarrow \text{products} \tag{65}$$

The second important shift is that chain branching steps become of increasing importance. In this temperature regime, the $O(^3P)$ atom is still not significant. The important chain branching step is

$$HO_2 + RH \rightarrow R + H_2O_2{}^* \rightarrow 2HO \tag{66}$$

Third, hydrogen atom abstraction from olefins by radicals becomes important, especially with HO radicals:

$$HO + C_2H_4 \rightarrow H_2O + C_2H_3 \tag{47c}$$

$$HO + C_2H_2 \rightarrow H_2O + C_2H \tag{67}$$

Thus the carbon–hydrogen ratio in the organic material is increased and soot formation is promoted in the fuel-rich region. The radicals thus formed undergo oxidation, although the detailed steps are not known. The probable steps are

$$C_2H_3 + O_2 \rightarrow C_2H_2 + HO_2 \tag{68}$$

$$C_2H + O_2 \rightarrow HCO + CO \tag{69}$$

In the temperature regime >450°C, CO plays an important role as an intermediate. An important equilibrium reaction is

$$CO + HO \leftrightharpoons CO_2 + H \qquad (70,-70)$$

When this reaction goes in the forward direction, chain branching is enhanced, but the reverse reaction reduces the chain branching steps. In addition CO promotes termination via

$$O(^3P) + CO(+M) \rightarrow CO_2(+M) \tag{71}$$

Another shift is that the CH_3 radical oxidizes mainly via the relatively slow reaction (52), because reaction (53), which produces CH_3O_2, is reversed.

Cracking

At the temperatures of rapid oxidations, the hydrocarbons can undergo pyrolysis, which leads mainly to the production of lower-molecular-weight species for alkane oxidation. The pyrolysis follows a simple Rice–Herzfeld mechanism, which for C_2H_6 at temperatures between 550 and 1150°C is (Minkoff and Tipper, 1962)

$$C_2H_6 \rightarrow 2CH_3 \tag{72}$$

$$CH_3 + C_2H_6 \rightarrow CH_4 + C_2H_5 \tag{73}$$

$$C_2H_5 \rightarrow H + C_2H_4 \tag{74}$$

$$H + C_2H_6 \rightarrow H_2 + C_2H_5 \tag{75}$$

$$H + C_2H_5 \rightarrow H_2 + C_2H_4 \tag{76}$$

Thus C_2H_6 is converted to CH_4, H_2, and C_2H_4.

For the somewhat more complex i-C_4H_{10}, Konar *et al.* (1973) confirmed the mechanism to be (500–580°C).

$$i\text{-}C_4H_{10} \rightarrow 2 \text{ radicals} \tag{77}$$

$$i\text{-}C_4H_9 \rightarrow CH_3 + C_3H_6 \tag{78}$$

$$t\text{-}C_4H_9 \rightarrow H + i\text{-}C_4H_8 \tag{79}$$

$$CH_3 + i\text{-}C_4H_{10} \rightarrow CH_4 + C_4H_9 \tag{80}$$

$$H + i\text{-}C_4H_{10} \rightarrow H_2 + C_4H_9 \tag{81}$$

$$2CH_3 \rightarrow C_2H_6 \tag{82}$$

Thus the major products were C_3H_6, i-C_4H_8, CH_4, and H_2.

Olefin pyrolysis is somewhat more complex and leads to both cracking and the production of some heavier-molecular-weight species. Ethylene, at 480–580°C, gives C_2H_2, H_2, and 1,3-C_4H_6 as major products, with some butenes and C_2H_6. Also minor amounts of CH_4 and C_3H_6 are formed. The mechanism was discussed in some detail by Benson and Haugen (1967), and the major products are explained by

$$2C_2H_5 \rightarrow C_2H_5 + C_2H_3 \tag{83}$$

$$C_2H_5 \leftrightharpoons C_2H_4 + H \tag{84,−84}$$

$$C_2H_5 + H_2 \rightarrow C_2H_6 + H \tag{85}$$

$$C_2H_3 + M \leftrightharpoons C_2H_2 + H + M \tag{86,−86}$$

$$C_2H_3 + C_2H_4 \leftrightharpoons C_4H_6 + H \tag{87,−87}$$

$$C_2H_3 + C_2H_2 \leftrightharpoons C_4H_5 \tag{88,−88}$$

$$C_4H_5 + H_2 \leftrightharpoons 1,3\text{-}C_4H_6 + H \tag{89,−89}$$

with radical–radical reactions terminating the chain.

Pyrosynthesis

The nonoxidative reactions of C_2H_2 lead to pyrosynthesis. C_2H_2 thermolysis starts at temperatures >350°C, and low polymers are produced. The mechanism might be

$$2C_2H_2 \rightarrow C_2H_3 + C_2H \tag{90a}$$

$$\rightarrow C_4H_3 + H \tag{90b}$$

$$C_2H_3 + nC_2H_2 \rightarrow \text{polymer} \tag{91}$$

$$C_2H + nC_2H_2 \rightarrow \text{polymer} \tag{92}$$

$$C_4H_3 \rightarrow C_4H_2 + H \tag{93}$$

$$H + C_2H_2 \rightarrow C_2H_3 \tag{94}$$

At temperatures about 400–500°C, benzene and condensed aromatics are produced. The possible chemical routes involved in these processes have been considered by Badger *et al.* (1966). Their results, as summarized by Olsen and Haynes (1969) are illustrated in Fig. V-10. The lower polyunsaturated hydrocarbons, principally C_2H_2 and 1,3-C_4H_6, can combine by free radical mechanisms to produce benzene and substituted benzenes. These are the precursors to the polynuclear aromatic compounds, which may be reduced even further, in a fuel-rich mixture, to soot.

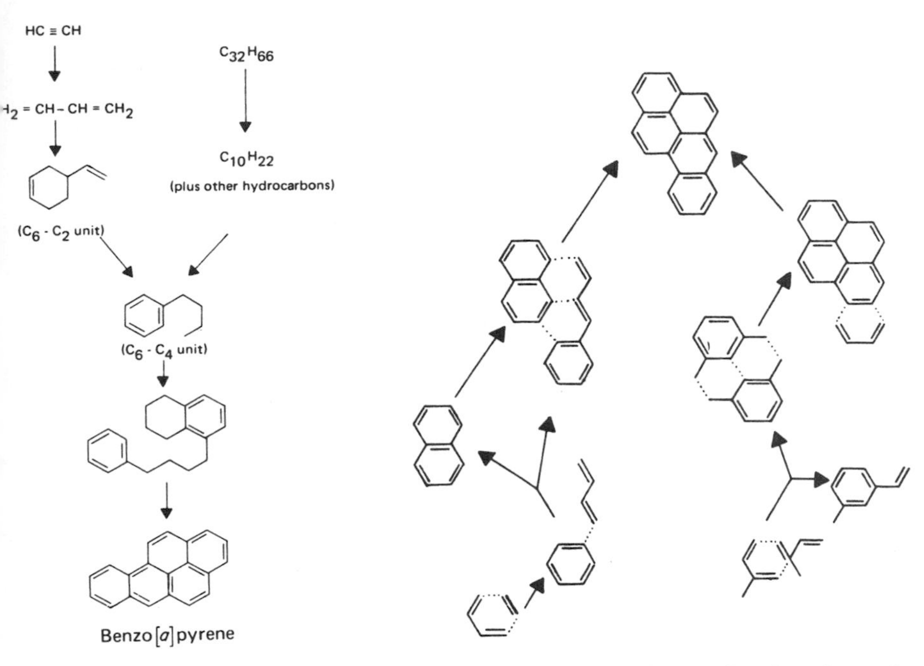

Fig. V-10. Possible pathways for the synthesis of benzo[*a*]pyrene based on the work of Badger *et al.* (1966). From Olsen and Haynes (1969) with permission of the U.S. Department of Health, Education, and Welfare.

Ignition

As the temperature continues to increase, the H_2O_2 thermal decomposition become more and more important relative to other removal processes, and ultimately chain branching becomes more important than termination; the radical concentrations cannot be maintained at steady state, and ignition occurs. This manifests itself either as a flame or an explosion. As an illustration of a typical rapid oxidation, we will consider C_4H_{10} at 900°K, without distinguishing between i-C_4H_{10} and n-C_4H_{10} or among the various C_4H_9 radicals. The initiating reaction is

$$C_4H_{10} + O_2 \rightarrow HO_2 + C_4H_9 \tag{95}$$

Termination by radical–radical processes cannot compete with chain branching, because the dominant radical is HO_2 and it is removed mainly by the chain branching reaction

$$HO_2 + C_4H_{10} \rightarrow 2HO + C_4H_9 \tag{96}$$

since the intermediate H_2O_2 thermally decomposes to 2HO radicals in

$\sim 10^{-2}$ sec. Furthermore the interaction of two HO_2 radicals no longer leads to chain termination, but to HO production, since reaction 11 is immediately followed by reaction 21. Thus ignition results.

The reactions of significance in product formation are (14a), (14b), (52), and

$$C_4H_9 + O_2 \rightarrow C_4H_8 + HO_2 \tag{97}$$

$$CH_3 + C_4H_{10} \rightarrow CH_4 + C_4H_9 \tag{98}$$

The H atoms produced in reaction (14a) are removed mainly by reactions (16) and (17) with a minor fraction forming H_2 via

$$H + C_4H_{10} \rightarrow H_2 + C_4H_9 \tag{99}$$

The O atom and HO radicals are removed almost exclusively via

$$O(^3P) + C_4H_{10} \rightarrow HO + C_4H_9 \tag{100}$$

$$HO + C_4H_{10} \rightarrow H_2O + C_4H_9 \tag{101}$$

The reactions determining the relative product distribution are (97), (14a), (14b), (52), and (98). Their rate coefficients are listed in Table V-4. The rate coefficient given for reaction (97) is that for the addition reaction:

$$C_4H_9 + O_2 \rightarrow C_4H_9O_2 \tag{102}$$

since it is more rapid than direct abstraction. It is followed by an internal abstraction to give a radical that decomposes to $C_4H_8 + HO_2$.

TABLE V-4

Some Rate Coefficients for Reactions of Importance in the Initial Stages of the Oxidation of C_4H_{10} at 900°K

Reaction	A[a] (cm^{3n}/sec)	E_a (kcal/mole)	Reference
$C_4H_9 + O_2 \rightarrow C_4H_8 + HO_2$	2.0×10^{-12}	0.0	Estimated by Knox (1968)
$C_4H_9 \rightarrow H + C_4H_8$	1.0×10^{14}	48.0	Estimate
$C_4H_9 \rightarrow CH_3 + C_3H_6$	1.0×10^{15}	39.0	Estimate
$CH_3 + O_2 \rightarrow CH_2O + HO$	1.0×10^{-12}	9.0	Clark *et al.* (1971b) and estimate[b]
$CH_3 + C_4H_{10} \rightarrow CH_4 + C_4H_9$	1.0×10^{-11}	14.6	Konar *et al.* (1973)[c]

[a] Units are cm^3/sec for second-order reactions and sec^{-1} for first-order reactions.
[b] Rate coefficient of Clark *et al.* at 1100–1400°K and estimated Arrhenius parameters.
[c] Average parameters for *i*-C_4H_{10} and *t*-C_4H_{10}.

TABLE V-5

Relative Importance of Major Products in the Initial Stages of the Oxidation of C_4H_{10} *at 900°K*[a]

	$[O_2]/[C_4H_{10}]$			
Product	6.50	3.25	1.625	1.00
C_4H_8	0.974	0.949	0.903	0.851
C_3H_6	0.026	0.051	0.097	0.149
H_2O	0.992	0.984	0.970	0.954
CH_2O	0.018	0.036	0.068	0.104
CH_4	0.008	0.016	0.030	0.046

[a] $[C_4H_{10}] = 1.0 \times 10^{18}$ molecules/cm³. Product fractions relative to C_4H_{10} consumed.

The product distribution, relative to C_4H_{10} consumed, computed for initial stages of the reaction is shown in Table V-5. Very little H_2 is formed, not only because the rate of reaction (99) is slower than those of reactions (16) and (17), but primarily because the formation of H atoms via reaction (14a) is only a relatively minor path for C_4H_9 removal.

Table V-5 indicates that for stoichiometric O_2–C_4H_{10} mixtures ($[O_2]/[C_4H_{10}] = 6.5$) to quite fuel-rich mixtures the principal products are C_4H_8 and H_2O. However, as the $[O_2]/[C_4H_{10}]$ ratio drops, more of the cracking products, C_3H_6 and CH_4, are produced. It is interesting to note that production of the only oxygen-bearing carbon compound formed in the initial stages of the reaction, i.e., CH_2O, is promoted by decreasing the $[O_2]/[C_4H_{10}]$ ratio and moving into the fuel-rich region. The reason is that CH_2O is produced through the CH_3 radical, which is produced by thermal cracking.

Very High Temperatures

At very high temperatures, more endothermic reactions play a role. It is also necessary to include the reverse of several reactions. In particular, HO_2 becomes less important, because of its rapid decomposition:

$$HO_2 + M \rightarrow H + O_2 + M \qquad (-16)$$

Now reaction (17) can become of considerable importance and the $O(^3P)$ atom becomes a significant chain branching carrier.

Above 2000°K the reverse of the terminating reactions must be considered:

$$M + H_2O \rightarrow H + OH + M \quad (-62)$$

$$M + CO_2 \rightarrow M + CO + O(^3P) \quad (-71)$$

$$M + O_2 \rightarrow M + 2O(^3P) \quad (-63)$$

Thus the oxidation cannot be turned off, and the internal combustion engine operates at peak temperatures >2000°C for this reason. Unfortunately since air is usually the oxidant, NO is produced by the endothermic reaction

$$O(^3P) + N_2 \rightarrow NO + N \quad (103)$$

followed by

$$N + O_2 \rightarrow NO + O(^3P) \quad (104)$$

These reactions do not interfere with the oxidation cycle since the $O(^3P)$ atom is regenerated. They tend to be more important in air-rich mixtures, since the $O(^3P)$ atoms are then more likely to react with the N_2 than with the hydrocarbon.

Soot Formation

The pyrolysis of C_2H_2 in fuel-rich mixtures leads to cracking above 600°C, and at temperatures of 1000 to 1500°C, H_2, CH_4, and soot (slightly hydrogenated carbon) is formed. The exact details of soot formation have never been elucidated definitively. There are two general theories for soot formation. In one theory, C_2H_2 undergoes pyrosynthesis to larger and larger polynuclear aromatic compounds, which have an increasing carbon–hydrogen ratio. This mechanism should be more important at relatively lower temperatures, where cracking is less important. At more elevated temperatures (>1000°C) there is more possibility of producing small free radicals with a high carbon content, say C_2H, and these may coalesce. Once the soot nucleus is formed, hydrocarbons may decompose on the surface, resulting in polyacetylenes and, ultimately, soot.

That polyacetylenes are present in flames has been demonstrated many times. A typical concentration profile in a flat C_2H_2–O_2 flame is shown in Fig. V-11. The species C_8H_2 is already close to the empirical formula of C_8H for soot.

Flame temperatures for C_2H_2 range from 2000 to 3000°C. In this temperature regime the oxidation is initiated mainly by branched chain reactions, once the flame has started. In the early stages before the radical concentrations have become sufficiently large for radical–radical and radical–

TABLE V-6

Rate Coefficients for Radical-Molecule Reactions of Importance in the Initial Stages of the High-Temperature Oxidation of C_2H_2

Reaction	A (cm³/sec)	E_a (kcal/mole)	Temp. range (°K)	Reference
$HO + C_2H_2 \rightarrow H_2O + C_2H$	1.0×10^{-11}	7.0	1000–1700	Williams and Smith (1970)
$C_2H + C_2H_2 \rightarrow C_4H_2 + H$	1.7×10^{-11}	4.0	—	Estimate
$C_2H + O_2 \rightarrow 2CO + H$	1.7×10^{-11}	7.0	1000–1700	Williams and Smith (1970)
$H + C_2H_2 \rightarrow H_2 + C_2H$	3.3×10^{-10}	19.0	1000–1700	Williams and Smith (1970)
$H + O_2 \rightarrow HO + O(^3P)$	3.7×10^{-10}	16.9	700–2500	Baulch *et al.* (1972)
$O(^3P) + C_2H_2 \rightarrow CH_2 + CO$	8.3×10^{-12}	2.5	1000–1700	Williams and Smith (1970)
$O(^3P) + C_2H_2 \rightarrow HO + C_2H$	3.3×10^{-10}	14.0	1000–1700	Williams and Smith (1970)
$CH_2 + O_2 \rightarrow 2H + CO_2$	1.7×10^{-11}	0.0	—	Estimate

product reactions to become important, the significant reactions are the radical–reactant reactions (17), (67), and

$$C_2H + C_2H_2 \rightarrow C_4H_2 + H \qquad (105)$$

$$C_2H + O_2 \rightarrow 2CO + H \qquad (106)$$

$$H + C_2H_2 \rightarrow H_2 + C_2H \qquad (107)$$

$$O(^3P) + C_2H_2 \rightarrow CH_2 + CO \qquad (51a)$$

$$\rightarrow HO + C_2H \qquad (51b)$$

$$CH_2 + O_2 \rightarrow 2H + CO_2 \qquad (108)$$

The rate coefficients for these reactions are listed in Table V-6. Reactions (17), (51b), and (108) are the branched-chain reactions that can promulgate the oxidation. All the species accumulate, and in the early part of the reaction their rates of accumulation, relative to C_2H_2 removal, are given in Table V-7 for a stoichiometric $[O_2]/[C_2H_2]$ mixture (2.5:1).

The C_2H and C_4H_2 species can be considered as a measure of soot formation, even though some fraction of these species is oxidized later in the

TABLE V-7

Relative Importance of Production Rate of Important Species in the Early Stages of the High-Temperature Oxidation of a Stoichiometric Mixture of C_2H_2 *and* O_2[a]

Product	Temperature (°K)		
	2000	2500	3000
H	0.133	0.082	0.059
$O(^3P)$	0.099	0.084	0.073
CH_2	0.001	0.001	0.008
HO	0.522	0.548	0.561
C_2H	0.132	0.155	0.170
C_4H_2	0.234	0.227	0.219
CO	0.676	0.702	0.723
H_2	0.106	0.114	0.119
CO_2	0.116	0.072	0.051
H_2O	0.258	0.258	0.258

[a] Rates relative to C_2H_2 consumption rate. $[O_2]/[C_2H_2] = 2.5$.

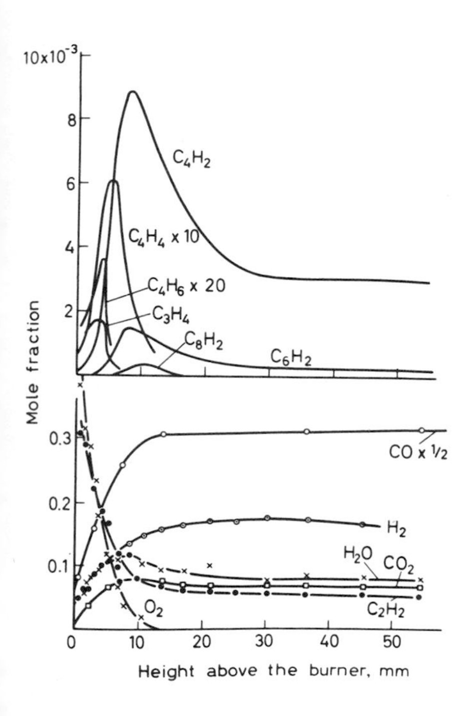

Fig. V-11. Concentration profiles in a flat C_2H_2–O_2 flame at the visual limit of carbon formation, C/O = 0.95/20 Torr burning pressure. From Homann (1967) with permission of the Combustion Institute.

flame. However, in the initial stages they account for over 50% of the carbon atoms at 2000–3000°K.

REFERENCES

Allara, D. L., Mill, T., Hendry, D. G., and Mayo, F. R. (1968). Oxidation of Organic Compounds II, *Advan. Chem. Ser. No. 76* p. 40, "Low-Temperature Gas- and Liquid-Phase Oxidations of Isobutane."

Badger, G. M., Donnelly, J. K., and Spotswood, T. M. (1966). *Aust. J. Chem.* **19,** 1023, "The Formation of Aromatic Hydrocarbons at High Temperatures. XXVII. The Pyrolysis of Isoprene."

Baldwin, R. R., and Walker, R. W. (1969). *Trans. Farad. Soc.,* **65,** 806 "Oxidation of Propionaldehyde in Aged Boric-Acid-coated Vessels. Part 2—Analytical Results."

Baldwin, R. R., Langford, D. H., Matchan, M. J., Walker, R. W., and Yocke, D. A. (1971a). *Symp. (Int.) Combust., 13th* 251, "The High-Temperature Oxidation of Aldehydes."

Baldwin, R. R., Matchan, M. J., and Walker, R. W. (1971b). *Trans. Faraday Soc.* **67,** 3521, "Oxidation of Propionaldehyde in KCl-Coated Vessels."

Baldwin, R. R., Fuller, A. R., Longthorn, D., and Walker, R. W. (1974). *J. Chem. Soc. Faraday Trans. I* **70,** 1257, "Oxidation of Formaldehyde in KCl-Coated Vessels."

Barat, P., Cullis, C. F., and Pollard, R. T. (1971). *Symp. (Int.) Combust., 13th* 179, "Studies of the Combustion of Branched-Chain Hydrocarbons."

Bass, A. M., and Laufer, A. H. (1973). *Int. J. Chem. Kinet.* **5**, 1053, "The Methyl Radical Combination Rate Constant as Determined by Kinetic Spectroscopy."

Baulch, D. L., Drysdale, D. D., Horne, D. G., and Lloyd, A. C. (1972). "Evaluated Kinetic Data for High Temperature Reactions," Vol. 1, "Homogeneous Gas Phase Reactions of the H_2–O_2 System." Butterworths, London.

Benson, S. W. (1968). Oxidation of Organic Compounds II, *Advan. Chem. Ser. No. 76* p. 143, "Some Current Views on the Mechanism of Free Radical Oxidations."

Benson, S. W. (1972). *Proc. Mater. Res. Symp., 4th* The Mechanisms of Pyrolysis, Oxidation and Burning of Organic Materials. Nat. Bur. Std. Spec. Publ. 357, p. 121, "Some Current Problems in Oxidation Kinetics."

Benson, S. W., and Haugen, G. R. (1967). *J. Phys. Chem.* **71**, 1735, "Mechanism for Some High-Temperature Gas-Phase Reactions of Ethylene, Acetylene and Butadiene."

Berry, T., Cullis, C. F., Saeed, M., and Trimm, D. L. (1968). Oxidation of Organic Compounds II, *Advan. Chem. Ser. No. 76*, p. 86, "Formation of O-Hetercycles as Major Products of the Gaseous Oxidation of *n*-Alkanes."

Clark, T. C., Izod, T. P. J., and Kistiakowsky, G. B. (1971a). *J. Chem. Phys.* **54**, 1295, "Reactions of Methyl Radicals Produced by the Pyrolysis of Azomethane or Ethane in Reflected Shock Waves."

Clark, T. C., Izod, T. P. J., and Matsuda, S. (1971b). *J. Chem. Phys.* **55**, 4644, "Oxidation of Methyl Radicals Studied in Reflected Shock Waves Using the Time-of-Flight Mass Spectrometer."

Donovan, R. J., and Husain, D. (1972). *Ann. Rep. 1971* **68A**, 123, "Reactions of Atoms and Small Molecules, Studied by Ultraviolet, Vacuum-Ultraviolet, and Visible Spectroscopy."

Fish, A. (1967). *Proc. Int. Oxidat. Symp., San Francisco* p. I-431, "Chain Propagation in the Oxidation of Alkyl Radicals."

Gál, D., Galiba, I., and Szabó, Z. (1958). *Acta Chim. Acad. Sci. Hung.* **16**, 39, "Slow and Cold Flame Oxidation of Acetaldehyde and Effect of Ethane on this Oxidation."

Garvin, D., and Hampson, R. F. (1974). Nat. Bur. Std. Rep. NBSIR-74-430, "Chemical Kinetics Data Survey. VII. Tables of Rate and Photochemical Data for Modelling of the Stratosphere" (Revised).

Hay, J. M., and Norrish, R. G. W. (1965). *Proc. Roy. Soc.* **A288**, 17, "The Oxidation of Acetylene."

Heicklen, J. (1968). Oxidation of Organic Compounds II, *Advan. Chem. Ser. No. 76* p. 23, "Gas-Phase Reactions of Alkylperoxy and Alkoxy Radicals."

Hikida, T., Eyre, J. A., and Dorfman, L. M. (1971). *J. Chem. Phys.* **54**, 3422, "Pulse Radiolysis Studies. XX. Kinetics of Some Addition Reactions of Gaseous Hydrogen Atoms by Fast Lyman-α Absorption Spectrophotometry."

Hoare, D. E. (1965). *In* "Low Temperature Oxidation" (W. Jost, ed.), p. 125. Gordon and Breach, New York, "The Combustion of Methane."

Homann, K. H. (1967). *Combust. Flame* **11**, 265, "Carbon Formation in Premixed Flames."

Karmilova, L. V., Enikolopyan, N. S., and Nalbandyan, A. B. (1960a). *Russ. J. Phys. Chem.* **34**, 261 (Engl. Transl.), "The Kinetics and Mechanism of the Oxidation of Methane. I. Fundamental Macrokinetic Laws."

Karmilova, L. V., Enikolopyan, N. S., and Nalbandyan, A. B. (1960b). *Russ. J. Phys. Chem.* **34**, 470 (Engl. Transl.), "Kinetics and Mechanism of Methane Oxidation. II. Kinetics of the Accumulation of Intermediate Products."

Knox, J. H. (1968). Oxidation of Organic Compounds II, *Advan. Chem. Ser. No. 76* p. 1, "Rate Constants in the Gas Phase Oxidation of Alkanes and Alkyl Radicals."
Knox, J. H., and Kinnear, C. G. (1971). *Symp. (Int.) Combust., 13th* 217, "The Mechanism of Combustion of Pentane in the Gas Phase between 250° and 400°C."
Konar, R. S., Marshall, R. M., and Purnell, J. H. (1973). *Int. J. Chem. Kinet.* **5**, 1007, "The Self-Inhibited Pyrolysis of Isobutane."
Lloyd, A. C. (1974). *Int. J. Chem. Kinet.* **6**, 169, "Evaluated and Estimated Kinetic Data for Gas Phase Reactions of the Hydroperoxyl Radical."
Minkoff, G. J., and Tipper, C. F. H. (1962). "Chemistry of Combustion Reactions." Butterworths, London.
Nalbandyan, A. B. (1967). *Proc. Int. Oxidat. Symp., San Francisco,* p. I–397, "The Photochemical Oxidation of Alkanes in the Gas Phase."
Olsen, D., and Haynes, J. L. (1969). U.S. Dept. of Health, Education and Welfare Publ. APTD 69-43, "Preliminary Air Pollution Survey of Organic Carcinogens. A Literature Review."
Ray, D. J. M., and Waddington, D. J. (1971). *Symp. (Int.) Combust., 13th* 261, "Co-oxidation of Acetaldehyde and Alkanes in the Gas Phase."
Scheer, M. D. (1955). *Symp. (Int.) Combust., 5th* 435, "Kinetics of the Gas Phase Oxidation of Formaldehyde."
Wang, H. Y., Eyre, J. A., and Dorfman, L. M. (1973). *J. Chem. Phys.* **59**, 5199, "Activation Energy for the Gas Phase Reaction of Hydrogen Atoms with Carbon Monoxide."
Williams, A., and Smith, D. B. (1970). *Chem. Rev.* **70**, 267, "The Combustion and Oxidation of Acetylene."

Chapter VI

PHOTOCHEMICAL SMOG

PRODUCTION

Since a chief source of pollutants is automobile emissions, it might be expected that pollutant levels could be correlated with automobile traffic, at least to some extent. Figure VI-1 shows the average hourly concentrations of various pollutants during the day of July 19, 1965, in Los Angeles. The CO and NO levels rise to a maximum at about 7:00 A.M. coincident with the morning rush hour traffic. A smaller peak is seen in the evening. Usually the evening peak is much larger than for this particular day. The hydrocarbons, although not shown, follow similar behavior. Of particular interest is the fact that the NO_2 morning peak is delayed by about 3 hr and the O_3 peak by about 5 hr. The evening peak for NO_2 is barely discernible and there is none for O_3.

The explanation for the delay in the NO_2 and O_3 peaks is that they are not primary emissions, but are produced photochemically in the atmosphere. Their production is important only during daylight, and thus the morning traffic emissions are mainly responsible for them. The evening rush hour traffic does not produce much smog, because the sunlight is weak and soon disappears. Nights are generally smog-free.

The photochemical effect has been reproduced many times in laboratory experiments. The time history of such an irradiation is shown in Fig. VI-2 for a C_3H_6–NO mixture. Initially, NO_2 and aldehydes are produced as the reactants are consumed. As the NO is exhausted, the NO_2 concentration passes through a maximum, and as the NO_2 is consumed, O_3 and other oxidants such as peroxyacetylnitrate are produced. Compounds such as HNO_3 and nitrates also must be present to account for the nitrogen balance.

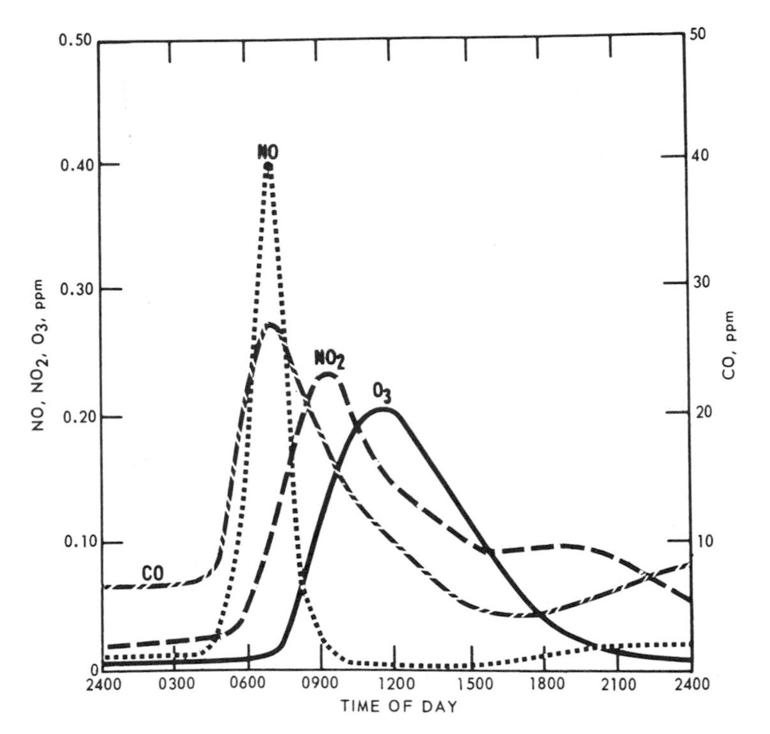

Fig. VI-1. Average 1-hr concentrations of selected pollutants in Los Angeles, California, July 19, 1965. From Environmental Protection Agency Document AP-84 (1971) with permission of the Environmental Protection Agency.

The most interesting feature of the above results is that the reaction is imperceptibly slow in the dark, yet none of the initial gases (NO, C_3H_6, O_2, N_2) absorb the ultraviolet radiation ($\lambda > 2900$ Å). In fact, the absorbing species is NO_2, and trace amounts of it must be present for the reaction to proceed. These may be produced from the well-known, but slow, reaction

$$2NO + O_2 \rightarrow 2NO_2 \tag{1}$$

The photochemical reaction is autocatalytic, and the NO_2 concentration rises at an accelerating rate as long as sufficient NO is present.

Once NO_2 is present, the production of O_3 follows from straightforward photochemical decomposition of NO_2:

$$NO_2 + h\nu \rightarrow NO + O(^3P) \tag{2}$$

$$O(^3P) + O_2 + M \rightarrow O_3 + M \tag{3}$$

where $h\nu$ represents sunlight at wavelengths <4000 Å (the photodissoci-

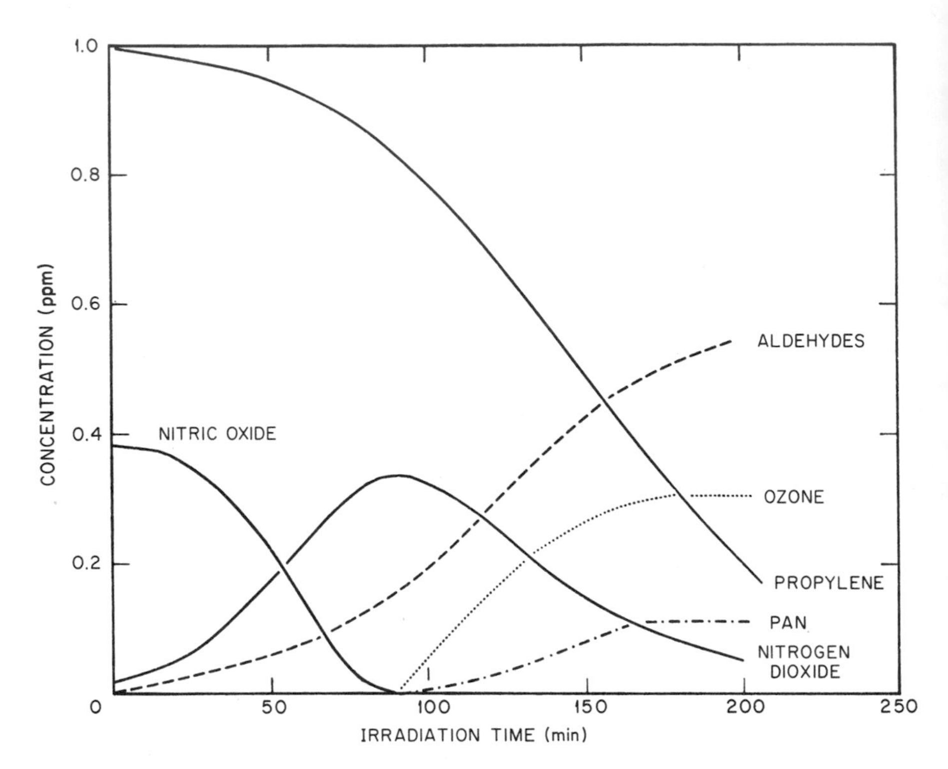

Fig. VI-2. Typical concentration changes during ultraviolet irradiation of a C_3H_6–NO mixture in air (PAN = peroxyacetyl nitrate). From Agnew (1968) with permission of the Royal Society.

ation limit of NO_2) and M is a chaperone gas, i.e., primarily N_2 or O_2 in the atmosphere. The complex chemistry of photochemical smog formation is that of the first phase, i.e., the conversion of NO to NO_2.

NO can be converted to NO_2 by rapid reaction with ozone:

$$NO + O_3 \rightarrow NO_2 + O_2 \tag{4}$$

However, this reaction leads to no net production of oxidant since one molecule of O_3 is removed for each NO_2 produced. In fact, it is known that hydrocarbons are needed for photochemical smog (i.e., oxidant) to be produced. This can be seen from Fig. VI-3, where oxidant production is shown to depend on both NO_x and nonmethane hydrocarbon concentration. In general, if both are increased so are the maximum oxidant concentrations. However, it should be observed that at constant $[NO_x]$ an increase in nonmethane hydrocarbon concentration first increases and then decreases

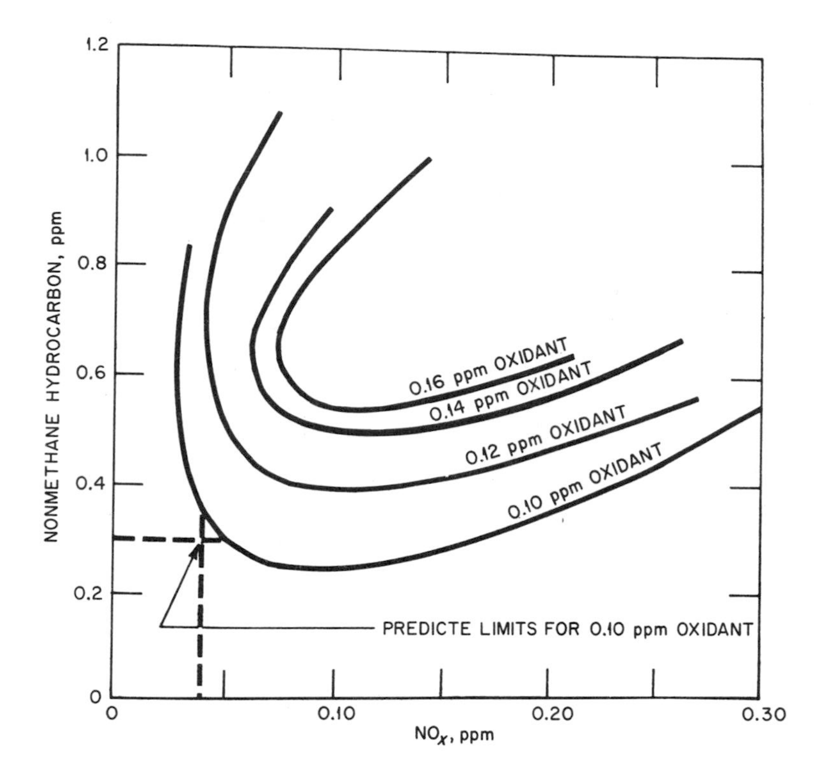

Fig. VI-3. Approximate isopleths for selected upper-limit maximum daily 1-hr average oxidant concentrations, as a function of the 6 to 9 A.M. averages of nonmethane hydrocarbons and total nitrogen oxides in Philadelphia, Washington, D.C., and Denver, June–August, 1966–1968. From Environmental Protection Agency Document AP-84 (1971) with permission of the Environmental Protection Agency.

oxidant production. The same effect is observed at constant hydrocarbon concentration when $[NO_x]$ is enhanced.

This effect was first reported by Haagen-Smit and Fox (1954) and has been substantiated with many hydrocarbons since then. In one of the more recent studies, Glasson and Tuesday (1970a) reported this effect for several hydrocarbons. In particular, they found that for propylene and *m*-xylene the maximum rate of oxidation occurred for hydrocarbon-to-NO ratios of $\sim$6 and $\sim$1.5, respectively.

Perhaps the most dramatic demonstration of the NO inhibition was shown by Dimitriades (1970, 1972), who examined the reactivity of automobile exhaust in an irradiated atmosphere. Since several hydrocarbons were present, the effective hydrocarbon concentration (HC) was computed on a per carbon basis by summing over the products of the concentrations and relative reactivity of each hydrocarbon. The measure of reactivity was the oxidant dosage, i.e., the product of the oxidant concentration and its time of existence. The effect for mixtures at a constant hydrocarbon-to-NO_x ratio is shown in Fig. VI-4. For ratios below about 4, the oxidant dosage actually drops as the pollutants increase! The implication is that if an automobile exhaust consists of hydrocarbons and oxides of nitrogen in an effective ratio <4:1, reducing the exhaust pollutants will actually increase the oxidant dosage! Clearly the chemistry of such systems must be somewhat complex to explain these observations.

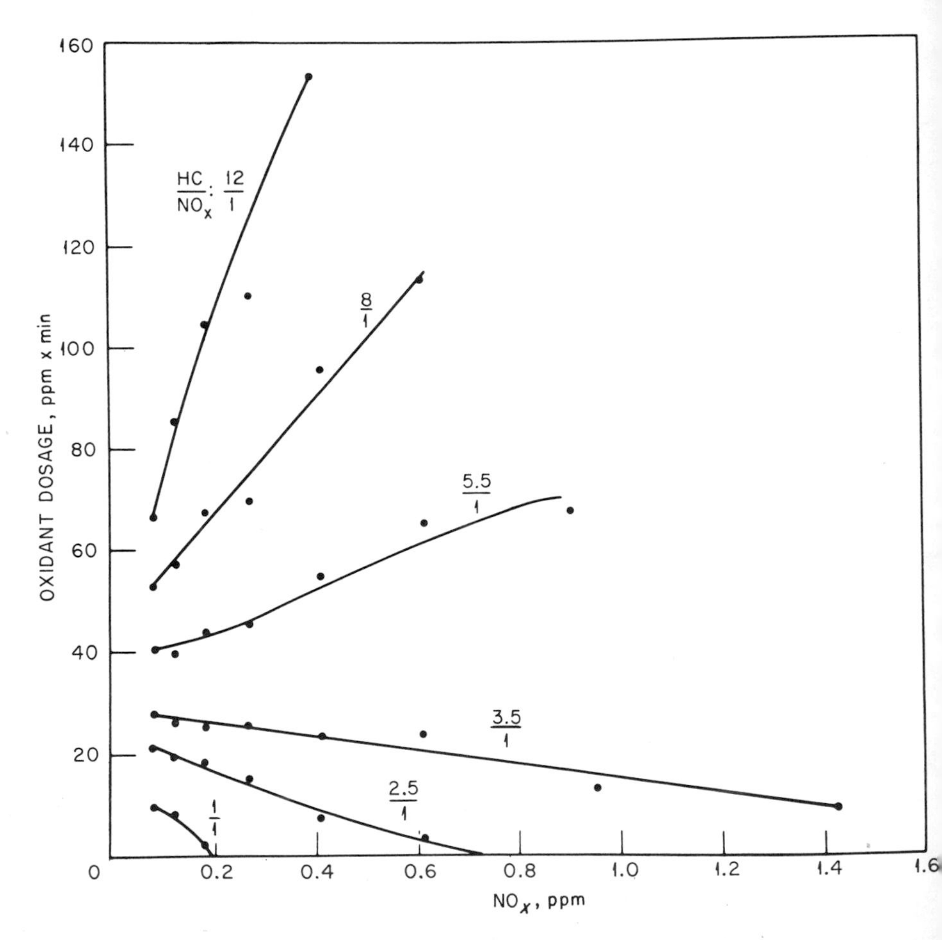

Fig. VI-4. Oxidant dosage of exhaust as a function of NO_x concentration at various hydrocarbon-to-NO_x concentration ratios. From Dimitriades (1970) with permission.

HYDROCARBON REACTIVITIES

The concept of hydrocarbon reactivity was mentioned above. In order to obtain hydrocarbon reactivities, the hydrocarbon–NO photooxidation has been studied in simulated atmospheres for several hydrocarbons and aldehydes (Altshuller and Cohen, 1963; Dimitriades and Wesson, 1972; Glasson and Tuesday, 1970b; Kopczynski, 1964; Schuck and Doyle, 1959). Measures of hydrocarbon reactivities have been obtained based on the rates of both hydrocarbon consumption and NO_2 production. An extensive review was made by Altshuller and Bufalini (1971). For hydrocarbons and aldehydes reported to be present in urban atmospheres (Garner, 1972), the relative reactivity by both methods is listed in Table VI-1. Arbitrarily i-C_4H_8 is considered to have a relative reactivity of 1.0.

TABLE VI-1

Reactivities of Some Hydrocarbons Identified in Urban Air Sampling[a]

Compound[b]	Reactivity based on		Relative rate constant	
	Hydro-carbons	NO oxidation	k_O	k_{OH}
Paraffins				
CH_4	0.0[c]	—	6.2×10^{-7} [d]	0.00016[e]
C_2H_6	—	—	2.9×10^{-5} [d]	0.0047[f]
C_3H_8	—	—	0.0072[g]	0.022[f]
n-C_4H_{10}	—	—	0.0100[h]	0.049,[e] 0.063[i]
i-C_4H_{10}	—	—	—	0.034[f]
n-C_5H_{12}	0.002[c]	—	—	—
i-C_5H_{12}	0.002[c]	—	0.050[c]	—
c-C_5H_{10}	—	—	—	—
n-C_6H_{14}	—	—	—	—
2-Methylpentane	0.024[c]	<0.1[j]	—	—
3-Methylpentane	0.012[c]	<0.1[j]	—	—
2,2-Dimethylbutane	—	—	—	—
2,3-Dimethylbutane	—	—	—	—
c-C_6H_{12}	0.004[c]	—	—	0.147[e]
Methycyclopentane	—	—	—	—
2-Methylhexane	—	—	—	—
3-Methylhexane	—	—	—	—
2,3-Dimethylpentane	—	—	—	0.138[e]
2,4-Dimethylpentane	—	—	—	—
Isooctane	0.30[c]	0.15[j]	0.44[h]	0.072[e]

TABLE VI-1—Continued

Compound[b]	Reactivity based on: Hydrocarbons	Reactivity based on: NO oxidation	Relative rate constant: k_O	Relative rate constant: k_{OH}
Aromatics				
Benzene	—	0.15[j]	0.0016[k]	—
Toluene	0.30[l]	0.2[j]	0.0057[k]	—
o-Xylene	—	0.4,[j] 0.45[m]	—	0.29[i] (o-, m-, p-Xylene)
m-Xylene	0.56[n]	1.0,[j] 0.90[m]	—	
p-Xylene	0.152,[c] 0.21[n]	0.4,[j] 0.45[m]	—	
m-Ethyltoluene	—	0.90[m]	—	—
p-Ethyltoluene	—	—	—	—
1,2,4-Trimethylbenzene	—	0.6,[f] 1.8[m]	—	—
1,3,5-Trimethylbenzene	—	1.2,[j] 2.4[m]	—	—
sec-Butylbenzene	—	—	—	—
β-Methylstyrene	—	2.94[m]	—	—
Olefins				
Acetylene	—	0.1[j]	0.0095[g]	0.0029[o]
Ethylene	0.045,[c] 0.1[n]	0.4,[k] 0.49,[p] 0.54[m]	0.049,[q] 0.029[r]	0.079[s]
Allene	—	—	—	—
Propylene	0.57,[c] 0.73[n]	1,[j] 1.0[p]	0.19[q]	0.27[r]
Propyne	—	—	—	—
1-Butene	0.44[c]	0.83[p]	0.25[q]	0.60[i]
cis-2-Butene	2.43,[c] 4.2[h]	2.0[f]	0.95[k]	0.94[i]
trans-2-Butene	4.05[c]	2,[d] 3.2[p]	1.13[k]	1.1[i]
Isobutene	1.0[a]	1.0[a]	1.0[a]	1.0[a]
1,3-Butadiene	0.80[c]	1.23[p]	0.97[k]	—
1-Pentene	0.50[c]	0.60[p]	—	—
cis-2-Pentene	—	1.54[p]	0.90[k]	1.4[i] (cis- and trans-2-Pentene)
trans-2-Pentene	—	2.2[p]	—	
2-Methyl-1-butene	—	0.97[p]	—	—
2-Methyl-2-butene	2.70[c]	5.4[p]	3.17[k]	1.84[i]
3-Methyl-1-butene	—	0.77[p]	—	—
2-Methyl-1,3-butadiene	1.5[c]	1.05[p]	—	—
Cyclopentene	—	6.6[p]	1.19[k]	—
1-Hexene	—	0.49[p]	0.26[k]	—
cis-2-Hexene	—	—	—	—
trans-2-Hexene	—	1.71[p]	—	—
cis-3-Hexene	3.5[n]	—	—	—

TABLE VI-1—Continued

	Reactivity based on		Relative rate constant	
Compound[b]	Hydro-carbons	NO oxidation	k_O	k_{OH}
Olefins				
trans-3-Hexene	—	1.68[p]	—	—
2-Methyl-1-pentene	—	0.66[p]	—	—
4-Methyl-1-pentene	—	—	—	—
cis-4-Methyl-2-pentene	—	0.92[p]	—	—
trans-4-Methyl-2-pentene	—	1.28[p]	—	—
Aldehydes				
CH_2O	—	0.80[m]	0.023[t]	0.24[l]
CH_3CHO	—	1.34[m]	0.025[k]	0.24[r]
Furfural	—	—	—	—
Acrolein	—	0.86[m]	—	—
Propionaldehyde	—	2.26[m]	—	0.47[i]
Butyraldehyde	—	1.48[m]	—	—
Crotonaldehyde	—	1.20[m]	—	—
Benzaldehyde	—	0.15[m]	—	—
o-Tolualdehyde	—	6.06[m]	—	—
m-Tolualdehyde	—	0.23[m]	—	—
p-Tolualdehyde	—	0.18[m]	—	—

[a] From Heicklen (1973), based on a relative reactivity of 1.0 for i-C_4H_8.
[b] The compounds reported to be present in urban air. Garner (1972).
[c] Schuck and Doyle (1959).
[d] Herron (1969), based on $k = 4.0 \times 10^9\ M^{-1}\ \mathrm{sec}^{-1}$ for 1-C_4H_8 (Huie *et al.*, 1971).
[e] Greiner (1970a), assuming relative rate coefficient $= 0.022$ for C_3H_8.
[f] Greiner (1967).
[g] Saunders and Heicklen (1966), assuming relative $k = 0.25$ for 1-C_4H_8.
[h] Herron and Huie (1969), based on $k = 4.0 \times 10^9\ M^{-1}\ \mathrm{sec}^{-1}$ for 1-C_4H_8 (Huie *et al.*, 1971).
[i] Morris and Niki (1971b).
[j] Altshuller and Cohen (1963).
[k] Cvetanović (1963).
[l] Kopczynski (1964), assuming C_6H_6 reactivity $= 0.15$.
[m] Dimitriades and Wesson (1972), assuming C_3H_6 reactivity $= 1.0$.
[n] Stephens and Scott (1962).
[o] Breen and Glass (1970).
[p] Glasson and Tuesday (1970b).
[q] Atkinson and Cvetanović (1972).
[r] Morris *et al.* (1971).
[s] Griener (1970b).
[t] Morris and Niki (1971a), assuming relative $k = 0.025$ for CH_3CHO.

It is readily apparent that both scales give similar results and that the reactivity can be associated with structure. The paraffins are the least reactive. Except for isooctane, their relative reactivity is <0.1 and often much less. The aromatics have reactivities from 0.15 for benzene to 2 for highly substituted benzenes. The olefins show reactivities of about 0.5 to 2.2, although cyclopentene has a value of 6.6. The relatively large value of 2.94 for β-methylstyrene can be attributed to the fact that it is both aromatic and olefinic. The aldehyde reactivities cover the same range as the aromatics except for the surprisingly high reactivity of *o*-tolualdehyde.

It seems reasonable to assume that the reactivity of a hydrocarbon is related to its rate coefficient for reaction with some active species in photochemical smog that reacts readily with hydrocarbons. The likely species are $O(^3P)$ and HO radicals. Other conceivable intermediates such as peroxy and alkoxy radicals are removed so much faster by other routes than by hydrocarbon attack that their role can be only minor—likewise for photoexcited aldehydes and ketones. NO and NO_2 react too slowly with hydrocarbons to be important. Photoexcited NO_2 may be quenched readily by hydrocarbons, but no report of chemical interaction has appeared.

Singlet O_2 is produced from the reaction of photoexcited NO_2 with O_2 and does react with hydrocarbons. Likewise, O_3 is a product of the NO_2 photodissociation and also reacts with hydrocarbons. However, in both cases reaction with hydrocarbons does not appear to lead to reactive free radicals as major products. A full discussion of the reactions of these two species will be given in Chapter VII, but for the time being they can be neglected.

The relative reactivities for the compounds in Table VI-1 with $O(^3P)$ and HO are also listed. All available data on any compound are not included, but only the most likely value. For $O(^3P)$ the results from Cvetanović's and Herron's laboratories have been preferred for two reasons. Both have made extensive studies and therefore are the most experienced. Also, any errors in values are likely to be internally consistent, and therefore the relative values from a given laboratory may be correct, even if the absolute values are not. This is not to imply that other studies are less accurate. In fact, the differences in values among several laboratories are usually $<50\%$. For HO radicals, the values of Greiner are preferred for the same reasons given above for $O(^3P)$ atoms.

For both $O(^3P)$ and HO, the reactivity of the hydrocarbon follows the same trends as the photochemical reactivity, although the detailed fit is not all that one might have expected. For CH_4 and C_2H_6 the reactivities are very small, but the other paraffin reactivities with $O(^3P)$ and HO, although still small, are larger than the photochemical reactivities. With the aromatics, the reverse seems to be the case, although the measurements are sparse. For the olefins, reasonably good matches are obtained in all cases except for cyclopentene, which has an abnormally large photochemical reactivity. With the aldehydes, the data are again sparse, but the HO and $O(^3P)$ reactivities are, respectively, about factors of 4 and 40 lower than the photochemical reactivities. Thus it can be concluded that, in a general way, photochemical reactivity is associated with the reactivity toward $O(^3P)$ and/or HO, but that the detailed correlation must be considerably more complex. Further evidence that some species in addition to $O(^3P)$ and HO are attacking hydrocarbons comes from the work of Bufalini and Altshuller (1967) and others (Altshuller *et al.*, 1966; Nicksic *et al.*, 1964), who noticed synergistic effects when two hydrocarbons were photooxidized simultaneously. Furthermore, the aldehydes absorb ultraviolet radiation to photodecompose and photooxidize, thus enhancing their reactivity over that expected from radical attack alone.

CHEMISTRY OF THE CONVERSION OF NO TO NO_2

A mechanism for the photochemical conversion of NO to NO_2 in urban atmospheres has been proposed by Heicklen *et al.* (1971) in which HO radical is the important chain carrier. With alkanes the initiating reaction is HO radical attack to remove an H atom. For example, with C_4H_{10}, the reaction sequence proposed was

$$C_4H_{10} + HO\cdot \rightarrow C_4H_9\cdot + H_2O \tag{5}$$

$$C_4H_9\cdot + O_2 \rightarrow C_4H_9O_2\cdot \tag{6}$$

$$C_4H_9O_2\cdot + NO \rightarrow C_4H_9O\cdot + NO_2 \tag{7}$$

$$C_4H_9O\cdot + O_2 \rightarrow C_3H_7CHO + HO_2\cdot \tag{8}$$

$$HO_2\cdot + NO \rightarrow HO\cdot + NO_2 \tag{9}$$

The overall reaction is

$$C_4H_{10} + 2NO + 2O_2 \rightarrow C_3H_7CHO + H_2O + 2NO_2 \tag{10}$$

Note that HO· is regenerated in reaction (9) and can reinitiate the reaction sequence.

With olefins the scheme is slightly modified because HO· can add to the double bond as well as abstract a hydrogen atom (Meagher and

Heicklen, 1975). Thus for $i\text{-}C_4H_8$ the reaction sequence for addition might be

$$i\text{-}C_4H_8 + HO\cdot \rightarrow HOC_4H_8\cdot \quad (11)$$

$$HOC_4H_8\cdot + O_2 \rightarrow HOC_4H_8O_2\cdot \quad (12)$$

$$HOC_4H_8O_2\cdot + NO \rightarrow HOC_4H_8O\cdot + NO_2 \quad (13)$$

$$HOC_4H_8O\cdot \rightarrow HOCH_2\cdot + (CH_3)_2CO \quad (14)$$

$$HOCH_2\cdot + O_2 \rightarrow HOCH_2O_2\cdot \quad (15)$$

$$HOCH_2O_2\cdot + NO \rightarrow HOCH_2O\cdot + NO_2 \quad (16)$$

$$HOCH_2O\cdot \rightarrow HO\cdot + CH_2O \quad (17)$$

The overall reaction would be

$$i\text{-}C_4H_8 + 2NO + 2O_2 \rightarrow (CH_3)_2CO + CH_2O + 2NO_2 \quad (18)$$

which conforms to the findings of Schuck and Doyle (1959), who observed that CH_2O and $(CH_3)_2CO$ were the major products of the photochemical oxidation of $i\text{-}C_4H_8$ in the presence of NO. For abstraction of an H atom (say from C_2H_4)

$$HO + C_2H_4 \rightarrow H_2O + C_2H_3 \quad (19a)$$

the fate of the C_2H_3 species is not known. However, the chain mechanism must be operative so that the overall reaction presumably is

$$C_2H_3 + 2O_2 + NO \rightarrow CH_2O + CO + NO_2 + HO \quad (20)$$

For aromatic compounds, the cycle is probably similar to that for the olefins. The aldehydes should behave like the alkanes, but with a slight modification. For example, for CH_3CHO the first few steps would become

$$CH_3CHO + HO\cdot \rightarrow CH_3\dot{C}O + H_2O \quad (21)$$

$$CH_3\dot{C}O + O_2 \rightarrow CH_3O\cdot + CO_2 \quad (22)$$

and then the sequence would proceed as in reactions (8) and (9). Reaction (22) is not a fundamental reaction but occurs in several steps.

The implication of the above schemes is that CO must also oxidize NO via reaction (9) and

$$HO\cdot + CO \rightarrow H\cdot + CO_2 \quad (23)$$

$$H\cdot + O_2 + M \rightarrow HO_2\cdot + M \quad (24)$$

Confirmation of the influence of CO has been given several times in recent years. The first report was by Glasson (1971), who examined an atmosphere with 2 ppm C_2H_4 and 1 ppm NO. The addition of 400 ppm CO doubled the rate of NO_2 production. Since the rate constant for $HO\cdot$ attack on

C_2H_4 is (1–3) $\times 10^{10}$ M^{-1} sec^{-1} (Greiner, 1970b; Morris *et al.*, 1971) and that for HO· attack on CO is 8.0×10^7 M^{-1} sec^{-1} (Baulch *et al.*, 1968), the relative reactivity between C_2H_4 and CO should be between 125 and 375. Since 200 times as much CO as C_2H_4 was used in Glasson's experiment, the expected rate of NO oxidation on the addition of CO would be between 1.6 and 2.9 times that in its absence, in excellent agreement with the observed result. Confirmation of the effect of CO with other hydrocarbons has subsequently occurred (Westberg *et al.*, 1971; Dodge and Bufalini, 1972).

Simultaneously with the report of Heicklen *et al.* (1971), Weinstock *et al.* (1971) were using a very similar mechanism to do a complete computer study on C_3H_6 photooxidation. They estimated the chain length to be about 280 and the lifetime of the HO radical to be 56 sec. Detailed computer studies have also been made by Westberg and Cohen (1970) and Hecht and Seinfeld (1972).

The reaction steps outlined above were in some cases well established, but in some cases speculative. Reactions (5), (6), (21), (23), and (24) were certain. Some evidence existed for reaction (8) with CH_3O and C_2H_5O radicals (Heicklen, 1968) and for reaction (9) at elevated temperatures (Tyler, 1962; Ashmore and Tyler, 1962). Reactions (11)–(17) were all speculative, and except for reaction (11), still are. The details, or even the validity, of reaction (20) are still unknown. Reaction (22) involves the addition of O_2 to CH_3CO (Weaver *et al.*, 1975) and probably proceeds via the sequence

$$CH_3CO + O_2\ (+M) \rightarrow CH_3C(O)O_2\ (+M) \tag{25}$$

$$CH_3C(O)O_2 + NO \rightarrow CH_3C(O)O + NO_2 \tag{26}$$

$$CH_3C(O)O \rightarrow CH_3 + CO_2 \tag{27}$$

$$CH_3 + O_2\ (+M) \rightarrow CH_3O_2\ (+M) \tag{28}$$

$$CH_3O_2 + NO \rightarrow CH_3O + NO_2 \tag{29}$$

Reaction (11) was first verified for C_2H_4 and C_3H_6 by Morris *et al.* (1971). Studies have been undertaken by Heicklen and his collaborators to verify reactions (7)–(9). For CH_3O radicals they have found that at 25°C, the relative reactivity of $CH_3O\cdot$ with O_2, NO_2, and NO is 4.7×10^{-5}, 0.83, and 1.0, respectively (Wiebe *et al.*, 1973). Thus under atmospheric conditions, the reaction with O_2 is at least five times as important as that with the oxides of nitrogen. Wiebe *et al.* (1973) also found that the principal fate of HO_2 was reaction (9). The rate constant for this reaction has been reported to be 9.7×10^{-13} cm^3/sec (Simonaitis and Heicklen, 1976). Reaction (9) must be the dominant fate of $HO_2\cdot$ in the atmosphere.

Reaction (29) [the analog to reaction (7)] has now been confirmed by Pate *et al.* (1974) and Simonaitis and Heicklen (1974b). According to Pate *et al.*, the reaction proceeds by O atom transfer 100% of the time, but according to Simonaitis and Heicklen about 80% of the time, the remainder of the time the products are presumably CH_2O and HONO.

INITIATION

The steps that initiate the photochemical smog cycle are photodecompositions that can be classified in four groups:

1. photolysis of NO_2 and O_3 to produce $O(^3P)$,
2. photolysis of aldehydes and ketones,
3. photolysis of HONO,
4. photolysis of nitrates, nitrites, and peroxides.

$O(^3P)$ Production

NO_2 and O_3 photodecompose at ground level via reaction (2) and

$$O_3 + h\nu \rightarrow O_2 + O(^3P) \tag{30}$$

with photodissociation coefficients of 8.0×10^{-3} and 3.8×10^{-4} sec^{-1}, respectively, for an overhead sun. Usually this process does not initiate the smog cycle since the $O(^3P)$ atoms generally combine with O_2 via reaction (3).

However, occasionally $O(^3P)$ reacts with aliphatic hydrocarbon or aldehydes to abstract an H atom,

$$O(^3P) + RH \rightarrow HO + R \tag{31}$$

$$O(^3P) + RCH(O) \rightarrow HO + RCO \tag{32}$$

or with an olefin to give a complex set of products.

The olefin reactions have been studied most thoroughly by Cvetanović and he has reviewed their chemistry (Cvetanović, 1963). The general scheme of olefin reactions is illustrated with C_2H_4:

$$O(^3P) + C_2H_4 \rightarrow \overline{CH_2CH_2O}^* \tag{33}$$

$$\overline{CH_2CH_2O}^* + M \rightarrow \overline{CH_2CH_2O} + M \tag{34}$$

$$\overline{CH_2CH_2O}^* \rightarrow CH_3CHO^* \tag{35}$$

$$CH_3CHO^* + M \rightarrow CH_3CHO + M \tag{36}$$

$$CH_3CHO^* \rightarrow CH_3 + HCO \tag{37}$$

In the case of C_2H_4, stabilization of the energetic intermediates (denoted by asterisks) does not occur to any appreciable extent even at 1 atm pressure, although some $\overline{CH_2CH_2O}$ and CH_3CHO were found. In fact, the HCO radical produced also has excess energy and may decompose to H + CO part of the time.

With the higher olefins, stabilization of the energetic intermediates occurs more easily. In the case of C_3H_6 at 1 atm pressure (of C_3H_6) the products are about 50% propylene oxide and 50% acetone plus propionaldehyde, but small amounts of the fragmentation products are observed. Presumably with atmospheric gases (O_2 and N_2), which are less efficient stabilizers than C_3H_6, a higher percentage of fragmentation may occur, but it will still be the minor process. With the C_4 and higher olefins, fragmentation should not be important under atmospheric conditions.

Two chlorinated olefins, C_2Cl_4 and C_2Cl_3H, are of particular importance because they are so widely used as solvents. Sanhueza and Heicklen (1974a,b) studied their reactions with $O(^3P)$ in the presence of O_2 and found that free radical production is important via the sequence of steps illustrated for C_2Cl_3H:

$$O(^3P) + C_2Cl_3H \rightarrow \text{adduct} \tag{38a}$$

$$\rightarrow CCl_2 + CClH(O) \tag{38b}$$

$$CCl_2 + O_2 \rightarrow Cl + CO + ClO \tag{39}$$

where reaction (38b) occurs 23% of the time with C_2Cl_3H and 19% of the time with C_2Cl_4 independent of pressure at 25°C. The Cl and ClO radicals add to the olefin, giving acid chlorides and regenerating chlorine atoms, so that a long-chain process occurs.

Photolysis of aldehydes and ketones

Of most importance in this group of initiating processes is the photolysis of CH_2O and CH_3CHO. The other aldehydes and ketones do absorb energy >2900 Å, and some do photodecompose. However, the presence of 150 Torr of O_2 markedly inhibits their photodecomposition, rendering it unimportant.

For CH_2O, the extinction coefficients and photodissociation yields have been summarized by Calvert *et al.* (1972). At 3130 Å and 25°C in dry air at 1 atm pressure, Osif (1975) has found quantum yields of 0.20 and 0.80 for reactions (40a) and (40b), respectively. Assuming no wavelength dependence allows us to estimate the ground level photodissociation coefficients for an overhead sun to be 1.9×10^{-5} and 7.8×10^{-5} sec^{-1}, respectively, for the processes

$$CH_2O + h\nu \rightarrow H_2 + CO \tag{40a}$$

$$\rightarrow H + HCO \tag{40b}$$

For CH_3CHO, the possible photodecomposition paths are

$$CH_3CHO + h\nu \rightarrow CH_4 + CO \tag{41a}$$

$$\rightarrow CH_3 + HCO \tag{41b}$$

$$\rightarrow CH_3CO + H \tag{41c}$$

Presumably all of these paths take place, but the quantum yield dependencies as a function of wavelength are not well known. Furthermore, Calvert and Hanst (1959) showed that the yields were reduced (but not to zero) in the presence of O_2. Process (41a) was not affected by replacing N_2 with O_2; process (41b) was reduced by about a factor of 5; and process (41c), by about a factor of 3. The photodissociation of CH_3CHO is important, but at this time (1975) we do not have good estimates for photodissociation coefficients.

Photolysis of HONO

Possibly the most important of all the initiating steps is

$$HONO + h\nu \rightarrow HO + NO \tag{42}$$

but unfortunately almost no reliable quantitative information exists. HONO absorbs radiation between 3000 and 4000 Å; the absorption spectrum is shown in Fig. VI-5. Extinction coefficients are not known, but preliminary work in Johnston's laboratory (see Hampson, 1973) indicates an approximate cross section of 10^{-19} cm^2/molecule over the wavelength range 3000–4000 Å. Since the total intensity in this wavelength regime is $\sim 2 \times 10^{16}$ photons/cm^2 sec for an overhead sun $J\{HONO\} \sim 6.45 \times 10^{-4}$ sec^{-1} if the quantum efficiency for HO production is unity (Garvin and Hampson, 1974).

The concentration of HONO in the atmosphere is also an unknown quantity. It has an equilibrium with NO, NO_2, and H_2O,

$$2HONO \leftrightharpoons NO + NO_2 + H_2O \tag{43,−43}$$

and the equilibrium constant is 0.72 atm at 25°C (Waldorf and Babb, 1963, 1964).

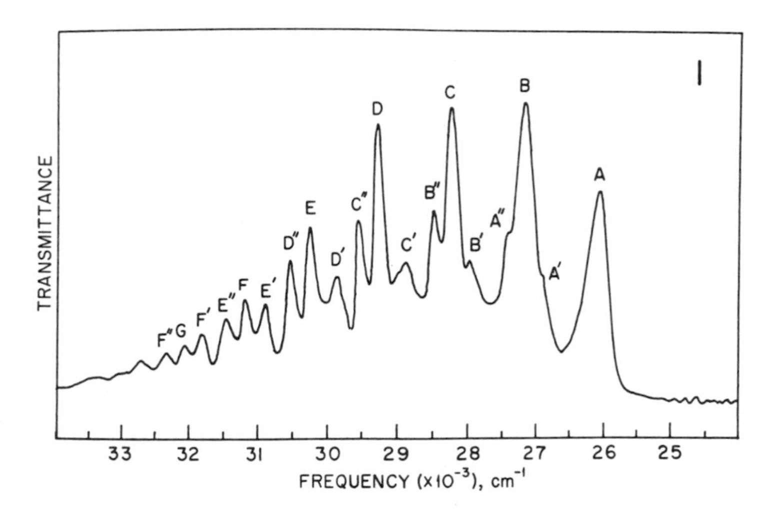

Fig. VI-5. Absorption spectrum of HONO at 25°C. From King *et al.* (1962) with permission of the National Research Council of Canada.

Photolysis of nitrates, nitrites, and peroxides

All of the following processes can initiate the free radical process

$$RONO + h\nu \rightarrow RO + NO \tag{44}$$

$$RONO_2 + h\nu \rightarrow RO + NO_2 \tag{45}$$

$$RO_2NO_2 + h\nu \rightarrow RO_2 + NO_2 \tag{46}$$

$$ROOR + h\nu \rightarrow 2RO \tag{47}$$

$$ROOH + h\nu \rightarrow RO + OH \tag{48}$$

In most cases the extinction coefficients, quantum yields, and concentrations of the species are not known, so that it is difficult to estimate the importance of the above. Generally the concentrations of the above species are small, so that initiation by these processes is not important. Furthermore, these compounds are formed by the reverse reactions of the photodecompositions (except for ROOH, which probably comes from $RO_2 + HO_2 \rightarrow ROOH + O_2$, but this does not alter the argument), which are terminations. Thus, these reverse processes tend to offset each other, making the photodecomposition steps only decelerators of termination.

TERMINATION

The important terminating processes are reaction (2) and

$$HO + HO_2 \rightarrow H_2O + O_2 \tag{49}$$

$$2HO_2 \rightarrow H_2O_2 + O_2 \tag{50}$$

$$HO_2 + NO_2 \rightarrow HONO + O_2 \tag{51}$$

$$HO + NO(+M) \rightarrow HONO(+M) \tag{52}$$

$$HO + NO_2(+M) \rightarrow HONO_2(+M) \tag{53}$$

In addition, termination can occur through the less important steps

$$RO + NO \rightarrow RONO \tag{54a}$$

$$RO + NO_2 \rightarrow RONO_2 \tag{55a}$$

$$\rightarrow R'O + HONO \tag{55b}$$

$$RO_2 + NO_2 \rightarrow RO_2NO_2 \tag{56a}$$

$$\rightarrow R'O + HONO_2 \tag{56b}$$

$$RO_2 + HO_2 \rightarrow RO_2H + O_2 \tag{57}$$

$$RO_2 + NO \rightarrow R'O + HONO \tag{58}$$

where R′O is an aldehyde or ketone. These reactions are offset somewhat as terminating reactions, because their products photodecompose to regenerate radicals (see above). However, reactions (53b) and (58) may lead to termination all the time if HONO is in equilibrium with NO, NO_2, and H_2O. However, if the product R′O photodecomposes to free radicals, then this tends to reduce the effective termination. Finally we point out that there is an alternative path to reaction (54a):

$$RO + NO \rightarrow R'O + HNO \tag{54b}$$

However, this path never leads to termination since, in the atmosphere it is always followed by

$$HNO + O_2 \rightarrow HO_2 + NO \text{ (or } HO + NO_2) \tag{59}$$

In fact, if R′O is capable of photodecomposition, reaction (54b) is an initiating rather than a terminating reaction. The most important example of reaction (54b) is with CH_3O, since CH_2O is produced. In the CH_3O case, path (54b) occurs 14.5% of the time (Wiebe and Heicklen, 1973), so that the CH_3O + NO reaction may actually have the net effect of enhancing rather than reducing the chain length.

Another pair of reactions that lead to termination are those removing HONO, viz.,

$$HO + HONO \rightarrow H_2O + NO_2 \quad (60)$$

$$O(^3P) + HONO \rightarrow HO + NO_2 \quad (61)$$

However, these terminations are mitigated if HONO is reformed from its equilibrium with NO, NO_2, and H_2O. Reactions (53) and (56b), as well as any oxidation of NO_2 or HONO by O_3 or H_2O_2, are the paths to $HONO_2$ production, and represent the ultimate fate of the oxides of nitrogen (except for possible further reaction with NH_3 to give NH_4NO_3).

MODEL CALCULATION

It is useful to see how well the preceding mechanism can fit the observations shown in Figs. VI-1 and VI-2. In order to compute a model system, we use the simplest olefin, C_2H_4, as the hydrocarbon. The problem can be considered in three parts: (1) the $O(^3P)$ chain, which is slow, but which is needed to initiate the oxidation and produce O_3 and CH_2O, (2) the HO chain sequence, which accounts for the bulk of the oxidation, and (3) the HONO cycle, which is really a subcycle of the HO chain.

$O(^3P)$ Chain

For the $O(^3P)$ chain, initiation is due to photolysis of NO_2 and, later on in the reaction, O_3, via reactions (2) and (30). The chain-carrying steps can be simplified as follows for C_2H_4:

$$O(^3P) + C_2H_4 \rightarrow HCO + CH_3 \quad (62)$$

$$HCO + O_2 \rightarrow HO_2 + CO \quad (63)$$

$$CH_3 + O_2\ (+M) \rightarrow CH_3O_2\ (+M) \quad (28)$$

$$CH_3O_2 + NO \rightarrow CH_3O + NO_2 \quad (29a)$$

$$CH_3O + O_2 \rightarrow CH_2O + HO_2 \quad (64)$$

$$O(^3P) + CH_2O \rightarrow HO + HCO \quad (65)$$

For simplicity, we have omitted the reaction

$$CH_3O_2 + NO \rightarrow CH_2O + HONO \quad (29b)$$

since the HONO photodissociates, so that no termination occurs by this process; only the ratio $[HO]/[HO_2]$ is altered, which would reduce the chain length slightly, if this reaction path proceeds to any significant extent.

The termination reaction is (3). This termination step, which is the O_3-producing step, can be followed by several reactions that remove the oxides of nitrogen:

$$NO + O_3 \rightarrow NO_2 + O_2 \tag{4}$$

$$NO_2 + O_3 \rightarrow NO_3 + O_2 \tag{66}$$

$$NO + NO_3 \rightarrow 2NO_2 \tag{67}$$

$$NO_2 + NO_3 \rightarrow N_2O_5 \tag{68}$$

In addition, O_3 can react with the olefin

$$O_3 + C_2H_4 \rightarrow CH_2O + CH_2O_2 \tag{69}$$

The fate of the zwitterion, CH_2O_2, under atmospheric conditions is not known exactly. It could give CH_2O, HCOOH, or $CO + H_2O$, and probably gives all of these products under various conditions. The formation of polyoxides may also be possible. Again for simplicity, we shall assume that $CO + H_2O$ is produced all the time. This will not affect the computation in a qualitative way, but will only tend to decelerate the oxidation of NO slightly.

HO Chain

The HO chain is initiated by CH_2O photolysis via reactions (40a) and (40b). The chain steps are assumed to be

$$HO + CH_2O \rightarrow H_2O + HCO \tag{70}$$

$$HO + C_2H_4 \rightarrow H_2O + C_2H_3 \tag{19a}$$

$$\rightarrow HOC_2H_4 \tag{19b}$$

The detailed fate of C_2H_3 and HOC_2H_4 is not known. However, from the earlier discussion we can see that the overall processes can be represented by reaction (20) and

$$HOC_2H_4 + 2O_2 + 2NO \rightarrow 2CH_2O + 2NO_2 + HO \tag{71}$$

The work of Meagher and Heicklen (1975) suggests that $k_{(19a)}/k_{(19b)} \sim \frac{1}{3}$.

The H atom produced in reaction (40b) adds to O_2:

$$H + O_2 + M \rightarrow HO_2 + M \tag{72}$$

Termination occurs by reaction (53).

HONO Cycle

There are four other reactions involving HO, HO_2, and HONO that need to be included. The first of these is the important chain carrying reaction (9). There are also the two apparent chain-terminating reactions (51) and (52). However, these reactions do not really terminate the chain because they are always followed by the chain initiating reaction (42). Since $J_{(42)} = 6.45 \times 10^{-4}$ sec^{-1}, the lifetime of HONO is 1500 sec, which is relatively short compared to the radiation times involved in the photochemical oxidation ($\sim 10^4$ sec). Thus HONO can be considered to be in its steady state at all times and

$$I\{HONO\} = R_{(51)} + R_{(52)} \tag{73}$$

where $R_{(51)}$ and $R_{(52)}$ are the rates of reactions (51) and (52), respectively.

Rate Expressions

With the above reactions, the steady-state expressions can be used to compute $[O(^3P)]$, $[HO]$, $[HO_2]$, and $[NO_3]$: These are

$$[O(^3P)] = \frac{I_{(2)}\{NO_2\} + I_{(30)}\{O_3\}}{k_{(3)}[O_2][M] + k_{(62)}[C_2H_4] + k_{(65)}[CH_2O]} \tag{74}$$

$$[HO] = \frac{k_{(9)}[HO_2][NO] + I_{(42)}\{HONO\} + k_{(65)}[O(^3P)][CH_2O]}{k_{(52)}[NO] + k_{(53)}[NO_2] + k_{(70)}[CH_2O]} \tag{75}$$

$$[HO_2] = \frac{2I_{(40b)}\{CH_2O\} + (2k_{(62)}[C_2H_4] + k_{(65)}[CH_2O])[O(^3P)] + k_{(70)}[HO][CH_2O]}{k_{(9)}[NO] + k_{(51)}[NO_2]} \tag{76}$$

$$[NO_3] = \frac{k_{(66)}[NO_2][O_3]}{k_{(67)}[NO] + k_{(68)}[NO_2]} \tag{77}$$

Thus if the concentrations of NO, NO_2, O_3, C_2H_4, and CH_2O are known, the concentrations of the free-radical intermediates can be computed. Also [HONO] can be computed from its steady-state expression, Eq. (73).

The rates of reactions of the pollutant species are

$$-d[C_2H_4]/dt = R_{(19)} + R_{(62)} + R_{(69)} \tag{78}$$

$$-d[NO]/dt = R_{(4)} + R_{(9)} + R_{(19a)} + 2R_{(19b)} + R_{(52)} + R_{(62)} + R_{(67)} - I_{(2)}\{NO_2\} - I_{(42)}\{HONO\} \tag{79}$$

$$d[CH_2O]/dt = R_{(19a)} + 2R_{(19b)} + R_{(62)} + R_{(69)} - I_{(40)}\{CH_2O\} - R_{(65)} - R_{(70)} \tag{80}$$

$$d[NO_2]/dt = R_{(4)} + R_{(9)} + R_{(19a)} + 2R_{(19b)} + R_{(62)} + 2R_{(67)} - I_{(2)}\{NO_2\} - R_{(51)} - R_{(53)} - R_{(66)} - R_{(68)} \quad (81)$$

$$d[O_3]/dt = R_{(3)} - R_{(4)} - I_{(30)}\{O_3\} - R_{(66)} - R_{(69)} \quad (82)$$

$$d[CO]/dt = R_{(19a)} + I_{(40)}\{CH_2O\} + R_{(62)} + R_{(65)} + R_{(69)} + R_{(70)} \quad (83)$$

$$d[HONO_2]/dt = R_{(53)} \quad (84)$$

$$d[N_2O_5]/dt = R_{(68)} \quad (85)$$

$$d[H_2O]/dt = R_{(19a)} + R_{(69)} + R_{(70)} \quad (86)$$

$$d[H_2]/dt = I_{(40a)}\{CH_2O\} \quad (87)$$

Thus all the rates can be computed and new concentrations estimated for a small time interval. In this way, a step-by-step calculation can give computed concentration profiles as a function of irradiation time. The pertinent reactions and their rate coefficients at 300°K and atmospheric pressure are listed in Table VI-2.

Concentration Profiles

The concentration profiles throughout the day can be computed using the above rate expressions if the solar intensity is known throughout the day. Since at the surface of the earth the sun's radiation is reduced because of scattering by the atmospheric molecules and particulates, the intensity is highest at noon and weakest at sunrise and sunset. The exact intensity function is not known, and in fact varies with atmospheric conditions. We consider a day with an overhead sun and assume that the intensity at all wavelengths is constant throughout the day. This approximation is not too bad, since significant reduction occurs only in the very early morning and very late afternoon.

Furthermore, as the photochemical smog is produced, the chains become so long that the above mechanism is no longer adequate. As the NO and NO_2 are removed, the HO and HO_2 concentrations increase, the dominant reactions of HO and HO_2 will become

$$HO_2 + O_3 \rightarrow HO + 2O_2 \quad (88)$$

$$HO + O_3 \rightarrow HO_2 + O_2 \quad (89)$$

and termination will shift to

$$HO + HONO \rightarrow H_2O + NO_2 \quad (60)$$

$$2HO \rightarrow H_2O + O(^3P) \quad (90)$$

$$HO_2 + HO \rightarrow H_2O + O_2 \quad (91)$$

$$2HO_2 \rightarrow H_2O_2 + O_2 \quad (92)$$

TABLE VI-2

Rate Coefficient at 300°K for Reactions of Pertinence in Model Photochemical Smog Calculation

Reaction	Reaction no.	J (sec^{-1}) or k (cm^{3n}/sec)[a]	Reference
$NO_2 + h\nu \rightarrow NO + O(^3P)$	(2)	8.0×10^{-3}	See Chapter II
$O_3 + h\nu \rightarrow O_2 + O(^3P)$	(30)	3.8×10^{-4}	See Chapter II
$O(^3P) + C_2H_4 \rightarrow HCO + CH_3$	(62)	8.4×10^{-13}	Garvin and Hampson (1974)
$O(^3P) + CH_2O \rightarrow HO + HCO$	(65)	1.6×10^{-13}	Garvin and Hampson (1974)
$O(^3P) + O_2 + M \rightarrow O_3 + M$	(3)	5.9×10^{-34}	Garvin and Hampson (1974)
$NO + O_3 \rightarrow NO_2 + O_2$	(4)	1.7×10^{-14}	Garvin and Hampson (1974)
$NO_2 + O_3 \rightarrow NO_3 + O_2$	(66)	3.1×10^{-17}	Garvin and Hampson (1974)
$NO + NO_3 \rightarrow 2NO_2$	(67)	8.7×10^{-12}	Garvin and Hampson (1974)
$NO_2 + NO_3 \rightarrow N_2O_5$	(68)	3.8×10^{-12}	Garvin and Hampson (1974)
$O_3 + C_2H_4 \rightarrow CH_2O + CH_2O_2$	(69)	2.7×10^{-18}	Garvin and Hampson (1974)
$CH_2O + h\nu \rightarrow H_2 + CO$	(40a)	1.9×10^{-5}	Calvert *et al.* (1972) and
$\rightarrow H + HCO$	(40b)	7.8×10^{-5}	Osif (1975)
$HO + CH_2O \rightarrow H_2O + HCO$	(70)	1.5×10^{-11}	Garvin and Hampson (1974)
$HO + C_2H_4 \rightarrow H_2O + C_2H_3$	(19a)	3.3×10^{-12}	Garvin and Hampson (1974)
$\rightarrow HOC_2H_4$	(19b)	1.7×10^{-12}	and Meagher and Heicklen (1975)
$HO + NO_2 \rightarrow HONO_2$	(53)	8.0×10^{-12}	Garvin and Hampson (1974)
$HO_2 + NO \rightarrow HO + NO_2$	(9)	9.7×10^{-13}	Simonaitis and Heicklen (1976)
$HO_2 + NO_2 \rightarrow O_2 + HONO$	(51)	1.8×10^{-13}	Simonaitis and Heicklen (1974a, 1976)
$HO + NO \rightarrow HONO$	(52)	8.0×10^{-12}	Sie *et al.* (1976)
$HONO + h\nu \rightarrow HO + NO$	(42)	6.4×10^{-4}	Garvin and Hampson (1974)
$HO + HONO \rightarrow H_2O + NO_2$	(60)	6.8×10^{-12}	Garvin and Hampson (1974)
$HO_2 + O_3 \rightarrow HO + 2O_2$	(88)	1.5×10^{-15}	Garvin and Hampson (1974)
$HO + O_3 \rightarrow HO_2 + O_2$	(89)	5.6×10^{-14}	Garvin and Hampson (1974)
$2HO \rightarrow H_2O + O(^3P)$	(90)	1.6×10^{-12}	Garvin and Hampson (1974)
$HO_2 + HO \rightarrow H_2O + O_2$	(91)	6.0×10^{-11}	Garvin and Hampson (1974)
$2HO_2 \rightarrow H_2O_2 + O_2$	(92)	3.3×10^{-12}	Garvin and Hampson (1974)

[a] Photodissociation coefficients for an overhead sun; $n = 0, 1$, or 2, for first-, second-, or third-order reactions, respectively.

With these reactions included, the concentration profiles obtained from the step-by-step (5-sec steps) computation using the above rate expressions are shown in Fig. VI-6 for a mixture initially containing 2.5×10^{13} molecules/cm^3 C_2H_4 (1.0 ppm), 1.25×10^{13} molecules/cm^3 NO (0.50 ppm), and 1.25×10^{12} molecules/cm^3 NO_2 (0.05 ppm). The pattern of results is quite similar to those in Fig. VI-2 with C_3H_6 as the hydrocarbon.

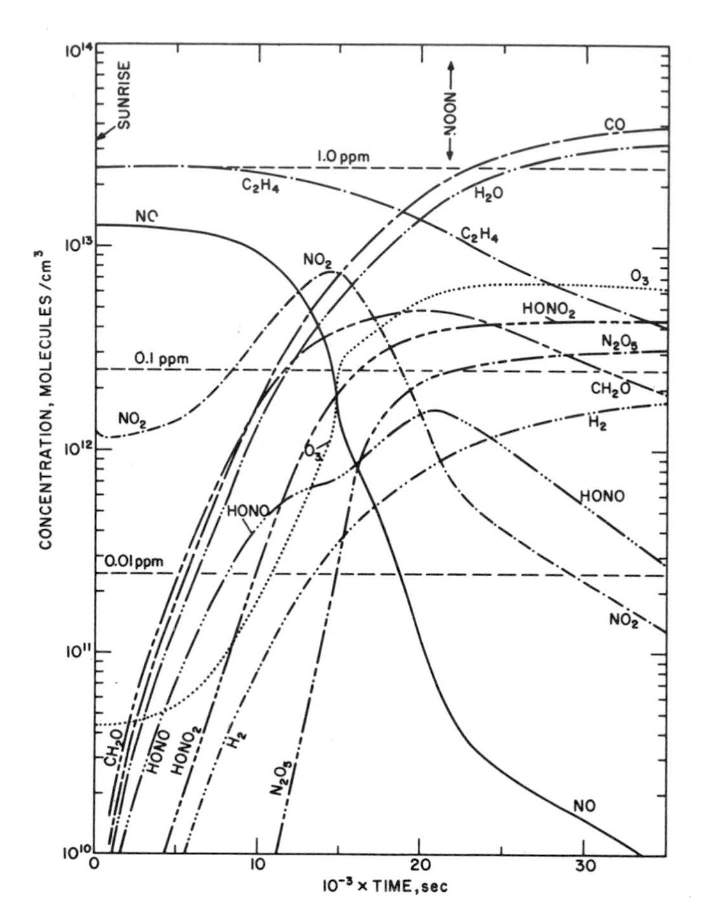

Fig. VI-6. Computed concentrations of pollutants in air at 300°K versus the time of day for a day with an overhead noonday sun, assuming that at sunrise the only pollutants are C_2H_4 at 1, NO at 0.5, and NO_2 at 0.05 ppm.

In the early morning hours the NO is removed slowly by the $O(^3P)$ chain. However, as the day progresses the sunlight intensity increases, CH_2O is produced, and the HO chain becomes important; the rate of NO disappearance is markedly accelerated. As the NO disappears, NO_2 is produced and passes through a maximum at 10:00 A.M. corresponding to about 60% of the initial NO concentration. At the maximum in [NO_2], the [NO] and [O_3] curves cross, and shortly after noon [O_3] reaches a maximum value corresponding to 54% of the initial NO pressure, as the NO_2 is replaced by HONO, $HONO_2$, and N_2O_5. Initially the C_2H_4 is not rapidly consumed, but the consumption rate increases noticeably after the [NO_2] maximum. At the time of the [O_3] maximum, most of the C_2H_4 is consumed. These results exactly parallel those shown in Fig. VI-2 except that the CH_2O concentration drops after the peak O_3 concentration is reached in Fig. VI-6. The reason for the discrepancy will be discussed later, although ultimately both the aldehyde and ozone concentrations must drop with prolonged irradiation.

Figure VI-6 can also be compared to actual atmospheric observations (Fig. VI-1). The relative maximum values for the NO, NO_2, and O_3

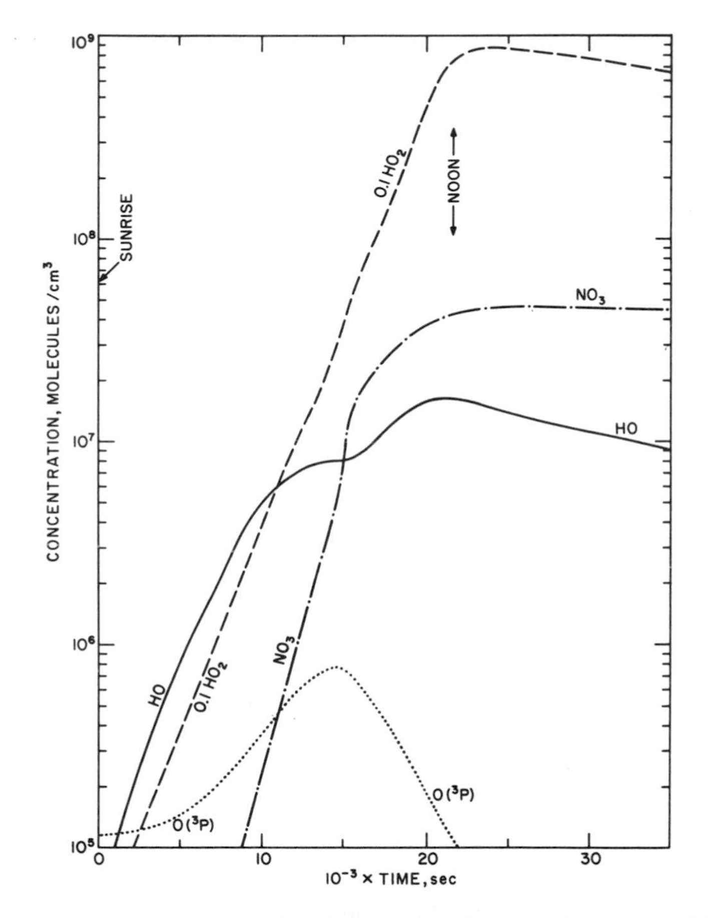

Fig. VI-7. Computed concentrations of free radical intermediates at 300°K versus the time of day for a day with an overhead noonday sun, assuming that at sunrise the only pollutants are C_2H_4 at 1, NO at 0.5, and NO_2 at 0.05 ppm.

concentrations are well matched in Figs. VI-1 and VI-6, but the time scales are not. In Fig. VI-1 the times to reach the maxima in $[NO_2]$ and $[O_3]$ are $2\frac{1}{2}$ and 5 hr, respectively, whereas in Fig. VI-6, the corresponding times are 5 and 7 hr, respectively. This discrepancy is probably due to the difference in relative reactivity between C_2H_4 and the hydrocarbon mixture present in the atmosphere with HO and $O(^3P)$ radicals, as well as the fact that in the atmosphere NO and hydrocarbons are continually being introduced.

The HONO concentration is also plotted in Fig. VI-6. Until the $[NO_2]$ maximum is reached, it is $<7 \times 10^{11}$ molecules/cm^3, i.e., $<6\%$ of the NO_x concentration. However, after the $[NO_2]$ maximum, when the NO_x concentration drops rapidly, the HONO and N_2O_5 concentrations rise steeply. This is a reflection that the chain-terminating step, reaction (53), is being reduced in importance, the rate of $HONO_2$ production decreases, and the chains are becoming longer and longer.

The effect of longer chains is seen even more dramatically in Fig. VI-7, in which the free radical intermediate concentrations are plotted. The $O(^3P)$ concentration follows the NO_2 profile since it is produced mainly from NO_2 photodissociation. However, the HO_x radicals increase in concentration as NO_2 is removed, the HO_2 radical very dramatically. In fact, $[HO_2]$ increases by over four orders of magnitude during the irradiation. Under these conditions, the photochemical process is essentially the photooxidation of CH_2O in the presence of O_3, and both CH_2O and O_3 should be removed with continued irradiation, as indicated in Fig. VI-6.

OXIDANT DOSAGE

The amount of ozone produced in the atmosphere depends on both the rate of NO oxidation and the maximum value of O_3 that can be achieved. The rate of NO oxidation is given by Eq. (79). Application of the steady-state assumption to [HONO] and the sum of $[HO] + [HO_2]$ gives the result

$$-d[NO]/dt \simeq R_{(19a)} + 2R_{(19b)} + 2I_{(40b)}\{CH_2O\} + R_{(70)} \quad (93)$$

where the contribution of the $O(^3P)$ chain reactions (62) and (65) have been neglected, since the bulk of the NO oxidation occurs through the HO cycle. Furthermore, with the same assumptions

$$[HO] \simeq 2I_{(40b)}\{CH_2O\}/k_{(53)}[NO_2] \quad (94)$$

so that

$$\frac{-d[\mathrm{NO}]}{dt} = 2I_{(40b)}\{\mathrm{CH_2O}\}\left(1 + \frac{(k_{(19a)} + 2k_{(19b)})[\mathrm{C_2H_4}] + k_{(70)}[\mathrm{CH_2O}]}{k_{(53)}[\mathrm{NO_2}]}\right) \tag{95}$$

of which the dominant term gives the approximation

$$\frac{-d[\mathrm{NO}]}{dt} \gtrsim \frac{2J_{(40b)}\{\mathrm{CH_2O}\}[\mathrm{CH_2O}](k_{(19a)} + 2k_{(19b)})[\mathrm{C_2H_4}]}{k_{(53)}[\mathrm{NO_2}]} \tag{96}$$

The CH_2O concentration increases approximately proportionately with the C_2H_4 concentration and faster than proportionately with the irradiation time in the morning. Thus the NO oxidation is autocatalytic, i.e., it becomes faster and faster as time goes on because of the accumulation of CH_2O. The rate of NO oxidation also increases proportionately with the square of the hydrocarbon concentration but inversely with the NO_2 concentration. Thus increasing hydrocarbons tends to promote oxidation (i.e., ultimate O_3 formation), but increasing $[NO_2]$ retards O_3 production.

On the other hand, ozone dosage depends on the maximum amount of O_3 produced. At the maximum, the O_3 concentration is given by

$$[\mathrm{O_3}]_{\max} = J_{(2)}\{\mathrm{NO_2}\}[\mathrm{NO_2}]/(k_{(4)}[\mathrm{NO}] + k_{(66)}[\mathrm{NO_2}] + k_{(69)}[\mathrm{C_2H_4}]) \tag{97}$$

Under the usual conditions, $[NO] \ll [NO_2] \ll [C_2H_4]$ at the $[O_3]$ maximum. From Fig. VI-1 and the rate coefficients in Table VI-2, it can be seen that $k_{(69)}[C_2H_4]$ is the dominant term after $[O_3]_{max}$ is reached. Thus $[O_3]_{max}$ increases with $[NO_2]$ and decreases as the hydrocarbon concentration is raised. Since $[NO_2]_{max}$ is more or less proportional to the initial NO concentration $[NO]_0$, $[O_3]_{max}$ is approximately proportional to $[NO]_0$, and most experiments give $[O_3]_{max} \sim (0.3 - 0.7)[NO]_0$.

Thus we have the situation that the rate of oxidation increases strongly with the hydrocarbon concentration, but $[O_3]_{max}$ decreases with increasing olefin concentrations. Thus increasing olefin concentration tends to increase oxidant at first, but large excesses of olefin will then reduce oxidant levels. Exactly the same thing happens with NO_x, but for the opposite reasons; the rate of oxidation is inhibited by NO_x, but $[O_3]_{max}$ increases with $[NO_x]$. Consequently the results depicted in Fig. VI-3 are explained.

If the ratio of olefin to NO_x concentrations is held constant, then $[O_3]_{max}$ is constant, but the rate of NO oxidation increases with $[NO_x]$, leading to an increase in oxidant dosage. This prediction agrees with the findings in Fig. VI-4 for high olefin-to-NO_x concentration ratios, where the chain lengths are long. However at low olefin-to-NO_x ratios, the results in Fig. VI-4 indicate that increasing NO_x decreases the oxidant dosage. In order to explain this phenomenon, it is necessary to examine the rate law for CH_2O production, Eq. (80). Ignoring the unimportant reactions (62), (65), and (69) in Eq. (80) gives

$$\frac{d[CH_2O]}{dt} = \frac{([k_{(19a)} + 2k_{(19b)}][C_2H_4] - k_{(7)}[CH_2O])2I_{(40b)}\{CH_2O\}}{k_{(53)}[NO_2]} - I_{(40)}\{CH_2O\} \tag{98}$$

When $[O_3]$ is increasing from 0.1 to 0.3 ppm in Fig. VI-6, the value for $[C_2H_4]/[CH_2O] \sim 3.5$. With this value and the rate coefficients in Table VI-2, $d[CH_2O]/dt$ is computed from Eq. (98) to be positive if $[C_2H_4]/[NO_2] > 2.2$, but negative if $[C_2H_4]/[NO_2] < 2.2$. Thus for high olefin-to-NO_x concentration ratios, $d[CH_2O]/dt$ is positive, CH_2O grows during the irradiation, and oxidant production is promoted. However, if the $[C_2H_4]/[NO_2]$ ratio drops below 2.2, $d[CH_2O]/dt$ becomes negative, and increasing the NO_2 pressure, even at constant olefin-to-NO_x concentration ratios, removes CH_2O and retards oxidant production. This occurs, of course, only if CH_2O is initially present, as it was in the experiments depicted in Fig. VI-4. If aldehydes are not initially present to any extent, then their concentration remains small, and the oxidation must be carried through the $O(^3P)$ chain, which is very slow. For C_2H_4, the general ratio $[C_2H_4]/[NO_2]$ for which $d[CH_2O]dt$ goes negative is $0.75/(1 - 2.3[CH_2O]/[C_2H_4])$. This value cannot be compared directly to those used in Fig. VI-4, because the hydrocarbon composition includes more than C_2H_4 and the CH_2O concentration is not known. However, a value of about 4 ppm C/ppm NO_x (corresponding to $[C_2H_4]/[NO_x] \sim 2$) found in that system is not out of line with the values expected from the above computation.

Finally, it is interesting to note that the absolute peak value that $[O_3]$ can ever attain is given by Eq. (97) for very high $[NO_2]/[NO]$ and $[NO_2]/[C_2H_4]$ values. This peak maximum O_3 concentration is

$$[O_3]_{\text{peak}} = J_{(2)}\{NO_2\}/k_{(66)} = 2.6 \times 10^{14} \text{ molecules/cm}^3 = 10 \text{ ppm}$$

This is a high level indeed and can cause serious physiological effects while it lasts (see Chapter IV). However, it will not cause death or permanent damage, and apparent good health will return after the episode is finished. Thus O_3 will never be an acute problem, but only a chronic one.

A further demonstration of the complex effect of NO_2 was shown by Bufalini and Altshuller (1967), who photolyzed NO_2 in air in the presence of 9.6 ppm of 1-C_4H_8. At low [NO_2], an increase in [NO_2] decreased the half-life for 1-C_4H_8 removal, as expected, since the rate of CH_2O production increases with intensity (i.e., NO_2 pressure) through the O(^{3}P) atom cycle. However, as the [NO_2]/[1-C_4H_8] ratio increased past 0.25, the lifetime for 1-C_4H_8 removal increased, showing the inhibition predicted by Eq. (98).

REFERENCES

Agnew, W. G. (1968). *Proc. Roy. Soc.* **A307,** 153, "Automotive Air Pollution Research."

Altshuller, A. P., and Bufalini, J. J. (1971). *Environ. Sci. Tech.* **5,** 39,"Photochemical Aspects of Air Pollution: A Review."

Altshuller, A. P., and Cohen, I. R. (1963). *Int. J. Air Water Pollut.* **7,** 787, "Structural Effects on the Rate of Nitrogen Dioxide Formation in the Photo-oxidation of Organic Compound–Nitric Oxide Mixtures in Air."

Altshuller, A. P., Klosterman, D. L., Leach, P. W., Hindaini, I. J., and Sigsby, J. E. (1966). *Int. J. Air Water Pollut.* **10,** 81, "Products and Biological Effects from Irradiation of Nitrogen Oxides with Hydrocarbons or Aldehydes under Dynamic Conditions."

Ashmore, P. G., and Tyler, B. J. (1962). *Trans. Faraday Soc.* **58,** 1108, "Reaction of Hydrogen Atoms with Nitrogen Dioxide."

Atkinson, R., and Cvetanović, R. J. (1972). *J. Chem. Phys.* **56,** 432, "Activation Energies of the Addition of O(^{3}P) Atoms to Olefins."

Baulch, D. L., Drysdale, D. D., and Lloyd, A. C. (1968). Dept. of Phys. Chem., Leeds Univ., "High Temperature Reaction Rate Data," No. 1.

Breen, J. E., and Glass, G. P. (1970). *Int. J. Chem. Kinet.* **3,** 145, "The Reaction of the Hydroxyl Radical with Acetylene."

Bufalini, J. J., and Altshuller, A. P. (1967). *Environ. Sci. Tech.* **1,** 133, "Synergistic Effects in the Photooxidation of Mixed Hydrocarbons."

Calvert, J. G., and Hanst, P. L. (1959). *Can. J. Chem.* **37,** 1671, "The Mechanism of the Photooxidation of Acetaldehyde at Room Temperature."

Calvert, J. G., Kerr, J. A., Demerjian, K. L., and McQuigg, R. D. (1972). *Science* **175,** 751, "Photolysis of Formaldehyde as a Hydrogen Atom Source in the Lower Atmosphere."

Cvetanović, R. J. (1963). *Advan. Photochem.* **1,** 115, "Addition of Atoms to Olefins in the Gas Phase."

Dimitriades, B. (1970). U.S. Bur. of Mines Rep. on Investigations 7433, "On the Function of Hydrocarbon and Nitrogen Oxides in Photochemical-Smog Formation."

Dimitriades, B. (1972). *Environ. Sci. Tech.* **6,** 253, "Effects of Hydrocarbon and Nitrogen Oxides on Photochemical Smog Formation."

Dimitriades, B., and Wesson, T. C. (1972). *J. Air Pollut. Contr. Ass.* **22,** 33, "Reactivities of Exhaust Aldehydes."

Dodge, M. C. and Bufalini, J. J. (1972). *Advan. Chem. Ser.* **113,** 232, "The Role of Carbon Monoxide in Polluted Atmospheres."

Environmental Protection Agency Document AP-84 (1971). "Air Quality Criteria for Nitrogen Oxides."

Garner, J. (1972). Environmental Protection Agency Rep., "Hydrocarbon Emission Effects Determination: A Problem Analysis."

Garvin, D., and Hampson, R. F. (1974). Nat. Bur. of Std. Rep. NBSIR 74-430, "Chemical Kinetics Data Survey VII. Tables of Rate and Photochemical Data for Modelling of the Stratosphere (Revised)."

Glasson, W. A. (1971). *In* "Chemical Reactions in Urban Atmospheres," (C. Tuesday, ed.), p. 59. Elsevier, Amsterdam.

Glasson, W. A., and Tuesday, C. S. (1970a). *Environ. Sci. Tech.* **4,** 37, "Inhibition of Atmospheric Photooxidation of Hydrocarbons by Nitric Oxide."

Glasson, W. A., and Tuesday, C. S. (1970b). *Environ. Sci. Tech.* **4,** 916, "Hydrocarbon Reactivities in the Atmospheric Photooxidation of Nitric Oxide."

Greiner, N. R. (1967). *J. Chem. Phys.* **46,** 3389, "Hydroxyl-Radical Kinetics by Kinetic Spectroscopy. II. Reactions with C_2H_6, C_3H_8, and iso-C_4H_{10} at 300°K."

Greiner, N. R. (1970a). *J. Chem. Phys.* **53,** 1070, "Hydroxyl Radical Kinetics by Kinetic Spectroscopy. VI. Reactions with Alkanes in the Range 300–500°K."

Greiner, N. R. (1970b). *J. Chem. Phys.* **53,** 1284, "Hydroxyl Radical Kinetics by Kinetic Spectroscopy. VII. The Reaction with Ethylene in the Range 300–500°K."

Haagen-Smit, A. J., and Fox, M. M. (1954). *Air Repair* **4,** 105, "Photochemical Ozone Formation with Hydrocarbons and Automobile Exhaust."

Hampson, R. F. (1973). Nat. Bur. of Std. Rep. NBSIR 73-207, "Chemical Kinetics Data Survey: 6, Photochemical and Rate Data for Twelve Gas Phase Reactions of Interest for Atmospheric Chemistry, Interim Report."

Hecht, T. A., and Seinfeld, J. H. (1972). *Environ. Sci. Tech.* **6,** 47, "Development and Validation of a Generalized Mechanism for Photochemical Smog."

Heicklen, J. (1968). *Advan. Chem. Ser. No. 76* **2,** 23, "Gas Phase Reactions of Alkylperoxy and Alkoxy Radicals."

Heicklen, J. (1973). Center for Air Environment Studies Publ. No. 299-73, Penn State Univ., "Photochemical Smog: Its Cause and Cure."

Heicklen, J., Westberg, K., and Cohen, N. (1971). *In* "Chemical Reactions in Urban Atmospheres" (C. Tuesday, ed.), p. 55. Elsevier, Amsterdam.

Herron, J. T. (1969). *Int. J. Chem. Kinet.* **1,** 527, "An Evaluation of Rate Data for the Reactions of Atomic Oxygen (O^3P) with Methane and Ethane."

Herron, J. T., and Huie, R. E. (1969). *J. Phys. Chem.* **73,** 3327, "Rates of Reaction of Atomic Oxygen. II. Some C_2 to C_8 Alkanes."

Huie, R. E., Herron, J. T., and Davis, D. D. (1971). *J. Phys. Chem.* **75,** 3902, "Absolute Rate Constants for the Reaction of Atomic Oxygen with 1-Butene over the Temperature Range of 259–493°K."

King, G. W., and Moule, D. (1962). *Can. J. Chem.* **40,** 2057, "The Ultraviolet Absorption Spectrum of Nitrous Acid in the Vapor State."

Kopczynski, S. L. (1964). *Int. J. Air Water Pollut.* **8,** 107, "Photooxidation of Alkylbenzene–Nitrogen Dioxide Mixtures in Air."

Meagher, J., and Heicklen, J. (1975). Center for Air Environment Studies Publication No. 386-75, Penn State Univ., "The Reaction of HO with C_2H_4."

Morris, E. D. Jr., and Niki, H. (1971a). *J. Chem. Phys.* **55,** 1991, "Mass Spectral Study of the Reaction of Hydroxyl Radical with Formaldehyde."

Morris, E. D. Jr., and Niki, H. (1971b). *J. Phys. Chem.* **75,** 3640, "Reactivity of Hydroxyl Radicals with Olefins."

Morris, E. D. Jr., Stedman, D. H., and Niki, H. (1971). *J. Amer. Chem. Soc.* **93,** 3570, "Mass Spectrometric Study of the Reactions of the Hydroxyl Radical with Ethylene, Propylene, and Acetaldehyde in a Discharge-Flow System."

Nicksic, S. W., Harkins, J., and Fries, B. A. (1964). *J. Air Pollut. Contr. Ass.* **14,** 158, "A Radiotracer Study of the Production of Formaldehyde in the Photooxidation of Ethylene in the Atmosphere. Part I. Method and Procedures."

Osif, T. (1975). Unpublished work at Penn State Univ., "Photooxidation of CH_2O."

Pate, C. T., Finlayson, B. J., and Pitts, J. N. Jr. (1974). *J. Amer. Chem. Soc.* **96,** 6554, "A Long Path Infrared Spectroscopic Study of the Reaction of Methylperoxy Free Radicals with Nitric Oxide."

Sanhueza, E., and Heicklen, J. (1974a). *Int. J. Chem. Kinet.* **6,** 553, "The Reaction of $O(^3P)$ with C_2HCl_3."

Sanhueza, E., and Heicklen, J, (1974b). *Can. J. Chem.* **52,** 3870, "The Reaction of $O(^3P)$ with C_2Cl_4."

Saunders, D., and Heicklen, J. (1966). *J. Phys. Chem.* **70,** 1950, "Some Reactions of Oxygen Atoms. I. C_2F_4, C_3F_6, C_2H_2, C_2H_4, C_3H_6, i-C_4H_8, C_2H_6, c-C_3H_6, and C_3H_8."

Schuck, E. A., and Doyle, G. J. (1959). Air Pollution Foundation, Rep. No. 29, San Marino, California, "Photooxidation of Hydrocarbons in Mixtures Containing Oxides of Nitrogen and Sulfur Dioxide."

Sie, B. K. T., Simonaitis, R., and Heicklen, J. (1976). *Int. J. Chem. Kinet.* **8,** 99, "The Reaction of OH with NO."

Simonaitis, R., and Heicklen, J. (1974a). *J. Phys. Chem.* **78,** 653, "Reactions of HO_2 with NO and NO_2."

Simonaitis, R., and Heicklen, J. (1974b). *J. Phys. Chem.* **78,** 2417, "Reactions of CH_3O_2 with NO and NO_2."

Simonaitis, R., and Heicklen, J. (1976). *J. Phys. Chem.* **80,** 1, "The Reactions of HO_2 with NO and NO_2 and of OH with NO."

Stephens, E. R., and Scott, W. E. (1962). *Proc. Amer. Petrol. Inst.* **42** (III), 665, "Relative Reactivity of Various Hydrocarbons in Polluted Atmospheres."

Tyler, B. J. (1962). *Nature (London)* **195,** 279, "Reaction of Hydrogen Peroxide and Nitric Oxide."

Waldorf, D. M., and Babb, A. L. (1963). *J. Chem. Phys.* **39,** 432, "Vapor-Phase Equilibrium of NO, NO_2, H_2O and HNO_2."

Waldorf, D. M., and Babb, A. L. (1964). *J. Chem. Phys.* **40,** 1165, "Erratum: Vapor-Phase Equilibrium of NO, NO_2, H_2O and HNO_2."

Weaver, J., Meagher, J., Shortridge, R., and Heicklen, J. (1975). *J. Photochem.* **4,** 341 "The Oxidation of Acetyl Radicals."

Weinstock, B., Daby, E., and Niki, H. (1971). *In* "Chemical Reactions in Urban Atmospheres" (C. Tuesday, ed.), p. 54. Elsevier, Amsterdam.

Westberg, K., and Cohen, N. (1970). *AIAA Fluid Plasma Dynam. Conf., 3rd, Los Angeles* AIAA Paper No. 70-753, "The Chemical Kinetics of Photochemical Smog as Analyzed by Computers," also see (1969). Aerospace Corp. Rep. No. ATR-70(8107)-1.

Westberg, K., Cohen, N., and Wilson, K. W. (1971). *Science* **171,** 1013, "Carbon Monoxide: Its Role in Photochemical Smog Formation."

Wiebe, H. A., and Heicklen, J. (1973). *J. Amer. Chem. Soc.* **95,** 1, "Photolysis of Methyl Nitrite."

Wiebe, H. A., Villa, A., Hellman, T. M., and Heicklen, J. (1973). *J. Amer. Chem. Soc.* **95,** 7, "Photolysis of Methyl Nitrite in the Presence of Nitric Oxide, Nitrogen Dioxide, and Oxygen."

Chapter VII

REACTIONS OF O_3 AND SINGLET O_2

NO_2 PHOTOLYSIS

Two important oxidant species in photochemical smog are O_3 and singlet O_2, the singlet O_2 being primarily in the $^1\Delta_g$ excited electronic state. Both species are produced as the result of NO_2 photolysis. The absorption spectrum of NO_2 is shown in Fig. VII-1. The absorption has a maximum at about 4000 Å with an absorption cross section of about 6.2×10^{-19} cm²/molecule.

At wavelengths below about 3300 Å, the absorption spectrum is continuous and photodissociation must occur. Above ~3300 Å, the absorption of NO_2 shows considerable structure and electronically excited bound states must exist. However, these cross to a repulsive state since the photodissociation coefficient is unity up to 3979 Å (Lee and Uselman, 1972; Jones and Bayes, 1973b), which corresponds to the bond dissociation enthalpy in NO_2 at 0°K:

$$NO_2 + h\nu\ (\leq 3979\ \text{Å}) \rightarrow NO + O(^3P) \quad (1)$$

The $O(^3P)$ immediately reacts with O_2 to produce O_3

$$O(^3P) + O_2 + M \rightarrow O_3 + M \quad (2)$$

Above 3979 Å, the quantum yield for reaction (1) drops until it is almost zero at 4050 Å. The extra energy needed for photodissociation above 3979 Å has been attributed to the thermal rotational energy by Pitts *et al.* (1964), who studied the temperature dependence at 4047 Å, and by Lee and Uselman (1972), who studied the wavelength dependence at 25°C. The rotational energy accounts for most, but not all, of the photodissociation between 3979 and 4050 Å. Jones and Bayes (1973b) found

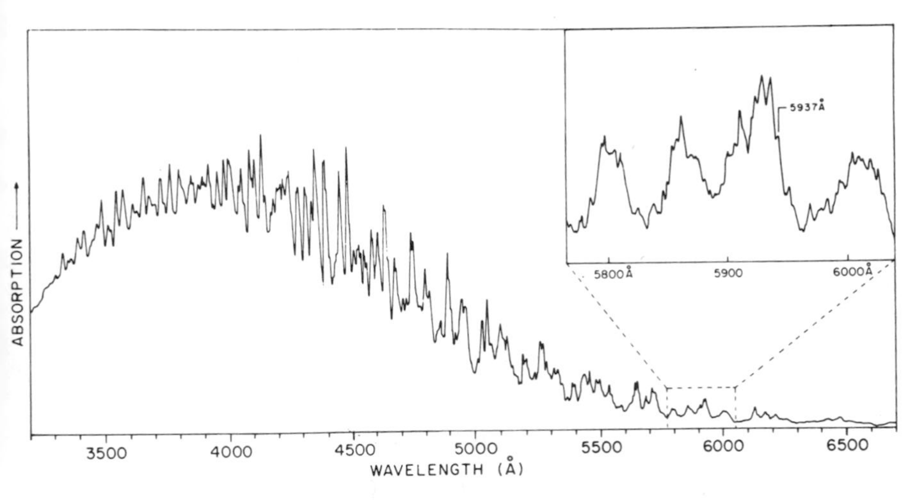

Fig. VII-1. Low-resolution absorption spectrum of NO_2. Insert shows medium resolution spectrum of the 5800–6000 Å region. From Stevens *et al.* (1973) with permission of the North-Holland Publ. Co.

that a small production of NO persists even at wavelengths in excess of 4300 Å, and they attributed this to the reaction

$$NO_2^* + NO_2 \rightarrow 2NO + O_2 \tag{3}$$

where NO_2^* represents some unspecified electronically excited state of NO_2. Reaction (3) is of no consequence in the atmosphere.

From the product of the photodissociation coefficients, the absorption cross sections (Fig. VII-1), and the solar radiation intensity (Fig. I-13), the photodissociation coefficient is obtained. For an overhead sun $J\{O(^3P)\} = 8 \times 10^{-3}$ sec^{-1} from NO_2 photolysis.

Above ~3300 Å, the absorption of NO_2 shows considerable structure. The absorption is complex and a complete spectroscopic analysis has not been made yet. Apparently there are two excited states ($^2B_1$ and $^2B_2$) involved in absorption, and there may be more states involved in the photophysics before deactivation occurs. The emission decay is not exponential, so that at least two states are involved in emission.

Neuberger and Duncan (1954) observed the same radiative lifetime of $\tau = 4.4 \times 10^{-5}$ sec for three different incident wavelengths (3950, 4300, and 4650 Å). This value was more than two orders of magnitude larger than that of 2.6×10^{-7} sec calculated from the integrated extinction coefficient. To account for the anomaly, Neuberger and Duncan proposed the participation of two excited electronic states, one formed in absorption and the other emitting the radiation. Eleven years later Douglas and

Huber (1965) showed that the $^2B_1$ state is responsible for the discrete absorption bands in the region from 3700 to 4600 Å. Douglas (1966) rejected the two-excited-states model and felt that the most likely explanation for the long radiative lifetime of NO_2 is the interaction of the excited electronic state with upper vibrational levels of the ground electronic state $^2A_1$.

Myers *et al.* (1965) studied the relative quenching efficiencies of several gases on the fluorescence. They found linearity in the Stern–Volmer plots and concluded that only one excited electronic state was responsible for the light emission. Further support for the one-excited-state model came from the work of Schwartz and Johnston (1969), who found the radiative lifetime to vary between 55 and 90 μsec, depending on the incident wavelength, and the work of Keyser *et al.* (1971), who found the radiative lifetime to be 55 μsec independent of exciting wavelength.

The two-excited-state mechanism of Neuberger and Duncan was revived by work at Penn State University in the early 1970s. Abe *et al.* (1971) demonstrated that emission is from a single vibrational level of the $^2B_2$ state when excitation is at 5145 Å. Braslavsky and Heicklen (1973) excited NO_2 with 4047, 4358, or 4800 Å radiation. They measured the fluorescence quenching by foreign gases over a very extended pressure range (up to 30 Torr) and found deviations in the Stern–Volmer plots. From this they concluded that the principal state emitting the radiation could not be the principal state formed in absorption. In fact, their results indicated that not even a two-excited-state model could explain the quenching behavior.

At the same time as the Penn State Studies, Sackett and Yardley (1972) examined the lifetime of decay for incident radiation at 4515–4605 Å. They found a nonexponential decay with an approximate lifetime of 62 to 75 μsec for emission >5200 Å. They interpreted their results in terms of absorption to both the $^2B_1$ and $^2B_2$ states, with the absorption to the latter state being more pronounced. Their results also seemed to indicate some unresolved quasi-continuous absorption underlying the bands in the visible region.

Final confirmation of the two-excited-state mechanism has been given by Stevens *et al.* (1973), who used a tunable laser to excite NO_2 at 5934–5940 Å. They observed absorption to both the $^2B_1$ and $^2B_2$ states. Two fluorescence lifetimes were obtained of 30 ± 5 and 115 ± 10 μsec, the former associated with the $^2B_2$ state, and the latter most likely belonging to the $^2B_1$ state.

A much shorter lifetime of 3.39 ± 0.36 μsec for the $^2B_2$ state has been reported by Solarz and Levy (1974), who measured the linewidth of the microwave optical double resonance spectrum. This lifetime is still an order of magnitude greater than that estimated from the integrated extinction coefficient.

Theoretical computations of the energy levels of the electronically excited states of NO_2 have been made in recent years (Fink, 1971; Gangi and Burnelle, 1971a; Hay, 1973). It is probable that the $^2B_1$, $^2A_2$, and $^4B_2$ states all have energies below that of the $^2B_2$ state. Gangi and Burnelle (1971a,b) suggested that absorption of light between 4000 and 6000 Å is mainly from the ground $^2A_1$ state to the $^2B_2$ state. They computed the radiative lifetime for the unperturbed $^2B_2$ state to be 0.1248 μsec, similar to the value computed by Neuberger and Duncan (1954). Thus the longer radiative lifetimes measured for the $^2B_2$ state support the argument of Douglas (1966) that the state formed in absorption is perturbed by other electronic states.

It is clear that the nature and number of the excited electronic states of NO_2 has not been resolved yet. For simplicity we include all the excited states together, and represent the process as

$$NO_2 + h\nu\ (>4050\ \text{Å}) \rightarrow NO_2^* \tag{4}$$

where NO_2^* is the sum of the excited electronic states, some of which are produced even at radiation down to 3300 Å. From the extinction coefficients and solar radiation intensity, $J\{NO_2^*\}$ can be computed to be 2.8×10^{-2} sec^{-1} for an overhead sun.

The excited NO_2^* is readily quenched by both N_2 and O_2 with about equal efficiency (Braslavsky and Heicklen, 1973). For that fraction which reacts with O_2, Frankiewicz and Berry (1973) found that $7.5 \pm 2.5\%$ gives $O_2(^1\Delta)$ and $0.14 \pm 0.02\%$ gives $O_2(^1\Sigma_g^+)$. On the other hand Jones and Bayes (1973a) found only about 4.5% of the collisions with O_2 gave singlet O_2 with 4000 Å incident radiation, and that this efficiency dropped linearly on both sides of the maximum to about 0% as the wavelength increased to 6000 Å and 2.5% as the wavelength decreased to 3300 Å. These results correspond to an average efficiency about one-third that of Franciewicz and Berry. Thus the NO_2^* removal reactions are

$$NO_2^* \rightarrow N_2 \rightarrow NO_2 + N_2 \tag{5}$$

$$NO_2^* + O_2 \rightarrow NO_2 + O_2 \tag{6a}$$

$$\rightarrow NO_2 + O_2\ (\text{singlet}) \tag{6b}$$

The percentage of NO_2^* that gives singlet O_2 in the atmosphere is 1.6%, using Frankiewicz and Berry's value or about one-third that from Jones and Bayes' value. However, Jones and Bayes' data give $O_2(^1\Delta)$ for wave-

TABLE VII-1

Rate Coefficients (M^{-1} sec^{-1}) at 300°K for Reactions with Hydrocarbons Present in Polluted Air

Hydrocarbon	$k\{HO\}^a$	$k\{O(^3P)\}^a$	$k\{O_2(^1\Delta_g)\}^b$	$k\{O_3\}^{b,d}$	$k\{O_3\}^{d,e}$
Paraffins					
CH_4	3.9×10^6	1.0×10^4	—	0.85×10^{-3}	—
C_2H_6	1.15×10^8	4.6×10^5	—	6.6×10^{-3}	—
C_3H_8	5.4×10^8	1.15×10^8	—	4.0×10^{-3}	—
n-C_4H_{10}	1.3×10^9	1.6×10^8	—	5.9×10^{-3}	—
i-C_4H_{10}	8.3×10^8	—	—	12.2×10^{-3}	—
Olefins					
C_2H_2	7.1×10^7	1.5×10^8	—	47	—
C_2H_4	1.9×10^9	6.0×10^8	1.1×10^3	1.6×10^3	1.8×10^3
C_3H_6	6.6×10^9	3.0×10^9	1.3×10^3	5.0×10^3	5.1×10^3
i-C_4H_8	1.5×10^{10}	4.0×10^9	1.4×10^3	5×10^3	3.9×10^3
trans-2-C_4H_8	2.7×10^{10}	1.8×10^{10}	$(1.0 \text{ or } 3.0) \times 10^4$	$(17 \text{ or } 260) \times 10^3$	17×10^3
cis-2-C_4H_8	2.3×10^{10}	1.5×10^{10}	$(1.3 \text{ or } 7.5) \times 10^4$	$(13 \text{ or } 200) \times 10^3$	13×10^3
i-C_4H_8	2.4×10^{10}	1.6×10^{10}	—	3.6×10^3	3.6×10^3
3-CH_3-1-butene	—	—	—	3.0×10^3	4.0×10^3
2-CH_3-1-butene	—	—	$(1.4 \text{ or } 10) \times 10^4$	4.0×10^3	3.0×10^3
2-CH_3-2-butene	4.5×10^{10}	5.0×10^{10}	3.3×10^{4} [c]	$(12 \text{ or } 450) \times 10^3$	12×10^3
Tetramethylethylene	—	—	—	14×10^3	14×10^3
1,3-C_4H_6	—	1.55×10^{10}	7.8×10^3	4.2×10^3	4.2×10^3
1-C_5H_{10}	—	—	1.9×10^3	4.0×10^3	3.9×10^3
cis-2-C_5H_{10}	3.4×10^{10}	1.44×10^{10}	2.0×10^4	10×10^3	10×10^3
trans-2-C_5H_{10}	3.4×10^{10}	—	—	13×10^3	13×10^3
1-C_6H_{12}	1.2×10^{10}	4.2×10^{10}	—	6.1×10^3	4.6×10^3

α-Pinene	—	—	—	99×10^3	—
2-CH_3-1,3-butadiene	—	—	9.5×10^3	—	—
Styrene	—	—	—	18×10^3	—

[a] From relative rate coefficients in Table VI-1 and $k = 4 \times 10^9$ for $O(^3P)$ + 1-C_4H_8 and $k = 6.6 \times 10^9$ for HO + C_3H_6.

[b] From review by Murray (1972); where 2 values are reported for $k\{O_3\}$ the larger is from the work of Bufalini and Altshuller (1965) at very low reactant concentrations in a large excess of O_2. $k\{O_3\}$ also reported for 1-C_7H_{14} (4.9×10^3), 1-C_8H_{16} (4.9×10^3), 1-$C_{10}H_{20}$ (6.5×10^3), *c*-C_6H_{10} (14×10^3 or 35×10^3).

[c] From Huie and Herron (1973).

[d] $k\{O_3\}$ obtained in presence of O_2.

[e] From Vrbaski and Cvetanović (1960a).

lengths extending below the photodissociation limit. Thus for an overhead sun, $J\{O_2(^1\Delta)\} = 4.3 \times 10^{-4}$ or $\sim 3 \times 10^{-4}$ sec^{-1}, depending on the value adopted. We adopt a value of 3.6×10^{-4} sec^{-1}, so that the ratio of singlet O_2 to O_3 produced becomes 4–5%.

The principal removal step for O_2^* (singlet O_2) is deactivation by O_2:

$$O_2^* + O_2 \rightarrow 2O_2 \tag{7}$$

with a rate coefficient of 2.2×10^{-18} cm^3/sec at 300°K (Garvin and Hampson, 1974). Thus the ratio $[O_2^*]/[O(^3P)]$ [assuming all $O(^3P)$ from NO_2 photolysis] is ~ 300.

Once O_3 is produced in the atmosphere, its photodecomposition in the Huggins bands (2900–3400 Å) can produce singlet O_2. From the calculations of Kummler *et al.* (1964) the O_3 photodissociation coefficient to produce singlet O_2 becomes $\sim 6 \times 10^{-6}$ sec^{-1} for an overhead sun, a factor of 60 smaller than that for NO_2 photolysis.

RATE COEFFICIENTS

The room temperature rate coefficients of O_3 and $O_2(^1\Delta)$ with various hydrocarbons are listed in Table VII-1. For comparison the rate coefficients with HO and $O(^3P)$ are also listed.

With the paraffins, the rates of reaction of $O_2(^1\Delta)$ are so slow that rate coefficients have not been measured. The O_3–paraffin reactions are six orders of magnitude slower than the O_3–olefin reactions, and they are also of no importance.

For reactions with olefins, both $O_2(^1\Delta_g)$ and O_3 have room temperature rate coefficients of about 10^4 M^{-1} sec^{-1}. However, compared to the rate

TABLE VII-2

Rate Constants for Deactivation of $O_2(^1\Delta_g)$ by Some Air Pollutants at Room Temperature[a]

Compound	10^{19} k (cm^3/sec)
SO_2	0.39 ± 0.09
H_2	2.1 ± 0.2
Acetone	16 ± 4
CH_2O	23 ± 3
Ambient air	3.4 ± 0.4
Dry synthetic air	3.0 ± 0.1

[a] From Penzhorn *et al.* (1974).

coefficients for $O(^3P)$ reactions, they are six orders of magnitude smaller. The ratio $[O_3]/[O(^3P)]$ is typically $\sim 10^6$ in the atmosphere (see Figs. VI-6 and VI-7), so that reactions of olefins with O_3 and $O(^3P)$ are of comparable importance. However, the ratio $[O_2(^1\Delta_g)]/[O(^3P)]$ is only ~ 300. Thus reactions of $O_2(^1\Delta_g)$ are of no importance in the photochemical smog cycle or in removing olefins. However, they are of interest in that they produce unique products. These products, although present in very small amounts, may cause very unpleasant effects.

The deactivation coefficients for $O(^1\Delta_g)$ by some other atmospheric pollutants are listed in Table VII-2. They are comparable to the deactivation coefficients with the olefins, and thus these interactions are also of no practical importance in polluted urban atmospheres.

TABLE VII-3

Gas Phase Rate Constants for Removal of $O_2(^1\Delta_g)$ *at Room Temperature*

Compound	k (M^{-1} sec^{-1})	Reference
2,3-Dimethyl-2-butene	7.6×10^5	Huie and Herron (1973)
trans-3-Methyl-2-pentene	$\sim 9 \times 10^3$	Huie and Herron (1973)
2,3-Dimethyl-2-pentene	5.2×10^5	Huie and Herron (1973)
2,3-Dimethyl-2-hexene	5.0×10^5	Huie and Herron (1973)
1-Ethoxy-2-ethyl-1-hexene	$\sim 4 \times 10^5$	Huie and Herron (1973)
Cyclopentene	1.5×10^7	Huie and Herron (1973)
1-Methylcyclopentene	1.1×10^4	Huie and Herron (1973)
2,3-Dimethylcyclopentene	3.0×10^5	Huie and Herron (1973)
1,2-Dimethylcyclohexene	3.0×10^5	Huie and Herron (1973)
1,3-Cyclohexadiene	6.8×10^4	Huie and Herron (1973)
2-Methylfuran	1.0×10^5	Huie and Herron (1973)
2,5-Dimethylfuran	1.5×10^7	Huie and Herron (1973)
Diethyl sulfide	9.1×10^4	Huie and Herron (1973)
Diethyl sulfide	4.1×10^5	Ackerman *et al.* (1971)
$(CH_3)_2S$	1.6×10^5	Ackerman *et al.* (1971)
$(CH_3S)_2$	1.0×10^4	Ackerman *et al.* (1971)
Thiophene	3.6×10^4	Ackerman *et al.* (1971)
CH_3SH	2.3×10^3	Ackerman *et al.* (1971)
CH_3SH	0.6×10^3	Penzhorn *et al.* (1974)
$(C_2H_5)_3N$	1.4×10^6	Huie and Herron (1973)
$(C_2H_5)_3N$	2.0×10^6	Furukawa and Ogryzlo (1973)
$(CH_3)_3N$	1.9×10^6	Furukawa and Ogryzlo (1973)
$(C_2H_5)_2NH$	7.3×10^4	Furukawa and Ogryzlo (1973)
$(CH_3)_2NH$	5.6×10^4	Furukawa and Ogryzlo (1973)
$C_2H_5NH_2$	7×10^3	Furukawa and Ogryzlo (1973)
CH_3NH_2	8×10^3	Furukawa and Ogryzlo (1973)

TABLE VII-4

Quenching Rate Coefficients for $O_2(^1\Sigma_g^+, v' = 0)$ *by Various Atmospheric Gases at 298°K*[a]

Compound	k(cm³/sec)	Compound	k(cm³/sec)	Compound	k(cm³/sec)
CO_2	4.4×10^{-13}	CH_4	1.1×10^{-13}	NO_2	3.1×10^{-14}
CO	4.3×10^{-15}	C_2H_6	3.6×10^{-13}	H_2	1.1×10^{-12}
SF_6	5.7×10^{-16}	C_2H_4	3.1×10^{-13}	N_2	1.8×10^{-15}
O_2	4.5×10^{-16}	N_2O	7.0×10^{-14}	He, Ne	$\sim 1.0 \times 10^{-17}$
H_2O	3.3×10^{-12}	NO	4.1×10^{-14}	Ar	5.8×10^{-18}
NH_3	8.6×10^{-14}			Kr, Xe	$\sim 1.0 \times 10^{-16}$

[a] From Filseth *et al.* (1970).

Singlet O_2 is also quenched by a number of other gases that, although not generally present in polluted atmospheres, may be present in some locations under some conditions. Those for which quenching constants have been measured are listed in Table VII-3. In general, their room temperature rate coefficients range from $\sim 10^4$ to 10^6 M^{-1} sec^{-1}. However, 2,5-dimethylfuran has a rate coefficient of 1.5×10^7 M^{-1} sec^{-1}, which means that it will quench 0.1% of the $O_2(^1\Delta_g)$ molecules at a concentration of 18.5 ppb in air.

The quenching of $O_2(^1\Sigma_g^+)$ for some atmospheric gases has been measured by Filseth *et al.* (1970) and their values are listed in Table VII-4.

REACTIONS OF OZONE

Reactions with Alkenes

Ozone reacts readily with alkenes to produce a variety of products. In the gas phase these include hydrocarbons, aldehydes, ketones, and acids of lower carbon number, as well as aerosols, CO, CO_2, and H_2O. The most extensive studies were made by Vrbaski and Cvetanović (1960b), who examined the reaction of ozonized oxygen with 16 alkenes. As an example of the complexity of the products, in the 1-C_4H_8 reaction the major products included propionaldehyde, CO_2, and HCOOH. Minor products were CH_3CHO, C_2H_6, C_2H_4, CH_3OH, n-C_3H_7CHO, and C_3H_6 in decreasing importance. Traces of methyl formate were also found, and presumably CH_2O, CO, and H_2O were also produced. The reaction is quite exothermic, and the initially formed products may undergo secondary decompositions. Thus some of the products observed may not be primary products of the reaction.

The principal initial reaction of O_3 with olefins is generally accepted to be the Criegee mechanism,

$$O_3 + R_1R_2C{=}CR_3R_4 \rightarrow R_1R_2C{=}O + R_3R_4COO \quad (8)$$

where R_1, R_2, R_3, and R_4 represent hydrocarbon (or H) radical fragments. The reaction proceeds through an intermediate molozonide of one or more of the structures **1–3**. One of these intermediates, presumably the first,

1 **2** **3**

has been trapped at low temperatures in the condensed phase and its NMR and infrared spectrum reported (Greenwood, 1965; Durham and Greenwood, 1968a,b; Greenwood and Durham 1969; Hull *et al.*, 1972). One of the products of the reaction is R_3R_4COO and its structure is usually considered to be intermediate between the diradical **4** and zwitterion formulations **5**. A possible third structure is **6**, but this is usually not considered to be important.

$$R_3R_4\dot{C}{-}O{-}O\cdot \rightleftharpoons R_3R_4\overset{+}{C}{-}O{-}O^- \qquad R_3R_4C{=}O{\rightarrow}O$$

4 **5** **6**

In the vapor phase the R_3R_4COO species has not been captured or isolated. However, there is ample evidence for its existence in the liquid phase, where the products of its reaction with aldehydes and ketones, with itself, and with solvents have been found (Murray, 1972). These reactions are

$$2\,R_3R_4CO_2 \longrightarrow \text{cyclic } (R_3R_4C)_2(O{-}O)_2 \quad (9)$$

$$nR_3R_4CO_2 \longrightarrow \left(R_3R_4C{-}O{-}O{-}\right)_n \quad (10)$$

$$R_3R_4CO_2 + R_1R_2C{=}O \longrightarrow \text{ozonide } R_1R_2C(O)(O{-}O)CR_3R_4 \quad (11)$$

$$R_3R_4CO_2 + CH_3OH \longrightarrow R_3R_4C(OCH_3)(OOH) \quad (12)$$

Some evidence also exists for some of these products in the gas phase. The secondary ozonides [the product of reaction (11)] have been observed with alkenes of C_6 or larger. Vrbaski and Cvetanović (1960b) reported a positive KI test for peroxides in the reaction of O_3 with *i*-C_4H_8, 1-C_4H_8, *cis*-, and *trans*-2-C_5H_{10}.

In the gas phase the $R_3R_4CO_2$ intermediate, which is formed with excess energy, presumably undergoes isomerizations and decompositions in preference to reaction with other species. For example with CH_3CHO_2, the possible reactions suggested by Vrbaski and Cvetanović are

$$(CH_3)(H)CO_2 \rightarrow CH_3C(=O){-}OH \tag{13a}$$

$$\rightarrow CH_3OH + CO \tag{13b}$$

$$\rightarrow CH_4 + CO_2 \tag{13c}$$

$$\rightarrow H_2O + CH_2CO \tag{13d}$$

The higher analogs might also transfer an H-atom before decomposition to produce olefins:

$$C_2H_5CHO_2 \rightarrow CH_3\dot{C}HCH_2OO\cdot \rightarrow C_2H_4 + CH_2O_2 \tag{14}$$

The above reactions are speculative, and there is no direct evidence for them. In fact, reaction (13c) cannot quantitatively explain the results of Vrbaski and Cvetanović, since they found CO_2 to be produced in considerable excess of the alkanes. Most of the CO_2 must come from other routes.

There are other reactions of the $R_3R_4CO_2$ intermediates that might play a role in the atmosphere. For these intermediates in which neither R_3 nor R_4 is an H atom, isomerization and decomposition should be relatively slow, and reactions with other species must be considered. Possible reactions are

$$R_3R_4CO_2 + CO \rightarrow R_3R_4CO + CO_2 \tag{15}$$

$$+ NO \rightarrow R_3R_4CO + NO_2 \tag{16}$$

$$+ NO_2 \rightarrow R_3R_4CO + NO_3 \tag{17}$$

$$+ SO_2 \rightarrow R_3R_4CO + SO_3 \tag{18}$$

$$+ O_3 \rightarrow R_3R_4CO + 2O_2 \tag{19}$$

$$+ O_2 \rightarrow R_3R_4CO + O_3 \tag{20}$$

$$+ \text{olefin} \rightarrow \text{adduct} \tag{21}$$

None of these reactions has been proved or disproved in the laboratory, but there is considerable evidence that reactions (19) and (21) proceed with halogenated olefins. The adduct produced in reaction (21) is then oxidized further.

The mechanism outlined above can explain the main features of the O_3–alkene reaction, but does not explain all the observations. This becomes more apparent when the ozonolysis results in the presence and absence of O_2 are compared. Wei and Cvetanovič (1963) examined the reactions in O_2 and in N_2. They found that the amount of O_3 to olefin consumed was 1.0 in N_2, but 1.4–2.0 in O_2, and that the ratio of products formed to O_3 consumed was greater in O_2 than N_2. Furthermore, the relative rates of reaction of several olefins with O_3 were different in O_2 and N_2. Recent results (Herron and Huie, 1974; Finlayson *et al.*, 1974) confirm that the presence of O_2 retards the rate of reaction.

It is difficult to explain these results with the above mechanism. The simple competition, reactions (19) and (20), is not consistent with the facts. In fact, it predicts that the O_3 to olefin consumption should be reduced in the presence of O_2, exactly the opposite of what is observed.

Further evidence that other processes are occurring comes from emission studies. The results in Pitts' laboratory (Finlayson *et al.*, 1972b, 1974) shows that the O_3–olefin reaction leads to light emission from electronically excited OH, aldehydes, ketones, and α-dicarbonyl compounds. In particular, the 1-olefins give rise to CH_2O emission. Also the HO Meinel vibrational emission ($v \leq 9$) has been observed. Its emission very closely resembles that in the upper atmosphere. Presumably the source is the same, i.e.,

$$H + O_3 \rightarrow O_2 + HO\ (v \leq 9) \tag{22}$$

Possibly the excited aldehydes and ketones could arise from the sequence of reactions

$$R_3R_4CO_2 + R_1R_2C{=}CR_1R_2 \rightarrow R_1R_2CO + R_3R_4CO + CR_1R_2 \tag{23}$$

$$CR_1R_2 + O_3 \rightarrow R_1R_2CO^* + O_2 \tag{24}$$

but reaction (24) should be completely quenched in the presence of O_2 if the triplet diradical is involved. Furthermore, there must be a production term for H atoms to account for the HO Meinel band emission. Therefore other mechanisms must be sought.

Some monofree radicals (species with one unpaired electron) must be present in the system. They could react in a chain process that is either promoted or inhibited by O_2. Promotion seems more likely, since it is difficult to envision how the inhibition of the radical chain would lead to a ratio of O_3 to olefin consumption of 1.4 to 2.0. Further evidence for radical chain promotion by O_2 comes from an examination of the reaction rate coefficients (Table VII-I), which were measured in the presence of O_2. For *cis*- and *trans*-2-C_4H_8 and tetramethylethylene, Bufalini and Altshuller (1965) found rate coefficients 15–30 times larger than found by Vrbaski and Cvetanović (1960a). Since the Bufalini and Altshuller experiments were done at very much higher $[O_2]/[O_3]$ ratios than those of Vrbaski and Cvetanović, these results suggest radical chain promotion by O_2. If so, then the rate coefficients listed in Table VII-1 do not represent true rate coefficients at all, but only upper limits.

Based on some of the above observations, as well as other thermochemical arguments, O'Neal and Blumstein (1973) have proposed an additional free radical mechanism following initial molozonide formation. They assume the molozonide to be in equilibrium with the diradical structure

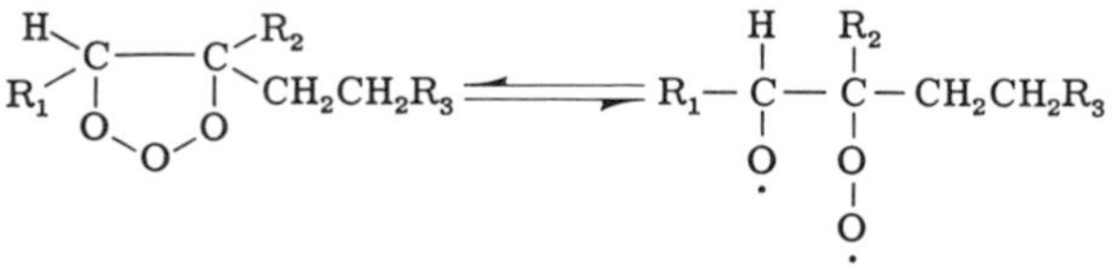

$$\text{(H)(R}_1\text{)C(–O–O–O–)C(R}_2\text{)CH}_2\text{CH}_2\text{R}_3 \rightleftharpoons \text{R}_1\text{–C(H)(O}\cdot\text{)–C(R}_2\text{)(OO}\cdot\text{)–CH}_2\text{CH}_2\text{R}_3 \qquad (25)$$

The diradical can undergo decomposition via the Criegee mechanism outlined above, or can first undergo intramolecular α, β, or γ H atom abstraction as illustrated in Fig. VII-2. O'Neal and Blumstein (1973) also estimated the energetics of the various reaction paths and concluded that for ethylene and propylene the diradical should decay mainly by the Criegee mechanism, while for 1-butene and other olefins with greater internal degrees of freedom internal hydrogen abstraction should dominate. Consequently, secondary ozonides should not be the major products from the latter ozonolysis reactions, particularly at higher total pressures. Low-temperature infrared studies (Hull *et al.*, 1972), on the other hand, revealed that not only ethylene and propylene but 2-butenes also gave secondary ozonides. Although O'Neal and Blumstein (1973) suggested that the reaction of molozonide with aldehydes may be the source of any unexpected secondary ozonides, the isotopic studies by Gillies and Kuczkowski (1972a, b) and Gillies *et al.* (1974) appear to rule out this possibility. It seems that the estimates of energetics of these ozonolysis reactions may not be completely valid or additional modifications of the reaction mechanism may be necessary.

A further shortcoming of the O'Neal–Blumstein mechanism is that internal β-hydrogen abstraction by the diradical was suggested to explain the chemiluminescence of O_3–olefin reactions. Of course, in the C_2H_4

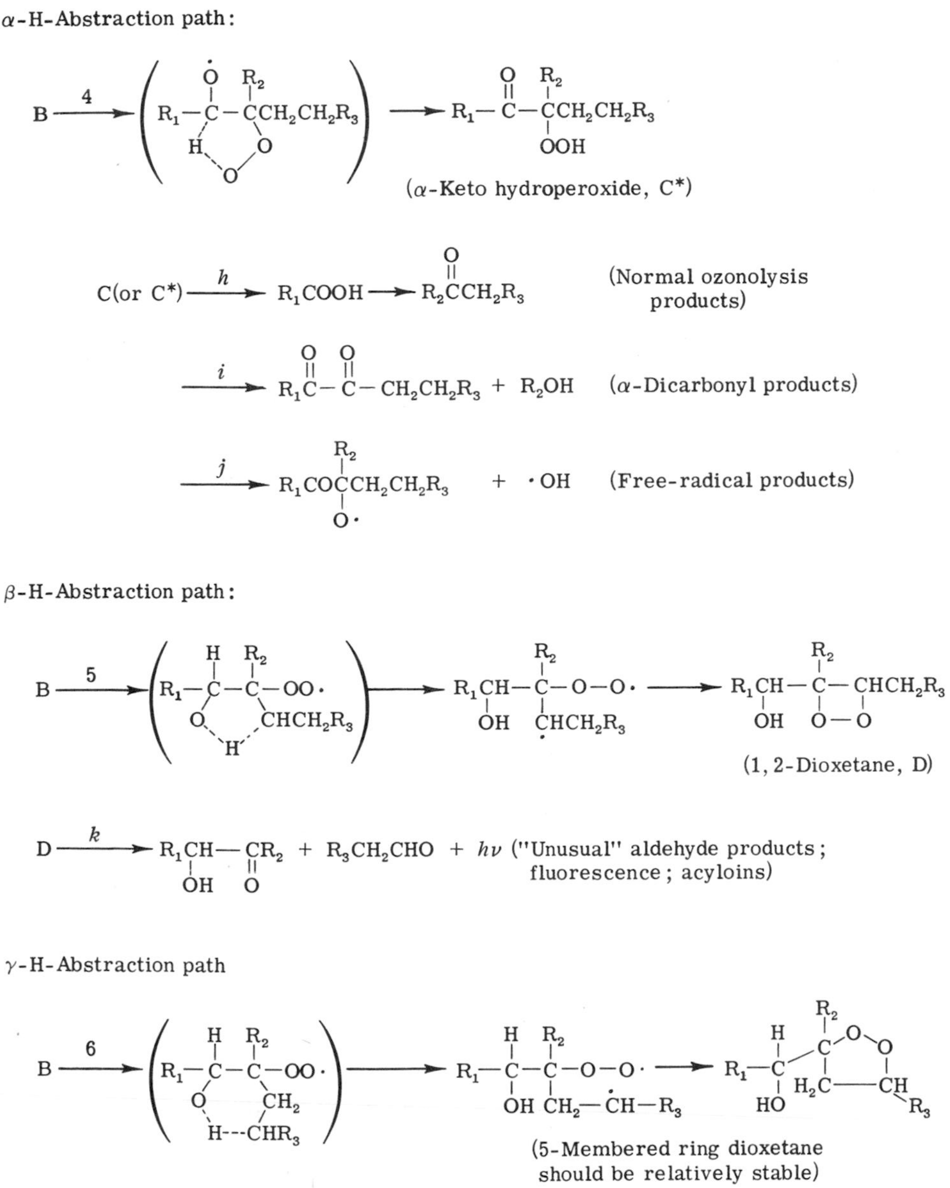

α-H-Abstraction path:

$$B \xrightarrow{4} \left(R_1-\overset{\dot{O}}{\underset{|}{C}}-\overset{R_2}{\underset{|}{C}}CH_2CH_2R_3 \ \text{(cyclic H}\cdots\text{O–O)} \right) \longrightarrow R_1-\overset{O}{\overset{\|}{C}}-\overset{R_2}{\underset{OOH}{C}}CH_2CH_2R_3$$

(α-Keto hydroperoxide, C*)

$$\text{C(or C*)} \xrightarrow{h} R_1COOH \longrightarrow R_2\overset{O}{\overset{\|}{C}}CH_2R_3$$ (Normal ozonolysis products)

$$\xrightarrow{i} R_1\overset{O}{\overset{\|}{C}}-\overset{O}{\overset{\|}{C}}-CH_2CH_2R_3 + R_2OH$$ (α-Dicarbonyl products)

$$\xrightarrow{j} R_1CO\overset{R_2}{\underset{O\cdot}{C}}CH_2CH_2R_3 + \cdot OH$$ (Free-radical products)

β-H-Abstraction path:

$$B \xrightarrow{5} \left(R_1-\overset{H}{\underset{O}{C}}-\overset{R_2}{C}-OO\cdot \ \ CHCH_2R_3 \ \text{(O}\cdots\text{H}\cdots\text{C)} \right) \longrightarrow R_1\underset{OH}{C}H-\overset{R_2}{\underset{\dot{C}HCH_2R_3}{C}}-O-O\cdot \longrightarrow R_1\underset{OH}{C}H-\overset{R_2}{\underset{O-}{C}}-\underset{O}{C}HCH_2R_3$$

(1, 2-Dioxetane, D)

$$D \xrightarrow{k} R_1\underset{OH}{C}H-\underset{O}{\overset{\|}{C}}R_2 + R_3CH_2CHO + h\nu$$ ("Unusual" aldehyde products; fluorescence; acyloins)

γ-H-Abstraction path

$$B \xrightarrow{6} \left(R_1-\overset{H}{\underset{O}{C}}-\overset{R_2}{C}-OO\cdot \ \ CH_2 \ \ H\text{---}CHR_3 \right) \longrightarrow R_1-\overset{H}{\underset{OH}{C}}-\overset{R_2}{\underset{CH_2-\dot{C}H-R_3}{C}}-O-O\cdot \longrightarrow R_1-\overset{H}{\underset{HO}{C}}-\overset{R_2}{C}(-O-O-)\underset{H_2C-CH-R_3}{}$$

(5-Membered ring dioxetane should be relatively stable)

Fig. VII-2. **O'Neal–Blumstein (1973) scheme for free radical formation by internal hydrogen abstraction in the ozonolysis of olefins. From O'Neal and Blumstein (1973) with permission of John Wiley and Sons.**

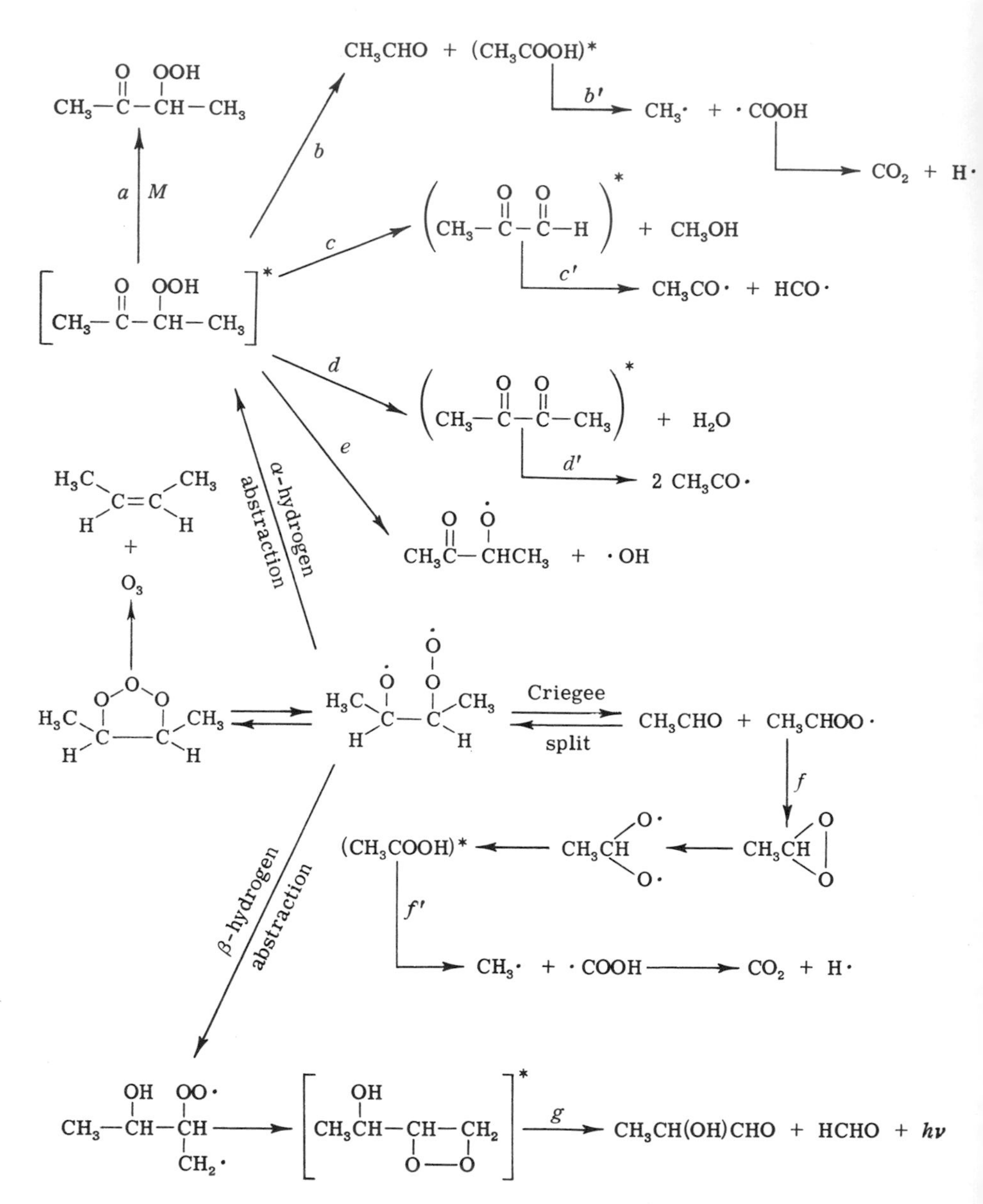

Fig. VII-3. Finlayson *et al.* (1974) modification of the O'Neal–Blumstein (1973) scheme to account for chemiluminescence in the ozonolysis of olefins. From Finlayson *et al.* (1974) with permission of the American Chemical Society.

system there is no β-hydrogen. Finlayson *et al.* (1974) modified the O'Neal–Blumstein mechanism as outlined in Fig. VII-3. They suggested that electronically excited CH_2O was produced by α-hydrogen abstraction,

$$\begin{array}{ll} O\diagdown H & \\ \;\;C\text{—} & CH_2 \\ \;\;| & | \\ H\diagdown_{O}\diagup & O \end{array} \longrightarrow HCOOH + CH_2O^* \qquad (26)$$

but this route seems unlikely from both energetic and steric considerations. The experimental evidence clearly requires the presence of free H atoms, and Finlayson *et al.* (1974) suggested two routes both of which are variations of sequential H atom splitting from the zwitterion:

$$H_2COO \rightarrow HCOOH^* \rightarrow H + HCOO \rightarrow H + CO_2 \qquad (27)$$

Again, these routes seem unlikely and Sanhueza *et al.* (1975) preferred hydrogen abstraction by the single O atom in the diradical:

$$\begin{array}{lcl} O\cdot & \cdots & H \\ | & & | \\ H_2C & \text{—} & CH \\ & & | \\ \cdot O & \text{—} & O \end{array} \longrightarrow H_2COH + HCOO \longrightarrow H + CO_2 \qquad (28)$$

The routes proposed by Finlayson *et al.* (1974) for electronically excited OH production were

$$O + H \rightarrow HO^* \qquad (29)$$

$$O + HCO^* \rightarrow OH^* + CO \qquad (30)$$

However, at this time all the proposed routes to chemiluminescence must be considered speculative.

Reactions with Haloalkenes

Some studies have been made of the gas phase reactions of the chloroethylenes with O_3 in the author's laboratory, and the results have been reviewed recently (Sanhueza *et al.*, 1975). For the molecules studied (C_2H_3Cl, CCl_2CH_2, CHClCHCl, and C_2Cl_4) there is a chain mechanism in the absence of O_2, that is quenched in the presence of O_2. Since O_2 is known to promote the chains of the monofree radical reaction, it was concluded that only diradical species were present, and that they cause a chain decomposition of the olefin in the absence of O_2. The major products were the one-carbon carbonyl products, but the mechanism is complex.

The second significant characteristic of the ozonolysis reactions concerns the origin of the biradical species. In the low-temperature infrared study of the ozonolysis of C_2H_4 (Hull *et al.*, 1973), both the primary and secondary ozonides were detected. Thus, the reaction in this case may be considered to proceed by the Criegee-type mechanism, where the primary ozonide is the source of the biradical species. The infrared study of the low-temperature ozonolysis of CH_2CCl_2 and CHClCHCl (Kolopajlo *et al.*, 1975), on the other hand, showed no formation of primary ozonides even though the reactions were taking place as evidenced by the appearance of the infrared bands of phosgene and formyl chloride, respectively. The source of the biradicals for these reactions must, therefore, be other than the primary ozonides. In the case of the low-temperature ozonolysis of CH_2CHCl, the formation of both formyl chloride and the primary ozonide was observed in a temperature range where the decomposition rate of the latter was negligible. Thus, two independent reaction paths appear to be available here, one involving the primary ozonide and the other channel bypassing this intermediate. These spectroscopic observations indicate, in fact, that the parallel-path reaction mechanism proposed by Williamson and Cvetanović (1968) may be correct in principle and, furthermore, it can now be modified to be consistent with the experimental observations.

Thus we see that the O_3 reactions proceed primarily by different paths for different ethylenes. We find that C_2H_4, CH_2CHCl, and the higher unhalogenated olefins form molozonides that decompose by the Criegee or the O'Neal–Blumstein mechanism to give a rate law first order in both olefin and O_3 over the entire pressure range. On the other hand, the higher chloroethylenes do not form molozonides but react with O_3 by a complex rate law that deviates from second order (first order in each reactant) at low reactant pressures, indicating reversibility of the initial reaction step. Furthermore, CHClCHCl undergoes geometrical isomerization, whereas the 2-butenes do not.

The higher chloroethylenes, and perhaps to some extent C_2H_3Cl, proceed through the π-complex path, which is best represented by the steps

$$RR' + O_3 \leftrightarrows RR'O_3 \quad (\pi\text{-complex}) \tag{31}$$

$$RR'O_3 + RR' \leftrightarrows R_2R_2'O_3 \tag{32}$$

$$R_2R_2'O_3 \rightarrow 2R'O + 2RO_2 \tag{33a}$$

$$\rightarrow 2RO + 2R'O_2 \tag{33b}$$

where R and R′ are the two parts of the olefin joined by the double bond. Both reactions (31) and (32) must be reversible, at least in the CHClCHCl system, to account for the deviation in the second-order kinetics observed at low reactant pressures. For a full discussion of the complete mechanism, rate laws, and rate constants, the reader is referred to Sanhueza *et al.* (1975).

The π-complex has been observed also in the low-temperature studies of all the lower nonhalogenated olefins except C_2H_4 (Hull *et al.*, 1973). Therefore, the π-complex path may play some role in the ozonolysis of these compounds as well. However, this path presumably is less important than the molozonide path.

The reactions of O_3 with the perfluoroolefins C_2F_4, C_3F_6, and 2-C_4F_8 has been examined and the results reviewed (Heicklen, 1969). At room temperature the products are the carbonyl fluorides. However, the reaction is complex since even 2-C_4F_8 produces CF_2O as the main product. At lower temperatures peroxidic polymers are produced in the ozonolysis of C_2F_4, indicating the stability of CF_2O_2 to decomposition.

Presumably the perfluoroolefin ozonolysis is similar to that for the higher chloroethylenes. No deviation was observed in the second-order rate law for C_2F_4 or C_3F_6 at low pressures, but the deviation did occur with 2-C_4F_8. The data for all three perfluoroolefins were interpreted (Heicklen, 1969) in terms of a changing rate law. However, it was believed that no CF_2O_2 diradicals were present because no $\overline{CF_2CF_2O}$ was observed as a product in the room temperature gas phase ozonolysis, yet $\overline{CF_2CF_2O}$ was a product when CF_2O_2 was produced in the C_2F_4–O_2–O system (Heicklen, 1969). If this inference is correct then the ozonolysis of C_2F_4 must be different than that for either the olefins or the chloroethylenes. One note of caution in this inference is that CF_2OO may exist as either a triplet or singlet species, which may react differently. Thus the triplet CF_2OO (presumably produced in the O–O_2–C_2F_4 system) might lead to $\overline{CF_2CF_2O}$, whereas the singlet CF_2OO (as expected in the O_3–C_2F_4 system) might not lead to $\overline{CF_2CF_2O}$. However, if CF_2OO is not produced, the π-complex mechanism might still explain the results by adding the reaction

$$R_2R_2'O_3 + O_3 \rightarrow 2RO + 2R'O + 2O_2 \tag{34}$$

Reactions with Alkynes

The reaction of O_3 with the alkynes C_2H_2, CH_3C_2H, $CH_3C_2CH_3$, and $C_2H_5C_2CH_3$ was studied at 21°C by DeMore (1971). He found the reaction to be about 5–6 times more rapid in the absence of O_2 than in its presence, indicating a chain mechanism. Presumably this chain is suppressed by O_2, and the rate coefficients in the presence of O_2, for O_2 and alkyne pressures in excess of O_3 pressures, were

Alkyne:	C_2H_2	CH_3C_2H	$CH_3C_2CH_3$	$C_2H_5C_2CH_3$
k (at 21°C) (M^{-1} sec^{-1}):	18 ± 3	13 ± 7	20 ± 3	24 ± 8

These rate coefficients are much smaller than for the O_3–alkene reactions. The reason for the decreased rate coefficient is due to the larger activation energies. For example C_2H_2 has an activation energy of 10.8 kcal/mole, compared to a value of 4.7 kcal/mole for C_2H_4. In fact, the pre-exponential factors for the alkyne reactions are about three orders of magnitude greater than those for the alkene reactions.

The products of the reaction with C_2H_2 were mainly CO and HCOOH, the CO being formed at about the same rate that O_3 was consumed. Also produced were CO_2 and $(HCO)_2$ at about one-fourth the rate of O_3 consumption. However, these less important products appeared to be associated with the accompanying slower heterogeneous reaction. Presumably the homogeneous reaction went essentially by

$$C_2H_2 + O_3 \rightarrow HCOOH + CO \tag{35}$$

The HCOOH may be produced with much excess energy and further decompose in part to either $H_2O + CO$ or $H_2 + CO_2$. For the higher analogs the main reaction paths are presumably

$$CH_3C_2H + O_3 \rightarrow CH_3COOH + CO \tag{36}$$

$$CH_3C_2CH_3 + O_3 \rightarrow CH_3COOH + CH_2CO \tag{37}$$

As in the case of C_2H_2 other products were found, these being CO_2, HCOOH, CH_2O, and CH_3COCHO with CH_3C_2H; and CO, CO_2, and $(CH_3CO)_2$ with $CH_3C_2CH_3$. Interestingly enough, CH_2CO was absent in the CH_3C_2H–O_3 reaction.

DeMore showed that the preexponential factors for the O_3–alkene reaction were consistent with the five-member ring intermediate. However, with alkynes the preexponential factors are three orders of magnitude greater, and a linear activated complex must be involved. The mechanism envisioned by DeMore was

$$O_3 + C_2H_2 \longrightarrow H\dot{C}{=}CH{-}O{-}O{-}\dot{O} \longrightarrow \text{(five-member ring: O–O–O bridging HC=CH)} \longrightarrow \dot{O}{-}O{-}HC{=}CH{-}\dot{O} \longrightarrow HC(=O){-}O{-}C(=O)H \longrightarrow CO + HCOOH \tag{38}$$

The diradical intermediate has a double bond, and thus is energetically unable to split into the fragments obtained in the Criegee mechanism for alkene reaction. With the higher analogs, the energetic anhydride produced decomposes via

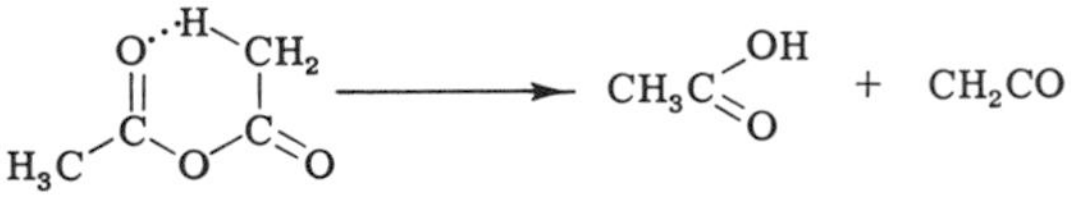

$$H_3C{-}C(=O\cdots H{-}CH_2{-})\,O{-}C{=}O \longrightarrow CH_3C(=O)OH + CH_2CO \qquad (39)$$

Reactions with Other Molecules

The reactions of O_3 with alkanes are very slow and not important in atmospheric processes. The products of the reactions are alcohols and O_2. Presumably the O_3 adds across C—H bonds, since the alcohols formed in solution studies show considerable retention of configuration (Hellman and Hamilton, 1974):

$$R{-}H + O_3 \longrightarrow \begin{matrix} R{-}{-}{-}{-}H \\ | \quad\quad | \\ O\diagdown_{O}\diagup O \end{matrix} \longrightarrow \begin{matrix} R \quad\quad H \\ | \;\; \diagup \;\; | \\ O\diagdown_{O}\diagup O \end{matrix} \longrightarrow ROH + O_2 \qquad (40)$$

The reactions of O_3 with aromatic rings produce π-complexes (Bailey *et al.*, 1971, 1974; Hull *et al.*, 1973) but this is a completely reversible process, and no net reaction has ever been measured. Bufalini and Altshuller (1965) have shown that reactant removal is immeasurably slow under atmospheric conditions.

The reaction of O_3 with aldehydes is apparently not rapid in the vapor phase, although the reaction has been studied in solution (White and Bailey, 1965; Erickson *et al.*, 1966; and references therein). Recent results in our laboratory indicate that the homogeneous gas phase reaction of O_3 with CH_2O has a rate coefficient $\leq 2.4 \times 10^{-24}$ cm^3/sec at 25°C (Braslavsky and Heicklen, 1975). However, Finlayson *et al.* (1972a) have reported the chemiluminescence of the HO Meinel bands in the O_3–CH_3CHO reaction at 3 Torr total pressure at room temperature, but found this chemiluminescence absent with other aldehydes. In the CH_3CHO system, a weaker visible emission was tentatively identified as CH_2O fluorescence.

The SO_2 luminescence has been seen in the reaction of O_3 with H_2S, CH_3SH, and $(CH_3)_2S$ at 0.3–1 Torr total pressure at room temperature (Akimoto *et al.*, 1971). In fact, the H_2S–O_3 reaction has been studied by Cadle and Ledford (1966) who found the stoichiometry to be

$$H_2S + O_3 \rightarrow H_2O + SO_2 \qquad (41)$$

However, the reaction mechanism was quite complex and proceeded in part heterogeneously.

The CS_2–O_3 and NH_3–O_3 reactions were studied by Olszyna and Heicklen (1970, 1972). These reactions were complex and proceeded in part heterogeneously. They are probably too slow to be of major importance under atmospheric conditions. The CS_2–O_3 reaction gave OCS, CO, CO_2, and SO_2, and was envisioned to proceed through the oxidation of CS:

$$CS + O_3 \longrightarrow \text{[C–S ring with O–O–O]} \longrightarrow OCS + O_2 \quad (42a)$$

$$\longrightarrow CO_2 + SO_2 \quad (42b)$$

$$\longrightarrow CO + SO_2 \quad (42c)$$

The NH_3–O_3 reaction proceeds by the overall stoichiometry

$$2NH_3 + 4O_3 \rightarrow 4O_2 + H_2O + NH_4NO_3 \quad (43)$$

although minor amounts of N_2 and N_2O were also produced. The initial step in the reaction is presumably the formation of the charge transfer complex, which has been trapped at low temperatures in condensed phase, along with other intermediates (Huston *et al.*, 1974). The sequence of events envisioned is

$$NH_3 + O_3 \rightarrow NH_3^+O_3^- \rightarrow NH_2 + HO + O_2 \quad (44)$$

$$HO + NH_3 \rightarrow H_2O + NH_2 \quad (45)$$

$$NH_2 + O_3 \rightarrow NH_2O + O_2 \quad (46)$$

$$NH_2O + O_3 \rightarrow HO + O_2 + HNO \quad (47)$$

$$HNO + 2O_3 \rightarrow HNO_3 + 2O_2 \quad (48)$$

$$HNO_3 + NH_3 \rightarrow NH_4NO_3 \quad (49)$$

Perhaps the most interesting feature of the CS_2–O_3 and NH_3–O_3 reactions was the unusual rate laws that were found. In the former system the rate was second order in $[O_3]$, whereas in the latter system it was first order in $[O_3]$. However, in both systems the apparent rate coefficients depended on the initial O_3 pressure $[O_3]_0$.

This peculiar rate law dependence was subsequently seen in the allene–ozone system (Toby and Toby, 1974), where the rate law was

$$-d[O_3]/dt = k[O_3][\text{allene}] + k'[O_3]^2[\text{allene}]/[O_3]_0 \quad (50)$$

The system was studied at 226–325°K in the gas phase. Initial O_3 pressures ranged from 0.01 to 0.7 Torr and initial allene pressures ranged from 0.05 to 6 Torr. The reaction was heterogeneous, at least in part, and the rate coefficients measured were

$$\log(k,\ M^{-1}\ \text{sec}^{-1}) = 6.0 \pm 0.7 - (5500 \pm 1000)/\Theta$$

$$\log(k',\ M^{-1}\ \text{sec}^{-1}) = 6.9 \pm 0.7 - (6200 \pm 800)/\Theta$$

where $\Theta \equiv 2.303RT$. At the higher initial O_3 pressures the most important product was O_2 followed by CO, H_2O, CO_2, and C_2H_4. The oxygen balance was good, but about one-half the carbon atoms were missing. Again the mechanism must be complex.

SINGLET O_2

As we have seen, the reactions of singlet O_2 are not as important as the reactions of $O(^3P)$ or O_3 in the chemistry of polluted atmospheres. Their interest lies in the fact that unique products are formed, some of which may have adverse biological effects. Singlet O_2 generally refers to the two lowest electronically excited states of O_2, i.e., the $^1\Delta_g$ and $^1\Sigma_g^+$ states. The former is present at considerably higher concentrations than the latter and is thus the more important state, even though the latter state is of higher energy and more reactive. Often the chemical studies have been done with mixtures of the two states, but in some cases the individual reactions have been separated. The results that might be pertinent to polluted atmospheres are discussed below.

In the vapor phase $O_2(^1\Delta_g)$ is known to react with olefins to give allylic hydroperoxides:

$$O_2(^1\Delta_g) + \;>C=C(-CH<)< \;\longrightarrow\; \text{[cyclic O–O···H–C, C=C, C transition state]} \;\longrightarrow\; >C(OOH)-C<=C< \qquad (51a)$$

The allylic hydroperoxide has been isolated in the gas phase reaction of $O_2(^1\Delta_g)$ with c-C_6H_{10} and 1,2-dimethylcyclohexene. With tetramethylethylene, the allylic hydroperoxide is the sole product of the reaction in both solution (Foote and Wexler, 1964; Winer and Bayes, 1966) and in the gas phase (Broadbent *et al.*, 1968; Gleason *et al.*, 1970). The reaction of $O_2(^1\Delta_g)$ with C_2H_4 takes place, but the product has not been isolated.

Besides forming allylic hydroperoxides, $O_2(^1\Delta_g)$ can add to olefins to form dioxetanes (Kopecky *et al.*, 1969; Kopecky and Mumford, 1969):

$$>C=C< + O_2(^1\Delta_g) \longrightarrow \begin{array}{c} O\!-\!O \\ |\quad\;\; | \\ -C\!-\!C- \\ |\quad\;\; | \end{array} \qquad (51b)$$

The dioxetane initially formed may be sufficiently "hot" to undergo cleavage to produce carbonyl compounds before being stabilized, and this would be the expected fate in the gas phase.

The reaction of $O_2(^1\Delta_g)$ with 2,5-dimethylfuran produces the ozonide (Broadbent *et al.*, 1968; Gleason *et al.*, 1970; Pitts, 1970)

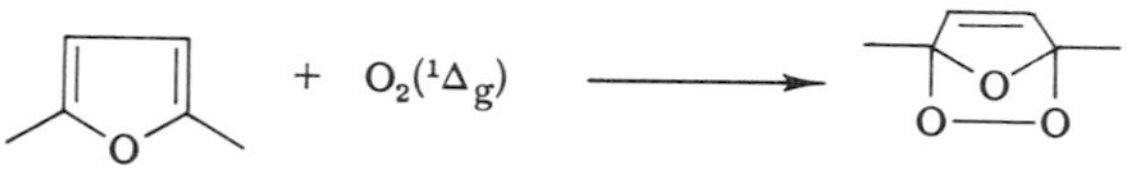

$+ \; O_2(^1\Delta_g) \longrightarrow$ (52)

This reaction was first shown to occur in solution (Foote and Wexler, 1964), but the product molecule is unstable in solution, and in the presence of CH_3OH solvent, adds the solvent to give

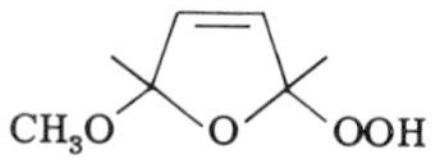

CH_3O O OOH

Other reactions of $O_2(^1\Delta_g)$ with olefins have been observed in solution. From the point of view of atmospheric chemistry, perhaps the most interesting are

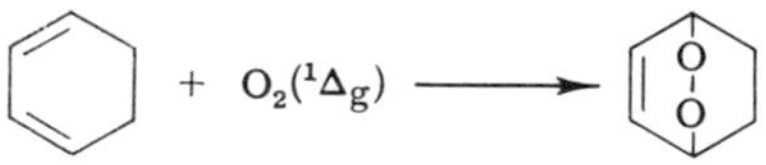

$+ \; O_2(^1\Delta_g) \longrightarrow$ (Foote and Wexler, 1964) (53)

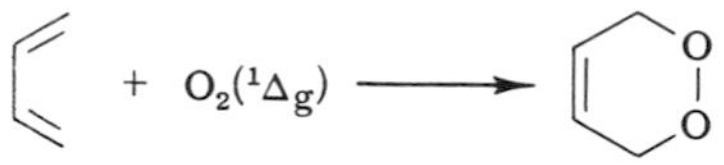

$+ \; O_2(^1\Delta_g) \longrightarrow$ (McKeown and Waters, 1966) (54)

With aromatic compounds the following interesting reactions in solution have been reported:

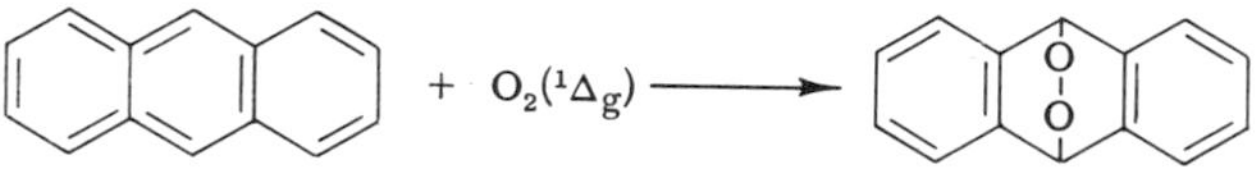

$+ \; O_2(^1\Delta_g) \longrightarrow$

(Corey and Taylor, 1964)
(McKeown and Waters, 1966) (55)

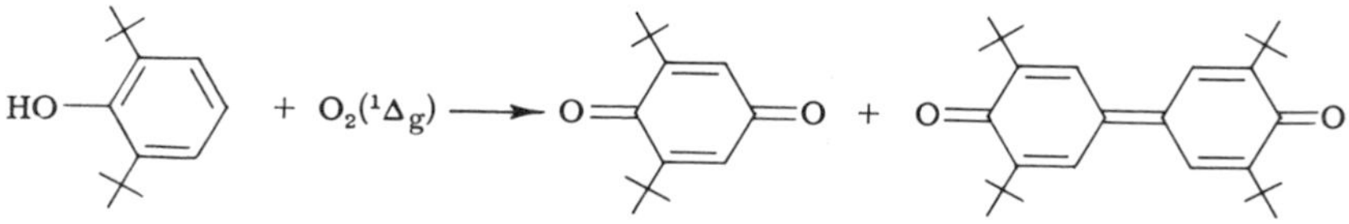

HO $+ \; O_2(^1\Delta_g) \longrightarrow$ O O + O O

(Matsura *et al.*, 1969) (56)

With alkynes there was one report of a gas phase reaction with 3-hexyne (Herron and Huie, 1970). However no products were identified. There are no reports of reactions of singlet O_2 with saturated hydrocarbons or terpenes.

Singlet O_2 is known to dissociate O_3 (see Chapter II):

$$O_2(^1\Delta_g \text{ or } ^1\Sigma_g^+) + O_3 \rightarrow 2O_2 + O(^3P) \quad (57)$$

It has also been shown that the reaction with methyl sulfide, ethyl sulfide, and methyl disulfide yields the sulfoxides (Murray, 1972)

$$2RSR + O_2(\text{singlet}) \rightarrow 2R\overset{\overset{\displaystyle O}{\|}}{S}R \quad (58)$$

$$2RS_2R + O_2(\text{singlet}) \rightarrow 2R\overset{\overset{\displaystyle O}{\|}}{S}SR \quad (59)$$

REFERENCES

Abe, K., Myers, F., McCubbin, T. K., and Polo, S. R. (1971). *J. Mol. Spectros.* **38,** 552, "Resonance Fluorescence Spectrum of Nitrogen Dioxide."

Ackerman, R. A., Rosenthal, I., and Pitts, J. N. Jr. (1971). *J. Chem. Phys.* **54,** 4960, "Singlet Oxygen in the Environmental Sciences. X. Absolute Rates of Deactivation of $O_2(^1\Delta_g)$ in the Gas Phase by Sulfur Compounds."

Akimoto, H., Finlayson, B. J., and Pitts, J. N. Jr. (1971). *Chem. Phys. Lett.* **21,** 199, "Chemiluminescent Reactions of Ozone with Hydrogen Sulphide, Methyl Mercaptan, Dimethyl Sulphide, and Sulfur Monoxide."

Bailey, P. S., Ward, J. W., and Hornish, R. E. (1971). *J. Amer. Chem. Soc.* **93,** 3552, "Complexes of Ozone with Carbon π Systems."

Bailey, P. S., Ward, J. W., Carter, T. P. Jr., Nieh, E., Fischer, C. M., and Khashab, A.-I. Y. (1974). *J. Amer. Chem. Soc.* **96,** 6136, "Studies Concerning Complexes of Ozone with Carbon π Systems."

Braslavsky, S., and Heicklen, J. (1973). *J. Photochem.* **1,** 203, "Quenching of the Fluorescence of NO_2."

Braslavsky, S., and Heicklen, J. (1975). Unpublished results, "The Gas Phase Reaction of O_3 with H_2CO."

Broadbent, A. D., Gleason, W. S., Pitts, J. N. Jr., and Whittle, E. (1968b). *Chem. Commun.* 1315, "The Reactions between $O_2(^1\Delta_g)$ and Tetramethylethylene and 2,5-Dimethylfuran in the Gas Phase."

Bufalini, J. J., and Altshuller, A. P. (1965). *Can. J. Chem.* **43,** 2243, "Kinetics of Vapor-Phase Hydrocarbon–Ozone Reactions."

Cadle, R. D., and Ledford, M. (1966). *Int. J. Air Water Pollut.* **10,** 25, "The Reaction of Ozone with Hydrogen Sulfide."

Corey, E. J., and Taylor, W. C. (1964). *J. Amer. Chem. Soc.* **86,** 3881, "A Study of the Peroxidation of Organic Compounds by Externally Generated Singlet Oxygen Molecules."

DeMore, W. B. (1971). *Int. J. Chem. Kinet.* **3,** 161, "Rates and Mechanism of Alkyne Ozonation."

Douglas, A. E. (1966). *J. Chem. Phys.* **45,** 1007, "Anomalously Long Radiative Lifetimes of Molecular Excited States."

Douglas, A. E., and Huber, K. P. (1965). *Can. J. Phys.* **43,** 74, "The Absorption Spectrum of NO_2 in the 3700–4600 Å Region."

Durham, L. J., and Greenwood, F. L. (1968a). *Chem. Commun.* 24, "The *cis*-Molozonide."

Durham, L. J., and Greenwood, F. L. (1968b). *J. Org. Chem.* **33,** 1629, "Ozonolysis. X. The Molozonide as an Intermediate in the Ozonolysis of *cis*- and *trans*-Alkenes."

Erickson, R. E., Bakalik, D., Richards, C., Scanlon, M., and Huddleston, G. (1966). *J. Org. Chem.* **31,** 461, "Mechanism of Ozonation Reactions. II. Aldehydes."

Filseth, S. V., Zia, A., and Welge, K. H. (1970). *J. Chem. Phys.* **52,** 5502, "Flash Photolytic Production, Reactive Lifetime, and Collisional Quenching of $O_2(b^1\Sigma_g^+, v' = 0)$."

Fink, W. H. (1971). *J. Chem. Phys.* **54,** 2911, "Calculations of the Ground and Excited State Energies of NO_2."

Finlayson, B. J., Gaffney, J. S., and Pitts, J. N. Jr. (1972a). *Chem. Phys. Lett.* **17,** 22, "The Ozone-Induced Chemiluminescent Oxidation of Acetaldehyde."

Finlayson, B. J., Pitts, J. N. Jr., and Akimoto, H. (1972b). *Chem. Phys. Lett.* **12,** 495, "Production of Vibrationally Excited OH in Chemiluminescent Ozone–Olefin Reactions."

Finlayson, B. J., Pitts, J. N. Jr., and Atkinson, R. (1974). *J. Amer. Chem. Soc.* **96,** 5356, "Low-Pressure Gas-Phase Ozone–Olefin Reactions. Chemiluminescence, Kinetics and Mechanisms."

Foote, C. S., and Wexler, S. (1964). *J. Amer. Chem. Soc.* **86,** 3879, "Olefin Oxidations with Excited Singlet Molecular Oxygen."

Frankiewicz, T. C., and Berry, R. S. (1973). *J. Chem. Phys.* **58,** 1787, "Production of Metastable Singlet O_2 Photosensitized by NO_2."

Furukawa, K., and Ogryzlo, E. A. (1973). *J. Photochem.* **1,** 163, "A Redetermination of the Rate Constants for the Quenching of Gaseous $O_2(^1\Delta_g)$ by Aliphatic Amines."

Gangi, R. A., and Burnelle, L. (1971a). *J. Chem. Phys.* **55,** 843, "Electronic Structure and Electronic Spectrum of Nitrogen Dioxide. II. Configuration Interaction and Oscillator Strengths."

Gangi, R. A., and Burnelle, L. (1971b). *J. Chem. Phys.* **55,** 851, "Electronic Structure and Electronic Spectrum of Nitrogen Dioxide. III. Spectral Interpretation."

Garvin, D., and Hampson, R. F. (1974). Nat. Bur. of Std. Rep. NBSIR 74-430, "Chemical Kinetics Data Survey. VII. Tables of Rate and Photochemical Data for Modelling of the Stratosphere (Revised)."

Gillies, C. W., and Kuczkowski, R. L. (1972a). *J. Amer. Chem. Soc.* **94,** 6337, "Mechanism of Ozonolysis. Microwave Spectrum, Structure, and Dipole Moment of Ethylene Ozonide."

Gillies, C. W., and Kuczkowski, R. L. (1972b). *J. Amer. Chem. Soc.* **94,** 7609, "Oxygen-18 Formaldehyde Insertion in the Ozonolysis of Ethylene and the Microwave Spectrum of Oxygen-18 Ethylene Ozonide."

Gillies, C. W., Lattimer, R. P., and Kuczkowski, R. L. (1974). *J. Amer. Chem. Soc.* **96,** 1536, "Microwave and Mass Spectral Studies of the Ozonolysis of Ethylene, Propylene, and *cis*- and *trans*-2-Butene with Added Oxygen-18 Formaldehyde and Acetaldehyde."

Gleason, W. S., Broadbent, A. D., Whittle, E., and Pitts, J. N. Jr. (1970). *J. Amer. Chem. Soc.* **92,** 2068, "Singlet Oxygen in the Environmental Sciences. IV. Kinetics

of the Reactions of Oxygen ($^1\Delta_g$) with Tetramethylethylene and 2,5-Dimethylfuran in the Gas Phase."
Greenwood, F. L. (1965). *J. Org. Chem.* **30,** 3108, "Ozonolysis VII. Factors Controlling the Stability of *cis*- and *trans*-Molozonides of Straight-Chain Alkenes. Role of Nucleophilic Solvents in Alkene–Ozone Reactions."
Greenwood, F. L., and Durham, L. J. (1969). *J. Org. Chem.* **34,** 3363, "Ozonolysis. XI. Kinetic Studies on the Conversion of the Molozonide into Ozonolysis Products."
Hay, P. J. (1973). *J. Chem. Phys.* **58,** 4706, "Theoretical Investigations of the Low-Lying Doublet States of NO_2."
Heicklen, J. (1969). *Advan. Photochem.* **7,** 57, "Gas Phase Oxidation of Perhalocarbons."
Hellman, T. M., and Hamilton, G. A. (1974). *J. Amer. Chem. Soc.* **96,** 1530, "On the Mechanism of Alkane Oxidation by Ozone in the Presence and Absence of $FeCl_3$."
Herron, J. T., and Huie, R. E. (1970). *Ann. N.Y. Acad. Sci.* **171,** 229, "Mass Spectrometer Studies on the Reactions of Singlet Oxygen in the Gas Phase."
Herron, J. T., and Huie, R. E. (1974). *J. Phys. Chem.* **78,** 2085, "Rate Constants for the Reactions of Ozone with Ethene and Propene, from 235.0 to 362.0°K."
Huie, R. E., and Herron, J. T. (1973). *Int. J. Chem. Kinet.* **5,** 197, "Kinetics of the Reactions of Singlet Molecular Oxygen ($O_2\ ^1\Delta_g$) with Organic Compounds in the Gas Phase."
Hull, L. A., Hisatsune, I. C., and Heicklen, J. (1972). *J. Amer. Chem. Soc.* **94,** 4856, "Low Temperature Infrared Studies of Simple Alkene–Ozone Reactions."
Huston, T., Hisatsune, I. C., and Heicklen, J. (1974). Unpublished work at Penn State Univ.
Jones, I. T. N., and Bayes, K. D. (1973a). *J. Chem. Phys.* **59,** 3119, "Formation of $O_2(a^1\Delta_g)$ by Electronic Energy Transfer in Mixtures of NO_2 and O_2."
Jones, I. T. N., and Bayes, K. D. (1973b). *J. Chem. Phys.* **59,** 4836, "Photolysis of Nitrogen Dioxide."
Keyser, L. F., Levine, S. Z., and Kaufman, F. (1971). *J. Chem. Phys.* **54,** 355, "Kinetics and Mechanism of NO_2 Fluorescence."
Kolopajlo, L., Hisatsune, I. C., and Heicklen, J. (1975). To be published.
Kopecky, K. R., and Mumford, C. (1969). *Can. J. Chem.* **47,** 709, "Luminescence in the Thermal Decomposition of 3,3,4-Trimethyl-1,2-dioxetane."
Kopecky, K. R., Van de Sande, J. H., and Mumford, C. (1968). *Can. J. Chem.* **46,** 25, "Preparation and Base-Catalyzed Reactions of Some β-Halohydroperoxides."
Kummler, R. H., Bortner, M. H., and Baurer, T. (1969). *Environ. Sci. Tech.* **3,** 248, "The Hartley Photolysis of Ozone as a Source of Singlet Oxygen in Polluted Atmospheres."
Lee, E. K. C., and Uselman, W. M. (1972). *Faraday Discuss. Chem. Soc.* No. 53, 125, "Molecular Predissociation of Nitrogen Dioxide Studied by Fluorescence Excitation Spectroscopy."
Matsuura, T., Yoshimura, N., Nishinaga, A., and Saito, I. (1969). *Tetrahedron Lett.* 1669, "Photoinduced Reactions. XXX. Hydrogen Abstraction from a Phenol by Singlet Oxygen."
McKeown, E., and Waters, W. A. (1966). *J. Chem. Soc. B* 1040, "The Oxidation of Organic Compounds by "Singlet" Oxygen."
Murray, R. (1972). Private communication
Myers, G. H., Silver, D. M., and Kaufman, F. (1966). *J. Chem. Phys.* **44,** 718, "Quenching of NO_2 Fluorescence."
Neuberger, D., and Duncan, A. B. F. (1954). *J. Chem. Phys.* **22,** 1693, "Fluorescence of Nitrogen Dioxide."

Olszyna, K. J., and Heicklen, J. (1970). *J. Phys. Chem.* **74**, 4188, "The Reaction of Ozone with Carbon Disulfide."

Olszyna, K. J., and Heicklen, J. (1972). *Advan. Chem. Ser. No. 113*, Amer. Chem. Soc., "Photochemical Smog and Ozone Reactions," p. 191 "The Reaction of Ozone with Ammonia."

O'Neal, H. E., and Blumstein, C. (1973). *Int. J. Chem. Kinet.* **5**, 397, "A New Mechanism for Gas Phase Ozone–Olefin Reactions."

Penzhorn, R.-D., Güsten, H., Schurath, U., and Becker, K. H. (1974). *Environ. Sci. Tech.* **8**, 907, "Quenching of Singlet Molecular Oxygen by Some Atmospheric Pollutants."

Pitts, J. N. Jr. (1970). *Ann. N.Y. Acad. Sci.* **171**, 239, "Singlet Molecular Oxygen and the Photochemistry of Urban Atmospheres."

Pitts, J. N. Jr., Sharp, J. H., and Chan, S. I. (1964). *J. Chem. Phys.* **40**, 3655, "Effects of Wavelength and Temperature on Primary Processes in the Photolysis of Nitrogen Dioxide and a Spectroscopic-Photochemical Determination of the Dissociation Energy."

Sackett, P. B., and Yardley, J. T. (1972). *J. Chem. Phys.* **57**, 152, "Dynamics of NO_2 Electronic States Excited by a Tunable Dye Laser."

Sanhueza, E., Hisatsune, I. C., and Heicklen, J. (1975). Center for Air Environment Studies Publ. No. 387-75, Penn State Univ., "Oxidation of Haloethylenes."

Schwartz, S. E., and Johnston, H. S. (1969). *J. Chem. Phys.* **51**, 1286, "Kinetics of Nitrogen Dioxide Fluorescence."

Solarz, R., and Levy, D. H. (1974). *J. Chem. Phys.* **60**, 842, "Lifetime of Electronically Excited NO_2: Evidence for a Short-lived State."

Stevens, C. G., Swagel, M. W., Wallace, R., and Zare, R. N. (1973). *Chem. Phys. Lett.* **18**, 465, "Analysis of Polyatomic Spectra Using Tunable Laser-Induced Fluorescence: Applications to the NO_2 Visible Band System."

Toby, F. S., and Toby, S. (1974). *Int. J. Chem. Kinet.* **6**, 417, "Reaction between Ozone and Allene in the Gas Phase."

Vrbaski, T., and Cvetanović, R. J. (1960a). *Can. J. Chem.* **38**, 1053, "Relative Rates of Reaction of Ozone with Olefins in the Vapor Phase."

Vrbaski, T., and Cvetanović, R. J. (1960b). *Can. J. Chem.* **38**, 1063, "A Study of the Products of the Reactions of Ozone with Olefins in the Vapor Phase as Determined by Gas–Liquid Chromatography."

Wei, Y. K., and Cvetanović, R. J. (1963). *Can. J. Chem.* **41**, 913, "A Study of the Vapor Phase Reaction of Ozone with Olefins in the Presence and Absence of Molecular Oxygen."

White, H. M., and Bailey, P. S. (1965). *J. Org. Chem.* **30**, 3037, "Ozonation of Aromatic Aldehydes."

Williamson, D. G., and Cvetanović, R. J. (1968). *J. Amer. Chem. Soc.* **90**, 4248, "Rates of Reactions of Ozone with Chlorinated and Conjugated Olefins."

Winer, A. M., and Bayes, K. D. (1966). *J. Phys. Chem.* **70**, 302, "The Decay of $O_2(a^1\Delta)$ in Flow Experiments."

Chapter VIII

SO_2 CHEMISTRY

SO_2 is oxidized in the atmosphere to give H_2SO_4, $(NH_4)_2SO_4$, and organic sulfur compounds. The reactions involved are of three main types: (1) photolysis of SO_2, (2) free radical reactions with SO_2, and (3) reaction on surfaces or dissolution of SO_2 into water droplets followed by reaction in the liquid phase.

PHOTOLYSIS OF SO_2

Primary Process

The absorption spectrum of SO_2 is shown in Fig. VIII-1. Three absorption bands are shown. The first is a very weak absorption and extends from 4000 to 3400 Å. Its most intense band has a peak extinction coefficient of 0.095 M^{-1} cm^{-1} (to base 10) at 3700 Å. This absorption produces the triplet state $^3B_1$ and phosphorescence from this state has been observed. In fact, much of its photophysics has been worked out by examining the phosphorescence.

The decay rate of the $^3B_1$ phosphorescence has been measured a number of times. The most recent and accurate measurement was made at 30°C by Strickler *et al.* (1974), who found the rate to be pressure dependent. The quenching coefficient by ground state SO_2 was $(4.52 \pm 0.18) \times 10^8$ M^{-1} sec^{-1}. The first-order decay coefficient, obtained by extrapolation to zero pressure, was $(1.12 \pm 0.20) \times 10^3$ sec^{-1}, very much larger than the radiative decay coefficient of 79 ± 5 sec^{-1}. These results agreed with several previous determinations and indicated that only 7% of the first-order decay was accompanied by emission. This is an unexpectedly small percentage for a small molecule like SO_2 and indicates considerable complexity in the excited state physics.

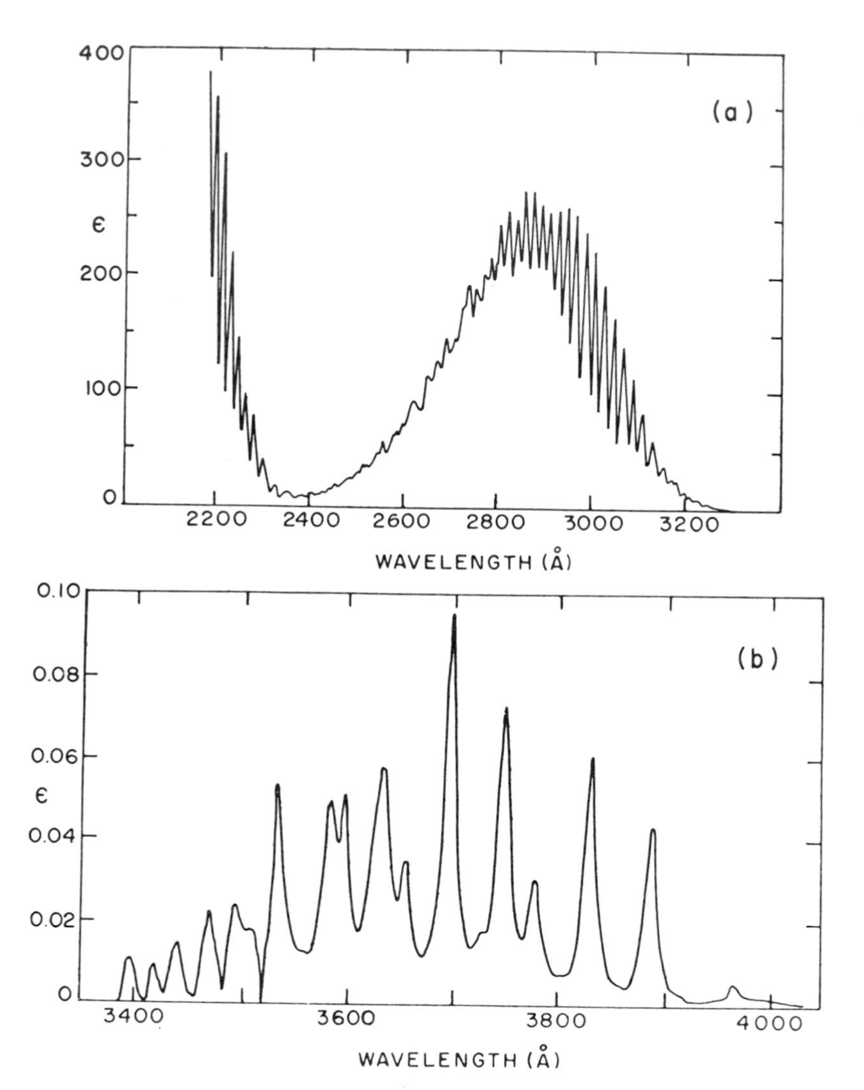

Fig. VIII-1. Absorption spectrum of SO_2. Extinction coefficients (to base 10) in units of M^{-1} cm^{-1}. From Sidebottom *et al.* (1972a) with permission of the American Chemical Society.

The third absorption starts at 2400 Å and becomes very intense as the wavelength drops. Presumably this absorption produces the $^1B_2$ state. This wavelength region is not important in the lower atmosphere, but it should be pointed out that there is a low-lying triplet state $^3B_2$ associated with B_2 symmetry that may be accessible at wavelengths >2900 Å.

The second absorption band starts at 3300 and extends to 2400 Å. There is noticeable vibrational structure above 2800 Å, but this structure is not apparent at lower wavelengths, probably because of pronounced overlap among the bands. The peak extinction coefficient is about 300 M^{-1} cm^{-1} (to base 10) at 2900 Å. The absorption into this band produces a singlet state. The structure of the absorption band is very complex for such a simple molecule, and two electronically excited states are energetically accessible, $^1A_2$ and $^1B_1$. Furthermore, Greenough and Duncan (1961) measured the emission lifetime and obtained a value of 42 μsec, a factor of 70 larger than the lifetime of 0.6 μsec computed from the integrated absorption coefficient (Strickler and Berg, 1962). In spite of these problems the absorption and fluorescence were interpreted with one excited singlet electronic state. Douglas (1966) discussed several possible mechanisms to account for the discrepancy in lifetime and ascribed the main effect to mixing of levels of the excited singlet state with some lower state.

Evidence that one excited singlet state could not be correct came from the work of Cehelnik *et al.* (1971), who showed that photochemical reactions could occur even if the light-emitting states were quenched. In spite of this, the one excited singlet state model lingered on. Brand and Nanes (1973) interpreted the absorption spectrum at 2900–3400 Å in terms of one excited singlet state, as did Dixon and Halle (1973) in their studies at 3150–3400 Å. The latter authors proposed that the absorption corresponded to a vibronically induced B_1–A_1 transitions of a $^1A_2 \leftarrow {}^1A_1$ forbidden electronic transition ($^1A_1$ is the ground state). The fluorescence spectrum was also interpreted in terms of one emitting state by Kent *et al.* (1974), but they designated the emitting state as $^1B_1$.

The first spectroscopic evidence (aside from the lifetime discrepancy) that the explanation was not so simple came from the work of Sidebottom *et al.* (1972b), who found nonexponential decay in the fluorescence excited by radiation <2900 Å. They considered that different vibrational levels of the excited singlet state had different decay times. However, it was Stockburger *et al.* (1973) who provided conclusive spectroscopic proof that one excited singlet state could not explain the pressure quenching of fluorescence; the principal state formed in absorption could not be the principal fluorescing state—the latter state is produced in a first-order process by internal conversion.

Brus and McDonald (1973, 1974) quickly established that both excited singlet states were emitting. The major emitting state $^1B_1$ had a variable lifetime of 80 to 600 μsec in conformance with the findings of Sidebottom *et al.* (1972b) and a fluorescence efficiency of near unity, i.e., it did not participate in radiationless transitions. The minor emitting state $^1A_2$ had a lifetime of ~50 μsec. Calvert (1973) then showed that all the earlier spectroscopic data were consistent with this two excited singlet state model, and subsequent work of Caton and Gangadharan (1974), who examined $^3B_1$ phosphorescence, was also consistent with this model.

Possibly the $^1A_2$ state is the principal state formed in absorption, as suggested by Dixon and Halle (1973). However, the lifetime of ~50 μsec is still much too large to be consistent with that computed from the integrated absorption coefficient. More recent data by Levine and Kaufman (1974), who used phase shift measurements of lifetimes, indicate lifetimes of 10 and 100 μsec for the two states (absorption at 2800–3200 Å). The latter value agrees with the Brus and McDonald data, but the former value is smaller than the shorter lifetime reported by Brus and McDonald and is approaching the lifetime computed from the integrated absorption coefficient. Presumably the remaining factor of 14 discrepancy is accounted for by absorption into another singlet state, designated SO_2^*.

In summary, we can conclude that in the second absorption band SO_2^* is produced $>90\%$ of the time, with $^1A_2$ production accounting for the rest of the absorption ($<10\%$). The fluorescence is principally from $^1B_1$, which is produced by interconversion from SO_2^*, but a minor contribution comes from $^1A_2$. There is not sufficient energy for dissociation of SO_2 with incident radiation >2180 Å. Thus the chemistry of the second absorption band is that of excited molecules.

The photoabsorption coefficients (i.e., the rate coefficients for radiation absorption) have been computed for atmospheric conditions by Sidebottom *et al.* (1972a) and they are listed in Table VIII-1. It can be seen that the absorption to the singlet states is 55–330 times as important as absorption to the triplet state.

TABLE VIII-1

Photoabsorption Coefficients for SO_2[a]

Solar zenith angle (deg)	$J\{2900\text{–}3000$ Å$\}$ ($\times 10^{-4}$ sec^{-1})	$J\{3400\text{–}4000$ Å$\}$ ($\times 10^{-6}$ sec^{-1})
0	3.16	0.95
20	2.86	0.91
40	1.99	0.80
60	0.73	0.57
80	0.093	0.17

[a] From Sidebottom *et al.* (1972a).

When the absorption is into the second band, not only are the singlet states produced, but the phosphorescence of $^3B_1$ is also observed. Thus in addition to the excited singlet states, at least one triplet state is also present to complicate the primary photophysics. The $^1B_1$ and $^3B_1$ emitting states are readily quenched by many gases including N_2, O_2, and H_2O. However, studies of photochemical products show little effect of quenching by foreign gases for absorption in the 3000–3300 Å region. A number of studies in the author's laboratory (Cehelnik *et al.*, 1971, 1973; Braslavsky *et al.*, 1972; Luria and Heicklen, 1974) led to the conclusion that there must be at least two additional excited electronic states (the Cehelnik–Spicer states) that do not emit radiation and are not quenched by many of the gases that quench the $^1B_1$ and $^3B_1$ emitting states (e.g., SO_2, N_2, CO), which are mainly responsible for the chemistry.

These two Cehelnik–Spicer states were the "nonemitting" singlet designated SO_2^*, and a nonemitting triplet designated SO_2^{**}. The singlet state is presumably the principal state formed in absorption. Further experiments by Fatta *et al.* (1974) led to the suggestion that the Cehelnik–Spicer triplet state might really be two states. This evidence was obtained by examining the SO_2 photosensitized emission of biacetyl.

The photosensitized emission followed kinetics suggesting two precursor SO_2 triplet states, but foreign gases such as N_2 and CO did not quench the emission, ruling out the $^3B_1$ emitting triplet state. Furthermore one precursor state was readily quenched by H_2O vapor while the other was not. Each precursor state accounted for about half of the sensitized emission. Fatta *et al.* thus added to the Cehelnik–Spicer triplet state another triplet designated $SO_2^{\ddagger}$, which is the one not quenched by H_2O vapor. Presumably these two nonemitting triplet states (SO_2^{**} and $SO_2^{\ddagger}$) are the $^3B_2$ and $^3A_2$ states, although it is not clear which is which. In fact, it has not been shown that the $^3B_2$ state is sufficiently low in energy to be accessible at wavelengths of 3000 to 3300 Å. Further work will be required to see if indeed two nonemitting triplet states exist in this region.

Support for at least one other low-lying triplet state in addition to the emitting one comes from the work of Wampler *et al.* (1973). They produced the $^3B_1$ state both by direct absorption at 3829 Å and indirectly by absorption into the singlet manifold. They examined the temperature dependence of the phosphorescence quenching for 21 gases. The results were independent of the method of excitation, and the Arrhenius parameters for quenching are listed in Table VIII-2. The results fall into three groups:

1. For the atmospheric gases O_2, Ar, CO_2, N_2, CO, and CH_4, the activation energies were identical within the experimental uncertainty, i.e., 2.8 ± 0.3 kcal/mole. This value corresponds to the vibrational level

TABLE VIII-2

Arrhenius Parameters for $SO_2(^3B_1)$ *Quenching Rate Coefficients for Various Gases*[a]

Quenching gas	log A (M^{-1} sec^{-1})	E_a (kcal/mole)
SO_2	10.05 ± 0.15	2.0 ± 0.2
O_2	10.35 ± 0.35	3.2 ± 0.5
Ar	9.78 ± 0.37	2.8 ± 0.6
CO_2	10.26 ± 0.44	2.8 ± 0.4
N_2	9.98 ± 0.28	2.9 ± 0.4
CO	9.98 ± 0.26	2.7 ± 0.4
CH_4	9.93 ± 0.14	2.5 ± 0.6
C_2H_6	10.43 ± 0.18	3.1 ± 0.3
C_3H_8	10.31 ± 0.19	2.4 ± 0.3
n-C_4H_{10}	10.73 ± 0.15	2.4 ± 0.2
iso-C_4H_{10}	10.18 ± 0.20	1.7 ± 0.3
neo-C_5H_{12}	10.48 ± 0.06	2.7 ± 0.1
cyclo-C_6H_{12}	10.56 ± 0.18	1.6 ± 0.3
NO	11.96 ± 1.02	1.4 ± 0.7
C_2H_4	11.36 ± 0.83	1.1 ± 0.6
CH_2CHF	11.22 ± 0.42	1.7 ± 0.4
C_3H_6	11.65 ± 0.85	0.9 ± 0.6
cis-2-C_4H_8	11.66 ± 0.50	0.7 ± 0.4
cyclo-C_5H_8	11.44 ± 0.79	0.4 ± 0.5
C_6H_6	11.38 ± 1.03	0.5 ± 0.7
C_6F_6	10.37 ± 0.94	0.7 ± 0.7

[a] From Wampler *et al.* (1973).

of either the symmetric (2.6 kcal/mole) or the asymmetric (2.8–3.1 kcal/mole) stretching vibration in the $^3B_1$ state and suggested to Wampler *et al.* that quenching was from one of these levels to another low-lying triplet state that intersected one of these vibrational levels of $^3B_1$.

2. The activation energies for the paraffins were lower than for the gases above and decreased as the molecule complexity increased, suggesting chemical quenching via H atom abstraction.

3. The rate coefficients for the aromatic and olefin hydrocarbons and NO were near collision frequency and suggested addition of $SO_2(^3B_1)$ to the π-systems of these molecules.

The detailed primary process when SO_2 absorbs into the second band has not been elucidated yet. However, the effective quantum yields of production of the various excited states under atmospheric conditions can be estimated. The chemical results with C_2F_4 (Cehelnik *et al.*, 1971) and

C_2H_2 (Luria and Heicklen, 1974) indicate that triplet states are produced with a yield $\phi\{\text{triplet } SO_2\} = 0.04$ for incident radiation at 3130 Å. This value is consistent with the value of 0.095 obtained with 2875 Å incident radiation and 0.082 with 2650 Å incident radiation by SO_2 photosensitization of biacetyl phosphorescence obtained in Calvert's laboratory (Horowitz and Calvert, 1972a). Calvert and his co-workers attributed all the triplet SO_2 to $^3SO_2$. However, the results of Fatta *et al.* (1974) indicate that the triplet yield is about half SO_2^{**} and about half $SO_2^{\ddagger}$, with $^3B_1$ accounting for <15% of the total. Thus the values for $\phi\{\text{triplet } SO_2\}$ reported by Horowitz and Calvert may need modification.

The singlet state yields are harder to estimate. Fortunately, under atmospheric conditions, very little $^1B_1$ is produced and that is readily quenched. The SO_2^* yield is presumably near unity, but only a small fraction of this reacts (at least with CO). The quantum yields for excited state production are summarized in Table VIII-3.

Some of the excited states of SO_2 can be removed by collisional quenching with atmospheric gases. The important atmospheric gases are O_2, N_2 and, H_2O. The quenching constants for these gases, compared to the first-order removal rates of the respective states, are listed in Table VIII-2. The

TABLE VIII-3

Rates of Formation and Removal of Excited States of SO_2 *for Absorption at 2900–3300 Å under Atmospheric Conditions*

Excited state	ϕ	$k\{N_2\}$[a] (cm³/molecule)	$k\{O_2\}$[a] (cm³/molecule)	$k\{H_2O\}$[a] (cm³/molecule)	$J\{O_2\}$[b] (sec⁻¹)
$^1SO_2(^1B_1)$	~0.01	4.3×10^{-16}	5.1×10^{-16}	1.5×10^{-15}	$\sim 0.7 \times 10^{-6}$
$^3SO_2(^3B_1)$	<0.006	7.4×10^{-17}	1.0×10^{-16}	6.0×10^{-16}	$<0.45 \times 10^{-6}$ [c]
SO_2^*	~1.0	0	0	0	0
SO_2^{**} [d]	0.02	0	2.1×10^{-19} (both SO_2^{**} and $SO_2^{\ddagger}$)	$>3 \times 10^{-17}$	~0
$SO_2^{\ddagger}$ [d]	0.02	0		0	3.2×10^{-6}

[a] Relative to first-order removal rate coefficient.

[b] $J\{O_2\}$ is the effective photo rate coefficient for the removal of excited SO_2 in the atmosphere at ground level by O_2 for an overhead sun. With $[H_2O]/[O_2] = 0.10$ the actual rate of SO_2 disappearance could be twice this value, because a ground state SO_2 molecule might be removed with each excited SO_2 via reactions (1a) and (2).

[c] The minimum total value is 0.22×10^{-6} sec⁻¹ because of the absorption into the band at 3400–4000 Å. For the same reason the maximum total value could be as high as 0.67×10^{-6} sec⁻¹.

[d] These two states are probably $^3A_2$ and $^3B_2$, but it is not clear which is which.

relative quenching constants for $^1B_1$ and $^3B_1$ with several gases, compared to SO_2, have been measured in three laboratories (Mettee, Calvert, and Heicklen laboratories) with good consistency (Mettee, 1969; Sidebottom *et al.*, 1972a; Rao *et al.*, 1966b; Horowitz and Calvert, 1972b; Stockburger *et al.*, 1973). The self-quenching and first-order decay constants were measured by Rao *et al.* (1969a). Combination of these data gives the results listed in Table VIII-3. The quenching constants for the nonemitting states come from the work of Cehelnik *et al.* (1971) for O_2 quenching and Fatta *et al.* (1974) for H_2O quenching.

Photooxidation

In the presence of O_2, the photolysis of SO_2 is known to lead to SO_3 and, if H_2O vapor is also present, to H_2SO_4 aerosol. The photooxidation has been studied in several laboratories under atmospheric conditions, and three separate studies (Bufalini, 1971) are consistent with an effective rate coefficient for H_2SO_4 production of 0.28×10^{-6} sec^{-1} for an overhead sun independent of relative humidity above 10%.

The mechanism for the photooxidation is not known. The reactions most often postulated are

$$SO_2 \text{ (electronically excited)} + O_2 \rightarrow SO_4 \quad (1a)$$

$$\rightarrow SO_2 + O_2 \quad (1b)$$

$$SO_4 + SO_2 \rightarrow 2SO_3 \quad (2)$$

$$SO_3 + H_2O \rightarrow H_2SO_4 \quad (3)$$

However, there is no direct evidence for the SO_4 intermediate. In addition to the question of the existence of SO_4, it is also not known which electronic states are involved and what fraction of reaction proceeds by chemical quenching, reaction (1a)

In regard to the excited states involved, they must be the triplet states under atmospheric conditions. Presumably the states quenched by H_2O are not involved, since the photooxidation proceeds readily in the presence of H_2O vapor. The photochemical rate coefficients $J\{O_2\}$ for the quenching of these states by O_2 under atmospheric conditions for an overhead sun with $[H_2O]/[O_2] = 0.10$ have been computed. They are listed in Table VIII-3 for absorption into the singlet band, i.e., at wavelengths of 2900 to 3300 Å. For absorption at 3400–4000 Å into the $^3B_1$ band, $J\{O_2\} = 0.22 \times 10^{-6}$ sec^{-1}. The values of $J\{O_2\}$ for the $SO_2^{\ddagger}$ state are independent of the value of $[H_2O]/[O_2]$. There is a moderate dependence on $[H_2O]/[O_2]$ for $J\{O_2\}$ for the $^3B_1$ state. In the absence of water, the values will be about

15% higher, and at $[H_2O]/[O_2] = 0.20$, i.e., ~100% relative humidity, the values will be 15% lower.

If every removal of electronically excited SO_2 by O_2 led to reaction via reactions (1a), (2), and (3), then the rate coefficient for SO_2 removal would be 6.4×10^{-6} sec^{-1} for $SO_2^{\ddagger}$ and $(0.44–1.34) \times 10^{-6}$ sec^{-1} for $^3B_1$. These values are larger than the observed values of 0.28×10^{-6} sec^{-1}, so that reaction (1b) must be dominant for $SO_2^{\ddagger}$ and occurs to some extent for $^3B_1$.

The quantum yield of SO_2 disappearance for absorption at 2900–3400 Å has been obtained in many laboratory studies of the photooxidation with widely varying results ranging from 0.5×10^{-3} to 10×10^{-3} for atmospheric conditions (Allen *et al.*, 1972). A more recent and presumably more accurate measurement comes from the work of Cox (1972), who found 0.3×10^{-3}. Work in our laboratory (Smith *et al.*, 1975) is consistent with this value. From this value the photooxidation coefficient of 9.5×10^{-8} sec^{-1} is obtained for absorption <3400 Å for an overhead sun. Thus two-thirds of the oxidation must occur from absorption at 3400–4000 Å. Since $^3B_1$ is also present for incident radiation <3400 Å, the photooxidation in this region probably also occurs through this state. The $SO_2^{\ddagger}$ state accounts for little if any of the oxidation, and it is removed almost exclusively by physical quenching with O_2, reaction (1b).

The Calvert group (Sidebottom *et al.*, 1972a) has examined the oxidation of $^3B_1$ by studying the photooxidation with a laser light source at 3828.8 Å. They have found that the photooxidation does proceed through this state, but quantum yields were not obtained.

Another complication in the photooxidation is the possibility of SO_3 photolysis as pointed out by Bufalini (1971). There is a weak absorption of SO_3 extending to 3100 Å, and the following reaction is energetically possible:

$$SO_3 + h\nu \rightarrow SO_2 + O(^3P) \tag{4}$$

This reaction would have the effect of reducing the rate of photooxidation of SO_2.

Photolysis of SO_2–Hydrocarbon Mixtures

The photolysis of SO_2 in the presence of several light paraffins and olefins in the absence of O_2 was studied by Dainton and Ivin (1950a,b). The sole product was an aerosol, which was mainly the sulfinic acid (RSO_2H) with some disulfinic and disulfonic acids for the paraffins. The product of the olefin reaction was not well characterized but resembled a polysulfone. The quantum yields were independent of SO_2 pressure, but did depend on hydrocarbon pressure, a result confirmed by Timmons (1970). This behavior is that expected if the precursor electronically excited states of SO_2 are the nonluminescing ones, i.e., those not quenched by SO_2.

This is not to imply that hydrocarbons do not quench the luminescing states. In fact, the Calvert group found quenching with large rate coefficients (see Table VIII-2). Cox (1973) found that excited SO_2 could induce the geometrical isomerization of the 2-butenes. Primarily the isomerization was induced by triplet SO_2, but there was evidence for the participation of singlet SO_2. Cox interpreted the results on the basis that the principal state of SO_2 involved was $^3B_1$. On this basis he deduced the quenching coefficients to be 1.62×10^{11} and 1.42×10^{11} M^{-1} sec^{-1}, respectively, for *cis*- and *trans*-C_4H_8-2.

The simplest mechanism to sulfinic acid formation is direct insertion

$$R{-}H + SO_2 \longrightarrow \begin{array}{c} R{-}H \\ \vdots \quad \vdots \\ O{=}S{-}O \end{array} \longrightarrow R{-}\overset{\overset{O}{\|}}{S}{-}OH \tag{5}$$

and almost all the work has been interpreted in this way. Recently, Penzhorn *et al.* (1975) examined the photolysis of SO_2 with radiation >2900 Å in the presence of n-C_4H_{10} and other small paraffin hydrocarbons. In the butane system, they found as products the isomeric (n-butane and *sec*-butane) sulfinic and sulfonic acids and various isomeric butyl thiosulfonates. The complexity of products suggests a free-radical mechanism proceeding through H atom abstraction:

$$SO_2 \text{ (electronically excited)} + RH \rightarrow HSO_2 + R \tag{6}$$

With paraffins, work has been done in the presence of dry air (Johnston and der Jain, 1960; Kopcyznski and Altshuller, 1962). With high SO_2 and paraffin pressures (1–50 Torr), aerosol formation was readily apparent, but as the concentrations of SO_2 and paraffin dropped toward 0.05–0.10 Torr, the aerosol production became unobservable. These observations are consistent with expectations for the nonemitting triplet states where the quenching coefficients for the hydrocarbons are 10^2–10^3 times that for O_2. Thus above 1 Torr pressure, the states SO_2^{**} and $SO_2^{\ddagger}$ are quenched primarily by the hydrocarbon. Below this pressure, the situation shifts and quenching is primarily by O_2; the aerosol production is reduced. Thus at hydrocarbon concentrations of 1 ppm, the photo rate coefficient for aerosol production should be about 3×10^{-9} sec^{-1} for an overhead sun. This is too small to convert any significant amount of SO_2 to aerosol. However, with 1 ppm SO_2, about 250 particles/cm^3 of 0.1 μm diameter could be produced in one hour.

Luria *et al.* (1974a,b) have studied the reaction with C_2H_2 and allene. With C_2H_2 the products were CO and a solid of composition $C_3H_4S_2O_3$. They visualized the reaction as occurring via

$$SO_2\ (\text{excited}) + C_2H_2 \rightarrow CO + CH_2SO \quad (7)$$

$$CH_2SO + C_2H_2 + SO_2 \rightarrow C_3H_4S_2O_3 \quad (8)$$

With allene the gas phase products were CO and C_2H_4, the C_2H_4 being about one-fourth as important as the CO. An aerosol was also formed of composition C_5H_8SO. They envisioned the reaction as proceeding via

$$SO_2\ (\text{excited}) + C_3H_4 \rightarrow CO + C_2H_4 + SO \quad (9a)$$

$$\rightarrow CO + C_2H_4SO \quad (9b)$$

$$C_2H_4SO + C_3H_4 \rightarrow C_5H_8SO \quad (10)$$

They determined the electronically excited states responsible for the reaction. For C_2H_2 only the SO_2^{**} and $SO_2^{\ddagger}$ states are involved. Since the quantum yield of CO production was about 0.04 (Luria and Heicklen, 1974) the chemical quenching process accounts for all the reaction and physical quenching is unimportant. The allene reaction was more complex and there was evidence that all of the excited SO_2 states (except possibly $^1B_1$) could be involved.

The only SO_2–aromatic system in which chemical products were monitored was done by Braslavsky and Heicklen (1972). They found products in the SO_2–thiophene system when irradiation was into the band at 2900–3400 Å, but none for irradiation at 3660 Å, thus eliminating $^3B_1$ as the photochemically active state. Two sets of products were found. One set included H_2 (the major product), CH_2CO, and SCO, and could be attributed to sensitization by the two singlet states $^1B_1$ and SO_2^*. The other set of products are the same as those produced in the direct photolysis of thiophene, i.e., C_2H_2, CH_2CCH_2, CH_3CCH, CS_2, CH_2CHCCH, and polymer. These could be attributed to sensitization by the nonluminescing singlet and triplet states.

FREE RADICAL REACTIONS

Reaction with O(^{3}P)

The photolysis of SO_2–NO_2 mixtures in air was first carried out by Gerhard and Johnstone (1955), who found the rate of oxidation of 10 to

20 ppm SO_2 to be unaffected by the addition of 1 to 2 ppm NO_2. However, Renzetti and Doyle (1960) found that 1 ppm NO_2 enhanced aerosol production in the photolysis of 0.14 ppm SO_2. Urone *et al.* (1968) found that the rate of photooxidation of 8 ppm SO_2 was increased about 100-fold, independent of relative humidity, in the presence of 15 ppm NO_2.

The explanation for the above observation is that the $O(^3P)$ atom produced in the photolysis of NO_2 reacts with SO_2:

$$O + SO_2 + M \rightarrow SO_3 + M \tag{11}$$

The rate coefficient for this reaction at 300°K is known to be 1.0×10^{-32} cm^6/sec with O_2 or N_2 as chaperone gases (Cohen and Heicklen, 1972). At atmospheric pressures, the reaction may be entering the second-order region, and the rate coefficient may be somewhat smaller. However, using this value and the value of 8×10^{-3} sec^{-1} for the photodissociation coefficient of NO_2 with an overhead sun, the rate coefficient for photooxidation of SO_2 by $O(^3P)$ atoms becomes $0.128[NO_2]/[O_2]$ sec^{-1}. The ratio of this coefficient to that for photooxidation through the excited electronic states of SO_2, 0.28×10^{-6} sec^{-1}, gives the ratio of SO_2 disappearance by the two processes:

$$J\{O(^3P)\}/J\{SO_2 \text{ photolysis}\} = 0.46 \times 10^6[NO_2]/[O_2] \tag{12}$$

For 15 ppm NO_2 used by Urone *et al.* (1968) this ratio becomes 33, in reasonably good agreement with the ~100-fold increase observed. For typical atmospheric conditions, when $[NO_2] \sim 0.2$ ppm, the $O(^3P)$ mechanism will be only 50% as important as direct photooxidation. The importance of this reaction will be reduced by the reaction

$$O(^3P) + SO_3 \rightarrow O_2 + SO_2 \tag{13}$$

but the rate coefficient for this reaction has not yet been established. Unless it is very large the reaction will not be important, since SO_3 will preferentially be removed by H_2O vapor to give H_2SO_4.

Renzetti and Doyle (1960) also found that if 1 ppm NO, rather than NO_2, was added to 0.5 ppm SO_2 in air (relative humidity = 50%) the aerosol production was retarded rather than enhanced during irradiation. They suggested that inhibition occurred by

$$SO_4 + NO \rightarrow SO_3 + NO_2 \tag{14}$$

Altshuller and Bufalini (1965) suggested alternative reactions for the inhibition:

$$SO_3 + NO \rightarrow SO_2 + NO_2 \tag{15}$$

$$SO_2 \text{ (excited)} + NO \rightarrow SO_2 + NO \tag{16}$$

Reaction (16) cannot play a significant role since the relative quenching coefficients for NO and O_2 are about 400 for all the triplet states of SO_2, but $[NO]/[O_2]$ is only 5×10^{-6}. There is no evidence for or against reactions (14) and (15).

Reactions with HO and HO_2

The reaction of HO with SO_2 occurs by

$$HO + SO_2\ (+M) \rightarrow HOSO_2\ (+M) \tag{17}$$

The limiting low- and high-pressure rate coefficients for this reaction are 2.5×10^{-31} cm^6/sec and 1.5×10^{-13} cm^3/sec at room temperature (Castleman, 1975). The reaction is entirely in the high-pressure second-order regime above 250 Torr pressure. The effect of reaction (17) would be as a terminating reaction, since it presumably would be followed by

$$HOSO_2 + HO \rightarrow H_2SO_4 \tag{18}$$

Thus its effect would be to promote H_2SO_4 aerosol production by termination of the chain oxidation of NO to NO_2. Reaction (17) is probably not important in HO removal since it competes with the dominant terminating reaction

$$HO + NO_2 \rightarrow HONO_2 \tag{19}$$

which has a large rate coefficient of 8×10^{-12} cm^3/sec (Garvin and Hampson, 1974).

With HO_2 the rate coefficient has been measured by Payne *et al.* (1973) to be 9×10^{-16} cm^3/sec. Presumably the reaction is

$$HO_2 + SO_2 \rightarrow HO + SO_3 \tag{20}$$

Again this reaction must compete with the HO_2 reactions with NO and NO_2, both of which have rate coefficients more than 100 times larger. The HO_2–SO_2 reaction is probably of no significance in the atmosphere in regard to HO_2 removal.

Reactions with RO and RO_2

Evidence in our laboratory indicates that the reactions of CH_3O (and presumably other RO radicals) with SO_2 are slow, too slow to be of any significance. However, Shortridge (1973) has evidence that $CH_3C(O)O_2$ reacts readily with SO_2 via

$$CH_3C(=O)O_2 + SO_2 \longrightarrow CH_3C(=O)O + SO_3 \tag{21}$$

Such a reaction in the atmosphere would carry the chain, and thus oxidation of the hydrocarbon would not be retarded (possibly slightly accelerated) and SO_2 would be oxidized. In fact, it is just these arguments that led Shortridge to propose the above reaction. He studied the oxidation of CH_3CHO and found that its rate was unaffected by the addition of SO_2, although the SO_2 was removed and an aerosol was produced.

In the atmosphere the above reaction would compete with

$$CH_3C\begin{matrix}{=}O\\ \diagdown O_2\end{matrix} + NO \longrightarrow CH_3C\begin{matrix}{=}O\\ \diagdown O\end{matrix} + NO_2 \qquad (22)$$

$$CH_3C\begin{matrix}{=}O\\ \diagdown O_2\end{matrix} + NO_2 \longrightarrow CH_3C\begin{matrix}{=}O\\ \diagdown O_2NO_2\end{matrix} \quad \text{(PAN)} \qquad (23)$$

Thus the rate of conversion of NO to NO_2 and the rate of PAN production would be reduced. Since PAN is a primary eye irritant, eye irritation should also be diminished. [Wilson *et al.* (1972a) have shown that H_2SO_4 is not an eye irritant at the levels produced.] A number of experiments on simulated atmospheres have been made and the results reviewed by Bufalini (1971). Many conflicting observations have been reported, but the bulk of the data indicates that the addition of SO_2 to olefin–NO_x air mixtures reduces the rate of photooxidation of NO to NO_2, reduces the oxidant levels and eye irritation, and increases the aerosol production rate. More recent work (Wilson *et al.*, 1972a,b) confirms these general conclusions.

In the early stages of the reaction, before much NO_2 is produced, the competition would be between the reactions

$$RO_2 + NO \rightarrow RO + NO_2 \qquad (24)$$

$$RO_2 + SO_2 \rightarrow RO + SO_3 \qquad (25)$$

The rate coefficients for these two reactions are not known, but we assume that they are equal. Then at equal concentrations of NO and SO_2 the rate of NO_2 production would equal the rate of SO_3 production, and the sum of these rates would equal the NO to NO_2 conversion rate in the absence of SO_2. This conclusion would only be true for that fraction of the chain carried through RO_2, and not HO_2, radicals, i.e., in the early stages of the olefin catalyzed conversion of NO to NO_2. In the later stages, or in the paraffin-catalyzed conversion, HO_2 plays a more important role.

The rate of SO_2 conversion to SO_3, $R\{SO_2 \rightarrow SO_3\}$, by the above mechanism for an olefin hydrocarbon is given approximately by the formula

$$R\{SO_2 \rightarrow SO_3\} \simeq I\{\text{rad}\}\phi\{\text{chain}\}f \qquad (26)$$

where $I\{\text{rad}\}$ is the photoproduction rate of radicals. Assume the photolysis of CH_2O is the predominant source; then for an overhead sun $I\{\text{rad}\} = 15.6 \times 10^{-5}[CH_2O]$. The chain length, $\phi\{\text{chain}\}$, is approximately given by

$$\phi\{\text{chain}\} \simeq k_{(28)}[\text{olefin}]/k_{(29)}[NO_2] \tag{27}$$

where reactions (28) and (29) are

$$HO + \text{olefin} \xrightarrow{O_2} RO_2 \tag{28}$$

$$HO + NO_2 \rightarrow HONO_2 \tag{29}$$

The fraction of the chain oxidizing SO_2, f, is

$$f \equiv k_{(25)}[SO_2]/(k_{(24)}[NO] + k_{(25)}[SO_2]) \tag{30}$$

Thus for an overhead sun

$$R\{SO_2 \rightarrow SO_3\} = 15.6 \times 10^{-5}[CH_2O]\frac{k_{(28)}[\text{olefin}]}{k_{(29)}[NO_2]}\,\frac{k_{(25)}[SO_2]}{(k_{(25)}[SO_2] + k_{(24)}[NO])} \tag{31}$$

For $f = \frac{1}{2}$, $\phi = 20$, and $[CH_2O] = 0.1$ ppm, $R\{SO_2 \rightarrow SO_3\}$ becomes 15.6×10^{-5} ppm/sec. This can be compared with the rate of photooxidation of 0.28×10^{-6} at 1 ppm SO_2 or 0.28×10^{-7} at 0.1 ppm. Even if $f = 0.1$, more photooxidation of SO_2 will proceed by this route than through either electronically excited SO_2 or $O(^3P)$.

Reaction in the O_3–Olefin System

Harkins and Nicksic (1965) apparently were the first to report that SO_2 could be oxidized in the O_3–olefin reaction. This reaction was studied in detail by Cox and Penkett (1972), who found that the addition of SO_2 to the O_3–olefin reaction in the <ppm range in air

(1) did not change the rate of O_3 consumption,
(2) reduced the olefin to O_3 consumption ratio,
(3) increased the amount of product aldehydes,
(4) produced H_2SO_4 aerosol at a rate increasing with SO_2 pressure, and
(5) the rate of H_2SO_4 production increased with added H_2O vapor.

Except for the fifth observation, which is difficult to explain, these results are all consistent with the oxidation of SO_2 by the zwitterion–diradical intermediate (Cox and Penkett also considered the initially formed molozonide as the possible oxidizing agent, but its lifetime is probably too short for it to be the reactive intermediate):

$$\begin{matrix} R_3 \\ \diagdown \\ C{-}O{-}O \\ \diagup \\ R_4 \end{matrix} + SO_2 \rightarrow R_3R_4C{=}O + SO_3 \tag{32}$$

This reaction would be in competition primarily with unimolecular disappearance reactions

$$\begin{matrix} R_3 \\ \diagdown \\ COO \\ \diagup \\ R_4 \end{matrix} \rightarrow \text{removal} \tag{33}$$

The values of $k_{(33)}/k_{(32)}$ found are listed in Table VIII-4. The effect of H_2O vapor might reflect a possible competition for SO_3 by H_2O and the olefin.

The importance of the oxidation of SO_2 by the zwitterion can be estimated under atmospheric conditions to be

$$R\{SO_2 \rightarrow SO_3\} = k_{(35)}[\text{olefin}][O_3]\frac{k_{(32)}[SO_2]}{k_{(32)}[SO_2] + k_{(33)}} \tag{34}$$

where reaction (35) is

$$O_3 + \text{olefin} \rightarrow \text{zwitterion–diradical} + \text{aldehyde or ketone} \tag{35}$$

For $k_{(33)}/k_{(32)} = \frac{1}{2}$ ppm, $k_{(35)} = 10^4\ M^{-1}\ \text{sec}^{-1}$, and [olefin] = $[O_3]$ =

TABLE VIII-4

Relative Importance of Reactions (33) and (32) in the SO_2–O_3–*Olefin System*[a]

Olefin	Relative humidity (%)	$k_{(33)}/k_{(32)}$ (ppm)
cis-C_4H_8-2	<10	0.30 ± 0.03
cis-C_4H_8-2	40	0.63 ± 0.10
cis-C_4H_8-2	75	1.44 ± 0.50
trans-C_4H_8-2	—	0.64 ± 0.13
cis-C_5H_{10}-2	—	0.48 ± 0.08
1-C_6H_{12}	—	0.41 ± 0.06
2-Methyl-1-pentene	—	0.29 ± 0.12

[a] From Cox and Penkett (1972).

$[SO_2] = 0.1$ ppm, this rate is 0.70×10^{-6} ppm/sec, about 35 times faster than the photooxidation for an overhead sun.

For the simulated atmospheric experiments of Harkins and Nicksic (1965), the concentrations used were 4 ppm *trans*-2-butene, 2 ppm SO_2, and 0.5–2.0 ppm NO. Thus if we take the maximum in $[O_3]$ to be ~ 0.5 ppm, the O_3 induced oxidation rate at the $[O_3]$ maximum would be 1.06×10^{-3} ppm/sec, whereas that for monoradical RO_2 oxidation would be smaller, although of the same order of magnitude. Consequently in these experiments aerosol should be produced before O_3 appears, but the aerosol rate should increase significantly after the NO disappears and the O_3 appears. This conforms to the findings of Harkins and Nicksic (1965).

RELATIVE IMPORTANCE OF HOMOGENEOUS SO_2 OXIDATION REACTIONS

It is useful at this point to examine the relative importance of the homogeneous paths to SO_2 oxidation. We shall examine how the importance of the paths changes during the course of the computed C_2H_4–NO–air run depicted in Fig. VI-6 with the assumption that the SO_2 pressure is so low that it has no influence on the course of the SO_2-free irradiation.

The oxidation rate coefficients for an overhead sun for four of the processes are given by the expressions

$$J\{\text{excited } SO_2\} = 0.14 \times 10^{-6} \text{ sec}^{-1}$$

$$J\{O(^3P)\} = 0.61 \times 10^{-6}[NO_2] \text{ sec}^{-1}$$

$$J\{\text{monoradical}\} = 9.8 \times 10^{-6} \frac{[CH_2O]}{[NO_x]} \frac{[C_2H_4]}{[NO_2]} \text{ sec}^{-1}$$

$$J\{O_3\} = 120 \times 10^{-6}[C_2H_4][O_3] \text{ sec}^{-1}$$

where the concentrations of all the species are in ppm. In obtaining the above expressions certain assumptions were made. For $J\{\text{excited } SO_2\}$, it was assumed that the oxidation rate in the presence of NO_x is $\frac{1}{2}$ that in its absence. For $J\{\text{monoradical}\}$ it was assumed that the RO_2–SO_2 reaction had a rate coefficient 0.1 that for the RO_2–NO_x reaction (i.e., $k_{SO_2} = 0.1\, k_{NO}$, $k_{NO} = k_{NO_2}$). For $J\{O_3\}$, the value $k_{(33)}/k_{(32)}$ was assumed $= 0.55$ ppm. For the reactions of SO_2 with HO and HO_2, the oxidation coefficient is found directly as the product of the radical concentration and the corresponding rate coefficient.

The values of the six oxidation coefficients are plotted versus time of day in Fig. VIII-2. $J\{\text{excited } SO_2\}$ and $J\{O(^3P)\}$ are always negligible. In the early part of the day, $J\{\text{monoradical}\}$ is the dominant term, but

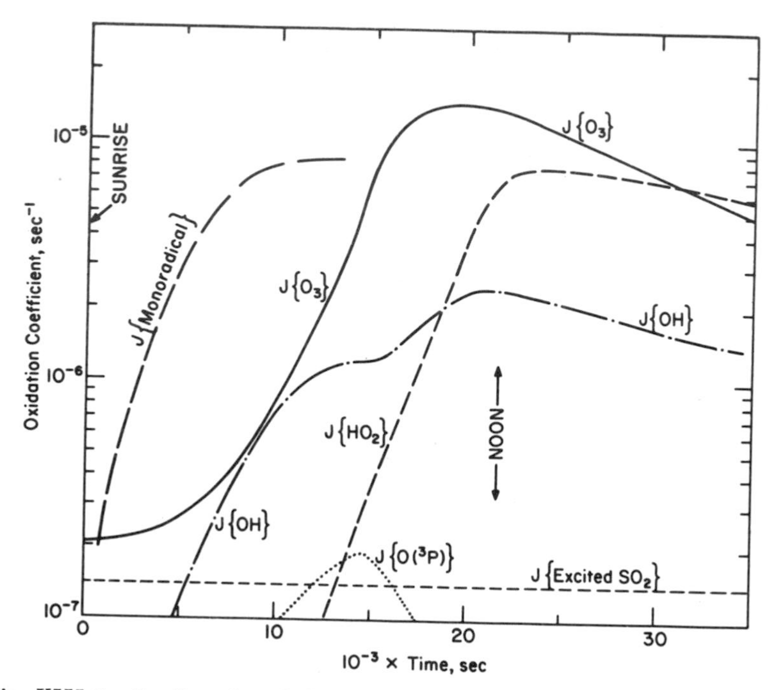

Fig. VIII-2. Semilog plot of the computed oxidation coefficients for SO_2 by the homogeneous routes versus the time of day for the conditions of Fig. VI-6.

$J\{O_3\}$ becomes at least as important before noon when both $[C_2H_4]$ and $[O_3]$ have significant values. Around noon and later, oxidation by HO and HO_2 can also become important. The overall oxidation coefficient is $\sim 10^{-5}$ sec^{-1} or larger during midday. This corresponds to an SO_2 oxidation rate $\gtrsim 3\%$ per hour.

OXIDATION IN H_2O SOLUTION

SO_2 can dissolve in H_2O, where it can be oxidized to sulfate. This oxidation was studied by Miller and dePena (1972). They found that their data, as well as the results of earlier investigators, could be explained by the mechanism

$$SO_2(g) + H_2O(l) \leftrightarrows H_2SO_3(aq) \qquad (36,-36)$$

$$H_2SO_3(aq) \leftrightarrows HSO_3^-(aq) + H^+(aq) \qquad (37,-37)$$

$$H_2SO_3(aq) + HO^-(aq) \leftrightarrows HSO_3^-(aq) + H_2O(l) \qquad (38,-38)$$

$$HSO_3^-(aq) \leftrightarrows SO_3^{-2}(aq) + H^+(aq) \qquad (39,-39)$$

$$SO_3^{-2}(aq) + \tfrac{1}{2}O_2 \rightarrow SO_4^{-2}(aq) \qquad (40)$$

They assumed that the equilibria between SO_2(g) and H_2SO_3(aq) and between HSO_3^-(aq) and SO_3^{-2}(aq) always held and that the reaction rate was determined by the other three reactions. These assumptions were consistent with all the data. Calculations of the rate of the SO_2(g)–H_2O(l) reaction rate coefficient show that under any conditions of interest the first equilibrium is obeyed.

A summary of the pertinent rate and equilibria data is given in Table VIII-5. The rate coefficients for reaction (37) were taken from Wang and Himmelblau (1964). For reaction (38) they assumed the rate coefficients to be the same as those for the analogous reaction of CO_2 found by Himmelblau and Bobb (1958). Miller and dePena measured sulfate production rates in the reaction of gaseous SO_2 in air with liquid H_2O, and found that the mechanism fitted their data with a pseudo-first-order rate constant 3×10^{-3} sec^{-1}. This value is for distilled H_2O. Other investigators found that metal salt catalysts in the H_2O accelerated the oxidation rate. Manganese salts were particularly effective in this regard.

Miller and dePena (1972) examined the rates of SO_2 removal from the gas and sulfate formation for a static system in which the dissolved anion concentrations increased as the experiments proceeded. These studies were done both in the presence and absence of NH_3 vapor (which influences the solution pH), and it was found that the rates were faster in the presence of NH_3. However, in all these experiments the rates of SO_2 removal were too slow to be of consequence in removing significant amounts of SO_2. Nevertheless, we can extrapolate their results to conditions of interest.

TABLE VIII-5

Reactions of Importance for SO_2 *Oxidation in Water Droplets*

Reaction	Reaction no.	Equilibrium constant	Units	Rate coefficient	Units
$SO_2(g) + H_2O(l) \leftrightharpoons H_2SO_3(aq)$	(36, −36)	1.24	M/atm	—	—
$H_2SO_3(aq) \leftrightharpoons HSO_3^-(aq) + H^+(aq)$	(37, −37)	1.28×10^{-2}	M	3.2×10^{-2}	sec^{-1}
$H_2SO_3(aq) + HO^-(aq) \leftrightharpoons HSO_3^-(aq) + H_2O(l)$	(38, −38)	1.26×10^{12}	M^{-1}	2.9×10^5	M^{-1} sec^{-1}
$HSO_3^-(aq) \leftrightharpoons SO_3^{-2}(aq) + H^+(aq)$	(39, −39)	6.24×10^{-8}	M	—	—
$SO_3^{-2}(aq) + (\frac{1}{2})O_2 \rightarrow SO_4^{-2}(aq)$	(40)	—	—	3×10^{-3} [a]	sec^{-1}

[a] At atmospheric conditions; $[O_2] = 0.21$ atm.

The reaction rate of SO_2 uptake is determined by reactions (37) and (38). It is only of importance when reactions (37) and (38) are far from equilibrium so that their reverse reactions are negligible. The expression for SO_2 removal then becomes

$$k\{SO_2\} \equiv \frac{-d[SO_2]/dt}{[SO_2]} = (k_{(37)} + k_{(38)}[HO^-])K_{(36,-36)}[H_2O]$$

$$= (3.2 \times 10^{-2} + 2.9 \times 10^5[HO^-])9.2 \times 10^{-22}[H_2O] \quad (41)$$

where $k\{SO_2\}$ is in sec^{-1}, $[HO^-]$ is the solution molar concentration, and $[H_2O]$ is the number of liquid H_2O molecules per cubic centimeter of air. This rate coefficient is only fast enough to be important in basic solution (i.e., in the presence of NH_3) and for high H_2O concentrations. Thus even at a pH of 8.0 and $[H_2O] = 10^{16}$ molecules/cm^3 (heavy fog), $k\{SO_2\}$ is only 2.9×10^{-6} sec^{-1}, which corresponds to a lifetime of 4 days. Unless the H_2O droplets are quite basic or there is heavy rain, SO_2 washout will not be important.

The rate of conversion of any HSO_3^- or SO_3^{2-} in solution to SO_4^{2-} will be rapid, however. The equilibrium is such that for acid droplets, HSO_3^- will be present, but in basic solution mostly SO_3^{2-}. The oxidation rates then are

$$d[SO_4^{-2}]/dt = k_{(40)}[SO_3^{-2}] \qquad (pH > 7.2) \quad (42)$$

$$d[SO_4^{-2}]/dt = k_{(40)}K_{(39,-39)}[HSO_3^-]/[H^+] \qquad (pH < 7.2) \quad (43)$$

Thus for $k_{(40)} = 3 \times 10^{-3}$ sec^{-1} for $K_{(39,-39)} = 6.24 \times 10^{-8}$ M, the latter expression gives a conversion rate coefficient for HSO_3^- to SO_4^{2-} of $1.87 \times 10^{-10}/[H^+]$, which equals 1.87×10^{-4} sec^{-1} for pH = 6 (i.e., a lifetime of 1.5 hr). The value of $k_{(40)} = 3 \times 10^{-3}$ sec^{-1} is for the uncatalyzed reaction and is thus a minimum value. If any metal ion catalysts are present, the rate coefficient will be larger. Thus in H_2O droplets, the oxidation of HSO_3^- and SO_3^{2-} to SO_4^{2-} in the atmosphere is rapid (lifetime less than 2 hr), but the uptake of SO_2 by H_2O droplets is not important, except possibly in a heavy rain.

In regard to the Mn^{2+} catalyzed reaction, Matteson *et al.* (1969) found that Mn^{2+} solutions cause a rapid uptake of SO_2 without giving H_2SO_4. This is attributed to most of the SO_2 being tied up as a complex. The reaction scheme envisioned was

$$SO_2 + Mn^{2+} \leftrightharpoons Mn \cdot SO_2^{2+} \quad (44)$$

$$2Mn \cdot SO_2^{2+} + O_2 \leftrightharpoons 2Mn \cdot SO_3^{2+} \quad (45)$$

$$Mn \cdot SO_3^{2+} + H_2O \leftrightharpoons Mn^{2+} + H_2SO_4 \quad (46)$$

with $k_{(44)} = 2.4 \times 10^5$ M^{-1} min^{-1} and $k_{(45)} = 36$ M^{-2} min^{-1}. However, for reaction (44) to double the uptake of SO_2 over that from reaction (36) alone would require a Mn^{2+} concentration of at least 7.7×10^{-3} M, an unreasonably high value.

REFERENCES

Allen, E. R., McQuigg, R. D., and Cadle, R. D. (1972). *Chemosphere* **1**, 25, "The Photo-oxidation of Gaseous Sulfur Dioxide in Air."

Altshuller, A. P., and Bufalini, J. (1965). *Photochem. Photobiol.* **4**, 97, "Photochemical Aspects of Air Pollution: A Review."

Brand, J. C. D., and Nanes, R. (1973). *J. Mol. Spectrosc.* **46**, 194, "The 3400–3000Å Absorption of Sulfur Dioxide."

Braslavsky, S., and Heicklen, J. (1972). *J. Amer. Chem. Soc.* **94**, 4864, "Photolysis of Sulfur Dioxide in the Presence of Foreign Gases II. Thiophene."

Brus, L. E., and McDonald, J. R. (1973). *Chem. Phys. Lett.* **21**, 283, "Collision Free, Time Resolved Fluorescence of SO_2 Excited Near 2900Å."

Brus, L. E., and McDonald, J. R. (1974). *J. Chem. Phys.* **61**, 97, "Time Resolved Fluorescence Kinetics and $^1B_1$ ($^1\Delta_g$) Vibronic Structure in Tunable Ultraviolet Laser Excited SO_2 Vapor."

Bufalini, M. (1971). *Environ. Sci. Tech.* **5**, 685, "Oxidation of Sulfur Dioxide in Polluted Atmospheres—A Review."

Calvert, J. G. (1973). *Chem. Phys. Lett.* **20**, 484, "Non-Radiative Decay Processes in Isolated SO_2 Molecules Excited Within the First Allowed Absorption Band (2500–3200Å)."

Castleman, A. W. Jr. (1975). Unpublished work at Brookhaven Nat. Lab.

Caton, R. B., and Gangadharan, A. R. (1974). *Can. J. Chem.* **52**, 2389, "Electronic and Vibrational Relaxation in Sulfur Dioxide Vapor."

Cehelnik, E., Spicer, C. W., and Heicklen, J. (1971). *J. Amer. Chem. Soc.* **93**, 5371 "Photolysis of Sulfur Dioxide in the Presence of Foreign Gases I. Carbon Monoxide and Perfluoroethylene."

Cehelnik, E., Heicklen, J., Braslavsky, S., Stockburger, L., III, and Mathias, E. (1973). *J. Photochem.* **2**, 31, "Photolysis of SO_2 in the Presence of Foreign Gases IV. Wavelength and Temperature Effects with CO."

Cohen, N., and Heicklen, J. (1972). *In* "Comprehensive Chemical Kinetics" (C. H. Bamford and C. F. H. Tipper, eds.), Vol. 6 Reactions of Non-Metallic Inorganic Compounds," p. 1. Elsevier, Amsterdam. "The Oxidation of Inorganic Non-Metallic Compounds."

Cox, R. A. (1972). *J. Phys. Chem.* **76**, 814, "Quantum Yields for the Photooxidation of Sulfur Dioxide in the First Allowed Absorption Region."

Cox, R. A. (1973). *J. Photochem.* **2**, 1, "The Sulphur Dioxide Photosensitized cis-trans Isomerization of Butene-2."

Cox, R. A., and Penkett, S. A. (1972). *J. Chem. Soc., Faraday Trans. I* **68**, 1735, "Aerosol Formation from Sulphur Dioxide in the Presence of Ozone and Olefinic Hydrocarbons."

Dainton, F. S., and Ivin, K. J. (1950a). *Trans. Faraday Soc.* **46**, 374, "The Photochemical Formation of Sulphinic Acids from Sulphur Dioxide and Hydrocarbons."

Dainton, F. S., and Ivin, K. J. (1950b). *Trans. Faraday Soc.* **46**, 382, "The Kinetics of the Photochemical Gas Phase Reactions between Sulphur Dioxide and n-Butane and 1-Butene Respectively."

Dixon, R. N., and Halle, M. (1973). *Chem. Phys. Lett.* **22,** 450, "The 3400Å Band System of Sulphur Dioxide."

Douglas, A. E. (1966). *J. Chem. Phys.* **45,** 1007, "Anomalously Long Radiative Lifetimes of Molecular Excited States."

Fatta, A. M., Mathias, E., Heicklen, J., Stockburger, L., III, and Braslavsky, S. (1974). *J. Photochem.* **2,** 119, "Photolysis of SO_2 in the Presence of Foreign Gases V. Sensitized Phosphorescence of Biacetyl."

Garvin, D., and Hampson, R. F. (1974). Nat. Bur. Std. Rep. NBSIR 74-430, "Chemical Kinetics Data VII. Tables of Rate and Photochemical Data for Modelling of the Stratosphere (Revised)."

Gerhard, E. R., and Johnstone, H. F. (1955). *Ind. Eng. Chem.* **47,** 972, "Photochemical Oxidation of Sulfur Dioxide in Air."

Greenough, K. F., and Duncan, A. B. F. (1961). *J. Amer. Chem. Soc.* **83,** 555, "The Fluorescence of Sulfur Dioxide."

Harkins, J., and Nicksic, W. S. (1965). Presented at Amer. Chem. Soc. Meeting, Div. of Petroleum Chem., Detroit, April, 1965. "Role of Hydrocarbon Photooxidation Rates in the Atmospheric Oxidation of Sulfur Dioxide."

Himmelblau, D. M., and Bobb, A. L. (1958). *J. Amer. Inst. Chem. Eng.* **4,** 143, "Kinetic Studies of Carbonation Reactions Using Radioactive Tracers."

Horowitz, A., and Calvert, J. G. (1972a). *Int. J. Chem. Kinet.* **4,** 175, "The SO_2-Sensitized Phosphorescence of Biacetyl Vapor in Photolyses at 2650 and 2875Å. The Intersystem Crossing Ratio in Sulfur Dioxide."

Horowitz, A., and Calvert, J. G. (1972b). *Int. J. Chem. Kinet.* **4,** 191, "A Study of the Intersystem Crossing Reaction Induced in Gaseous Sulfur Dioxide Molecules by Collisions with Nitrogen and Cyclohexane at 27°C."

Johnston, H. S., and dev Jain, K. (1960). *Science* **131,** 1523, "Sulfur Dioxide Sensitized Photochemical Oxidation of Hydrocarbons."

Kent, J. E., O'Dwyer, M. F., and Shaw, R. J. (1974). *Chem. Phys. Lett.* **24,** 221, "Single Vibronic Level Fluorescence of SO_2."

Kopczynski, S. L., and Altshuller, A. P. (1962). *Int. J. Air Water Pollut.* **6,** 133, "Photochemical Reactions of Hydrocarbons with Sulfur Dioxide."

Levine, S. Z., and Kaufman, F. (1974). Presented at the *Can. Chem. Conf., 57th* Chem. Inst. Canada, Regina, Sask., June, 1974, Abstract No. 294, "Radiative Lifetimes of NO_2 and SO_2."

Luria, M., and Heicklen, J. (1974). *Can. J. Chem.* **52, 3451,** "Photolysis of SO_2 in the Presence of Foreign Gases VI. Acetylene and Allene."

Luria, M., de Pena, R. G., Olszyna, K. J., and Heicklen, J. (1974a). *J. Phys. Chem.* **78,** 325, "Kinetics of Particle Growth III. Particle Formation in the Photolysis of Sulfur Dioxide–Acetylene Mixtures."

Luria, M., Olszyna, K. J., de Pena, R. G., and Heicklen, J. (1974b). *Aerosol Sci.* **5,** 435, "Kinetics of Particle Growth V. Particle Formation in the Photolysis of SO_2–Allene Mixtures."

Matteson, M. J., Stoeber, W., and Luther, H. (1969). *Ind. Eng. Chem. Fundam.* **8,** 677, "Kinetics of the Oxidation of Sulfur Dioxide by Aerosols of Manganese Sulfate."

Mettee, H. D. (1969). *J. Phys. Chem.* **73,** 1071, "Foreign Gas Quenching of Sulfur Dioxide Vapor Emission."

Miller, J. M., and de Pena, R. G. (1972). *J. Geophys. Res.* **77,** 5905, "Contribution of Scavenged Sulfur Dioxide to the Sulfate Content of Rain Water."

Payne, W. A., Stief, L. J., and Davis, D. D. (1973). *J. Amer. Chem. Soc.* **95,** 7614, "A Kinetics Study of the Reaction of HO_2 with SO_2 and NO."

Penzhorn, R.-D., Filby, W. G., Günther, K., and Streylitz, L. (1975). Private communication. "The Photoreaction of Sulphur Dioxide with Hydrocarbons. Part II. Chemical and Physical Aspects of the Formation of Aerosols with Butane."

Rao, T. N., Collier, S. S., and Calvert, J. G. (1969a). *J. Amer. Chem. Soc.* **91,** 1609, "Primary Photophysical Processes in the Photochemistry of Sulfur Dioxide at 2875Å."

Rao, T. N., Collier, S. S., and Calvert, J. G. (1969b). *J. Amer. Chem. Soc.* **91,** 1616, "The Quenching Reactions of the First Excited Singlet and Triplet States of Sulfur Dioxide with Oxygen and Carbon Dioxide."

Renzetti, N. A., and Doyle, G. J. (1960). *Int. J. Air Pollut.* **2,** 327, "Photochemical Aerosol Formation in Sulphur Dioxide-Hydrocarbon Systems."

Shortridge, R. (1973). Unpublished results at Penn State Univ.

Sidebottom, H. W., Badcock, C. C., Calvert, J. G., Rabe, B. R., and Damon, E. K. (1971). *J. Amer. Chem. Soc.* **93,** 3121, "Mechanism of the Photolysis of Mixtures of Sulfur Dioxide with Olefin and Aromatic Hydrocarbons."

Sidebottom, H. W., Badcock, C. C., Jackson, G. E., Calvert, J. G., Reinhardt, G. W., and Damon, E. K. (1972a). *Environ. Sci. Tech.* **6,** 72, "Photooxidation of Sulfur Dioxide."

Sidebottom, H. W., Otsuka, K., Horowitz, A., Calvert, J. G., Rabe, B. R., and Damon, E. K. (1972b). *Chem. Phys. Lett.* **13,** 337, "Vibronic Effects in the Decay of the Fluorescence Excited in SO_2 and NO_2."

Smith, R., de Pena, R. G., and Heicklen, J. (1975). *Int. J. Colloid Interface Chem.* **53,** 202, "Kinetics of Particle Growth VI. Sulfuric Acid Aerosol from the Photooxidation of SO_2 in Moist O_2-N_2 Mixtures."

Stockburger, L., III, Braslavsky, S., and Heicklen, J. (1973). *J. Photochem.* **2,** 15, "Photolysis of SO_2 in the Presence of Foreign Gases III. Quenching of Emission by Foreign Gases."

Strickler, S. J., and Berg, R. A. (1962). *J. Chem. Phys.* **37,** 814, "Relationship between Absorption Intensity and Fluorescence Lifetime of Molecules."

Strickler, S. J., Vikesland, J. P., and Bier, H. D. (1974). *J. Chem. Phys.* **60,** 664, "$^3B_1$-$^1A_1$ Transition of SO_2 Gas II. Radiative Lifetime and Radiationless Processes."

Timmons, R. B. (1970). *Photochem. Photobiol.* **12,** 219, "The Photochemically Induced Reactions of Sulphur Dioxide with Alkanes and Carbon Monoxide."

Urone, P., Lutsep, H., Noyes, C. M., and Parcher, J. F. (1968). *Environ. Sci. Tech.* **2,** 611, "Static Studies of Sulfur Dioxide Reactions in Air."

Wampler, F. B., Otsuka, K., Calvert, J. G., and Damon, E. K. (1973). *Int. J. Chem. Kinet.* **5,** 669, "The Temperature Dependence and the Mechanism of the SO_2 ($^3B_1$) Quenching Reactions."

Wang, J. C., and Himmelblau, D. M. (1964). *J. Amer. Inst. Chem. Eng.* **10,** 574, "A Kinetic Study of Sulfur Dioxide in Aqueous Solution with Radioactive Tracers."

Wilson, W. E. Jr., Levy, A., and McDonald, E. H. (1972a). *Environ. Sci. Tech.* **6,** 423, "Role of SO_2 and Photochemical Aerosol in Eye Irritation from Photochemical Smog."

Wilson, W. E. Jr., Levy, A., and Wimmer, D. B. (1972b). *J. Air Pollut. Contr. Ass.* **22,** 27, "A Study of Sulfur Dioxide in Photochemical Smog II. Effect of Sulfur Dioxide on Oxidant Formation in Photochemical Smog."

Chapter IX

AEROSOL CHEMISTRY

The generation of aerosols in polluted atmospheres has been investigated in several experiments (Goetz and Pueschel, 1967; Bricard *et al.*, 1968; Husar and Whitby, 1973). Husar and Whitby collected a sample of Pasadena air in 1969, removed all the particulate matter, and then irradiated the mixture. They measured total particle density for particles >25 Å diam. Bricard *et al.* (1968) performed a similar experiment with Paris air. Their results are shown in Fig. IX-1. Initially there is a sharp rise in particle count, reaching values of 10^5 to $10^7/cm^3$ in 10–20 min. Then the particle number density decreases, but the particles already present grow in size. This behavior has been seen in model laboratory experiments for both thermal and photochemical systems (dePena *et al.*, 1973; Luria *et al.*, 1974a,b; Olszyna *et al.*, 1974), and the effects have been explained quantitatively.

Initially irradiation produces particles by homogeneous nucleation. These particles then grow by condensation and coagulation. As the size of the particles increases, condensation of the vapor on the particles already present becomes more important, the vapor species concentrations drop, and nucleation becomes negligible. At the peak particle density, the nucleation and coagulation rates are equal. For longer times, after nucleation becomes unimportant, the particle number density decreases by coagulation until a value of $\sim 10^4/cm^3$ is reached. At these levels coagulation also becomes slow, and the number density remains more or less constant, while the particles grow in size. When the particles become very large and are removed by gravitational settling, then another burst of nucleation can occur and the cycle repeats. The detailed laboratory studies show that condensation and coagulation proceed with near unit efficiency, i.e., nearly every collision results in reaction.

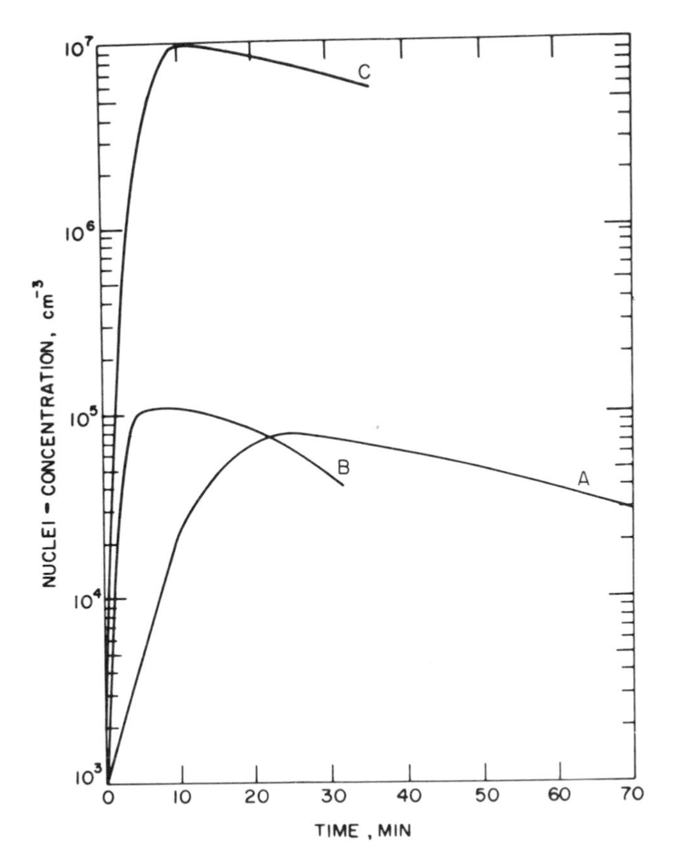

Fig. IX-1. Change in total number concentration in irradiated, initially particle-free, ambient air. (A) Paris (data from Bricard *et al.*, 1968), (B) Los Angeles smog, (C) Los Angeles smog with a generation rate of $10^5/cm^3$-sec. From Husar and Whitby (1973), with permission of the American Chemical Scoeity.

Further corroboration of the above comes from experiments on polluted atmospheres that are not filtered prior to irradiation (Husar and Whitby, 1973). In these experiments, homogeneous nucleation did not occur and the particle density did not rise. The particles already present grew, i.e., they acted as seed particles for the removal of low-volatile vapor material. These results confirmed the earlier findings of Goetz and Pueschel (1965) that particulate matter in air is of major significance in aerosol formation.

Samples of ambient air in several major U.S. cities in 1970 (Lee and Goranson, 1972) showed that typical particulate matter concentrations were 50 $\mu g/m^3$. The size distribution was log normal with cumulative percent mass, and yearly average mean diameters were $\sim$0.46–0.83 μm for the six cities. If the specific gravity of the particles were 1.0, and all the particles had a diameter of 0.46 μm, then the particle number density would be $10^3/cm^3$. Actually since $\frac{1}{2}$ the mass is in smaller particles, there are a much larger number of them, 10 times as many or more. Large cities typically have 10^5 and country areas 10^4 particles/cm^3. On the other hand, there can only be very few larger particles (e.g., $<100/cm^3$ of 1 μm or larger, $<10/cm^3$ of 2 μm or larger). These particle number densities are of the same order of magnitude expected from the laboratory and simulated atmosphere experiments.

There are numerous sources of particulate matter in the atmosphere, and it is of interest to examine the chemistry of each one.

DIRECT INTRODUCTION

From Mechanical Processes

Limestone, metals, their oxides and halides, and asbestos are examples of particulate material introduced directly into the air. These materials can act as catalysts for heterogeneous reactions, or they may undergo chemical reactions themselves. Thus the metals and metal salts will convert to oxides. Also for communities near oceans, sea sprays can bring in H_2O droplets carrying NaCl, sulfates, and nitrates.

Smith *et al.* (1969) found that submicron particles of Fe_3O_4, Al_2O_3, lead oxides and platinum effectively removed SO_2 in the dark. For Fe_3O_4, sorption of SO_2 reached 3% at $[SO_2] = 6.2$ ppm and 1.8×10^5 particles/cm^3 of Fe_3O_4. With Al_2O_3, 50% sorption of SO_2 occurred at 1 ppm SO_2. Lead oxides were the most effective, giving complete removal of SO_2 within five minutes after mixing.

Urone *et al.* (1968) also studied the removal of SO_2 by metal oxides. They found rapid SO_2 removal in the dark with Fe_2O_3, Fe_3O_4, Al_2O_3, PbO, PbO_2, and CaO. The removal was somewhat slower with $CaCO_3$, Cr_2O_3, and V_2O_5. NaCl did not promote the oxidation of SO_2. With moisture or irradiation in the NaCl or V_2O_5 experiments, the oxidation rates did not greatly increase. In the V_2O_5 experiments a significant amount of the SO_2 was adsorbed but not oxidized; it could be released by rinsing the particles.

From Sea Spray

Salts from ocean water are brought over mainland in the sea spray carried by winds. Sodium chloride is the most prevalent of these salts and it can enter reactions. Schroeder and Urone (1974) found that NaCl particles could react with 0.1–1.5% mixtures of NO_2 and SO_2 in air to produce NOCl. The reaction was rapid but surface area dependent and proceeded equally on dry or moist particles.

Other effects of sea spray aerosols would be to dissolve atmospheric gases, and then promote solution reactions through catalysis by the salt ions. An example would be metal–ion catalysis of the oxidation of dissolved SO_2, as discussed in Chapter VIII.

From Incomplete Combustion

Particulate matter is made during incomplete combustion and ejected into the atmosphere. This material is usually soot, i.e., carbon and polyacetylenic material, but also contains inorganic ash, complex tars, and resins. It has been estimated that an average grass fire extending over one acre produces 2×10^{22} fine particles. Also the concentration of condensation nuclei in a ventilated kitchen containing a large gas range was found to exceed $5 \times 10^5/cm^3$ (Cadle, 1966, p. 12). Most of these particles are very small Aitken nuclei (i.e., <0.1-μm radius); thus they have considerable surface area. In the atmosphere these species can undergo oxidation or catalyze surface reactions.

Activated charcoal is known to be an excellent catalyst for many reactions. There is evidence that it catalyzes the oxidation of SO_2 at elevated temperatures. It is known to catalyze the conversion of NO to NO_2 at room temperature (Heicklen and Cohen, 1968). One report indicates that certain types of charcoal can increase the rate 1000-fold over the gas phase reaction

$$2NO + O_2 \rightarrow 2NO_2 \tag{1}$$

If so the rate coefficient would become $\sim 10^7\ M^{-2}$ sec, which for 0.5 ppm NO corresponds to a lifetime for NO oxidation in air of 570 sec, short enough to be important in urban atmospheres.

This conclusion is consistent with reasonable adsorption coefficients of 10^{-8} cm^3/sec and particle concentrations of $10^5/cm^3$, which occur on heavily polluted days. Thus if adsorption of the atmospheric gas is the controlling process, then surface-catalyzed oxidations can proceed with lifetimes ~ 20 min. Thus for NO_2 and SO_2, the surface-catalyzed oxidations could reach 10 and 100 times the importance of the gas phase oxidations, respectively.

Another aspect of absorption is that even if no chemical reaction occurs, the absorption spectrum of the adsorbed species can be shifted markedly, thus altering the photochemistry. Daubendiek and McMillan (1973) found that 1,3-pentadiene adsorbed on Pyrex could be photoisomerized with 3660 Å radiation, even though in the gas phase 1,3-pentadiene does not absorb above 2900 Å.

PHOTOCHEMICAL SMOG

NH_4NO_3

In photochemical smog the end product of NO oxidation is HNO_3. The HNO_3 can react with NH_3 or other materials to give nitrate aerosols. NH_4NO_3 is an important constituent of the atmospheric aerosols.

The formation of particulate NH_4NO_3 has been studied extensively in our laboratory (dePena *et al.*, 1973; Olszyna *et al.*, 1974, 1976). Mixtures of O_3 and NH_3 were reacted in N_2 or air to produce HNO_3 *in situ*. The HNO_3 could then react with NH_3 in the vapor phase:

$$NH_3(g) + HNO_3(g) \leftrightarrows NH_4NO_3(g) \quad (2)$$

It is not yet known what fraction of the HNO_3 is free and what fraction becomes tied up as a gaseous NH_4NO_3. Originally (Olszyna *et al.*, 1974) the evidence suggested that the equilibrium was shifted entirely to the left, but more recently (Olszyna *et al.*, 1976), with the accumulation of more experimental data, the evidence favors the equilibrium being shifted to the right under the experimental conditions.

The position of the equilibrium is rather important, because it determines the conditions under which NH_4NO_3 precipitates. If the original interpretation is valid, then homogeneous nucleation of NH_4NO_3 commences when the product $[NH_3][HNO_3] = 5.8 \times 10^{27}$ molecules^2/cm^6 at 25°C. This would correspond to conditions when, if $[NH_3] = [HNO_3]$, their concentrations would be 3.0 ppm, much higher than would ever be realized in the atmosphere. However, this value is the particle nucleation pressure and not the vapor pressure, which is unknown and must be at least an order of magnitude lower, since NH_4NO_3 is quite nonvolatile. The nucleation reaction corresponding to this situation was found to be

$$qNH_3 + qHNO_3 \rightarrow (NH_4NO_3)_q \quad (3)$$

with $q = 8.0$ and $k_{(3)} = 6.2 \times 10^{-224}$ cm^{45}/min.

On the other hand, with the later interpretation, the equilibrium in reaction (2) is shifted to the right, and the critical $NH_4NO_3(g)$ pressure needed to initiate nucleation at 25°C was 1.0×10^{12} molecules/cm^3 (40 ppb), both in the presence and absence of H_2O, so that HNO_3 would never accumulate in the vapor phase (in the presence of NH_3). In this case the nucleation reaction is

$$qNH_4NO_3(g) \rightarrow (NH_4NO_3)_q \quad (4)$$

with $q = 9$ and $k_{(4)} = 3 \times 10^{-108}$ cm^{24}/min.

Regardless of which mechanism applies, once particulate NH_4NO_3 or any other material on which NH_4NO_3 can condense is present, removal of gaseous HNO_3 or NH_4NO_3, as the case may be, proceeds rapidly by condensation. If free HNO_3 is present, Olszyna *et al.* (1974) found that the condensation mechanism was

$$HNO_3 + C_n \leftrightharpoons C_n \cdot HNO_3 \tag{5a}$$

$$NH_3 + C_n \cdot HNO_3 \rightarrow C_{n+1} \tag{5b}$$

where C_n represents particulate NH_4NO_3 with n NH_4NO_3 molecules. Under their conditions (68–1300 ppm NH_3), reaction (5b) was always fast enough so that reaction (5a) was rate controlling. They found $k_{(5a)} = 1.24 \times 10^{-6}$ cm^3/min for very small particles (this value corresponds to reaction of every collision for 84 Å diameter particles). For typical size particles in the air (~5000 Å diameter) and typical particle concentrations of $10^4/cm^3$, the lifetime of HNO_3 would be about 1 sec. Of course, the actual lifetime might be larger either because, for particles other than NH_4NO_3, adsorption of HNO_3 may not occur with unit efficiency or because of the 1000-fold lower concentration of NH_3 in the atmosphere, reaction (5a) is not rate controlling. If NH_4NO_3 is associated in the gas phase, then the condensation coefficient is 1.24×10^{-6} cm^3/min, regardless of the NH_3 concentration. In either case it is clear that NH_4NO_3 precipitation must be an important process in the atmosphere.

Organic Nitrates

In photochemical smog, aerosols can be produced. Laboratory and atmospheric observations indicate that NO_2 is important in their formation and that these aerosols are more important when the hydrocarbons are larger olefins or aromatics.

It is well known that peroxy radicals $RO_2\cdot$ can add to NO_2 to produce peroxynitrates. These materials are of low volatility, and if the RO_2 radical is sufficiently large or bulky, perhaps they precipitate or react further with the olefins to give particulate organic nitrates. The aerosols formed under these conditions are eye irritants, which suggests that peroxynitrates are present.

There is laboratory evidence for the participation of RO_2NO_2 in aerosol formation. Mitschele and Heicklen (1972) found that NO_2 reacted with the 2-butenes to give a suspended oil, both in the absence and presence of

O_2. With O_2 present the reaction scheme presumably is

$$NO_2 + \rangle C{=}C\langle \longrightarrow O_2N{-}\rangle C{-}C\langle\cdot \xrightarrow{O_2} O_2N{-}\rangle C{-}C\langle{-}O_2\cdot \quad (RO_2) \tag{6}$$

$$RO_2 + NO_2 \longrightarrow RO_2NO_2 \tag{7}$$

The oil was never identified. It may be the RO_2NO_2 or some further reaction product of it. The NO_2–olefin reaction is too slow to be of importance in the atmosphere, but presumably other RO_2 radicals formed by other routes would behave similarly.

Volatile organic nitrates, such as peroxyacetylnitrate (PAN) hydrolyze readily in an aqueous medium to give NO_2^- (Hidy and Burton, 1975). Hidy and Burton (1975) have also suggested that in the presence of large concentrations of sulfuric acid, NO_2 may be absorbed to produce nitrosulfuric acid ($HNOSO_4$), which could then be neutralyzed by NH_3 to give nitrates.

Reaction of O_3 with Hydrocarbons

O_3 reacts with larger olefins, particularly cyclic olefins, to give aerosols. Presumably these are peroxy compounds that are produced by polymerization of the zwitterion or addition of it to the olefin. The importance of these reactions under atmospheric conditions is not known. However, they primarily might be responsible for the haze often observed over wooded areas, which naturally give terpenes. Ripperton and Lillian (1971) found that irradiation of mixtures of 0.1 ppm NO_2 and 0.5 ppm α-pinene (a natural terpene) in air produced condensation nuclei. Presumably the nuclei come from the reaction of the terpene with the O_3 produced in the photochemical smog cycle. In the presence of NO_2 or SO_2 any polymeric material may be modified due to incorporation of these compounds. In addition, SO_2 is converted to SO_3 in the presence of O_3 and olefins, as discussed in the previous chapter. Thus SO_3 may react with the organics to give organic sulfur aerosols or with H_2O to give H_2SO_4 aerosol.

FROM SO_2

SO_2 can react with organic material, under the appropriate circumstances, to produce organic aerosols. It can also react with NH_3 to give a variety of products, and be oxidized to give H_2SO_4 in the presence of water vapor. Not only are these compounds found in polluted urban atmospheres, as discussed in Chapter IV, but H_2SO_4 aerosols, and possibly ammoniated sulfur compounds, are found in the stratosphere.

There is a distinct aerosol layer in the lower stratosphere, starting at 17 km with a broad relative maximum at ~20km and persisting to 35 to 40km, known as the Junge layer. Scattering of sunlight from this layer is believed to be the cause of purple twilights. Particle concentrations in the upper atmosphere drop regularly as the altitude rises. For Aitken nuclei (radius $<0.1\ \mu m$) the number density is $\sim 300/cm^3$ at 7 km, which decreases to a flat plateau value of ~ 5–$10/cm^3$ at 20–30 km (Cadle, 1966). For particles $>0.15\ \mu m$ radius the lower stratospheric concentrations are a few tenths per cm^3 (Hofmann, 1974). Thus absolute particle concentrations decrease with increasing altitude, but in the lower stratosphere not as fast as the gravitational falloff would predict.

The most abundant constituent in the stratospheric aerosol is sulfate (SO_4^{2-}), but Cadle (1972) (see also Cadle and Grams, 1975) has reported that sometimes SiO_3^{2-} is as important as SO_4^{2-}. There has been some uncertainty whether the sulfate is primarily H_2SO_4 or $(NH_4)_2SO_4$. Recently Castleman (1974) has reexamined the data and concluded that in the tropics and high latitudes, the sulfate is all H_2SO_4, with a maximum of 13.4% $(NH_4)_2SO_4$ in the midlatitude stratosphere.

The sources of the stratospheric aerosol could be tropical upwellings, storms, and upward eddy diffusion. A large part appears to come directly from volcanic eruptions. Thus the sulfur content was $(1\text{–}4) \times 10^{-15}$ g/cm^3 in 1959 and rose 1000-fold by 1965 after the Agung (1963) and Taal (1965) eruptions. In 1966 stratospheric sulfur was $\sim 10^{-12}$ gm/cm^3 (Castleman, 1974).

Organic Sulfur Particles

Although no studies have been made in the presence of O_2, the photolysis of SO_2 in the presence of hydrocarbons leads to particulate matter in the absence of O_2. As discussed in the previous chapter the SO_2–alkane system produces sulfinilic acids. With C_2H_2 and allene replacing the alkane, Luria *et al.* (1974a,b) found particles of empirical formulas $C_3H_4S_2O_3$ and C_5H_8SO, respectively.

The nucleating reactions for the two systems examined by Luria *et al.* are

$$r(C_3H_4S_2O_3)_3 \rightarrow (C_3H_4S_2O_3)_{3r} \tag{8}$$

$$rC_5H_8SO \rightarrow (C_5H_8SO)_r \tag{9}$$

For the C_2H_2 system, the fundamental species undergoing condensation was considered to be the trimer of $C_3H_4S_2O_3$ from a molecular weight determination. The values of r in the two systems were 3.6 and 6.7, respectively, with $k_{(8)} = 3.4 \times 10^{-30}$ cm$^{7.8}$/min and $k_{(9)} = 3.0 \times 10^{-59}$ cm$^{17.1}$/min. Once particles were formed they grew by condensation with rate coefficients approaching unit efficiency.

Another possible route to organic sulfate formation is by solution of organic material into H_2SO_4 aerosol followed by reaction. However, we have no information on these reactions.

H_2SO_4

When SO_3 is produced (see Chapter VIII) it reacts readily with H_2O vapor to produce H_2SO_4:

$$H_2O(g) + SO_3(g) \leftrightarrows H_2SO_4(g) \quad (10)$$

with $k_{(10)} = 9.1 \times 10^{-13}$ cm^3/sec at room temperature (Castleman, 1975). At 29°C the equilibrium constant is 7.8×10^9 atm^{-1}, and it is larger at lower temperatures. Thus under all atmospheric conditions, no free SO_3 exists; it is quickly converted to H_2SO_4 vapor.

H_2SO_4 has a very low vapor pressure, especially in the presence of H_2O. Vapor pressure data for various concentration H_2SO_4 solutions at 20–100°C are presented in Table IX-1. The solution is highly nonideal and a high-boiling azeotrope exists at 98.3 wt % H_2SO_4, which greatly reduces the vapor pressure of that expected for ideality. Vapor pressure data are not available for temperatures below 0°C. However, they can be estimated from the thermodynamic functions listed in Table IX-2 for 298°C. It is readily apparent from the information in Tables IX-1 and IX-2 that H_2SO_4 will condense under all atmospheric conditions. Thus essentially all of the SO_2 that is oxidized will be present as H_2SO_4 aerosol.

NH_3 Compounds

NH_3 reacts with SO_2 in the absence of H_2O to give two compounds $NH_3 \cdot SO_2$ and $(NH_3)_2 \cdot SO_2$. These reactions are unaffected by the presence of either N_2 or O_2. The thermodynamic parameters have been found to be (Landreth *et al.*, 1975) $\Delta H = 18.4$ kcal/mole and $\Delta S = 45.1$ cal/mole-°K for the 1:1 adduct and $\Delta H \simeq 33$ kcal/mole and $\Delta S \simeq 87$ cal/mole-°K for the 2:1 adduct. At 200°K, the equilibrium constants for the reactions

$$NH_3 \cdot SO_2(s) \leftrightarrows NH_3(g) + SO_2(g) \quad (11)$$

$$(NH_3)_2SO_2(s) \leftrightarrows 2NH_3(g) + SO_2(g) \quad (12)$$

can be computed to be 5.61×10^{-11} atm^2 and 8.95×10^{-18} atm^3, respec-

TABLE IX-1

Vapor Pressure (Torr) above H_2SO_4 *Solutions at Various Temperatures*[a]

H_2SO_4 (wt %)	Temperature (°C)				
	20°	40°	60°	80°	100°
0	17.54	54.4	146.6	352	760
10	16.25	50.0	134.5	323	704
20	14.9	45.9	123.6	297	650
25	14.0	43.2	116.4	280	610
30	13.0	40.0	107.9	260	567
40	10.4	32.2	87.5	211.8	465
50	7.22	24.5	62.4	152.7	338
55	5.40	17.25	48.0	118.7	265
60	3.69	12.05	34.2	86.1	195.5
65	2.25	7.55	22.0	56.9	132.2
70	1.19	4.22	12.8	34.1	81.6
75	0.53	1.93	6.33	17.7	44.2
80	0.18	0.72	2.49	7.48	19.9
85	0.04	0.19	0.73	2.43	7.06
90	0.004	0.022	0.10	0.39	1.28
94	0.00024	0.00016	0.009	0.04	0.15
96	0.00005	0.00004	0.0024	0.012	0.05
98.3	0.00003	0.00025	0.0015	0.008	0.003
100	0.00035	0.0025	0.014	0.07	0.27

[a] From Luchinskii (1956).

TABLE IX-2

Thermodynamic Data for the H_2SO_4 *System*[a]

Molecule	$\Delta H_f°\{298\}$	$\Delta G_f°\{298\}$	$S°\{298\}$
$SO_3(g)$[b]	−94.58	−88.69	61.34
$H_2SO_4(l)$	−194.55	−164.94	37.50
$H_2SO_4(g)$	−177.00	−156.81	69.13
$H_2SO_4 \cdot H_2O(l)$	−269.51	−227.18	51.35
$H_2SO_4 \cdot 2H_2O(l)$	−341.09	−286.77	66.06
$H_2SO_4 \cdot 3H_2O(l)$	−411.19	−345.18	82.55
$H_2SO_4 \cdot 6\text{-}\frac{1}{2}H_2O(l)$	−653.26	−546.40	140.51
$H_2O(l)$	−68.32	−56.69	16.71
$H_2O(g)$	−57.80	−54.63	45.10

[a] From Stull and Prophet (1971). [b] From Wagman *et al.* (1968).

tively. These values are too large for the solid compounds to be present under any atmospheric conditions.

Scargill (1971) has examined the reaction of NH_3 with SO_2 in the presence of H_2O vapor and finds that both ammonium sulfite, $(NH_4)_2SO_3$, and ammonium pyrosulfite, $(NH_4)_2S_2O_5$, are produced:

$$(NH_4)_2SO_3(s) \leftrightarrows 2NH_3(g) + SO_2(g) + H_2O(g) \tag{13}$$

$$(NH_4)_2S_2O_5(s) \leftrightarrows 2NH_3(g) + 2SO_2(g) + H_2O(g) \tag{14}$$

He found $\Delta H = 64.832$ and 81.00 kcal/mole, respectively, and $\Delta S =$ 152.22 and 191.66 cal/mole-°K, respectively, for the two reactions over the temperature range 273–298°K. Extrapolation to 200°K gives equilibrium constants of $10^{-42.0}$ atm^4 and $10^{-46.6}$ atm^5, respectively. Just above the tropopause, the H_2O concentration is (2 ± 1) ppm (see Chapter I) but the NH_3 and SO_2 concentrations are not known. If we assume that $[NH_3] \sim [SO_2]$, then at 200°K $(NH_4)_2S_2O_5$ will precipitate at SO_2 pressures of $\sim 10^{-10}$ atm, but $(NH_4)_2SO_3$ will precipitate first at concentrations of $\sim 2 \times 10^{-12}$ atm! Significant amounts of gaseous SO_2 and NH_3 cannot coexist under these conditions.

In the presence of O_2, mixtures of moist NH_3 and SO_2 do not produce $(NH_4)_2SO_3(s)$ at room temperature, but rather $(NH_4)_2SO_4(s)$ (McLaren *et al.*, 1974). In our laboratory we have evidence that this is so even at very low temperatures (Hisatsune and Heicklen, 1975). Since SO_2 is presumably present in excess of NH_3 above the tropopause, any NH_3 rising to the tropopause should precipitate as $(NH_4)_2SO_4$. Possibly this heterogeneous oxidation is an important oxidation route for SO_2 in the stratosphere.

FOG FORMATION

When the atmosphere becomes supersaturated in H_2O vapor, usually because of a temperature drop, condensation of H_2O occurs, producing a fog. A fog typically contains particles of 7 to 15 μm diameter with a number density of 50 to 100/cm^3 for a light fog, and 500 to 600/cm^3 for a heavy fog, i.e., 10^{15}–10^{17} liquid H_2O molecules/cm^3 of air (Amelin, 1966).

Since even clean air contains 10^3 particles/cm^3 and urban air often contains 10^4–10^5 particles/cm^3, there are sufficient particles so that condensation will be heterogeneous, i.e., occur on the particles already present. The equilibrium vapor pressure of H_2O depends on the drop size, but it is only 1% higher for droplets of 0.1 μm diameter than for infinitely large particles. Since the mean diameters of particles in urban air are 0.4–0.8 μm, the supersaturation will be negligible before condensation occurs. In fact, if the condensation nuclei are soluble in H_2O, then condensation will occur below the vapor pressure of H_2O, because the soluble species reduce the vapor pressure. If Raoult's law is obeyed (i.e., ideal solution) then, for example, condensation nuclei can grow to 90 mole % H_2O at only 90% relative humidity.

However, if the temperature drop is sufficiently sharp, the heterogeneous nucleation will not be sufficiently fast, the supersaturation will rise, and homogeneous nucleation will occur. This can give rise to $>10^5$ water droplets/cm^3, but they will be relatively small. For example, if 2.5 Torr of H_2O vapor condenses into 10^5 particles/cm^3, the mass average particle diameter is 3 μm, most of which will be <1 μm in diameter.

The rate of H_2O vapor removal can be computed from condensation coefficients. They are diffusion controlled and increase proportionately with the diameter of the particle, reaching 2×10^{-5} cm^3/sec for a particle of 1 μm diameter. For a H_2O vapor pressure of 30 Torr (10^{18} molecules/cm^3) (vapor pressure at 29°C) the lifetime required for a particle to grow to a droplet containing 3×10^{13} H_2O molecules (i.e., $\sim$10 μm diameter) would be $\sim$50 sec for 1% supersaturation. At lower H_2O vapor pressures (i.e., lower temperatures) and lower supersaturations, the times involved could reach one hour or more.

REFERENCES

Amelin, A. G. (1966). "Theory of Fog Condensation" (English transl., 1967). Monson Wiener Bindery, Ltd., Jerusalem, Israel.

Bricard, J., Billard, F., and Madelaine, G. (1968). *J. Geophys. Res.* **73,** 4487, "Formation and Evolution of Nuclei of Condensation that Appear in Air Initially Free of Aerosols."

Cadle, R. D. (1966). "Particles in the Atmosphere and Space." Van Nostrand-Reinhold, Princeton, New Jersey.

Cadle, R. D. (1972). Climatic Impact Assessment Program, *Proc. Surv. Conf.* p. 130, "Composition of the Stratospheric Sulfate Layer."

Cadle, R. D., and Grams, G. W. (1975). *Rev. Geophys. Space Phys.* **13,** 475, "Stratospheric Aerosol Particles and Their Optical Properties."

Castleman, A. W. Jr. (1974). *Space Sci. Rev.* **15,** 547, "Nucleation Processes and Aerosol Chemistry."

Castleman, A. W. Jr. (1975). Unpublished work at Brookhaven Nat. Lab.

Daubendiek, R. L., and McMillan, G. R. (1973). *J. Amer. Chem. Soc.* **95,** 1374, "Photochemistry of the Gas-Solid Interface. The System 1,3-Pentadiene-Pyrex."

de Pena, R. G., Olszyna, K., and Heicklen, J. (1973). *J. Phys. Chem.* **77,** 438, "Kinetics of Particle Growth I. Ammonium Nitrate from the Ammonia-Ozone Reaction."

Goetz, A., and Pueschel, R. F. (1965). *J. Air Pollut. Contr. Ass.* **15,** 90, "The Effect of Nucleating Particulates on Photochemical Aerosol Formation."

Goetz, A. and Pueschel, R. F. (1967). *Atmos. Environ.* **1,** 287, "Basic Mechanisms of Photochemical Aerosol Formation."

Heicklen, J., and Cohen, N. (1968). *Advan. Photochem.* **5,** 157, "The Role of Nitric Oxide in Photochemistry."

Hidy, G. M., and Burton, C. S. (1975). *Int. J. Chem. Kinet.* Symp. No. 1, "Chemical Kinetics Data for the Upper and Lower Atmosphere," p. 509. "Atmospheric Aerosol Formation by Chemical Reactions."

Hisatsune, I. C., and Heicklen, J. (1975). *Can. J. Chem.* **53,** 2646 "Infrared Spectroscopic Study of Ammonia-Sulfur Dioxide-Water Solid State System"

Hofmann, D. J. (1974). *Can. J. Chem.* **52,** 1519, "Stratospheric Aerosol Determination."

Husar, R. B., and Whitby, K. T. (1973). *Environ. Sci. Tech.* **7,** 241, "Growth Mechanisms and Size Spectra of Photochemical Aerosols."

Landreth, R., de Pena, R. G., and Heicklen, J. (1975). Unpublished work at Penn State Univ., "Redetermination of the Thermodynamics of the Reactions $(NH_3)_n \cdot SO_2(s) \leftrightarrows nNH_3(g) + SO_2(g)$."

Lee, R. E. Jr., and Goranson, S. (1972). Environmental Protection Agency Rep. AP-108, "Cascade Impactor Network."

Luchinskii, G. P. (1956). *Zh. Fiz. Khim.* **30,** 1208, "Physical-Chemical Study of the H_2O-SO_3 System."

Luria, M., de Pena, R. G., Olszyna, K. J., and Heicklen, J. (1974a). *J. Phys. Chem.* **78,** 325, "Kinetics of Particle Growth III: Particle Formation in the Photolysis of Sulfur Dioxide-Acetylene Mixtures."

Luria, M., Olszyna, K. J., de Pena, R. G., and Heicklen, J. (1974b). *Aerosol Sci.* **5,** 435, "Kinetics of Particle Growth-V: Particle Formation in the Photolysis of SO_2-Allene Mixtures."

McLaren, E., Yencha, A. J., Kushnir, J. M., and Mohnen, V. A. (1974). *Tellus* **26,** 291, "Some New Thermal Data and Interpretations for the System SO_2-NH_3-H_2O-O_2."

Mitschele, J., and Heicklen, J. (1972). Presented at 163rd Amer. Chem. Soc. Nat. Meeting, Boston, "Gas Phase Reaction of NO_2 with *trans*-Butene-2."

Olszyna, K. J., de Pena, R. G., Luria, M., and Heicklen, J. (1974). *Aerosol Sci.* **5,** 421, "Kinetics of Particle Growth-IV: NH_4NO_3 from the NH_3-O_3 Reaction Revisited."

Olszyna, K. J., de Pena, R. G., and Heicklen, J. (1976). *Int. J. Chem. Kinet.* **8,** 357 "Kinetics of Particle Growth VII. NH_4NO_3 from the NH_3-O_3 Reaction in the Presence of Air and Water Vapor."

Ripperton, L. A., and Lillian, D. (1970). *J. Air Pollut. Contr. Ass.* **21,** 629, "The Effect of Water Vapor on Ozone Synthesis in the Photo-oxidation of Alpha-Pinene."

Scargill, D. (1971). *J. Chem. Soc. A* 2461, "Dissociation Constants of Anhydrous Ammonium Sulphite and Ammonium Pyrosulphite Prepared by Gas-phase Reactions."

Schroeder, W. H., and Urone, P. (1974). *Environ. Sci. Tech.* **8,** 756, "Formation of Nitrosyl Chloride from Salt Particles in Air."

Smith, B. M., Wagman, J., and Fish, B. R. (1969). *Environ. Sci. Tech.* **3,** 558, "Interaction of Airborne Particles with Gases."

Stull, D. R., and Prophet, H. (1971). "JANAF Thermochemical Tubles," 2nd ed. Nat. Bur. of Std., Washington, D.C.

Urone, P., Lutsep, H., Noyes, C. M., and Parcher, J. F. (1968). *Environ. Sci. Tech.* **2,** 611, "Static Studies of Sulfur Dioxide Reactions in Air."

Wagman, D. D., Evans, W. H., Parker, V. B., Halow, I., Bailey, S. M., and Schumm, R. H. (1968). Nat. Bur. Std. Tech. Note 270-3, "Selected Values of Chemical Thermodynamic Properties."

Chapter X

CONTROL METHODS

The main sources of atmospheric pollution are:

1. motor vehicles and other sources of transportation,
2. fuel consumption, particularly in industry and power plants, but also in commercial and domestic uses,
3. industrial processes,
4. waste disposal by incineration, both public and private, and
5. solvent emissions from industrial, commercial, and domestic uses.

In this chapter we shall discuss the methods used to control each of these sources and examine their effectiveness.

STATIONARY SOURCES

The stationary sources of emission include fuel consumption in power plants; industrial, commercial, and domestic heating; solvent emissions: industrial, domestic, and commercial (e.g., dry cleaning); and incineration, both public and private. The pollutants include hydrocarbons, CO, oxides of nitrogen, particulates, and SO_2.

Hydrocarbons

Hydrocarbon control from stationary sources has been effective where applied. The methods used are (NAPCA Publication No. AP-64, 1970)

1. evaporation control,
2. combustion of escaping vapors,
3. adsorption, principally by activated charcoal,

4. absorption into liquid solvents,
5. condensation of vapors by cooling,
6. substituting photochemically nonreactive materials for reactive hydrocarbons.

The difficulty of control is not its effectiveness, but the lack of use. It is difficult to enforce solvent release at dry cleaning establishments. Private home enforcement is even more difficult. Large amounts of solvents are emitted in spray cans used in homes. The solvents are generally the chemically unreactive and water-insoluble perhaloalkanes. However, their very inertness is a problem in that they are accumulating in the atmosphere. One consequence is that they rise into the lower stratosphere and photodissociate to give chlorine atoms, which attack the protective ozone layer, as discussed in Chapter II.

CO

In flames CO emissions are lowest at the stoichiometric air–fuel composition (NAPCA Publication No. AP-62, 1970). Under fuel-rich conditions, combustion is not complete, and CO emissions are enhanced. For fuel-lean conditions, the flame temperature is decreased and the residence times are shorter; this also leads to increased CO. To minimize CO, combustors should have high turbulence, sufficient residence time, high temperatures, and use a stoichiometric air–fuel ratio.

NO_x

Oxides of nitrogen are emitted from oil- and gas-filled boilers. These emissions can be reduced by two-stage combustion, using less excess air, or by furnace modifications to reduce the peak temperatures and alter the time–temperature history (APCO Publication No. AP-84, 1971).

An approach that is feasible in principle is to use catalytic conversion of NO to N_2 in exhaust gases (APCO Publication No. AP-84, 1971). There are copper catalysts that will promote the reaction

$$NO + CO \rightarrow CO_2 + \tfrac{1}{2}N_2 \tag{1}$$

Unfortunately CO is needed, so that the flame must be fuel rich, which gives rise to CO and hydrocarbon emissions.

Particles

Particles can be removed from stack gases by electrostatic precipitators with >99% efficiency. However, the efficiency is reduced at the normal operating temperature of 270 to 300°F by reducing the sulfur content of the coal, because it is more difficult to charge the particles. Thus at 0.5% sulfur content, the precipitators may be only 50% efficient. This problem is reduced by operating at higher temperatures. High-temperature (580–610°F) precipitators have been developed with efficiencies 99.6% independent of the sulfur content of the coal (Amer. Chem. Soc. Rep., 1969).

H_2S and SO_2

SO_2 is produced primarily from the burning of sulfur-containing fuels. However, it is also an important waste in metal refining, where the sulfides are often converted to oxides:

$$MS + \tfrac{3}{2}O_2 \rightarrow MO + SO_2 \qquad (2)$$

This process is important for M = Cn, Zn, Fe, and Pb. Other industrial processes can give SO_2 (sulfuric acid manufacture) or H_2S (paper mills).

Bituminous coal, an important industrial and power plant fuel, can contain up to 7% sulfur. Anthracite coal contains $<0.7\%$ sulfur, but it only accounts for 4% of coal consumption. The sulfur in coal is present as pyrites (FeS), organic compounds, and sulfates.

The methods of reducing SO_2 emissions from fuels include (NAPCA Publication No. AP-52, 1969): (1) fuel substitution, (2) dispersion from tall stacks, (3) modification of combustion processes, (4) desulfurization of fuel, and (5) flue gas desulfurization.

1. In fuel substitution, a high-sulfur fuel is replaced by a low-sulfur fuel. This is always expensive and will become more so as the energy crisis becomes more serious.
2. Dispersion of flue gases from tall stacks does not reduce emissions, but only deposits them where they cause less harm. Stacks up to 1000 ft tall have been constructed. Some of the emissions from these stacks still reach the earth, but at considerably diluted levels.
3. Modification of combustion processes includes heat recovery, improving generating system efficiency, high-pressure combustion, magnetohydrodynamics, electrogas dynamics, and two-step combustion. In two-step combustion, the first step is a gasification procedure that produces H_2S, which can be removed, and the second step is the combustion process.
4. Desulfurization of coal involves mechanical removal of some pyrites by either wet or dry settling, and liquefaction and gasification processes. Desulfurization of oil is often done by treatment with H_2 over a catalyst to produce H_2S.
5. A typical flue gas from an electric power plant burning 2.4% sulfur coal contains 1500 ppm SO_2, 10 ppm SO_3, 8.5 gm/m^3 particulates, and 400 ppm NO_x. There are 60 or 70 processes for flue gas desulfurization.

These include sorption on activated carbon, metal oxide sorption, aqueous solution sorption, catalytic oxidation, and limestone reactions. The four most important processes are: (a) alkalized alumina process, (b) limestone reaction, (c) catalytic oxidation, and (d) Beckwell SO_2 recovery process.

a. In the alkalized alumina (partly Al_2O_3, partly Na_2O) process, the oxide spheres absorb SO_2 in the flue gas, giving up to 90% recovery. The SO_2-saturated spheres are then sent to a regenerating chamber where SO_2 is regenerated at 300–650°F. The SO_2 is then converted to H_2S by reaction with H_2 and CO at 1200°F.

b. In the limestone–dolomite recovery process, the limestone–dolomite is placed directly into the boiler to produce metal oxides:

$$CaCO_3 \xrightarrow{\Delta} CaO + CO_2 \tag{3}$$

$$MgCO_3 \xrightarrow{\Delta} MgO + CO_2 \tag{4}$$

These oxides then react with SO_2 and SO_3 in the flue gas to produce sulfites and sulfates, which can be removed by electrostatic precipitation. The SO_3 reacts completely, but only about 25% of the SO_2 is removed in the vapor phase. Complete SO_2 removal occurs in the aqueous phase in a scrubber before the electrostatic precipitator. This method has the disadvantage that there are then large amounts of sulfates to dispose.

c. In catalytic oxidation, the SO_2 is converted to SO_3 over a V_2O_5 catalyst. The SO_3 then reacts with H_2O vapor to give H_2SO_4, which is either condensed by cooling or reacted with NH_3 to produce $(NH_4)_2SO_4$. Particulate matter interferes with the catalyst, so that it must be removed first.

d. In the Beckwell recovery process, SO_2 is scrubbed with K_2SO_4 to give $KHSO_4$ and finally a pyrosulfite precipitate that can be converted to sulfur.

The H_2S produced in coal gasification, liquid fuel desulfurization, the alkalized alumina recovery process, or in papermill effluent can be removed. Often this is done by the Claus process, which produces marketable elemental sulfur via

$$2H_2S + 3O_2 \rightarrow 2SO_2 + 2H_2O \tag{5}$$

$$2H_2S + SO_2 \rightarrow S_2 + 2H_2O \tag{6}$$

New York City: A Case History

In 1970, M. Eisenbud, the first administrator of the Environmental Protection Administration of the City of New York, summarized (Eisenbud, 1970) the effects of New York's control policies. In 1966 the New York City Council passed a law that stated:

1. Sulfur content of all fuels would be limited to 1% by the 1969–70 heating season.
2. No incinerators could be installed in newly constructed buildings.
3. All existing apartment house incinerators were to be closed or improved according to a specific timetable.
4. Emission controls were to be installed on all municipal incinerators.
5. All open burning of leaves, refuse, and building demolition materials were to be banned within city limits.

As a result, the SO_2 emissions in 1969 were reduced by 56% over those in 1965. The annual maximum SO_2 level of 2.2 ppm SO_2 dropped to 0.8 ppm in 1969. Figure X-1 shows the number of hours various SO_2 levels

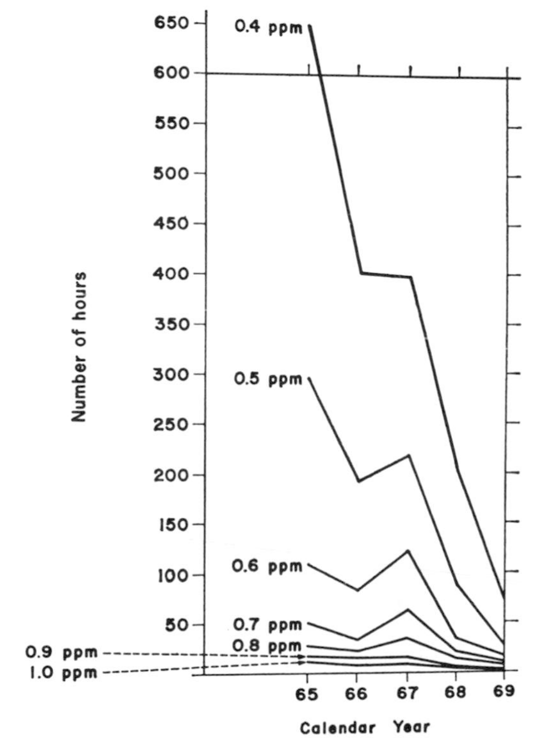

Fig. X-1. Number of hours per year the concentration of SO_2 exceeded the indicated level, 1965–1969. From Eisenbud (1970) with permission of the American Association for the Advancement of Science.

were exceeded each year, and a dramatic drop occurred as the law was enforced.

Unfortunately, the picture has not been completely positive. It was observed that as SO_2 levels dropped, oxidant levels increased, as would be expected from laboratory experiments. Also the energy crisis made it difficult to enforce the first provision of the law. During the 1972–1973 winter, Mayor Lindsay permitted high-sulfur fuels to be burned on occasion to provide heat.

Particulates are the other important pollutant in New York City. In 1969, 69,120 tons of them were emitted, about one-third from space heating, and over one-third from incineration. In 1969, three of the city's eleven incinerators were closed down.

MOTOR VEHICLES

Motor vehicles emit hydrocarbons, partially oxidized hydrocarbons, CO, Pb, and oxides of nitrogen (principally NO). All the emissions except the hydrocarbon fuel are products of combustion and are emitted entirely in the exhaust gases. However, the hydrocarbon fuel can be emitted through evaporation from the fuel tank and carburetor and crankcase blowby. For a car with no control devices the hydrocarbon emissions come approximately 65% from the exhaust gases, 20% from crankcase blowby, and 15% from evaporation from the fuel tank and carburetor (NAPCA Publication No. AP-64, 1970).

Crankcase and Evaporation Loss Control

Crankcase hydrocarbon losses were essentially eliminated by installing positive crankcase ventilation systems on cars starting in the U.S. with 1963 models. These systems recycle crankcase ventilation air and blowby gases instead of venting them to the atmosphere.

Fuel evaporation was eliminated in cars starting with 1971 models by storing fuel vapors in the crankcase or in charcoal cannisters that absorb hydrocarbons. Fuel evaporation could also be reduced by using lower volatile fuels. This would have the advantage of being generally applicable to all cars, not only those with evaporation control devices. Also evaporation losses during transfer at the refinery or gas station would be reduced.

Operation of the Internal Combustion Engine

Before discussing exhaust gases, it is necessary to understand the combustion process in the internal combustion engine. The oxidation is carried on through free radical attack on hydrocarbons in a branched chain explosion (see Chapter V). The key chain branching step is

$$O(^3P) + RH \rightarrow HO + R \tag{7}$$

Thus $O(^3P)$ atoms are the essential free radical intermediate. They are produced readily at elevated temperatures through H atom attack on O_2.

Unfortunately, air is used as the oxidant and consists mainly of N_2. In fact, the gasoline–air mixture at its stoichiometric ratio of 1/14.5 is 73% N_2. At elevated temperatures the energetically unfavorable reaction

$$O(^3P) + N_2 \rightarrow NO + N \tag{8}$$

can occur, followed by

$$N + O_2 \rightarrow NO + O(^3P) \tag{9}$$

Thus $O(^3P)$ is not consumed, but NO is produced. Reaction (8) is 75 kcal/mole endothermic, its rate coefficient being given by the Arrhenius expression: $k_{(8)} = 5 \times 10^{10} \exp\{-75{,}500/RT\}\ M^{-1}\ \text{sec}^{-1}$ (Glick *et al.*, 1957). The competing reaction (7) has a rate coefficient of $1.2 \times 10^{10} \exp\{-4100/RT\}\ M^{-1}\ \text{sec}^{-1}$ if RH is $n\text{-}C_4H_{10}$ (Marsh and Heicklen, 1967). Thus the fraction of NO produced to hydrocarbon consumed is highly dependent on temperature. For various fuel–air ratios and temperatures, the relative importance of reactions (7) and (8) is given in Table X-1 for $n\text{-}C_4H_{10}$ as the fuel.

TABLE X-1

Ratio of $O(^3P)$ *Atoms Reacting with* N_2 *and* $n\text{-}C_4H_{10}$ *at Various Temperatures for Various* $[n\text{-}C_4H_{10}]/[Air]$ *Volume Ratios*

Temperature (°K)	$[Air]/[n\text{-}C_4H_{10}]$				
	21	26	31[a]	36	41
2000	1.07×10^{-6}	1.33×10^{-6}	1.59×10^{-6}	1.84×10^{-6}	2.10×10^{-6}
2200	5.50×10^{-6}	6.81×10^{-6}	8.12×10^{-6}	9.43×10^{-5}	1.07×10^{-5}
2400	2.15×10^{-5}	2.66×10^{-5}	3.17×10^{-5}	3.68×10^{-5}	4.19×10^{-5}
2600	6.79×10^{-5}	8.41×10^{-5}	1.00×10^{-4}	1.16×10^{-4}	1.33×10^{-4}
2800	1.82×10^{-4}	2.26×10^{-4}	2.69×10^{-4}	3.12×10^{-4}	3.56×10^{-4}
3000	4.29×10^{-4}	5.31×10^{-4}	6.33×10^{-4}	7.35×10^{-4}	8.37×10^{-4}
3200	9.06×10^{-4}	1.12×10^{-3}	1.34×10^{-3}	1.55×10^{-3}	1.77×10^{-3}
3400	1.75×10^{-3}	2.17×10^{-3}	2.59×10^{-3}	3.01×10^{-3}	3.43×10^{-3}
3600	3.16×10^{-3}	3.91×10^{-3}	4.66×10^{-3}	5.41×10^{-3}	6.16×10^{-3}
3800	5.34×10^{-3}	6.61×10^{-3}	7.88×10^{-3}	9.15×10^{-3}	1.04×10^{-2}

[a] Stoichiometric.

From the table it can be seen that NO formation is significant above 2600°K, but that it can be reduced by lowering the temperature or the air–fuel ratio. Each O(^{3}P) reacting with N_2 gives two NO molecules, whereas several O(^{3}P) atoms may be needed to oxidize the hydrocarbon. Thus as much as 0.1% of the oxidation might be of N_2 at temperatures >2600°K. Peak temperatures above 2800°K are reached in an efficiently running internal combustion engine. Over the years numerous attempts to reduce peak operating temperatures have resulted in failure, i.e., the power loss was too great a penalty.

The NO formation can be significantly reduced by running on fuel-rich mixtures. Under these circumstances, unburned hydrocarbons and CO increase; thus increasing their exhaust emissions as well as decreasing fuel efficiency. Figure X-2 shows the mole percent CO in auto exhaust as a function of air–fuel ratio. Slightly above the stoichiometric weight ratio of 14.5/1, and greater, CO emission is constant, but small. As the air–fuel ratio is lowered, the mole percent CO rises sharply. The amount of CO seems to depend almost completely on this ratio and is not markedly influenced by the fuel composition or other engine operating parameters (NAPCA Publication No. AP-62, 1970).

In order for the combustion to proceed smoothly in the internal com-

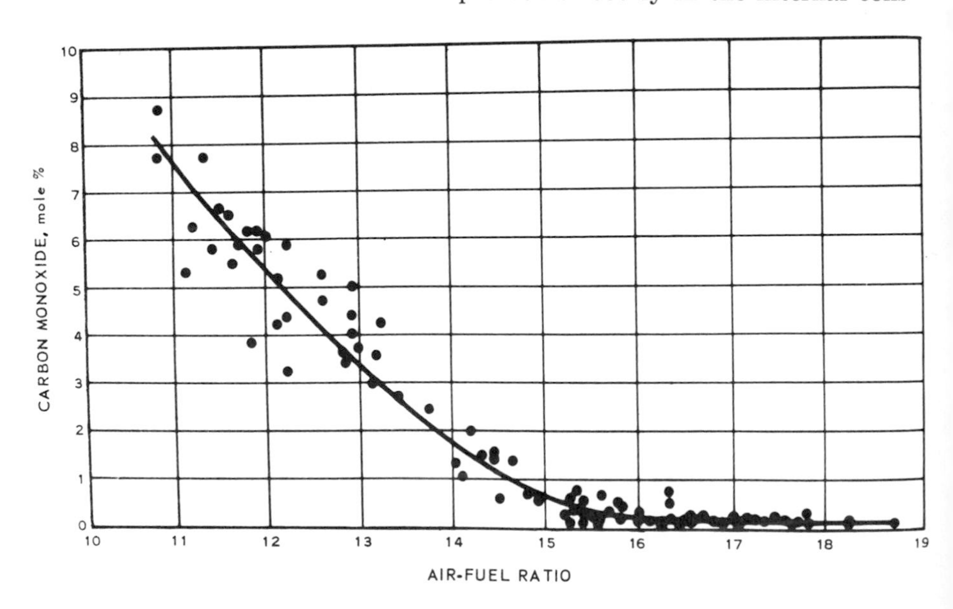

Fig. X-2. Effect of air–fuel ratio on exhaust gas CO concentrations from three test engines. From Hagen and Holiday (1964) with permission of the Society of Automotive Engineers, Inc.

bustion engine, antiknock agents such as $Pb(C_2H_5)_4$ are needed. They act as radical scavengers and prevent uncontrolled chain branching from causing localized unwanted explosions.

CO and Hydrocarbon Control

The 1970 U.S. Federal exhaust emission standards for hydrocarbons were 2.2 gm (measured as hexane)/vehicle mile; and for CO, 23 gm/vehicle mile. There are three possible ways to control exhaust emissions: (1) change the fuel composition, (2) change the engine's operating parameters, or (3) further oxidize the gases after leaving the engine, but before being exhausted to the atmosphere.

A change in fuel composition, by itself, has not proven particularly effective. At first thought it might appear that removing the photochemically active olefins and replacing them by the less reactive alkanes should help. However, this has little effect, because the alkanes are first oxidized to olefins in the combustion process (see Chapter V).

The second method of changing engine operating parameters means, in essence, operating fuel lean to increase combustion. This was the method adopted in the 1960s, and in fact was very successful. Unfortunately it was accompanied by an increase in NO_x emissions.

The third method is afterburning of the hydrocarbons in the exhaust manifold near the exhaust valves where the temperature is still high. In this method, additional air is injected into the exhaust gases and combustion occurs in a reactor of one of two types. The first is a thermal reactor that operates on the principle of permitting the exhaust gases enough time at high temperature to complete the oxidation of hycrocarbons and CO. To do this an insulated manifold of larger than normal volume is needed. The shortcoming of this device is both the larger volume of the manifold and the higher temperatures in it. At the higher temperatures, corrosion is a serious problem. The second type of afterburner is a catalytic convertor. The catalyst must be able to warm quickly, withstand high temperatures, and operate under a variety of conditions. With both types of afterburners, lead in the exhaust gases poses a problem. In the thermal reactor it promotes corrosion, and in the catalytic reactor it tends to poison the catalyst.

NO_x Control

In 1971, California standards required NO_x emission to be <4 gm/vehicle mile. In order to meet this standard several control methods were suggested. The most obvious was using a fuel-rich combustion mixture, but this increases hydrocarbon and CO emissions. Other suggestions included retarded spark timing, which reduces all three emissions, H_2O injection, and lower compression ratios. Unfortunately all these proposals have accompanying drawbacks.

One method that has proved successful on test engines is recycling part of the exhaust gases to reduce peak combustion temperatures. An 80% reduction in NO_x was achieved in this way. Unfortunately, in addition to the loss in engine power, the variables in the recycle system are too delicately balanced for easy mass production (Amer. Chem. Soc. Rep., 1969).

The method currently being adopted is catalytic conversion with noble metal catalysts. The principle is to convert NO to NO_2 through either of reactions (1) or

$$2NO \rightarrow N_2 + O_2 \qquad (10)$$

The method based on reaction (1) requires operating the engine fuel-rich to produce the CO (and reduce NO at the same time). Then after the exhaust has passed through the NO catalytic converter, air would be injected to oxidize the hydrocarbons and CO in a second converter. Needless to say, this double converter arrangement would be expensive.

The presence of lead in the exhaust tends to poison the NO catalytic converters even more severely than the hydrocarbon catalytic converters. Thus for these converters, the $Pb(C_2H_5)_4$ would have to be completely removed from the fuel. Furthermore, the variable air–fuel ratio in the engine and temperature fluctuation in the catalyst bed deteriorates the catalysts, and it is difficult to keep them operating efficiently for the required 50,000 mi.

Pb Control

At present there is no proven technology for removing lead from auto exhausts. In the middle 1960s gasoline contained an average of 2.4 gm Pb/gal gas. Two-thirds of this was exhausted and 25–50% of that became airborne (Amer. Chem. Soc. Rep., 1969). Lead is undesirable mainly because it poisons the catalysts that are being used in exhaust systems to remove CO, hydrocarbons, and NO_x. However it also poses a possible health and ecological hazard in itself, although this has not been proven.

In the late 1960s and early 1970s, efforts were made to remove lead compounds from gasoline. Without the lead, the octane rating of the fuel drops and efficiency is decreased about 25% (Gintzon, 1973). Thus a considerable economic penalty results. In order to compensate for the removal of the lead compounds, other antiknock agents have been incorporated in some gasolines to raise their octane rating. In particular, aromatic compounds have been used, and new high-aromatic gasolines appeared on the market in the early 1970s.

Unfortunately the aromatic hydrocarbons pose serious problems also. The higher the aromatic content of the fuel, the higher the aromatic content of the exhausts. If afterburning is not effected in the exhaust, these aromatics and their aldehydes will enter the atmosphere. Aromatic compounds form aerosols more easily than the straight-chain hydrocarbons. They also have interesting effects on the photochemical cycle.

In particular, benzaldehyde is a significant component of exhaust aromatics. It retards oxidant formation (Dimitriades and Wesson, 1972; Kuntz *et al.*, 1973; Gitchell *et al.*, 1974a) because of its free radical scavenging properties:

$$C_6H_5CHO + R \rightarrow C_6H_5\dot{C}O + RH \quad (11)$$

However, the radical produced in the scavenging process adds O_2 and then NO_2 to produce peroxybenzoyl nitrate, $C_6H_5C(O)O_2NO_2$, an effective eye irritant—200 times more potent than CH_2O (Heuss and Glasson, 1968). Thus Dimitriades and Wesson (1972) have shown that photochemical oxidation is reduced, but eye irritation is more severe, when benzaldehyde is present.

Heuss *et al.* (1974) tested the effect of aromatic content in gasoline on eye irritation. Adding aromatics to a special low-aromatic gasoline increased eye irritation, but had little or no effect on O_3 or NO_2 formation. Isopropylbenzene had the greatest effect, followed by *o*-xylene, *n*-propylbenzene, ethylbenzene, toluene, and benzene. Comparable tests with commercial gasolines did not show the correlation between eye irritation and total aromatic content, presumably because the commercial gasolines all gave high levels (compared to the special low-aromatic gasoline) of eye irritation anyway.

Diesel Engines

Diesel engines use less volatile fuel than internal combustion engines. They emit unburned hydrocarbons and NO_x, but since only about 1% of vehicles use diesel engines, this is not a serious problem. The major problems with diesel engines are the odor and soot content of the exhaust. Very little is known about odor control.

One method of controlling particulate levels in diesel exhausts is with 0.25% barium additives to the diesel fuel (Amer. Chem. Soc. Rep., 1969). It is not clear how these additives work, but they can reduce smoke emission by 50%. Possibly they inhibit dehydrogenation of hydrocarbons to carbon or promote oxidation of carbon particles. About 75% of the barium is exhausted as $BaSO_4$, the exact percentage depending on the sulfur content of the fuel. $BaSO_4$ is insoluble and harmless to humans. However, soluble Ba salts, some of which are formed in the exhaust, are toxic, but not at the levels present in the atmosphere.

Los Angeles: A Case Study

Hamming *et al.* (1973) reviewed the effects of motor vehicle control on Los Angeles air quality. Starting in 1959 air quality goals were established in California. As a result in 1960 a standard of 70% control of crankcase emissions (since raised to 100%) was set. In 1964, a standard of 90% control of evaporation losses was adopted and more stringent goals were set in 1965. Also in 1965, a goal of 65% control of NO_x emissions was set.

To bring about the desired goals the following programs were instituted:

1961 Voluntary crankcase control devises
1963 Mandatory crankcase control devices
1966 Control of hydrocarbon exhaust emissions
1970 Control devices for hydrocarbon evaporation losses
1971 NO_x control devices

The crankcase and evaporation control devices were extremely effective from the date of inception. Hydrocarbon exhaust emissions were lowered either by using increased air–fuel ratios with suitable engine modifications or by thermal reactor afterburners. The former method had the undesirable property of increasing NO_x exhaust emissions. Neither method gave the desired degree of control. The average 50,000 mi emissions from 1966–1969 U.S.-made motor vehicles exceeded the hydrocarbon standard by more than 40% and the CO standard by over 55%.

The effect of control devices on motor vehicles is shown in Fig. X-3. This figure shows the percent control of hydrocarbon and NO_x emissions from motor vehicles since 1960 with projected estimates beyond 1973. Starting in 1975 the estimates are based on both the California standards and more stringent Federal standards, which were to be inaugurated in that year. (As of this writing, March 1975, the Federal standards for HC and CO emissions have been increased from 0.41 to 1.5 and 3.4 to 15.0 gm, respectively, per vehicle mile for 1975 model cars.) Starting in 1960 when hydrocarbon control was initiated, the hydrocarbon levels decreased in Los Angeles. In 1966, when hydrocarbon exhaust emissions control started, NO_x levels increased since most cars achieved their lowered hydrocarbon emission by increasing the air–fuel ratio. In 1971, with the installation of NO_x control devices, NO_x levels also began to fall.

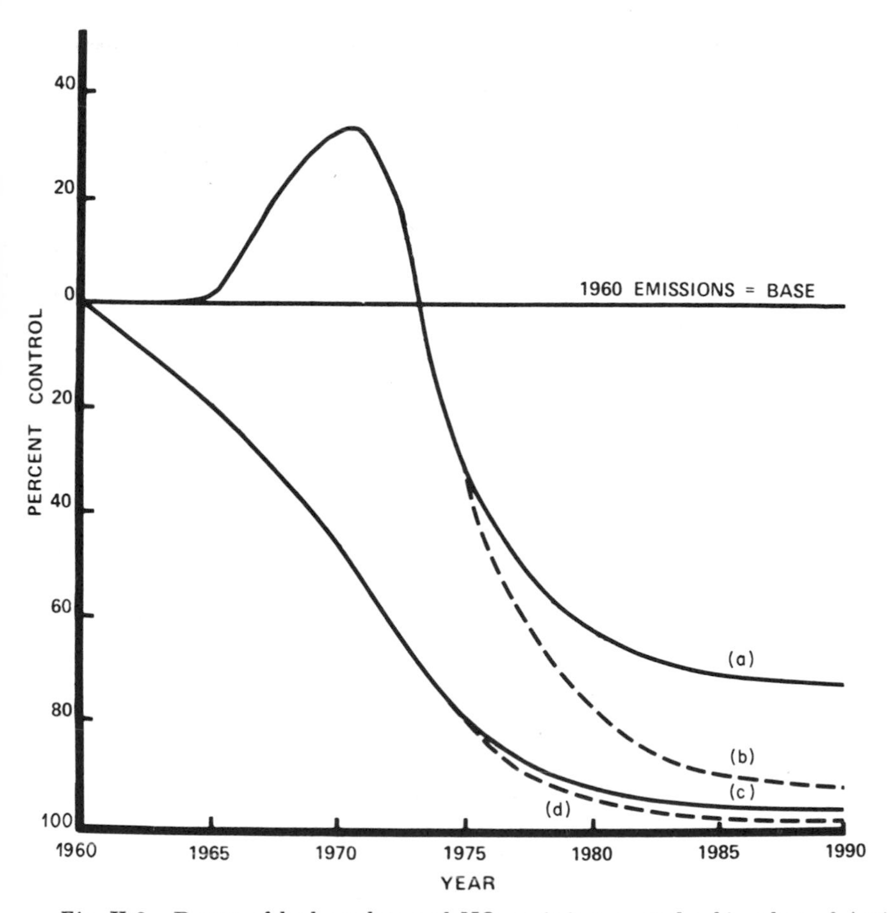

Fig. X-3. Degree of hydrocarbon and NO_x emissions control achieved as of April 1973 and forecast after 1974 using California and Federal Emission Standards. Oxides of nitrogen: (a) California, (b) Federal Standards. Hydrocarbons: (c) California, (d) Federal Standards. From Hamming *et al.* (1973) with permission of the Air Pollution Control Association.

The effect of hydrocarbon and NO_x control on air quality is shown in Figs. X-4–X-6. The measures of air quality used are maximum O_3 concentration (Fig. X-4), eye irritants (Fig. X-5), and steady-state NO_2 concentration (Fig. X-6). The level of these three parameters as a function of initial NO_x and hydrocarbon (expressed as hexane) concentrations is estimated from static smog chamber experiments and the Los Angeles County Air Pollution Control District atmospheric data. On the graphs are points labeled 1960, 1965, 1970, 1971, and 1972. These correspond to the maximum NO_x and hydrocarbon concentration during 6 to 9 A.M. in

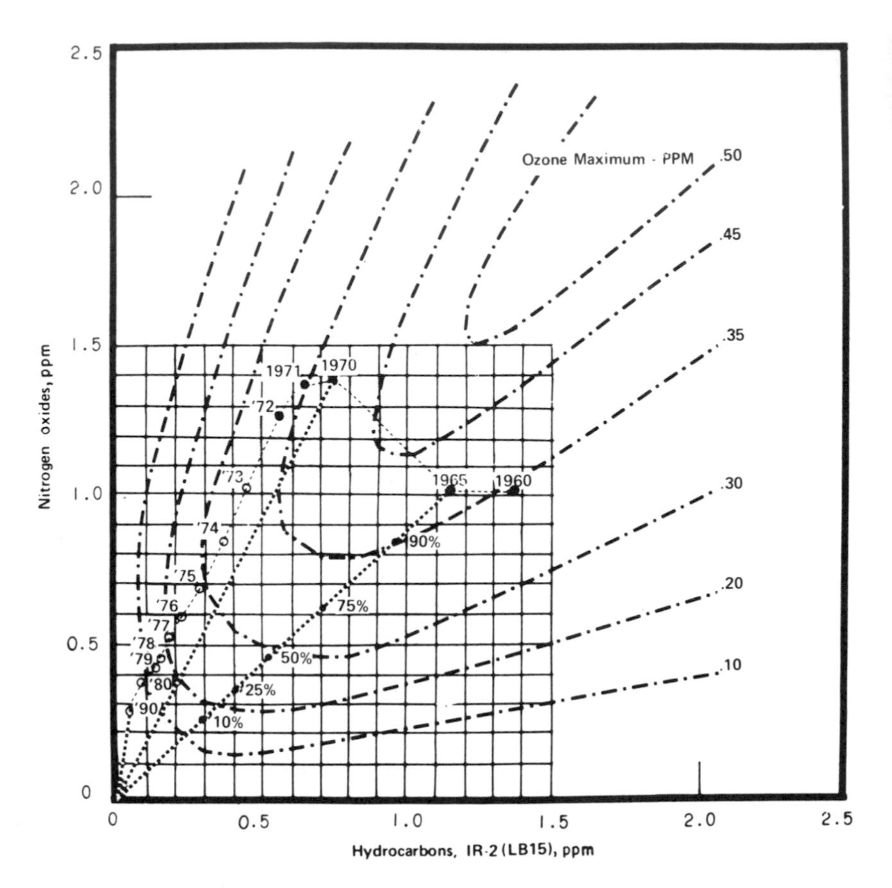

Fig. X-4. Effect of California's auto exhaust control program on maximum ozone concentration. From Hamming *et al.* (1973) with permission of the Air Pollution Control Association.

those years. Also shown are values projected for 1973 to 1990 using California standards. For the 1965 year, the percentage of days in which the HC–NO_x maxima were less than certain values are also specified (e.g., 75% of the days had peak NO_x <0.6 ppm and HC <0.7 ppm).

From 1960 to 1965, NO_x levels were not much affected, but peak hydrocarbon levels dropped 17%. This "control" had the effect of slightly increasing maximum O_3 levels, slightly increasing eye irritation, and leaving steady-state NO_2 levels unaffected. From 1965 to 1970 maximum HC was further reduced to about 50% of 1960 values, but NO_x rose almost 40%. Maximum O_3 levels rose, reaching a maximum in 1967 and 1968, and then dropped back to their 1965 levels by 1970. However, eye irritation was

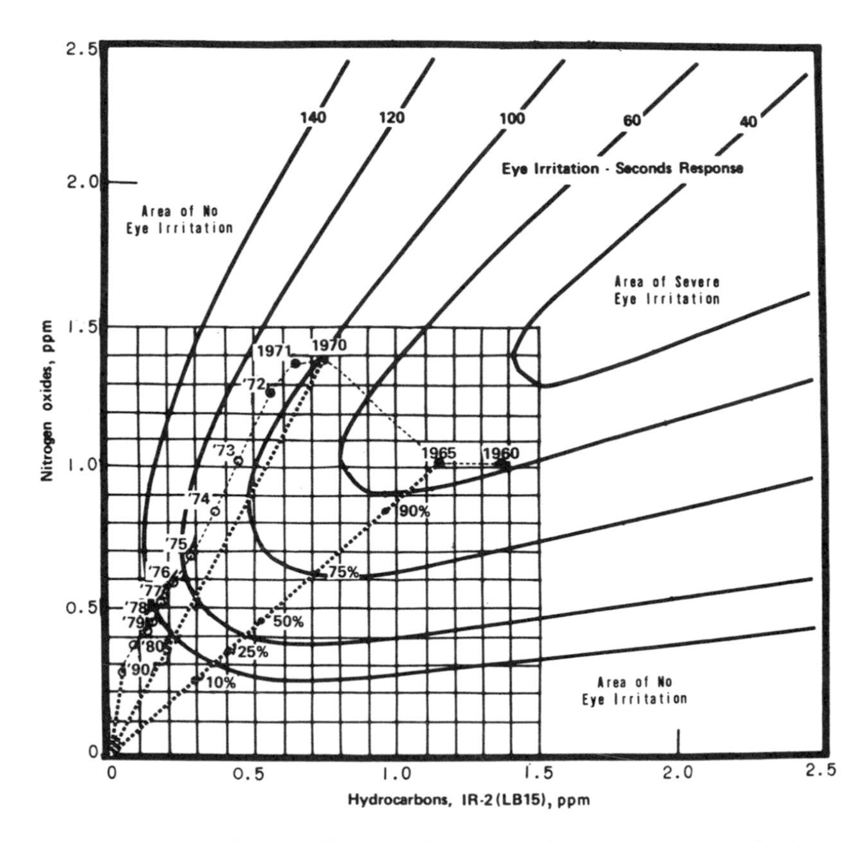

Fig. X-5. Effect of California's auto exhaust control program on atmospheric concentrations of eye irritants. From Hamming *et al.* (1973) with permission of the Air Pollution Control Association.

noticeably reduced, although steady-state NO_2 concentrations of course increased. Starting in 1971, NO_x control was instituted and the quality of air improved by all three measures. The projections for 1973 and beyond show increasing yearly improvement, with federal air quality standards being met by 1980 or soon thereafter.

More detailed information on O_3 levels was given by Tiao *et al.* (1975), who reported the relative frequency of daily maximum hourly readings at various levels for seven locations in the Los Angeles basis. The data are in Fig. X-7. In west and downtown Los Angeles, as well as Burbank, there has been a steady improvement from 1960 when controls were first started. However, in Pasadena and Azusa there was no improvement from 1965 to 1970, but in 1971 and 1972 the O_3 was reduced.

The question that remains is whether the envisioned degree of control

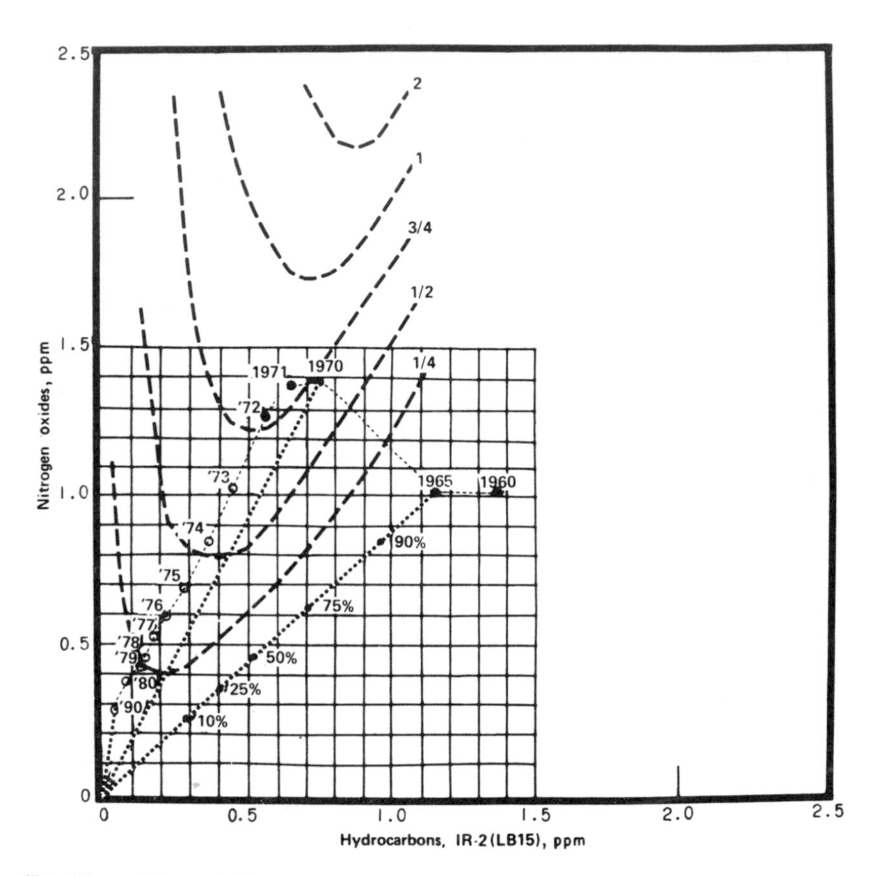

Fig. X-6. Effect of California's auto exhaust control program on steady-state NO_2 concentration. Numbers on curves indicate NO_2 concentration in parts per million. From Hamming *et al.* (1973) with permission of the Air Pollution Control Association.

can be realized. The estimates in the figures are based on motor vehicle control. They assume that all hydrocarbons and NO_x can be controlled by controlling motor vehicle emissions. In 1969, motor vehicles accounted for almost all the CO emissions, about 70% of hydrocarbons (90% of reactive hydrocarbons), and 65% of oxides of nitrogen (Mosher *et al.*, 1969). Of the hydrocarbons, about 90% of the olefins, 60% of the paraffins, and 45% of the aromatics in downtown Los Angeles come from vehicular emissions (Lonneman *et al.*, 1974). Haagen-Smit (1972) estimates that by 1990 non-motor vehicle sources will account for NO_x emissions >70% of 1960 total NO_x emissions in the South Coastal Basin of California. Consequently maximum NO_x levels <0.8 ppm may never be achieved unless stationary source controls are also installed.

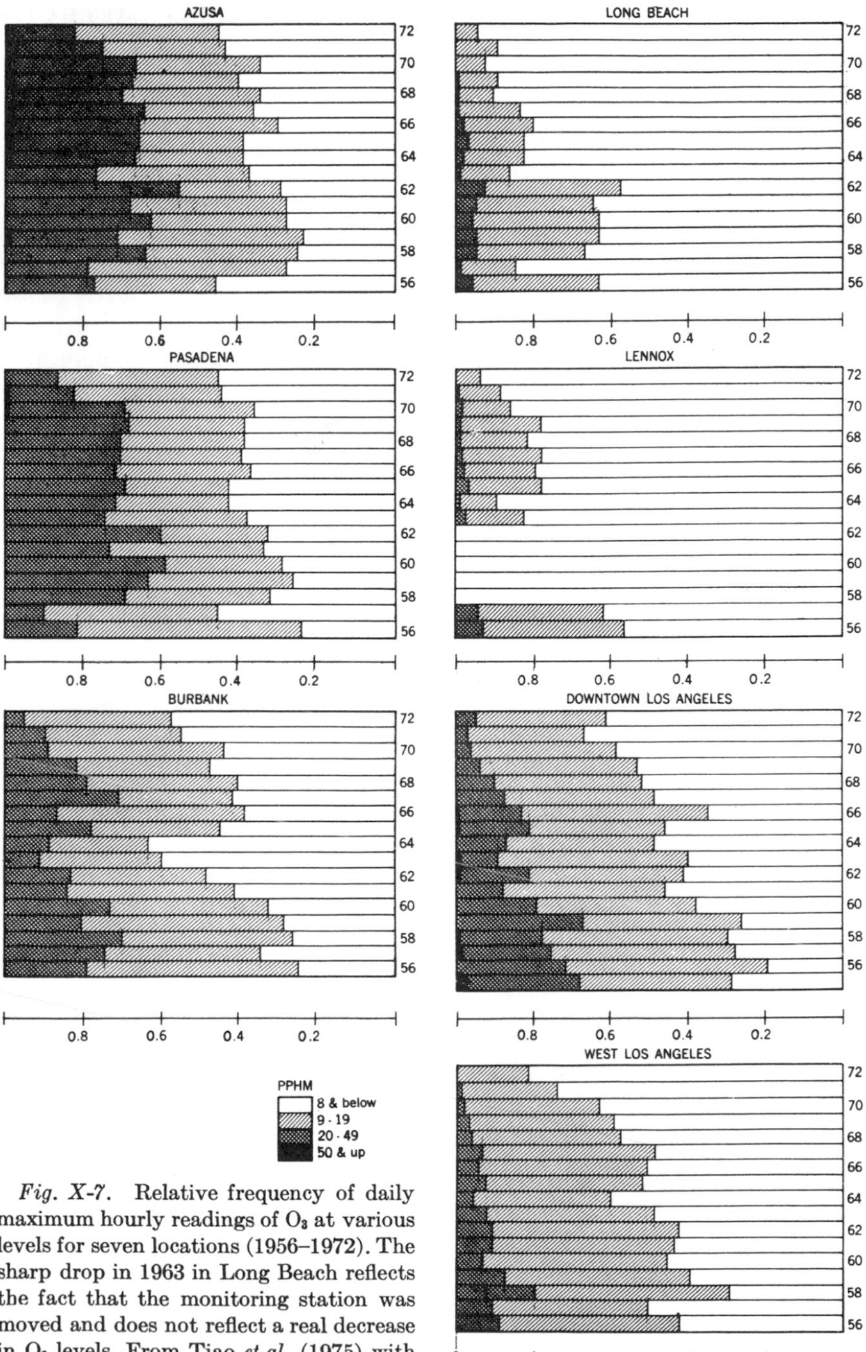

Fig. X-7. Relative frequency of daily maximum hourly readings of O_3 at various levels for seven locations (1956–1972). The sharp drop in 1963 in Long Beach reflects the fact that the monitoring station was moved and does not reflect a real decrease in O_3 levels. From Tiao *et al.* (1975) with permission of the Air Pollution Control Association.

Finally can one predict what level of control is needed to provide a desired improvement? In Fig. VI-6 a model calculation was done for a C_2H_4–NO–NO_2 mixture to compute the daytime concentration profiles. For a day with an overhead noonday sun, an initial mixture of 1.0 ppm C_2H_4, 0.5 ppm NO, and 0.05 ppm NO_2 produced 0.265 ppm O_3 shortly after noon. The concentration profiles are reproduced in Fig. X-8. A comparable computation is done with a fourfold reduction in each of the initial reactants, and the results also displayed in Fig. X-8. The peak O_3 concentration is delayed in appearing until later in the day, but it reaches 0.161 ppm, i.e.,

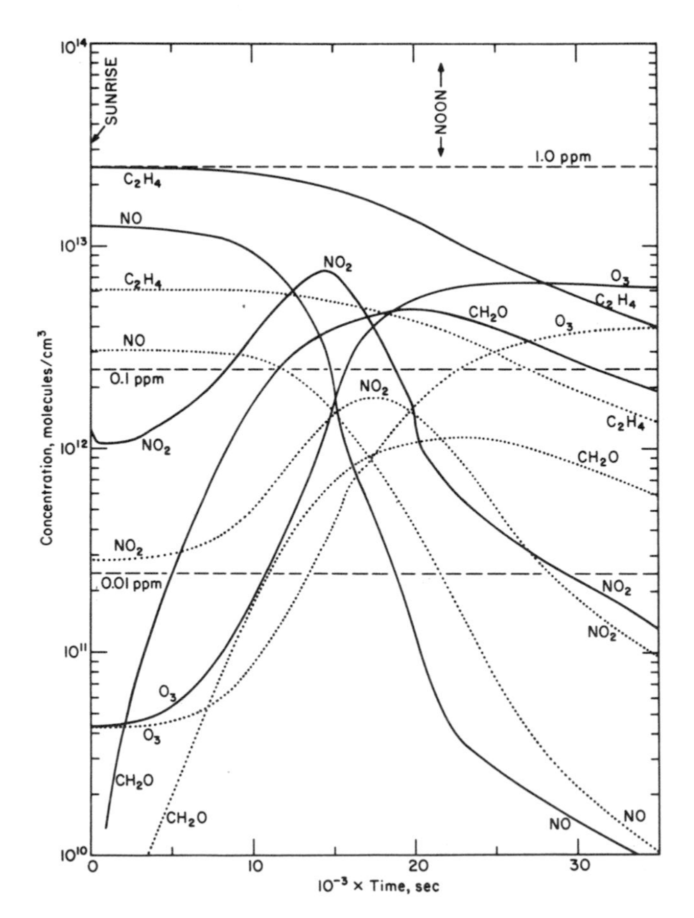

Fig. X-8. Calculated concentrations of some atmospheric species versus time of day for a day in which the noonday sun is overhead starting with C_2H_4, NO, and NO_2, based on the reaction scheme in Chapter VI corresponding to Fig. VI-6. —, – – –, two different sets of starting concentrations.

61% of the peak concentration for the original computation. On the other hand, a further reduction in the C_2H_4 concentration to 0.125 ppm, without any further reduction in NO_x, so delays the appearance of the O_3 concentration maximum that the standards are not violated; $[O_3]$ never exceeds 0.046 ppm. Thus we see that for high pollution levels, a reduction in NO_x or hydrocarbons does not reduce the O_3 levels very much, but if the pollutant levels are sufficiently reduced, there is more than a corresponding decline in O_3.

Projected Motor Vehicle Controls

In addition to setting control standards, one needs to consider whether the projected motor vehicle control standards can be reached, and what they will cost. This was the subject of the Report by the Committee on Motor Vehicle Emissions of the U.S. National Academy of Sciences (Gintzon, 1973).

The U.S. Federal standards set for cars manufactured in the 1970s are given in Table X-2. In developing systems to meet the standards of 1975, the engines must also be compatible with those of 1976. The greater control of NO_x required by the initial 1976 standards can be achieved by increasing exhaust gas recycle, adding a catalytic converter for NO to the exhaust system, or most likely, both.

TABLE X-2

U.S. Federal Standards for Automobile Emissions

	Emission (gm/vehicle mile)		
Car model	HC	CO	NO_x
1972	3.4	39.0	—
1973–1974	3.4	39.0	3.0
1975[a]	1.5[b]	15.0[c]	3.0
1976[a]	1.5[b]	15.0[c]	3.0[d]
1977[a]	1.5[b]	15.0[c]	2.0[d]

[a] In March 1975, EPA Director Train recommended that 1975 standards be retained through 1979 models and that 1980 and 1981 models use 1975 California standards of 0.9 gm/mile for HC, 9.0 gm/mile for CO, and 2.0 gm/mile for NO_x.

[b] Originally set at 0.41 gm and then relaxed.

[c] Originally set at 3.4 gm and then relaxed.

[d] Originally set at 0.41 gm and then relaxed.

The National Academy of Sciences Report concludes that four systems can meet the initially set 1975 standards and five systems the initially set 1976 standards. They are

Initially Set 1975 Standards

1. Modified conventional engine equipped with an oxidation catalyst.
2. Carbureted stratified charge engine.
3. Wankel engine equipped with an exhaust thermal reactor.
4. Diesel engine.

Initially Set 1976 Standards

1. Modified conventional engine equipped with dual catalysts.
2. Modified conventional engine equipped with dual catalysts plus thermal reactor.
3. Modified conventional engine equipped with two thermal reactors and a reduction catalyst.
4. Modified conventional engine equipped with three-way catalyst and electronic fuel injection.
5. Stratified charge engine employing fuel injection and equipped with an oxidation catalyst.

All these systems will cost the consumer money; the estimated expenses are listed in Table X-3. The system being developed by most U.S. automobile manufacturers is the dual catalyst system. It is the most disadvantageous with respect to initial cost, fuel economy, maintainability, and durability. When it was realized that there are approximately 10^8 cars in the U.S., the total annual cost for this system was estimated at \$23.5 billion, based on \$0.40/gal of gasoline. With 1975 gasoline prices of \$0.55/gal, this estimate should be raised to \$27.3 billion. Even this may be an underestimate, since Prud'homme (1974) has pointed out that refinery capital costs were not adequately figured in the National Academy of Sciences Report (Gintzon, 1973).

The carbureted stratified charge engine has been developed by Honda and achieves charge stratification with a prechamber and dual carburetor. The spark plug is located in the prechamber, which operates with a fuel-rich mixture to minimize NO_x formation. It also ensures good ignition. The mixture at the prechamber exit is approximately stoichiometric and propagates the flame into the fuel-lean mixture in the main chamber. The main chamber operates at lower temperature than the maximum in the prechamber so that not much NO_x is produced. Because of the fuel-lean mixture in the main chamber, unburned hydrocarbons are minimized. Emissions of NO_x, CO, and hydrocarbons are all lower than for a conventional engine at the same lean air–fuel ratio. In three tests without exhaust gas

TABLE X-3

Total Annual Cost to Customer of Emission Controls For Various Body and Engine Combinations[a]

	Sub-compact	Com-pact	Inter-mediate A[b]	Inter-mediate B[c]	Standard	Standard luxury	Luxury
Dual-catalyst system	127	138	215	243	263	285	361
Diesel with EGR	18	−45[d]	30	−18[d]	36	41	14
Wankel	86	117	120	105	133	198	148
Stratified-charge 3-valve	65	77	67	—	—	—	—
Feedback-controlled with electronic fuel injection	51	50	85	52	87	97	114

[a] Compared to cost of 1970 base-line car and amortized over five years. Includes increments in fuel and maintenance costs. Fuel cost estimates were based on $0.40/gal for all fuels and all years. From Gintzon (1973).

[b] Intermediate A bodies are those intermediates that currently use 6-cylinder engines.

[c] Intermediate B bodies are those intermediates that currently use 8-cylinder engines.

[d] The diesel 4-cylinder is used in compact cars and the 6-cylinder is used in intermediate B cars.

recycle or exhaust treatment devices, the average emissions were 0.25 gm/mile HC, 2.5 gm/mile CO, and 0.43 gm/mile NO_x, i.e., below or close to the initially set U.S. 1977 standards. The Honda motor has only been tested on small cars, and based on them the annual cost of using these motors can be estimated to be about $7.5 billion (1972 prices).

The Wankel rotary engine is a small, smooth-running, light-weight, relatively inexpensive engine. For the uncontrolled engine, its fuel economy is lower and its CO, HC, and NO_x emissions are higher than a conventional internal combustion engine. NO_x can be controlled by exhaust gas recycle, and hydrocarbons and CO by a thermal reactor afterburner. However, because of the fuel penalty, this system would cost ~$12 billion annually (1972 prices).

The diesel engine has the advantage of greater fuel economy than conventional engines and may actually be cheaper to run for some models. However, the disadvantages are that it is a heavy engine, and technology for use on small cars is limited and scattered. Also there is no indication that the diesel engine can meet the initially set 1977 standards for NO_x.

Another possibility of control is the feedback control system for air–fuel ratio to be used with conventional engines. In such a system an O_2 sensor detects the O_2 level in exhaust gases and feeds back a signal to an electronic fuel injector control unit, which adjusts the fuel–air supply and maintains control of the ratio. This careful control minimizes unwanted emissions. The annual cost of such a system is estimated to be ~\$8 billion.

Other systems that might be used but have not yet been shown to be feasible are gas turbines, Stirling engines, electronically driven vehicles, the Rankine engine, and other engines.

The annual costs of motor vehicle control programs are staggering, and one wonders if, or even why, the public should pay them. They will affect measurably the gross national product and the foreign balance of payments. Are these costs justified? Could the money be spent more usefully elsewhere? There will be a great increase in petroleum consumption, aggravating the energy crisis. The catalyst systems could pose a heavy drain on the world supply of noble metals.

CHEMICAL CONTROL OF THE ATMOSPHERE

Since the photochemical smog reaction is a long-chain free-radical process, the possibility of retarding the reaction by adding free-radical scavengers to the atmosphere is possible. This idea was apparently first proposed by Levine (1961), who added I_2 to a simulated atmosphere and observed that O_3 levels were reduced. This effect was confirmed by Stephens *et al.* (1962), who also found that aldehyde, PAN, aerosol production, and olefin consumption were retarded. It is not clear how I_2 works in this system. It is a well-known radical scavenger, but it scavenges the same radicals that O_2 does, so that radical scavenging is too inefficient to account for the results. Presumably the I_2 photodissociates and reacts with O_3 and NO_2:

$$I_2 + h\nu \rightarrow 2I \quad (12)$$

$$I + O_3 \rightarrow IO + O_2 \quad (13)$$

$$I + NO_2 \rightarrow ? \quad (14)$$

Stephens *et al.* also tried many other known radical scavengers but found them to be ineffective. However, in our laboratory we have tested some of these as well as others and found marked inhibition in model systems with C_3H_6 and NO at 16 and 8 ppm, respectively (Gitchell *et al.*,

1974a, b; Jayanty *et al.*, 1974). We reasoned that an effective free-radical terminator must not only scavenge the radicals but also terminate the chain. Thus the scavenging reaction is

$$RH + OH \rightarrow R + H_2O \tag{15}$$

where RH is the additive. However, the product radical must not have an H atom on the α-carbon position or the chain will be propagated:

$$R_1R_2CH + O_2 \longrightarrow R_1R_2CHO_2 \tag{16}$$

$$R_1R_2CHO_2 + NO \rightarrow R_1R_2CHO + NO_2 \tag{17}$$

$$R_1R_2CHO + O_2 \rightarrow R_1R_2C{=}O + HO_2 \tag{18}$$

Reaction (18) produces HO_2 and continues the chain. However, if there is no H on the α-carbon, then reaction (18) is too slow to compete with

$$RO + NO \rightarrow RONO \tag{19}$$

$$RO + NO_2 \rightarrow RONO_2 \tag{20}$$

and termination occurs. However, caution must be used to be sure that a free radical with an H on the α-carbon is really not produced. Thus two obvious additives that appear to be suitable, i-C_4H_{10} and CH_3CHO, actually promote rather than retard the chain process. With i-C_4H_{10} the scheme is

$$i\text{-}C_4H_{10} + HO \rightarrow t\text{-}C_4H_9 + H_2O \tag{21}$$

$$t\text{-}C_4H_9 + O_2 \rightarrow t\text{-}C_4H_9O_2 \tag{22}$$

$$t\text{-}C_4H_9O_2 + NO \rightarrow t\text{-}C_4H_9O + NO_2 \tag{23}$$

$$t\text{-}C_4H_9O \rightarrow (CH_3)_2CO + CH_3 \tag{24}$$

The resulting CH_3 radical contains H atoms on the α-carbon and continues the chain. With CH_3CHO the chain cycle is

$$CH_3CHO + HO \rightarrow CH_3CO + H_2O \tag{25}$$

$$CH_3CO + O_2 \rightarrow CH_3C(O)O_2 \tag{26}$$

$$CH_3C(O)O_2 + NO \rightarrow CH_3C(O)O + NO_2 \tag{27}$$

$$CH_3C(O)O \rightarrow CH_3 + CO_2 \tag{28}$$

and again CH_3 is produced.

Gitchell *et al.* (1974a, b) tested C_6H_5OH, C_6H_5CHO, and naphthalene, and found them all to be effective in retarding the conversion of NO to NO_2 in mixtures of 16 ppm C_3H_6 and 8 ppm NO in about 90 Torr O_2. The respective reactions are

$$C_6H_5OH + HO \rightarrow C_6H_5O + H_2O \tag{29}$$

$$C_6H_5CHO + HO \rightarrow C_6H_5\dot{C}O + H_2O \tag{30}$$

$$C_6H_8 + R \longrightarrow \text{[bicyclic radical with R]} \tag{31}$$

All three radicals have no H on the α-carbon and thus terminate the chain.

Jayanty *et al.* (1974) examined a series of amines. They found no inhibition by NH_3 or $C_2H_5NH_2$, fairly good inhibition by $(C_2H_5)_2NH$ and $(C_2H_5)_3N$, and strong inhibition by $(C_2H_5)_2NOH$ and *N*-methylaniline. The results with the last two compounds are shown in Fig. X-9. About 15% of the additive (compared to C_3H_6) essentially doubles the time to reach the maximum conversion of NO to NO_2. Retardation is initially even more effective, reducing the rate by over a factor of 10 with $(C_2H_5)_2NOH$, and after the additive is nearly consumed the reaction proceeds as in the uninhibited case (Stockburger *et al.*, 1976).

It is not exactly clear why some amines work and some do not. Probably the N—H bond strength plays an important role. However Jayanty *et al.* (1974) pointed out that the two unsuccessful inhibitors NH_3 and $C_2H_5NH_2$ would give rise to radicals with an H atom on the N atom, after abstraction of one H. The other molecules would not. Possibly by ánalogy with the α-carbon argument, the mechanism is similar. The apparent weakness of this argument is that Gitchell *et al.* (1974a) found previously $C_6H_5NH_2$ to be as good an inhibitor as $(C_2H_5)_2NOH$, and it gives rise to a radical with an H on the N atom. However, in this case resonance stabilization may occur

$$C_6H_5\text{—}\dot{N}H \rightleftharpoons \cdot C_6H_5\text{=}NH \tag{32}$$

and this equilibrium might stabilize the radical.

One of the problems with the free-radical inhibitors is that they may promote aerosol formation. Stephens *et al.* (1962) found that phenol slightly decreased the aerosol formation, while aniline increased the aerosol formation. Stephens *et al.* used a stirred-flow reactor, which has been shown by Wilson *et al.* (1971) adversely to affect the aerosol formation. Spicer *et al.* (1974) found that both phenol and aniline produce a marked increase in light scattering due to aerosol formation. This work was done in a 610-ft^3 smog chamber where it is possible that wall contamination could be important (Bufalini *et al.*, 1972).

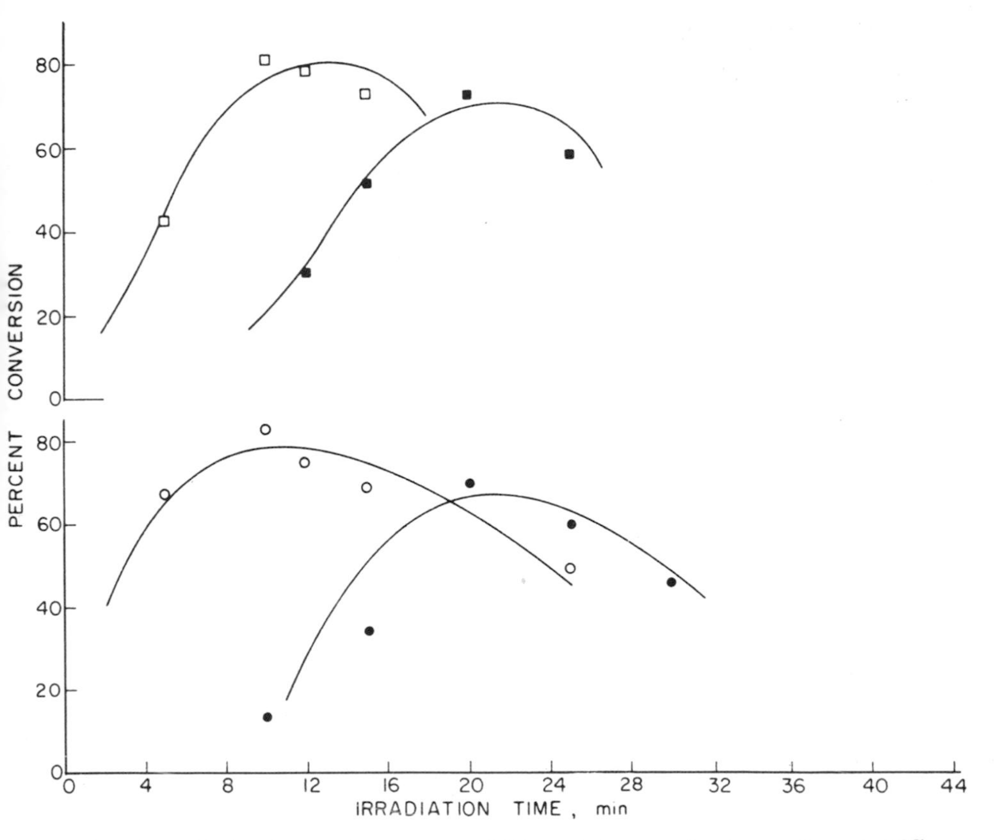

Fig. X-9. Plots of percent conversion of NO to NO_2 versus irradiation time at 25°C in the photolysis of mixtures initially containing 0.012 mTorr NO_2, 9.0 ± 2.5 mTorr NO, 15.3 ± 1.6 mTorr C_3H_6, and 100 Torr O_2 + CO_2 (mainly O_2). The experiments were done with (■) and without (□) 2.2 mTorr $(C_2H_5)_2NOH$ or with (●) and without (○) 2.4 mTorr *N*-methylaniline added. From Jayanty *et al.* (1974) with permission of Pergamon Press.

Stockburger and Heicklen (1976) irradiated mixtures of 10 Torr H_2O, 100 Torr O_2, and 650 Torr N_2 containing ppm concentrations of SO_2, C_3H_6, NO, and diethylhydroxylamine (DEHA). The particle number density N for particles with diameters >25 Å was monitored as a function of irradiation time. N depended strongly on the cell wall conditions. For a freshly cleaned cell and relatively small concentrations of SO_2, C_3H_6, NO, and DEHA, no particles were produced. After the cell was conditioned by repeated experiments particles were produced. The time histories, maximum values of N, or the times to reach the maxima were not markedly altered (less than a factor of 2) by adding the photochemical inhibitors DEHA or aniline (ϕNH_2) at 5–25% of the C_3H_6 concentrations, in either the absence or presence of SO_2. Thus at the present time the importance of aerosol production by inhibitors has not been determined.

The most promising of the additives is $(C_2H_5)_2NOH$, because of its high effectiveness and the low toxicity of it and its products. The initial products of the reaction are principally CH_3CHO, C_2H_5OH, HONO, and N_2O, with other products including $C_2H_5NO_2$, $C_2H_5ONO_2$, H_2O, CO, CO_2, probably CH_2O, and HNO_3 (Stockburger *et al.*, 1976). All of these, except C_2H_5OH, $C_2H_5NO_2$, and N_2O, are produced in photochemical smog anyway and with the possible exception of HONO their concentration levels will be reduced. N_2O is already present in the air and is harmless. The slow reaction of DEHA with NO_2 produces the same products, and the O_3–DEHA reaction produces CH_3CHO and $C_2H_5NO_2$ (Olszyna and Heicklen, 1976). Apparently DEHA always starts to react by losing the labile H atom:

$$(C_2H_5)_2NOH + HO \rightarrow (C_2H_5)_2NO + H_2O \tag{33}$$

$$(C_2H_5)_2NOH + NO_2 \rightarrow (C_2H_5)_2NO + HONO \tag{34}$$

$$(C_2H_5)_2NOH + O_3 \rightarrow (C_2H_5)_2NO + HO + O_2 \tag{35}$$

The $(C_2H_5)_2NO$ radical is then oxidized to $CH_3CHO + C_2H_5NO_2$ (or in the presence of the oxides of nitrogen to N_2O) and other products.

The question that must now be answered is How safe are DEHA, CH_3CHO, $C_2H_5NO_2$, and any other species that may be formed? In the case of CH_3CHO this question is irrelevant because it is formed already in the photochemical smog reaction, and since this reaction will be inhibited, CH_3CHO levels will actually drop. C_2H_5OH is of no concern because it is known to be harmless. However, both DEHA and $C_2H_5NO_2$ will be new species of possible biological significance in urban atmospheres.

The toxicological information that we have on DEHA is the following: Ingestion studies have been done with rats, and officials of the Pennwalt Corp. (the manufacturer of DEHA) believe that they could obtain FDA approval for DEHA as a food additive, if they wanted to do so. Short-term (<24 hr) inhalation studies of massive doses of DEHA have shown no effects. In the ten years that Pennwalt has manufactured this chemical, there have been no health incidents related to DEHA, and no special precautions are taken in handling it.

DEHA is a substituted hydroxylamine and should have similar characteristics to the hydroxylamines. The simplest one, hydroxylamine, is safe in the vapor unless exposure is to massive amounts. In the form of a salt it is used medically for its favorable effect on the central nervous system (Baker, 1966). DEHA itself is being tested as a spray for plants to protect them against oxidation damage. If it is used for this purpose, then of course it will be present in the atmosphere where the spraying occurs.

Ingestion of nitroethane by white rats has shown an LD_{50} (lethal dose to 50% of the population) to be 1625 mg/kg (Martin and Baker, 1967). This level is considered acceptable for food additives. No deaths were reported in rabbits or monkeys at 500 ppm exposure for 140 hr. There were no deaths with rabbits or guinea pigs at 1000 ppm after 6 hr exposure, but one rabbit died after 12 hr exposure (Mackle *et al.*, 1940). Threshold limit values for inhalation have been set for 100 ppm to which presumably industrial workers can be continually exposed without physiological effect (Martin and Baker, 1967). We expect that atmospheric levels would never exceed 0.1 ppm. Furthermore, nitroethane metabolizes readily and passes through the system quickly. It is completely eliminated in 30 hr. Smog chamber tests have shown that nitropropane (an analog of nitroethane) causes no eye irritation (Baker, 1974).

The information that is still needed is long-term inhalation studies at low concentrations. These studies have been started at Penn State University. Rats have been exposed to 9 ppm each of DEHA and $C_2H_5NO_2$ for 12 hr a day, 6 days a week. As of this writing (November 1975) this dose has been administered for 800 hr with the level of DEHA reaching 17 ppm for two 12-hr periods. The 9 ppm level is $\sim$200 times the level that would be used in a city. So far the rats have received a total dosage greater than any urban population would receive in 80 years under the proposed program. Blood tests, complete autopsies, and mutagenic studies are planned over a two-year period to see if there will be any long-term physiological or mutagenic effects.

Should the toxicological tests be favorable, then DEHA could be introduced into a city in 1977 (or possibly 1978). The levels of DEHA needed in the atmosphere would be about 10% of the reactive hydrocarbon level, or <0.1 ppm. Since 10^{11} gal of gasoline are consumed in the U.S. each year and about 1.0% of this enters the air as unburned olefinic hydrocarbons, approximately 10% of which will need to be controlled, about 10^7 gal of $(C_2H_5)_2NOH$/yr should give the desired levels. At a cost of about \$12/gal, this gives an expense of \$120 million. Distributing DEHA into the atmosphere and monitoring it could cost another \$50 million, so that the annual expense of such a program should be $\simeq$\$200 million, i.e., $\simeq$\$2 per automobile.

COST OF AIR POLLUTION

From the above we have seen that air pollution control will be expensive, running into annual costs of billions of dollars for motor vehicle control.

Chemical control of the atmosphere through additives is relatively cheap, but still comes to ~\$200 million/yr, a substantial sum of money. The questions one must ask are How much does air pollution cost, and what expense is justified in rectifying it?

The cost of air pollution based on its effects on human health was estimated by Ridker (1967) to be about \$2 billion (1958 dollars) annually in the U.S. This cost is associated with the increase in the respiratory diseases: cancer of the respiratory system, chronic bronchitis, acute bronchitis, the common cold, pneumonia, emphysema, and asthma. Lave and Seskin (1970) updated this estimate and more or less supported Ridker's value. They considered only the cost associated with treatment and foregone wages due to increased morbidity and mortality. They estimated that a 50% decrease in air pollution (in 1970) would save the following:

Disease:	Respiratory	Cardiovascular	Cancer	
Reduction (%):	25	10	15	
$\$10^6$ saved/yr:	1222	468	390	Total: \$2080

The savings due to increased health would be about \$2 billion annually.

More recently the health costs due to air pollution and cigarette smoking were reevaluated by Williams and Justus (1974), who pointed out that previous estimates of the urban factor on health were attributed entirely to air pollution. They suggested that other factors might also be important, that there probably was a synergistic effect with cigarette smoking, and that the influence of air pollution had been overestimated. Their reanalysis of the data led them to conclude that the 1970 nationwide (U.S.) health cost due to air pollution was \$62–311 million, and due to cigarette smoking, \$4.23 billion.

When plant damage and material damage are included, an estimate of the order of \$5 billion or more may not be unreasonable. Eisenbud (1970) reports that the total economic loss due to air pollution in large urban areas is \$65 per person per year. The Environmental Protection Agency (1974) estimated the total annual cost of air pollution at \$12.3 billion for the U.S., but this includes the probably overinflated value of \$4.6 billion in visits to doctors and lost work days. Property damage was estimated at \$5.8 billion. It is difficult to assess esthetic values, but presumably a longer, healthier life in more pleasant surroundings is worth something also. Possibly \$10 billion/yr is not an unreasonable price to pay for improved air, but larger sums are difficult to justify. Certainly \$5 billion/yr or less is justifiable, and <\$2 billion/yr must be a worthwhile financial investment.

REFERENCES

American Chemical Society Report (1969). Cleaning Our Environment. The Chemical Basis for Action.

Baker, P. J. Jr. (1966). *Encyclo. Chem. Tech.* **11,** 493, "Hydroxylamine and Hydroxylamine Salts."

Baker, P. J. Jr. (1974). Private communication.

Bufalini, J. J., Kopczynski, S. L., and Dodge, M. C. (1972). *Environ. Lett.* **3,** 101, "Contaminated Smog Chambers in Air Pollution Research."

Dimitriades, B., and Wesson, T. C. (1972). *J. Air Pollut. Contr. Ass.* **22,** 33, "Reactivities of Exhaust Aldehydes."

Eisenbud, M. (1970). *Science* **170,** 706, "Environmental Protection in the City of New York."

Environmental Protection Agency (1974). "The Economic Damages of Air Pollution."

Gintzon, E. L. (1973). Nat. Acad. of Sci., "Report by the Committee on Motor Vehicle Emissions."

Gitchell, A., Simonaitis, R., and Heicklen, J. (1974a). *J. Air Pollut. Contr. Ass.* **24,** 357, "The Inhibition of Photochemical Smog I. Inhibition by Phenol, Benzaldehyde, and Aniline."

Gitchell, A., Simonaitis, R., and Heicklen, J. (1974b). *J. Air Pollut. Contr. Ass.* **24,** 772, "The Inhibition of Photochemical Smog II. Inhibition by Hexafluorobenzene, Nitrobenzene, Napthalene, and 2,6-Di-tert-butyl-4-methylphenol."

Glick, H. S., Klein, J. J., and Squire, W. (1957). *J. Chem. Phys.* **27,** 850, "Single-Pulse Shock Tube Studies of the Kinetics of the Reaction $N_2 + O_2 \leftrightarrows 2NO$ Between 2000–3000°K."

Haagen-Smit, A. J. (1972). *Amer. Chem. Soc. Advan. Chem. Ser.* No. 113, Photochemical Smog and Ozone Reactions, p. 169, "Abatement Strategy for Photochemical Smog."

Hagen, D. F., and Holiday, G. W. (1964). *Vehicle Emissions* **6,** part 1, 206, "The Effects of Engine Operating and Design Variables on Exhaust Emissions."

Hamming, W. J., Chass, R. L., Dickinson, J. E., and MacBeth, W. G. (1973). *Ann. Meeting Air Pollut. Contr. Ass., 66th, Chicago* Paper No. 73-73, "Motor Vehicle Control and Air Quality: The Path to Clean Air for Los Angeles."

Heuss, J. M., and Glasson, W. A. (1968). *Environ. Sci. Tech.* **2,** 1109, "Hydrocarbon Reactivity and Eye Irritation."

Heuss, J. M., Nebel, G. J., and D'Alleva, B. A. (1974). *Environ. Sci. Tech.* **8,** 641, "Effects of Gasoline Aromatic and Lead Content on Exhaust Hydrocarbon Reactivity."

Jayanty, R. K. M., Simonaitis, R., and Heicklen, J. (1974). *Atmos. Environ.* **8,** 1283, "The Inhibition of Photochemical Smog III. Inhibition by Diethylhydroxylamine, *N*-methylaniline, Triethylamine, Diethylamine, Ethylamine, and Ammonia."

Kuntz, R. L., Kopczynski, S. L., and Bufalini, J. J. (1973). *Environ. Sci. Tech.* **7,** 1119, "Photochemical Reactivity of Benzaldehyde-NO_x and Benzaldehyde-Hydrocarbon-NO_x Mixtures."

Lave, L. B., and Seskin, E. P. (1970). *Science* **169,** 723, "Air Pollution and Human Health."

Levine, M. (1961). Lockheed Aircraft Corp., Burbank, California, Rep. 15055, "Atmospheric Chemical Reactions—Air Pollution."

Lonneman, W. A., Kopczynski, S. L., Darley, P. E., and Sutterfield, F. D. (1974). *Environ. Sci. Tech.* **8,** 229, "Hydrocarbon Composition of Urban Air Pollution."

Mackle, W., Scott, E. W., and Treon, J. F. (1940). *J. Ind. Hyg. Toxicol.* **22,** 315, "The Physiological Response of Animals to Some Simple Mononitroparaffins and to Certain Derivatives of These Compounds."
Marsh, G., and Heicklen, J. (1967). *J. Phys. Chem.* **71,** 250, "Some Reactions of Oxygen Atoms II. Ethylene Oxide, Dimethyl Ether, n-C_4H_{10}, n-C_7H_{16}, and Isooctane."
Martin, J. L., and Baker, P. J. Jr. (1967). *Encyclo. Chem. Tech.* **13,** 864, "Nitroparaffins."
Mosher, J. C., MacBeth, W. G., Leonard, M. J., Mullins, T. P., and Brunelle, M. F. (1969). *Ann. Meeting Air Pollut. Contr. Ass., 62nd, New York* Paper No. 69-28, "The Distribution of Contaminants in the Los Angeles Basin Resulting from Atmospheric Reactions and Transport."
National Air Pollution Control Administration (now Environmental Protection Agency) (1969). Publ. No. AP-52, "Control Techniques for Sulfur Oxide Air Pollutants."
National Air Pollution Control Administration (now Environmental Protection Agency) (1970). Publ. No. AP-62, "Air Quality Criteria for Carbon Monoxide."
National Air Pollution Control Administration (now Environmental Protection Agency) (1970). Publ. No. AP-64, "Air Quality Criteria for Hydrocarbons."
National Air Pollution Control Administration (now Environmental Protection Agency) (1971). Publ. No. AP-84, "Air Quality Criteria for Nitrogen Oxides."
Olszyna, K., and Heicklen, J. (1976). *Sci. Total Environ.*, **5,** 223, "The Inhibition of Photochemical Smog VI. The Reaction of O_3 with Diethylhydroxylamine."
Prud'homme, R. K. (1974). *Amer. Sci.* **72,** 191, "Automobile Emissions Abatement and Fuels Policy."
Ridker, R. G. (1967). "Economic Costs of Air Pollution." Frederick A. Praeger, New York.
Spicer, C. W., Miller, D. F., and Levy, A. (1974). *Environ. Sci. Tech.* **8,** 1028, "Inhibition of Photochemical Smog Reactions by Free Radical Scavengers."
Stephens, E. R., Linnell, R. H., and Reckner, L. (1962). *Science* **138,** 831, "Atmospheric Photochemical Reactions Inhibited by Iodine."
Stockburger, L. III, and Heicklen, J. (1976). *Atmos. Environ.* **10,** 51, "The Inhibition of Photochemical Smog IV. Effect of Diethylhydroxylamine on Particle Production."
Stockburger, L. III, Sie, B. K. T., and Heicklen, J. (1976). *Sci. Total Environ.* **5,** 201, "The Inhibition of Photochemical Smog V. Products of the Diethylhydroxylamine Inhibited Reaction."
Tiao, G. C., Box, G. E. P., and Hamming, W. J. (1975). *J. Air Pollut. Contr. Ass.* **25,** 260, "Analysis of Los Angeles Photochemical Smog Data: A Statistical Overview."
Williams, J. R., and Justus, C. G. (1974). *J. Air Pollut. Contr. Ass.* **24,** 1063, "Evaluation of Nationwide Health Costs of Air Pollution and Cigarette Smoking."
Wilson, W. E., Jr., Merryman, E. L., Levy, A., and Taliaferro, H. R. (1971). *J. Air Pollut. Contr. Ass.* **21,** 128, "Aerosol Formation in Photochemical Smog I. Effect of Stirring."

INDEX

A

Aerosol
  concentration, 361
    in cities, 355
  production, 354, 358–365
    rate coefficient, 340–341
  size distribution, 355–356
  stratospheric, 361
Acrolein lachrymation, 177, 196
Airglow, 37
  daytime, 43–44
  green line, 42
  HO Meinel bands, 41
  Meinel emission, 37
  N_2^+ (first negative), 41
  N(I), 41
  Na D doublet, 41
  NO—O afterglow, 37, 41–42
  NO β bands, 41
  NO γ bands, 41
  nighttime, 37–43
  O_2, 41
    atmospheric system, 43
    Herzberg bands, 41
    infrared bands, 41
    Kaplan–Meinel bands, 41
  O(I) lines, 37, 41
Air pollution
  cost, 393–394
  episodes, 156, 220–221
Allene, 341
Ar, 7
Auto exhaust composition, 169

C

Ca, 44
Ca^{2+} resonance lines, 44
Carcinogens, 177
  benzo[*a*]pyrene, 177
    concentration, 184
    emissions, 182–183, 185
    pyrosynthesis of, 265
  polyaromatics, 185
CH_3
  oxidation, 96, 285
  rate coefficients, 76
CH_4
  combustion, 257–258
    concentrations, in 258–259
    rate coefficients, 260–261
    surface effects, 241
  concentration, 7, 28–29, 58, 92–95, 101, 103–107, 165
    Los Angeles, 173
  diurnal variation, 103–107
  effects, 178
  emissions, 165–166
  flux, 165
  lifetime, 165
  photodissociation coefficient, 57, 94–95
  production, 94
  rate coefficients, 66, 84, 308
    diffusion, 95
  reactivity, 279
  removal, 94
  scale height, 95

C_2H_2
oxidation, 255–256
concentrations in, 271
products, 270
rate coefficients, 269, 308
soot formation, 268, 270–271
pyrosynthesis, 264
reaction with electronically excited SO_2, 341
reactivity, 280
C_2H_4
effect on plants, 170
oxidation, 255–256, 286
production, 293
rate coefficients, 295, 308
reactivity, 280
C_2H_6
cracking, 263–264
rate coefficients, 308
reactivity, 279
C_3H_4, *see* Allene
C_4H_{10}
combustion
product yields, 248–249, 267
radical concentrations, 245
rate coefficients, 246–247, 266
rates, 250–251
cracking, 263–264
ignition, 265
rate coefficients, 308
reactivity, 279, 373
CH_3CHO
oxidation, 254
photo, 287–288
reactivity, 281
C_2H_5CHO
oxidation, 254
reactivity, 281
CH_3CO oxidation, 285
Chlorofluoromethanes, 113–119
concentration, 115
lifetime, 119
photodissociation coefficient, 117
production, 116
CH_2O
concentration, 99, 103–107, 300
diurnal variation, 103–107
lachrymation by, 177, 196
oxidation, 252–253
photo, 287–288
photodissociation, 96
coefficient, 57
production, 96, 102, 293, 300
rate coefficients, 65, 76, 84, 295, 323
reactivity, 281
removal, 96, 102, 300
standards, 177
steady state, 96
threshold value limit, 177
CH_3O, 291
CH_3O_2, 291
Cl, 115
Cl_2, effect on plants, 171, 208
ClO, 115
CO, 29
cigarette smoking, 163
concentration, 7, 96, 101, 103–107, 160–161, 374
cities, 161–162, 274–275
diurnal variation, 162
latitude distribution, 161
longitude distribution, 160
control of, 368–369, 375
diffusion, 160
emissions, 157–159
standards, 385
flux, 159
lifetimes, 160
oxidation, 263
physiological activity, 162
production, 97, 159, 294
rate coefficients, 65, 78, 84
removal, 159
standards, 163
steady state, 97
threshold limit value, 162
CO_2, 29
concentration, 7, 97, 163
emissions, 163–164
infrared, 43
global effects, 164
production, 163
rate coefficients, 135
CO_3^-
concentration, 152–153
electron affinity, 151
rate coefficients, 135
reactions, 150–151

CO_4^-
concentration, 152–153
rate coefficients, 135
reactions, 150–151
Combustion
alkoxy radicals, 249
alkyl radical oxidation, 242–243
cool flames, 252
ignition, 265
Composition of atmosphere, 7
Cosmic radiation, 127
ion-pair production by, 129, 131
CS_2, 324

D

Density, 3–4
Diffusion, 3, 52
coefficients, 5, 52
eddy, 3–6
horizontal, 6
molecular, 3, 5–6
Diurnal cycle, 100–107, 147, 153
D region, 6
reactions, 134–136

E

Eclipse photoionization rate, 148
Electron, 43
affinities, 151
concentration, 122–124, 130, 138–140, 144, 151
diurnal, 153
production, 131
rate coefficients, 132, 138–139
reactions, 134–135, 149, 151
removal, 131
steady state, 132, 139
Elemental balance, 107–111
E region, 6
reactions, 134
Excess deaths, 156
Eye irritation, 381
acrolein, 177, 196
CH_2O, 177, 196
peroxybenzoyl nitrate, 196

F

Flux, 52
Fog, 364–365
F region, 6
reactions, 134

G

Gasoline composition, 168
Gravitational separation, 6

H

H, 37
concentration, 27, 99, 101, 103–109
diurnal variation, 103–107
production, 88, 102
rate coefficients, 84, 135
removal, 88, 102
H_2
concentration, 7, 28, 90, 92, 101, 103–109
production, 90, 92, 294
rate coefficients, 66, 84
removal, 90, 92
Halocarbons, *see also* Chlorofluoromethanes
concentration in cities, 207
effects, 207
odor threshold, 206–207
oxidation, 287
production, 206–207
reaction with O_3, 319–321
solvents, 206
threshold limit values, 206–207
HCl, effect on plants, 171, 208
HCO_3^- concentrations, 152–154
He, 7
Heterosphere, 5
$H(H_2O)_2^+$, 143
concentration, 127
$H(H_2O)_n^+$, 144–149
concentration, 147–148
HNO_2, 30
absorption spectrum, 289
concentration, 103–107
diurnal variation, 103–107
photodissociation coefficient, 56
photolysis, 288
production, 85, 102

rate coefficients, 84, 295
removal, 102
steady state, 85
HNO_3
concentration, 7, 25–26, 58, 87, 99, 101, 103–107, 110
diffusional loss, 86
diurnal variation, 103–107, 110
lifetime, 86–87
photodissociation, 86
coefficient, 57, 82, 86–87, 99
production, 85, 294
rate coefficients, 84
removal, 86
threshold limit value, 194
HO, 26–27
concentration, 86–87, 91, 94, 99, 101, 103–110, 293
diurnal variation, 103–107, 110
Meinel bands, 37, 41
production, 88, 102
rate coefficients, 78, 84, 279, 295, 308, 343
reaction with SO_2, 343
removal, 88, 102
vibrationally excited, 37
HO^-, *see* OH^-
HO_2
concentration, 91, 99, 101, 103–110, 293
diurnal variation, 103–107, 110
production, 89, 102
rate coefficients, 84, 295, 343
reaction with SO_2, 343
removal, 89, 102
HO_x, 83, 88
H_2O
concentration, 28, 91–93, 101, 103–109
diurnal variation, 103–107
infrared emission, 43
photodissociation coefficient, 57, 91–92
production, 92–93, 294
rate coefficients, 66, 84, 134–135
removal, 92
scale height, 91
H_2O^+ concentration, 127
H_2O_2
concentration, 91, 99, 101, 103–110
diurnal variation, 103–107, 110
lifetime, 90–91
photodissociation coefficient, 57, 90–91
production, 89, 102
rate coefficients, 65–66, 84
removal, 89, 102
steady state, 89
H_3O^+, 143
concentration, 127
Homosphere, 5
HONO, *see* HNO_2
H_2S
concentrations, 211–212
emissions, 209–210
flux, 212
lifetimes, 212
H_2SO_4
effects, 219–220
production, 338, 362
thermodynamic data, 363
Hydrocarbons, 169
concentrations, 169–170
in cities, 170, 173–176
control of, 367–368, 375
cracking, 263–264
effects, 37, 171, 177–181
emissions, 165–168
control of, 379
standards, 385
oxidation, 255–256
endothermic reactions, 267–268
in photochemical smog, 286
reaction mechanism, 262–263
soot formation, 268, 270–271
production, 243
pyrosynthesis, 264
rate coefficients, 308–309, 311, 322
reaction with
electronically excited SO_2, 339, 341
O_3, 312–319, 321, 323
singlet O_2, 325–327
reactivity, 279–283
Hydroperoxide
decomposition, 248
production, 243

I

Incinerator emission composition, 168
Ionosphere, 6, 122–154

K

Kr, 7

L

Lead, *see* Pb
Li, 44
Li^+, 44
Lifetime, 53–55, *see also* individual molecules
Lyman α, 30–31, 33, 125

M

Mesopause, 2–3
Mesosphere, 2–4
 elemental balance, 109–110
Metal oxides, 356
Meteors, 44
Molozonide, 313
Motor vehicle emissions
 benzo[*a*]pyrene, 183
 CO, 159
  control of, 372, 377–385
 cost, 386–388
 hydrocarbons, 167–169
 nitrogen oxides, 187
 particulate matter, 224–227

N

N_2, 37
 absorption cross section, 30
 concentration, 7, 9–11
 photodissociation, 31
 photoionization, 30, 125
  rate, 130
 potential-energy curves, 39
 rate coefficients, 134
 reactions, 373
N_2^-, 39
N_2^+, 43, 130
 absorption, 43
 concentration, 126–127, 132, 138
 first negative emission, 41, 43
 potential-energy curves, 39
 production, 132
 rate coefficients, 134
 removal, 132
Na, 44
NaH, 43
NaO, 43
$Na(^2P)$, 43
 D doublet emission, 41, 43–44
$N(^2D)$
 doublet emission, 43
 N(I), 41
 production, 43, 141
Ne, 7
Negative ions, 149–153
 concentration, 149
  diurnal variation, 153
 positive ion annihilation coefficient, 143–146, 152
NH_2, 83
NH_3, 83
 concentration, 7
 photodissociation, 83
  coefficient, 85
 reaction with
  O_3, 324
  SO_2, 362, 364
NH_x, 83
NH_4NO_3, 358–359
NH_2O, 83
NH_2O_2, 83
Nitrate photolysis, 289
Nitrite photolysis, 289
NO, 37
 $A\,^2\Sigma^+$ state, 44
 β bands, 41
 concentration, 7, 21–25, 55, 57–58, 77, 82, 87, 99, 101, 103–107, 110, 142, 188
  in cities, 190–191, 274–275
 diurnal variation, 103–107, 110
 effects,
  on O_3, 113
  physiological, 194–195
 emissions, 185
 γ bands, 21, 41, 44
 lifetime, 82, 142–143
 oxidation, 292
 photodissociation coefficient, 56, 74–75, 82, 99
 photoionization, 125, 143
  rates, 130
 potential-energy curves, 40
 production, 71, 141, 293, 374
 rate coefficients, 65, 76, 78, 84, 134–135, 295
 removal, 141
 scale height, 142
 steady state, 143
 threshold limit values, 194
NO^-, potential-energy curves, 40

NO^+, 43, 130, 144–145
concentration, 122–124, 126–127, 139–140
diurnal, 146–148
potential-energy curves, 40
production, 138
rate coefficients, 134–135
removal, 138
steady state, 139
NO_2
absorption, 304–305
concentration, 7, 25, 55, 58, 77, 79, 86–87, 98, 101, 103–107, 110, 188
control of, 382
in cities, 192–193, 274–275
maximum, 299
diurnal variation, 103–107, 110
effects,
on plants, 170
physiological, 194–195
electronically excited, 37, 305–307
emission, 37, 42
NO—O afterglow, 41–42
oxidation, 292
photodissociation coefficient, 56, 75, 305
photolysis, 275–276, 304–305
production, 102, 294
rate coefficients, 65, 76, 78, 84, 295
removal, 102, 294
steady state, 75, 98
threshold limit value, 194
NO_2^-, 143
concentration, 152–153
electron affinity, 151
rate coefficients, 134–136
reactions, 150–151
NO_2^+, 143
rate coefficients, 134
NO_3
concentration, 77, 79, 110, 293
diurnal variation, 110
photodissociation coefficient, 56, 77
production, 77
rate coefficients, 76, 78, 295
removal, 77
steady state, 79
NO_3^-
concentration, 152–154
electron affinity, 151
reactions, 136, 150
NO_x, 79
concentration, 80–82, 86, 100, 103–107
Los Angeles, 188–189
control of, 368–369, 375–376
diffusional loss, 80
diurnal variation, 103–107
effects, 189
on O_3, 112–113
plant damage, 189
emissions, 186–188
control of, 379
standards, 385
flux, 80–81, 186–187
lifetime, 80–82, 186
photodissociation coefficient, 81
production, 79
scale height, 81
standards, 189
N_2O, 71
concentration, 7, 25–26, 55, 58, 71, 73, 99, 101, 103–107
flux, 74
infrared emission, 43
photodissociation, 71
coefficient, 56, 72–73, 99
production, 72
rate coefficients, 66, 72
removal, 71
scale heights, 73
N_2O_5, 79
concentration, 110
diurnal variation, 110
production, 77, 294
rate coefficients, 76, 78
removal, 77
$NO(CO_2)^+$, rate coefficients, 135
$NO(H_2O)_x^+$, 144–145
concentration, 147
rate coefficients, 134–135
$NO(N_2)^+$, rate coefficients, 135
$N(^4S)$, 10, 43
concentration, 10–11, 57, 141–142
lifetime, 142
rate coefficient, 65, 134
scale height, 142
steady state, 75, 141

O

O^-, 150
O^+
concentration, 123, 126, 130, 133, 138

production, 133
rate coefficients, 134
removal, 133
O_2, 7–9, 12, 64
absorption, 31, 33, 63
coefficients, 32, 60
cross section, 33–34
Herzberg continuum, 33–34
Schumann–Runge bands, 33, 59
Schumann–Runge continuum, 33, 58
spectrum, 31
electronically excited, 37, 41–42
Herzberg bands, 41
Kaplan–Meinel bands, 41
photodissociation, 33
photodissociation coefficient, 56, 58–59, 61
Herzberg continuum, 59
Lyman α line, 59
Schumann–Runge bands, 59, 61
Schumann–Runge continuum, 59
photodissociation rate, 60–61
photoionization, 58, 125
rate, 130
potential-energy curves, 38
rate coefficients, 66, 76, 84, 134–135
O_2^-
concentration, 152–153
electron affinity, 151
potential-energy curves, 38
rate coefficients, 135–136
reactions, 149, 151
O_2^+
concentration, 122–123, 125, 127, 137–138
potential-energy curves, 38
production, 137
rate coefficients, 134, 308
removal, 137
steady state, 137
O_3, 36–37, 42, 60, 62–64, 70
absorption, 34, 63
Chappuis band, 36, 60
cross section, 35–36
Hartley band, 34, 60
Huggins bands, 34, 60, 70
quantum efficiency, 36
concentration, 7, 12–20, 27, 55, 58, 67, 98–99, 103–109
calculated, 68–69, 196
diurnal variation, 103–107
in cities, 196–199, 274–275, 383
maximum, 299
control of, 380
effects, 196
on animals, 204
on humans, 205
on plants, 170, 203
infrared emission, 43
lifetimes, 68–69, 98
photodissociation, 60, 63–64
coefficients, 56, 62
π-complex, 321
production, 102, 294
rate coefficients, 65–66, 76, 84, 135, 295, 308–309, 322–323, 372
reaction
emission, 315, 318
hydrocarbons, with 312–319, 321–323
mechanism, 319
removal, 102, 115, 294
by chlorofluorocarbons, 118
by NO_x, 112–113
standards, 196
steady state, 66
threshold limit value, 196
O_3^-
concentration, 152–153
electron affinity, 157
rate coefficients, 135–136
reactions, 149
O_4^-
concentration, 152–153
rate coefficients, 135
reactions, 150–151
$O(^1D)$, 33–34, 36, 42–43, 62–64
concentration, 10, 20, 55, 57, 91, 94, 103–107
calculated, 69
diurnal variation, 103–107
electronic energy, 35
emission doublet, 42
lifetimes, 69
O(I) red, 41
production, 35, 58, 60, 62–63, 102
rate coefficients, 65–66
removal, 43, 102
Odd nitrogen
concentration, 142
lifetime, 141–142

scale height, 142
steady state, 141
Odd oxygen, 63–64, 97
concentration, 98, 100
calculated, 68–69
lifetime, 67–68, 98–100
production, 66–67, 97
removal, 66–67, 97, 115
steady state, 97
$O_2(^1\Delta)$, 34, 42, 60, 70
concentration, 9–10, 18–20, 55, 57, 62, 103–107
calculated, 69
diurnal variation, 103–107
electronic energy, 35
infrared bands, 41, 44
production, 62, 64, 69, 102
rate coefficients, 65, 70, 135, 308–311
reaction mechanisms, 325–327
removal, 102
steady state, 71
OH, *see* HO
OH^-
concentration, 152–153
electron affinity, 151
rate coefficients, 135
reactions, 150–151
O-heterocycle production, 240–241, 243, 255
$O(^3P)$, 37, 58–60, 62–65, 70
concentration, 8–10, 12–13, 20, 27, 55, 57, 67, 98–99, 101, 103–109, 293
calculated, 68–69
diurnal variation, 103–107
lifetime, 67–69
photoionization, 125
rate, 130
production, 35, 102, 286
rate coefficients, 65, 76, 78, 84, 279, 308, 134–135
reactions with
C_4H_{10}, 373
N_2, 373
SO_2, 342
SO_3, 342
removal, 102
Optical depth, 54–55
$O(^1S)$, 8–9
electronic energy, 35
green line, 42
O(I) green, 41
production, 35, 42
$O_2(^1\Sigma)$, 43, 60, 70
Atmospheric System, 43
concentration, 55
calculated, 69
electronic energy, 35
emission, 43
rate coefficients, 65, 70, 312
removal, 70
steady state, 71
Oxidant
concentration, 198–201, 277
dosage, 278, 298–301
effects, 201
Ozonide, 316, 320, *see also* Molozonide

P

Particulates
concentrations, 226, 228–229
in cities, 228, 230–231
effects
meteorological, 231–233
on health, 233–235
other, 235
emissions, 224–225, 227
standards, 235
Pb
concentrations, 230
in cities, 231
standards, 230
Peroxide photolysis, 289
Peroxybenzoyl nitrate, 196
Photoabsorption rate, 51
Photochemical smog
chemical control by
I_2, 388
radical scavengers, 388–393
concentrations, 295–298, 384
free-radical intermediates, 297
HO chain, 292
HONO cycle, 292–293
NO inhibition, 277
NO to NO_2 conversion, 283–285
effect of CO, 284–285
effect of hydrocarbons, 284–285
NO_2 photodecomposition, 286
O_3 photodecomposition, 286
$O(^3P)$ chain, 291

oxidant dosage, 278
rate expressions, 293
rates of reaction in, 293, 295
simulated, 276
termination processes, 290–291
Photodissociation coefficient, 56–57
Photodissociation rate, 55
Photoionization rate, 128, 132, 137. 140, 152
eclipse, for 148
Photoreaction
efficiency, 51
rate, 51
Positive ions, 132–140, 143–148
concentration, 144, 147
electron annihilation coefficient, 132, 138–139, 143, 156
Lyman α, by 143
negative ion annihilation coefficient, 143–146, 152
Pressure, 4

R

Radiation intensity, 30–32, 54–55
Lyman α, 30–31, 33
Rate coefficients, *see also* individual molecules
photon absorption, 51
Reaction rate, 51–52
Recombination coefficient, 138–140

S

Scale heights, 52, *see also* individual molecules
Skin cancer, 113–115
SO_2
absorption
coefficients, 332–334
spectrum, 331–333
concentration, 211–214
in cities, 213, 215–217, 371
control of, 369–372
New York City, 370–372
conversion to SO_3, 344–345
effects, 217–218
on animals, 219
on humans, 219–223
on plants, 171, 218–219
electronically excited, 331–341
lifetime, 333
production, 337
rate coefficients, 336
reaction with hydrocarbons, 334–341
removal, 337, 339
emissions, 209–211
flux, 212
lifetime, 212
oxidation
coefficients, 348
in H_2O solution, 348–351
photooxidation, 338–339
rate coefficients, 310, 343
reaction with,
ammonia, 362, 364
HO and HO_2, 343
O_3–olefin system, 345–347
$O(^3P)$, 341–343
organic material, 360–362
RO and RO_2, 343–344
reactions, relative importance, 347–348
standards, 223
vapor pressure, 363
SO_3
photolysis, 339
reactions, 342
thermodynamic data, 363
Sodium, *see* Na
Solar heating, 3
Solvents
halocarbon, 206
production, 172
usage, 172
Steady state, 53–55, *see also* individual molecules
Stratopause, 2–3
Stratosphere, 2–4
elemental balance, 110–111
Sulfates
aerosol, 212
effects on humans, 222–223
emissions, 209–210
flux, 212
lifetimes, 212
stratospheric, 361
Supersonic transport, 112–113

T

Temperature, 1–4, 63
Terpene emissions, 165–166

Thermosphere, 2–4
  elemental balance, 107–109
Tropopause, 2–3
Troposphere, 2–3, 6
  elemental balance, 111

X

Xe, 7

Z

Zwitterion, 313–314, 321

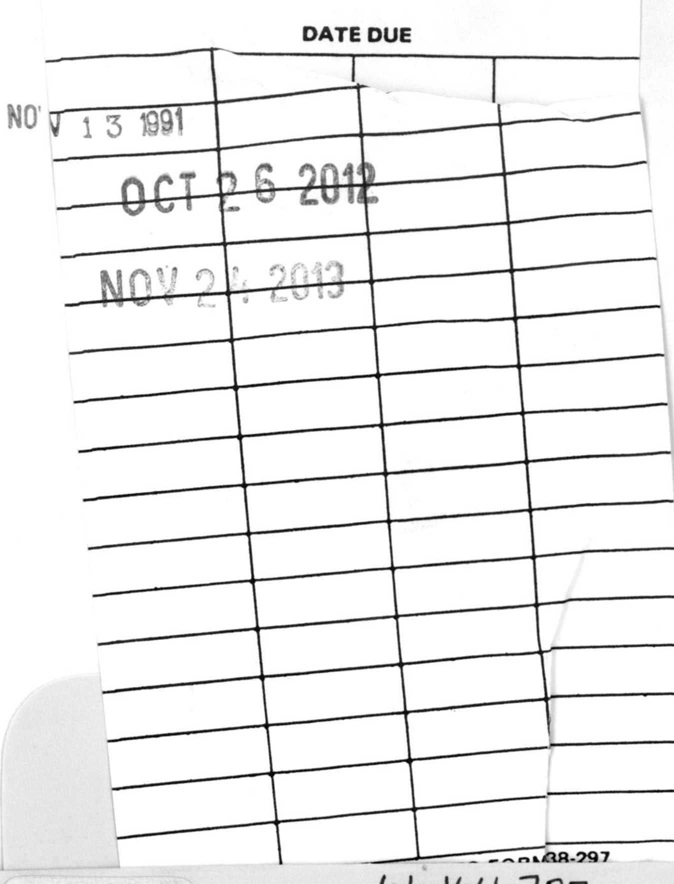
DATE DUE
NOV 13 1991
OCT 26 2012
NOV 24 2013